Insect Microbiomes

The microbiome of insects is intimately involved in their physiology, ecology, and evolution. This volume summarizes what is known of the microbiome of many different insect groups. For example, termites are able to digest cellulose thanks to their gut microbes. Honeybees make honey, in part, with the help of microbes. Chapters covering silkworms, honeybees, ticks, locusts, and mosquitos by leading researchers are all included.

Sunil Thomas is a Research Professor at the Lankenau Institute for Medical Research, Wynnewood, Pennsylvania. He earned his Ph.D. from Cochin University of Science and Technology and worked as a post-doctoral fellow at the Indian Institute of Science, Bangalore, and Mount Sinai School of Medicine, New York. His current research involves the development of immunotherapeutic and mRNA-based drugs for ulcerative colitis/colon cancer and Alzheimer's disease.

Insect Microbiomes

Edited by
Sunil Thomas

CRC Press
Taylor & Francis Group
Boca Raton London New York

CRC Press is an imprint of the
Taylor & Francis Group, an **informa** business

Designed cover image: Sunil Thomas

First edition published 2027
by CRC Press
2385 NW Executive Center Drive, Suite 320, Boca Raton FL 33431

and by CRC Press
4 Park Square, Milton Park, Abingdon, Oxon, OX14 4RN

CRC Press is an imprint of Taylor & Francis Group, LLC

ISBN: 978-1-041-07170-9 (hbk)
ISBN: 978-1-041-07171-6 (pbk)
ISBN: 978-1-003-63917-6 (ebk)

DOI: 10.1201/9781003639176

Typeset in Joanna
by Newgen Publishing UK

CONTENTS

PREFACE

Microorganisms represent the earliest and most enduring forms of life on Earth. Long before the emergence of complex multicellular organisms, microbial life shaped the planet's atmosphere, soils, waters, and biogeochemical cycles. Even today, all eukaryotic life—including plants, animals, and humans—remains fundamentally rooted in microbial ancestry. Modern eukaryotic cells themselves are conserved amalgamations of ancient microbial partnerships, a testament to the central role microorganisms have played in the evolution of life.

Higher organisms cannot exist in isolation from microorganisms. From plants drawing nutrients through symbiotic soil microbes to animals relying on gut bacteria for digestion, microorganisms are indispensable partners in survival. The collective community of microorganisms that inhabit a host or environment is known as the microbiome. These microbial communities are not passive inhabitants; rather, they form dynamic, interactive systems that profoundly influence the health, development, and resilience of their hosts.

The microbiome—comprising trillions of bacteria, archaea, fungi, viruses, and protozoa—acts much like a vital organ. In humans and animals, it supports digestion, synthesizes essential vitamins, regulates metabolism, trains the immune system to distinguish friend from foe, and influences inflammation, behavior, mood, and disease susceptibility. In plants, the microbiome enhances nutrient uptake, improves stress tolerance, and protects against pathogens. Microbial diversity is a defining feature of a healthy microbiome, conferring stability and adaptability to changing conditions.

Beneficial microorganisms produce a wide array of secondary metabolites that are critical for maintaining physiological balance and overall well-being. When this balance is disrupted—a condition known as dysbiosis—the consequences can be severe. Dysbiosis has been linked to infections, allergies, autoimmune diseases, metabolic disorders, reduced crop productivity, and even neurological and behavioral conditions. Understanding the mechanisms that govern microbial balance is therefore essential for advancing health, agriculture, and environmental sustainability.

Recent breakthroughs in high-throughput sequencing, metagenomics, and bioinformatics have revolutionized microbiome research. These technologies allow rapid identification and functional analysis of microbial communities associated with different hosts, diets, environments, and lifestyles. What once took decades to uncover can now be achieved in a fraction of the time, opening new frontiers in medicine, nutrition, agriculture, ecology, and biotechnology.

The six interconnected volumes listed below present a comprehensive exploration of microbial life across biological systems and ecosystems:

Volume 1: The Human Microbiome

Volume 2: Diet and the Microbiome

Volume 3: Animal Microbiomes

Volume 4: Plant Microbiomes

Volume 5: Insect Microbiomes

Volume 6: The Environmental Microbiome

Together, these volumes provide an integrated perspective on how microbial communities shape life on Earth.

I extend my sincere and heartfelt gratitude to all the authors who contributed chapters to these books. Their expertise, dedication, and scholarly insight form the foundation of these volumes. Each contribution reflects a commitment to scientific rigor and a shared passion for advancing knowledge in the rapidly evolving field of microbiome research.

The collective efforts of these authors have enriched the depth, breadth, and quality of these books, transforming it into a comprehensive and authoritative resource. Their willingness to share ideas, collaborate across disciplines, and invest their time and intellect has been invaluable. This work stands as a testament to their contributions, and I am deeply thankful for their role in bringing these books to fruition.

This year marks the 100th anniversary of the Lankenau Institute for Medical Research. The anniversary represents decades of unwavering commitment to scientific inquiry, education, and the advancement of medical knowledge. The Lankenau Institute for Medical Research has played a pivotal role in shaping and supporting my scientific endeavors. Its culture of intellectual rigor, encouragement, and dedication to research excellence has provided an environment in which curiosity could thrive, and ideas could evolve into meaningful scientific contributions.

This six-volume collection is presented as a tribute to the Lankenau Institute for Medical Research. It stands as a testament to the Institute's enduring legacy, its profound impact on generations of researchers, and its mission to advance science for the betterment of human health and society. Through this work, I honor its past achievements, acknowledge its present influence, and celebrate its continued inspiration for future scientific discovery.

CONTRIBUTORS

VEYSEL BAY
Faculty of Agriculture
Department of Animal Science
Ege University
Bornova-Izmir, Turkiye

YEHUDA BEN-SHAHAR
Department of Biology
Washington University in Saint Louis
Saint Louis, Missouri, USA

R. C. RAMCHARAN BHATTACHARYA
ICAR-National Institute for Plant Biotechnology
LBS Centre
Pusa Campus
New Delhi, India

YUN SANG CHO
Department of Animal and Plant Health Research
Animal and Plant Quarantine Agency
Kimcheon, Gyeongsangbuk-do, Republic of Korea

PITAMBAR DHAKAL
College of Veterinary Medicine
Henan Agricultural University
Zhengzhou, China

YIN FU
College of Veterinary Medicine
Henan Agricultural University
Zhengzhou, China

AYUSHI GUPTA
ICAR-National Institute for Plant Biotechnology,
LBS Centre, Pusa Campus
New Delhi, India

OLIVIA LUTZ
Lankenau Institute for Medical Research
Wynnewood, Pennsylvania, USA

ALEX MUPPASANI
Lankenau Institute for Medical Research
Wynnewood, Pennsylvania, USA

SURESH NAIR
Plant-Insect Interaction Group
International Centre for Genetic Engineering and Biotechnology (ICGEB)
New Delhi, India

BERRIN OZDIL
Faculty of Medicine
Department of Biophysics
Ege University
Bornova-Izmir, Turkiye

ZHANGRONG SONG
Department of Entomology and Nematology
University of Florida
Gainesville, Florida, USA

SUNIL THOMAS
Lankenau Institute for Medical Research
Wynnewood, Pennsylvania, USA

CHUN NIN ADAM WONG
Department of Entomology and Nematology
Genetics Institute
University of Florida
Gainesville, Florida, USA

LONGXIAN ZHANG
College of Veterinary Medicine
Henan Agricultural University
Zhengzhou, China

CHAPTER ONE

Insect-Associated Microbiomes and Their Diversity, Function, and Ecological Significance

Berrin Ozdil and Veysel Bay

1.1 INTRODUCTION

Insects harbour complex and functionally integrated microbial populations that occupy multiple body compartments, including the digestive tract, cuticle, reproductive organs, and specialised symbiotic tissues. These associations contain bacteria, fungi, archaea, and viruses, forming a composite biological unit in which host and microbes function as an ecological and evolutionary entity (Mondal et al., 2023). Within this holobiont framework, insect phenotypes emerge not only from host genetics but also from coordinated interactions between host physiology and microbial metabolism (Bennett & Moran, 2015) (Figure 1.1). Over evolutionary time, insect–microbe associations have expanded host metabolic capabilities, enabling exploitation of nutritionally imbalanced or chemically defended resources. Many insects rely on microbial partners to synthesise essential amino acids, vitamins, and cofactors, to degrade complex plant polymers, or to detoxify environmental toxins and pesticides (Douglas, 2015; Gupta & Nair, 2020; Jing et al., 2020; Wang et al., 2022). The stability and functional importance of these partnerships vary across insect taxa. Some insects maintain obligate, vertically transmitted symbioses that are indispensable for survival, whereas others exhibit flexible, environmentally acquired microbiomes that rapidly respond to diet and habitat conditions (Engel & Moran, 2013).

Classic examples of obligate mutualism include the association between aphids and their intracellular symbiont *Buchnera*, which compensates for amino-acid deficiencies in phloem sap diets (Jing et al., 2020; Wang et al., 2022). In contrast, insects such as flies and beetles typically acquire much of their microbiome from the environment, resulting in communities that are highly plastic and diet dependent (Engel & Moran, 2013). Social insects, including honeybees and termites, represent an intermediate strategy, maintaining relatively stable, host-specific microbiomes that are transmitted through social interactions and colony life (Engel & Moran, 2013). Beyond individual fitness, insect-associated microbiomes influence ecological processes at larger scales. Wood-feeding insects contribute to global carbon cycling through microbially mediated lignocellulose degradation (Liu et al., 2019; Tokuda et al., 2018). Pollinators depend on gut microbes that support nutrient assimilation and immune regulation, thereby affecting plant reproduction and ecosystem stability (Ricigliano et al., 2025; Sai Charan et al., 2025). In disease vectors, microbiomes modulate pathogen establishment and transmission, positioning microbial communities as critical determinants of human and animal health (Dennison et al., 2014; Weiss & Aksoy, 2011). The diversity and organisation of insect microbiomes are strongly constrained by host anatomy and

DOI: 10.1201/9781003639176-1

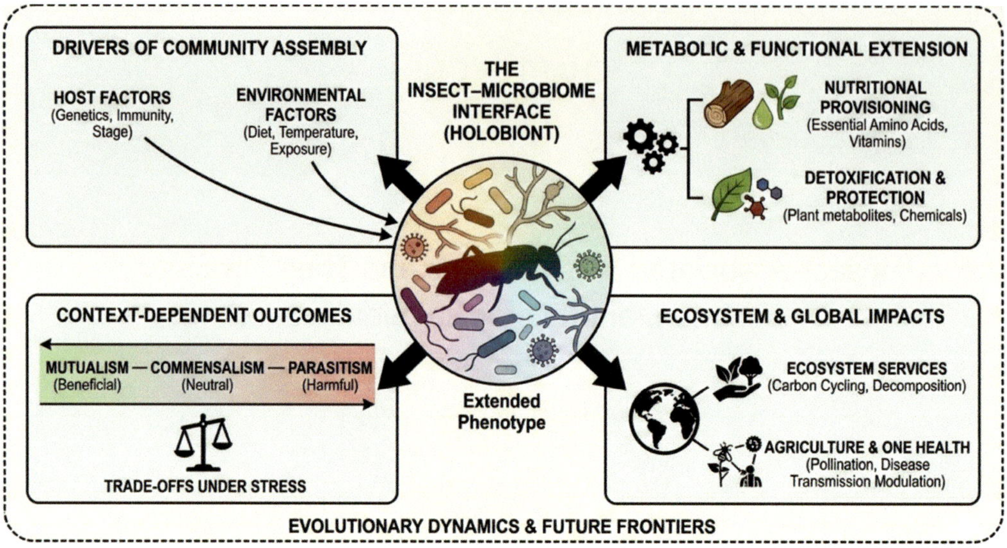

Figure 1.1 Conceptual framework of insect–microbiome interactions across biological and ecological scales. This schematic presents an integrative framework for understanding insect-associated microbiomes as components of the holobiont. Host and environmental factors drive microbial community assembly at the insect–microbiome interface, giving rise to an extended phenotype. Microbial partners expand host metabolic and functional capacity through nutritional provisioning, detoxification, and protection against stress. Outcomes of these associations are context dependent, ranging along a mutualism–commensalism–parasitism continuum, with trade-offs emerging under environmental stress. These interactions scale up to influence ecosystem processes, agricultural productivity, and One Health outcomes and are shaped over time by ongoing evolutionary dynamics and future research frontiers.

TABLE 1.1

Major modes of insect–microbiome association across ecological contexts

Association type	Transmission mode	Microbiome stability	Representative systems	Dominant functional roles
Obligate mutualism	Strict vertical	Very high	Aphids–*Buchnera*, tsetse–*Wigglesworthia*	Essential nutrient biosynthesis
Socially maintained	Horizontal (social)	High	Honeybees, termites	Digestion, immunity, colony health
Facultative symbiosis	Mixed	Intermediate	Beetles, stinkbugs	Detoxification, stress tolerance
Environmentally acquired	Horizontal (environmental)	Low	Flies, mosquitoes	Diet-dependent metabolism, plasticity

NOTE: This table summarises the principal types of insect–microbiome associations, classified by transmission mode and community stability. Obligate and socially maintained symbioses are characterised by high stability and essential nutritional or immune functions, whereas facultative and environmentally acquired microbiomes exhibit greater plasticity. These contrasting strategies reflect trade-offs between functional specialisation and ecological flexibility across insect taxa.

physiology. Although insects share a conserved body plan, variation in gut structure, feeding behaviour, and immune regulation creates distinct microbial habitats across species (Engel & Moran, 2013; Schmidt & Engel, 2021) (Table 1.1).

The insect digestive tract is divided into three major regions: foregut, midgut, and hindgut, each characterised by distinct physicochemical conditions that shape microbial colonisation (Engel & Moran, 2013; Mondal et al., 2023;

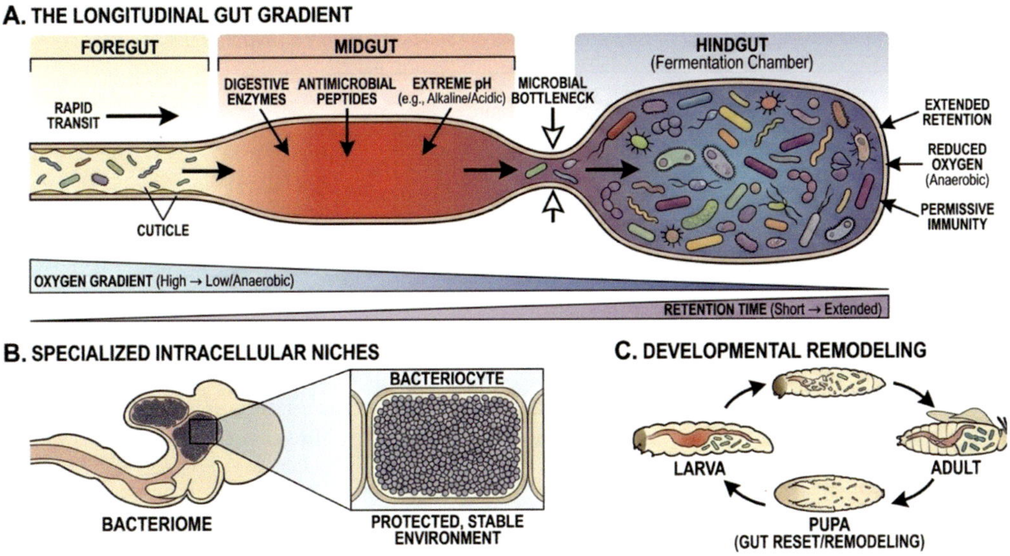

Figure 1.2 Host-derived anatomical and physiological filters shape insect microbiome assembly and structure. This diagram illustrates how the insect host actively regulates microbial communities through a hierarchy of spatial, chemical, and temporal constraints, rather than serving as a passive receptacle. (A) The Longitudinal Gut Gradient: The generalised digestive tract acts as a selective ecological sequence. The foregut serves primarily for transport with rapid transit times, supporting only transient microbes. The midgut functions as a restrictive bottleneck, where intense digestive enzymes, extreme pH (varying by taxa), and robust immune surveillance exclude most microorganisms. The hindgut often forms a specialised fermentation chamber characterised by reduced oxygen tension, extended retention times, and permissive immunity, fostering dense, complex microbial consortia responsible for metabolic functions. Physicochemical gradients (e.g., oxygen, retention time) dictate niche suitability along this axis. (B) Specialised Intracellular Niches: Some insects bypass gut dynamics by housing obligate symbionts within specialised cells called bacteriocytes (often aggregated into bacteriomes), providing a protected, stable environment isolated from gut immune fluctuations. (C) Developmental Remodelling: In holometabolous insects, metamorphosis involves significant restructuring or replacement of gut tissues during pupation, leading to a partial or complete 'reset' of the microbiome between larval and adult stages.

Siddiqui et al., 2022) (Figure 1.2, Table 1.2). The foregut is typically lined with a chitinous cuticle and exhibits rapid food passage, limiting opportunities for stable microbial attachment. As a result, microbial communities in this region are generally sparse and transient (Engel & Moran, 2013; Mondal et al., 2023; Siddiqui et al., 2022). The midgut represents the primary site of digestion and nutrient absorption. However, it is also a hostile environment for microbes due to intense proteolytic activity, the presence of the peritrophic matrix, and strong immune surveillance (Bai et al., 2021; Douglas, 2015; Engel & Moran, 2013; Jones et al., 2019; Mondal et al., 2023; Smith et al., 2017). Consequently, microbial abundance and stability in the midgut are often low, with many microbes passing through rather than establishing long-term residence.

In contrast, the hindgut frequently serves as the main site of microbial colonisation. Longer retention times, reduced oxygen availability, and favourable pH conditions allow dense and metabolically active microbial communities to develop (Ceja-Navarro et al., 2019; Douglas, 2015; Engel & Moran, 2013; Siddiqui et al., 2022). In termites and wood-feeding beetles, the hindgut functions as a fermentation chamber where bacteria and archaea cooperate to degrade lignocellulose and generate short-chain fatty acids that fuel host metabolism (Ceja-Navarro et al., 2019; Liu et al., 2023; Tokuda et al., 2018). In honeybees, the hindgut houses a highly specific bacterial consortium that supports pollen digestion and immune defence (De Beer et al., 2023; Zheng et al., 2019).

Although the digestive tract represents the primary site of microbial activity in many insects, functionally important microbial communities also colonise external and internal tissues beyond the gut. The insect cuticle supports diverse, environmentally acquired microbial assemblies that reflect habitat conditions, nesting environments, and social interactions (Birer et al., 2020). These cuticular

TABLE 1.2

Structural and physiological features of the insect gut and their influence on microbial stability

Region	Lining/barriers	Typical microbial stability	Citations
Foregut	Chitinous cuticle, rapid flow	Low, mostly transient	Engel and Moran (2013), Mondal et al. (2023), Siddiqui et al. (2022), and Engel and Moran (2013)
Midgut	Peritrophic matrix, strong immunity, and enzymes	Moderate–low, often replenished from diet	Bai et al. (2021), Douglas (2015), Engel and Moran (2013), Jones et al. (2019), Mondal et al. (2023), and Smith et al. (2017)
Hindgut	Cuticle with spines/plates, longer retention	High, often dense core communities	Ceja-Navarro et al. (2019), Douglas (2015), Engel and Moran (2013), and Siddiqui et al. (2022)

NOTED: This table compares the foregut, midgut, and hindgut in terms of anatomical barriers and typical microbial persistence. Rapid flow and a chitinous lining restrict stable colonisation in the foregut, while strong digestive and immune pressures limit microbial stability in the midgut. In contrast, the hindgut supports dense and stable core microbial communities due to longer retention times and more permissive conditions.

microbiomes can contribute to host defence by competitively excluding pathogens or by producing antimicrobial metabolites, thereby acting as a first line of protection against infection (Birer et al., 2020).

In several insect systems, specialised cuticular structures house defensive ectosymbionts with clearly defined protective roles. A well-characterised example is the beetle *Lagria villosa*, in which *Burkholderia* symbionts residing in dorsal cuticular invaginations synthesise the antifungal compound lagriamide. Experimental removal of these symbionts markedly reduces larval survival when exposed to pathogenic fungi, demonstrating the ecological importance of cuticle-associated microbes in offspring protection (Janke et al., 2022). Cuticular microbes have also been implicated in modulating insect responses to chemical stressors. Certain surface-associated bacteria and fungi can metabolise insecticides or induce host detoxification pathways, thereby contributing to increased tolerance to compounds such as deltamethrin and other synthetic chemicals (Lemoine et al., 2020).

Reproductive tissues constitute another critical microbial niche. Many insects harbour intracellular symbionts within specialised host cells known as bacteriocytes, which are organised into bacteriomes and transmitted vertically across generations. These endosymbionts are often nutritionally essential. In aphids, *Buchnera* synthesises essential amino acids absent from phloem sap, while in tsetse flies, *Wigglesworthia* provides B vitamins required for normal development and fertility (Douglas, 2015; Gupta & Nair, 2020). Elimination of these symbionts typically results in severe fitness defects, including impaired growth, sterility, or host mortality.

Some reproductive symbionts provide both nutritional and regulatory functions. *Wolbachia*, one of the most widespread intracellular symbionts in insects, occupies bacteriomes in hosts such as bed bugs and contributes to vitamin provisioning, particularly riboflavin and biotin, thereby supporting host growth and fecundity (Hosokawa et al., 2010; Ju et al., 2020). In addition to nutritional roles, *Wolbachia* and related symbionts can manipulate host reproduction by inducing cytoplasmic incompatibility, parthenogenesis, feminisation, or male killing. These effects profoundly influence host population structure and evolutionary dynamics (Al Darwish et al., 2025; Fricke et al., 2025). In whiteflies, interactions among co-resident bacteriocyte symbionts such as *Hamiltonella* and *Arsenophonus* further demonstrate how microbial communities within reproductive tissues shape sex ratios, fertility, and population persistence (Wang et al., 2020). Together, cuticular and reproductive microbiomes extend the functional reach of insect–microbe symbioses beyond digestion, contributing to defence, nutrient provisioning, reproductive success, and long-term host–symbiont coevolution.

Despite the extraordinary taxonomic diversity of insects and their associated microorganisms, insect microbiomes are often structured around a relatively limited set of recurrent microbial groups that perform conserved biological functions. The major microbial groups recurrently associated with insects and their convergent functional roles are summarised in Table 1.3. Bacterial communities dominate most insect microbiomes and are typically composed of members of the phyla Proteobacteria, Firmicutes, Actinobacteria, and Bacteroidetes (Chen et al., 2016; Yasika & Shivakumar, 2025). Proteobacteria are the most ubiquitous and functionally diverse group, encompassing both free-living gut bacteria and obligate intracellular symbionts. Members of this phylum contribute to amino acid and vitamin biosynthesis, degradation of plant-derived carbohydrates, and detoxification of allelochemicals and pesticides. Obligate nutritional symbionts such as *Buchnera*, *Wigglesworthia*, and *Ishikawaella* are characterised by reduced genomes that retain pathways essential for host nutrition and development (Jing et al., 2020; Wang et al., 2022). Firmicutes and Bacteroidetes play prominent roles in carbohydrate fermentation and energy extraction, particularly in herbivorous and detritivorous insects. These taxa contribute enzymes required for the breakdown of complex plant polysaccharides and generate short-chain fatty acids that support host metabolism and gut epithelial health (Jing et al., 2020; Yasika & Shivakumar, 2025). Actinobacteria, including *Bifidobacterium*, are especially abundant in pollinators and some larval stages, where they assist in the digestion, immune modulation, and production of antimicrobial compounds that protect hosts or brood environments (Dong et al., 2024).

Beyond bacteria, other microbial groups make important contributions in specific insect systems. Yeasts and filamentous fungi are frequent in flies, beetles, and social insects, where they aid nutrient assimilation, sterol provision, and detoxification (Gurung et al., 2019; Sibinga et al., 2025). In *Drosophila* and bark beetles, yeast and fungal volatiles shape host attraction, aggregation and parasitoid location, linking mycobiota directly to behaviour and multi-trophic interactions (Lemoine et al., 2025). In termites and cockroaches, gut consortia include abundant methanogenic archaea that convert H_2 and CO_2 to methane, stabilising redox conditions and closing the hydrogen sink in hindgut fermentation. This archaeal metabolism enhances the efficiency of cellulose and lignin degradation and contributes measurably to global methane fluxes (Brune, 2014; Holt et al., 2024). Viruses, including bacteriophages, regulate bacterial population dynamics and may contribute to microbiome stability, although their roles remain comparatively understudied. Across

TABLE 1.3

Major microbial groups associated with insects and their convergent functional roles

Microbial group	Typical insect systems	Dominant functions	Degree of redundancy
Proteobacteria	Flies, mosquitoes, bees	Metabolic flexibility, immune modulation	High
Firmicutes	Pollinators, herbivores	Fermentation, energy extraction	High
Actinobacteria	Beetles, solitary bees	Antimicrobial defence, nutrition	Moderate
Bacteroidetes	Pollen and detritus feeders	Complex polysaccharide degradation	Moderate
Obligate symbionts	Aphids, tsetse flies	Essential nutrient biosynthesis	Very low
Archaea	Termites, cockroaches	Hydrogen turnover, methanogenesis	Low
Viruses	Most insect systems	Community regulation, gene flow	Variable

NOTE: This table summarises the principal microbial taxa recurrently found in insect-associated microbiomes, their typical host systems, and dominant biological functions. Bacterial groups show high functional redundancy in metabolism and energy extraction, whereas obligate symbionts and archaea perform specialised roles with limited redundancy. Together, these patterns highlight functional convergence across taxonomically diverse insect–microbe associations.

these diverse taxa, strong functional convergence is evident. Insect-associated microbial communities repeatedly perform similar ecological roles, including nutrient provisioning, digestion of recalcitrant substrates, immune modulation, detoxification of xenobiotics, and manipulation of host reproduction (Gurung et al., 2019; Holt et al., 2024). These shared functional themes highlight the importance of evaluating insect microbiomes not only in terms of taxonomic composition but also through a functional and ecological lens.

By viewing insects as holobionts, it becomes clear that microbial communities are not passive passengers but active contributors to host metabolism, immunity, reproduction, and ecological performance. Host anatomy and physiology impose strong constraints on microbiome assembly, creating structured niches within the gut, on the cuticle, and within reproductive tissues. Across these niches, a limited number of microbial taxa repeatedly provide conserved functions that underpin insect survival and adaptation. With this conceptual and structural foundation in place, the next chapter will examine how insect microbiomes are dynamically assembled and reshaped by ecological and developmental processes.

1.2 DYNAMIC ASSEMBLY OF INSECT MICROBIOMES

Diet is one of the most powerful ecological forces shaping insect-associated microbiomes. Insects occupy an extraordinary diversity of feeding niches, ranging from strict herbivory and wood feeding to predation, omnivory, nectar feeding, and hematophagy. Each dietary strategy imposes distinct nutritional and chemical constraints that determine which microbial taxa can colonise the gut and which metabolic pathways dominate community function.

Plant-based diets often present insects with nutritionally imbalanced and chemically defended substrates. Herbivorous insects consuming leaves, wood, or phloem sap rely heavily on microbial partners to compensate for nutrient deficiencies and to degrade recalcitrant plant polymers. Termites and wood-feeding beetles host dense and metabolically complex hindgut communities that carry out lignocellulose degradation and fermentation, producing short-chain fatty acids that serve as a major energy source for the host (Liu et al., 2019; Tokuda et al., 2018). In phloem-feeding insects such as aphids and whiteflies, obligate intracellular symbionts provide essential amino acids and vitamins absent from plant sap, enabling the exploitation of otherwise inadequate diets (Bennett & Moran, 2015; Liu et al., 2023).

In contrast, predatory and omnivorous insects typically consume protein-rich diets and display more flexible, environmentally derived microbiomes (Table 1.4). Their gut communities are often less specialised and more variable, reflecting frequent exposure to diverse microbial sources through prey consumption and habitat

TABLE 1.4

Dietary niches and corresponding microbiome strategies in insects

Dietary niche	Nutritional constraints	Dominant microbiome strategy	Microbiome stability
Phloem sap	Amino acid deficiency	Obligate nutritional symbiosis	Very high
Wood/cellulose	Recalcitrant polymers	Fermentative consortia	High
Pollen/nectar	Complex carbohydrates	Specialised but flexible digestion	High
Predation/ omnivory	Protein-rich	Transient, environmentally acquired	Low
Blood	Oxidative stress, vitamin deficiency	Detoxification and immune modulation	Moderate
Generalist feeding	Variable	Rapid restructuring	Variable

NOTE: This table shows major insect feeding niches to their nutritional constraints, dominant microbiome strategies, and typical levels of community stability. Diet strongly shapes microbial function, with obligate symbioses supporting nutrient-poor diets and fermentative groups enabling the digestion of recalcitrant substrates. In contrast, predatory and generalist feeders exhibit more transient and flexible microbiomes that rapidly restructure in response to dietary change.

contact. Despite this variability, microbes still contribute to protein and lipid metabolism, nitrogen recycling, and immune defence.

Pollinators represent a distinct dietary niche. Honeybees and other bees consume pollen and nectar, which select for highly structured and host-specific gut microbiomes. Core bacterial taxa collaborate to degrade pollen walls, ferment plant-derived carbohydrates, regulate gut pH, and suppress pathogens, thereby enhancing foraging efficiency and colony health (De Beer et al., 2023; Zheng et al., 2019) (Figure 1.3).

Blood-feeding insects experience dramatic dietary shifts associated with hematophagy. Blood meals introduce large quantities of protein, iron, and heme, triggering profound changes in gut physiology and microbial composition. Some microbial taxa assist in detoxifying reactive oxygen species (ROS) generated during hemoglobin digestion, while others influence vitamin provisioning and pathogen establishment, with direct consequences for vector competence (Dennison et al., 2014; Weiss & Aksoy, 2011). Together, these examples illustrate how diet acts as a dominant ecological filter that structures insect microbiomes, shaping both community composition and functional potential (Figure 1.3).

Insect microbiomes are not static entities but dynamic assemblages that change across the host life cycle. Holometabolous insects often show pronounced microbiome shifts between larval and adult stages due to gut remodelling, immune activation, and gut purging during pupation (Chen et al., 2016; Manthey et al., 2023; Mason et al., 2023). Developmental transitions from egg to larva, pupa, and adult involve substantial alterations in diet, gut morphology, immune activity, and habitat, all of which influence microbial colonisation. Most insects hatch from eggs that are initially sterile or nearly sterile. Early microbial colonisation typically begins through the ingestion of microbes present on egg surfaces, surrounding substrates, or the first food encountered. Eggs are often weakly colonised or nearly sterile, with initial microbes derived from egg surfaces, oviposition sites, or brood structures (Jones et al., 2024; Yueqing et al., 2025). In some species, females actively inoculate eggs or brood environments with beneficial microbes, biasing early microbiome assembly towards advantageous taxa. Metamorphosis represents a major bottleneck for microbiome continuity. During pupation, the gut often undergoes extensive remodelling, including epithelial renewal and clearance of luminal contents. This process can eliminate much of the larval microbiome, effectively resetting microbial communities between life stages. As a result, larval and adult microbiomes may differ markedly in composition and function (Manthey et al., 2023).

Honeybees provide a well-studied example of post-metamorphic microbiome reassembly. Newly emerged adult workers possess nearly microbe-free guts and acquire their characteristic microbiome through social interactions within the hive (Vienna & S., 2021). Trophallaxis, grooming, and contact with nestmates and hive materials rapidly establish a stable adult gut community dominated by a small set of core bacterial taxa (De Beer et al., 2023; Zheng et al., 2019).

Other insects, such as mosquitoes, experience repeated microbiome restructuring driven by developmental and dietary shifts (Ragini et al., 2025). Aquatic larvae acquire microbes from water and detritus, whereas adults feeding on nectar or blood host distinct microbial assemblages shaped by adult physiology and immune responses (Weiss & Aksoy, 2011). These developmental dynamics highlight the importance of considering life stage when interpreting insect microbiome data and emphasise the transient nature of many host–microbe associations.

While many insect microbiomes are environmentally acquired, some symbiotic relationships are maintained through strict vertical transmission from mother to offspring. Vertical transmission ensures faithful inheritance of microbial partners that are essential for host survival, reproduction, or population persistence. Obligate nutritional symbionts are typically transmitted through the maternal germ line or specialised reproductive tissues (Björk et al., 2019; Bright & Bulgheresi, 2010). In aphids, *Buchnera* is transferred directly into developing embryos within the maternal body, guaranteeing immediate access to essential amino acids upon feeding (Bennett & Moran, 2015; Liu et al., 2023). In tsetse flies, *Wigglesworthia* is delivered to offspring via milk gland secretions during intrauterine larval development, playing a critical role in immune system maturation and fertility (Gupta & Nair, 2020).

Vertical transmission stabilises host–symbiont associations over evolutionary time but also constrains symbiont diversity and genome evolution. Many vertically transmitted symbionts

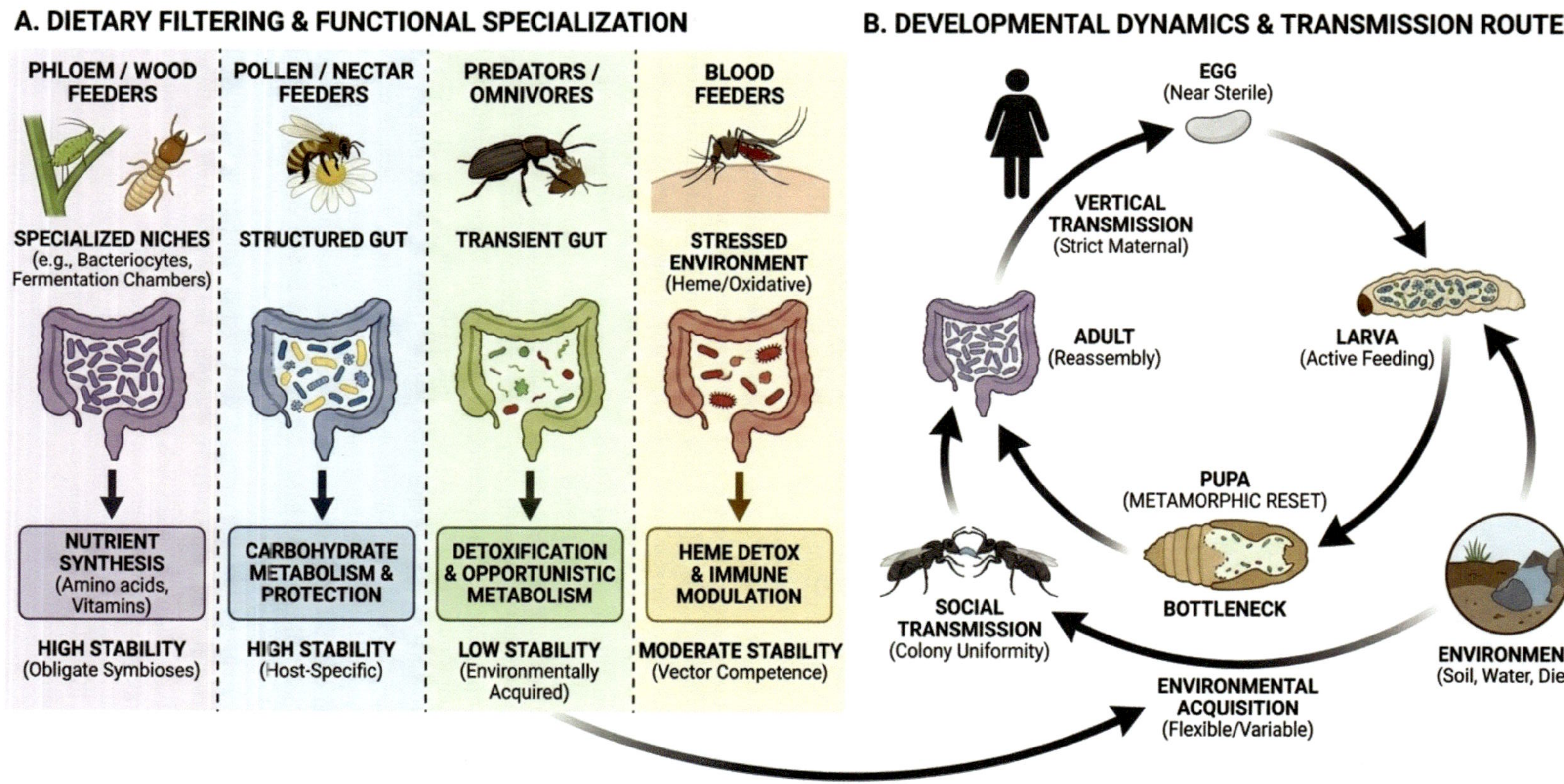

Figure 1.3 The dynamic landscape of insect microbiome assembly across diets and development. (A) Diet-driven microbiome specialization across insect feeding guilds. (B) Developmental and transmission-related routes shaping microbiome assembly.

exhibit reduced genomes and limited metabolic flexibility, reflecting long-term coevolution within a protected intracellular environment (Sudakaran et al., 2015). In contrast to strict maternal inheritance, many insects acquire their microbiomes horizontally from conspecifics or shared environments. Social interactions play a particularly important role in microbiome transmission among eusocial insects (Suenami et al., 2023). In honeybees, termites, and ants, trophallaxis, grooming, and close physical contact facilitate the transfer of beneficial microbes among colony members. These processes maintain colony-level microbiome uniformity and contribute to collective immunity and nutritional efficiency (De Beer et al., 2023; Zheng et al., 2019). Social transmission also buffers against microbiome loss during metamorphosis, ensuring rapid re-establishment of adult-specific communities. Horizontal transmission can also occur through shared food sources. Fruit flies acquire yeasts and bacteria from fermenting fruits, which influence larval development and adult behaviour. Predatory insects may obtain microbes directly from prey, leading to rapid turnover and diet-dependent community composition (Guilhot et al., 2023).

While horizontal transmission increases microbiome flexibility and ecological responsiveness, it also introduces variability and potential exposure to pathogens, necessitating robust immune regulation to maintain homeostasis.

Environmental acquisition represents a major route of microbiome assembly for insects with flexible diets or broad ecological niches. Soil-dwelling larvae ingest microbes from substrates rich in bacterial and fungal diversity, while aquatic larvae encounter microbial communities shaped by water chemistry and nutrient availability (Wang et al., 2025; Yun et al., 2014). Environmental sourcing allows insects to recruit microbes that enhance detoxification, nutrient assimilation, or stress tolerance in specific habitats. However, reliance on environmental acquisition also increases susceptibility to dysbiosis under rapidly changing conditions, such as habitat degradation, pesticide exposure, or climate stress. Generalist insects, including cockroaches and some dipteran larvae, exemplify extreme microbiome plasticity. Their gut communities shift rapidly in response to available food sources, enabling exploitation of diverse environments but also resulting in high interindividual variability. Diet acts as a primary ecological filter, selecting microbial taxa and functions suited to specific nutritional niches (Luo et al., 2021; Mugo-Kamiri et al., 2024). Developmental transitions, particularly metamorphosis, impose major restructuring events that can reset microbial communities. Transmission routes ranging from strict vertical inheritance to flexible environmental acquisition determine the stability, diversity, and evolutionary trajectory of insect–microbe associations (Huang et al., 2021; Sanaei et al., 2024). These dynamic processes set the stage for understanding how insects maintain microbial homeostasis in the face of continual microbial flux.

1.3 IMMUNE REGULATION AND MICROBIOME HOMEOSTASIS

Insects lack an adaptive immune system but possess a highly effective innate immune repertoire that enables them to recognise, regulate, and respond to microbial colonisation (Zhang et al., 2025). This immune system must simultaneously perform two seemingly opposing tasks: eliminating pathogenic microbes while tolerating beneficial symbionts that contribute to host fitness. Achieving this balance is essential for maintaining microbiome homeostasis and overall insect health (Bai et al., 2021; Zhang et al., 2021). Insect immunity relies on epithelial barriers, cellular defences, and conserved signalling pathways that detect microbial molecules. Pattern-recognition receptors (PRRs) recognise conserved microbial-associated molecular patterns such as peptidoglycan, lipopolysaccharides, and fungal β-glucans (Zhang et al., 2025). Recognition triggers downstream immune signalling cascades that regulate antimicrobial peptide (AMP) production, ROS generation, and epithelial renewal (Zhang et al., 2021). Two signalling pathways dominate systemic and gut-associated immune responses. The Toll pathway responds primarily to fungi and Gram-positive bacteria, whereas the immune deficiency (Imd) pathway is mainly activated by Gram-negative bacteria. Additional pathways, including Janus Kinase/Signal Transducer and Activator of Transcription (JAK/STAT), JNK, and Forkhead Box O (FOXO), modulate immune responses, stress signalling, and epithelial homeostasis, particularly within the gut environment (Bai et al., 2021; Wang et al., 2025).

To sustain stable microbiomes, insects must avoid excessive immune activation against resident microbes. Immune tolerance is achieved through spatial, molecular, and quantitative regulation of immune responses (Figure 1.4). In the insect gut, immune activity is often compartmentalised. The peritrophic matrix and epithelial mucus layers physically separate microbes from gut epithelial cells, reducing direct immune stimulation (Douglas, 2015; Khan et al., 2023). At the molecular level, specialised PRRs such as peptidoglycan recognition proteins (PGRPs) degrade bacterial cell wall fragments before they can trigger immune signalling. In *Drosophila*, PGRP-LB and PGRP-SC reduce peptidoglycan availability, thereby preventing chronic Imd pathway activation while still permitting pathogen detection (Zhang et al., 2021).

Social insects provide additional examples of immune tolerance. Core gut bacteria in honeybees, including *Snodgrassella alvi* and *Gilliamella apicola*, form stable, epithelium-associated communities that are generally tolerated and can even enhance host immunity rather than provoke damaging inflammation. Honeybees host a small, conserved microbiota dominated by about nine species clusters, with *S. alvi* forming a biofilm directly on the ileal epithelium and *G. apicola* growing on top of this layer. These bacteria are long-term, host-specialised symbionts that persist across individuals and colonies, indicating an evolved mutualistic relationship rather than a typical inflammatory response to epithelial colonisation (Horak et al., 2020; Powell et al., 2016). Host immune responses are finely tuned to allow colonisation by these beneficial taxa while retaining the capacity to respond rapidly to opportunistic pathogens (De Beer et al., 2023; Zheng et al., 2019). Honeybee colonies illustrate immune tolerance to tightly associated gut symbionts, *S. alvi* and *G. apicola*, form stable, epithelium-adjacent biofilms yet are largely non-pathogenic, instead modulating immunity and providing colonisation resistance and pathogen control. This regulated tolerance highlights that immune quiescence is not passive but actively maintained through host–microbe coevolution.

The intestinal epithelium is a dynamic barrier that both senses microbes and shapes their spatial organisation, keeping dense communities in the lumen while limiting contact with the mucosa and blood compartment (Soderholm & Pedicord, 2019). AMPs, ROS, and epithelial renewal collectively prevent microbial overgrowth and translocation into the hemocoel (Bai et al., 2021). ROS production, mediated by enzymes such as dual oxidase (Duox), plays a key role in restricting microbial proliferation. Low, regulated ROS levels suppress excessive bacterial growth without eliminating beneficial microbes. Excessive ROS production, however, can damage host tissues and disrupt microbiome composition, underscoring the importance of precise regulation (Bai et al., 2025; Xiao et al., 2017). Commensal microbes actively contribute to gut homeostasis. Many symbionts modulate gut pH, compete for nutrients and attachment sites, and produce antimicrobial compounds that inhibit invading pathogens. Short-chain fatty acids produced during microbial fermentation nourish epithelial cells and dampen inflammatory signalling, reinforcing barrier integrity (Douglas, 2015; Gupta & Nair, 2020). Disruption of these regulatory processes, whether through immune gene mutations, antibiotic exposure, or environmental stress, often leads to dysbiosis characterised by microbial imbalance and host pathology (Figure 1.4).

Beyond passive coexistence, insect microbiomes actively enhance host defence through multiple mechanisms.

Competitive Exclusion: Resident microbes occupy ecological niches within the gut, limiting resources and space available to invading pathogens. This competitive exclusion represents a primary defence layer that reduces infection risk without requiring immune activation (Zhang et al., 2021).

Antimicrobial Metabolite Production: Many insect-associated microbes produce antimicrobial compounds. Actinobacteria associated with solitary bees and beetles synthesise antibiotics that protect larvae and food stores from fungal and bacterial pathogens (Dong et al., 2024).

Immune Priming: Some symbionts induce low-level immune activation that prepares the host for subsequent pathogen exposure. In mosquitoes, specific gut bacteria stimulate immune pathways that suppress *Plasmodium* development, reducing malaria transmission potential (Weiss & Aksoy, 2011). Through these mechanisms, microbiomes function as integral components of the insect immune system rather than as independent entities.

Dysbiosis refers to a disruption of normal microbiome composition and function, often resulting from environmental stressors, dietary changes, pesticide exposure, or immune dysfunction. In insects, dysbiosis frequently leads to impaired digestion, excessive immune activation, increased susceptibility to pathogens, and reduced

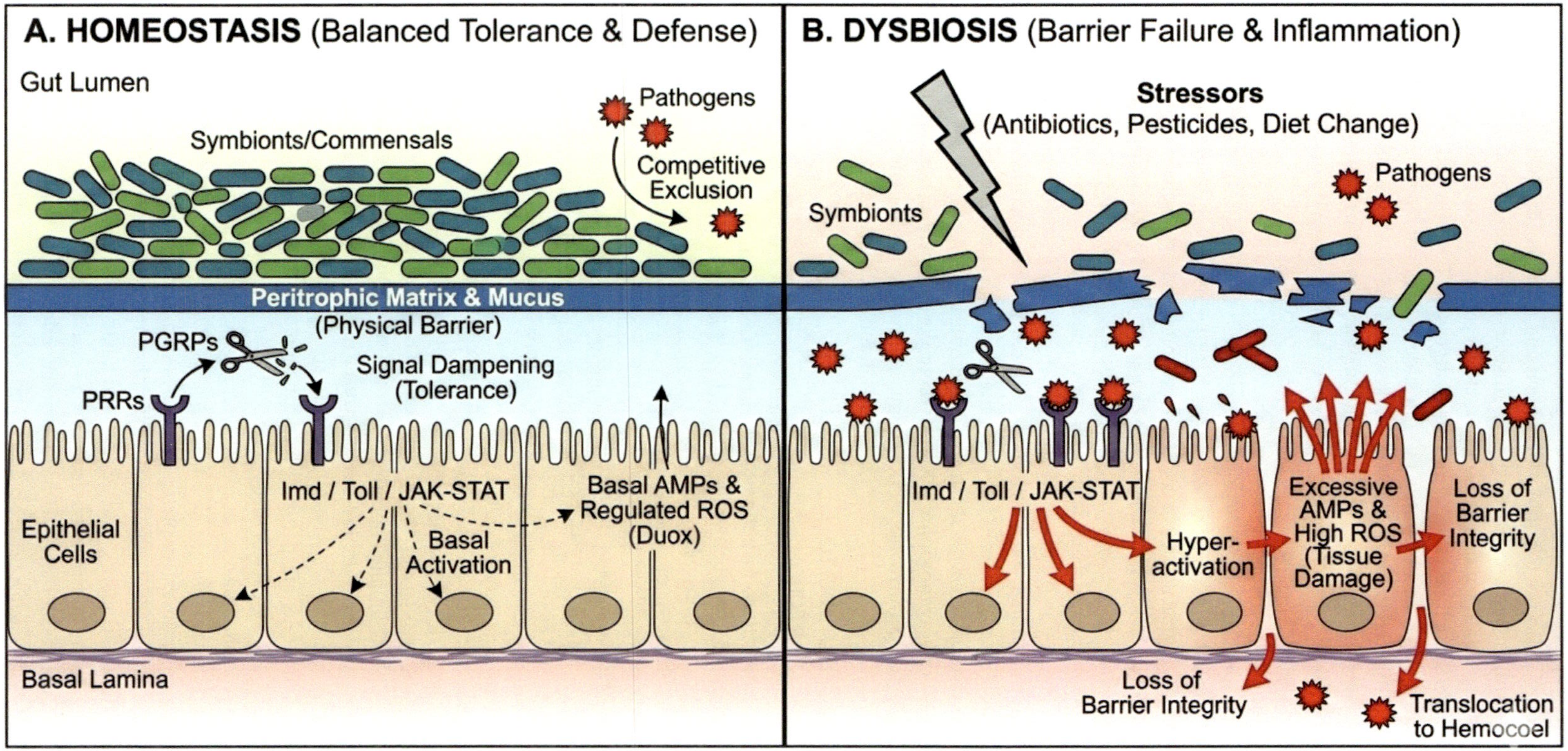

Figure 1.4 Schematic representation of immune-microbiome homeostasis (A) and dysbiosis (B) in the insect gut. Under homeostatic conditions, commensal and symbiotic microbes are spatially constrained by the peritrophic matrix and mucus layer, promote competitive exclusion of pathogens, and maintain controlled activation of innate immune pathways. In contrast, dysbiosis induced by stressors such as antibiotics, pesticides, or dietary change disrupts barrier integrity, leads to excessive immune activation, epithelial damage, and facilitates pathogen translocation into the hemocoel.

fitness (Zhang et al., 2021). Honeybees provide a well-documented example. Exposure to antibiotics or agricultural chemicals disrupts core gut microbiota, weakening immune stability and increasing vulnerability to pathogens such as *Nosema* and *Crithidia*. These effects are often accompanied by increased mortality and reduced colony performance. In agricultural pests and disease vectors, dysbiosis can alter feeding behaviour, reproductive success, and vector competence. Changes in microbial composition may either increase or decrease pathogen transmission, demonstrating that immune–microbiome interactions have broad ecological and epidemiological consequences (Dennison et al., 2014; Weiss & Aksoy, 2011).

Insects maintain microbial homeostasis through finely regulated immune mechanisms that balance tolerance and defence. Physical barriers, immune signalling pathways, and microbial contributions collectively stabilise gut communities while preventing pathogen invasion. When this balance is disrupted, dysbiosis can compromise host health, fitness, and ecological performance.

1.4 MICROBIOME CONTRIBUTIONS TO HOST FITNESS AND ADAPTATION

One of the most fundamental contributions of insect-associated microbiomes is their role in compensating for dietary limitations. Many insects exploit nutritionally imbalanced or chemically complex food sources that would be insufficient to sustain development in the absence of microbial partners (Al Darwish et al., 2025; Douglas, 2009). Symbiotic microorganisms expand host metabolic capacity by synthesising essential nutrients, providing digestive enzymes, and facilitating energy extraction from recalcitrant substrates (Douglas, 2015; Mondal et al., 2023). Together, these microbiome-mediated processes form an integrated fitness framework linking nutrition, development, reproduction, behaviour, and stress resilience (Figure 1.5).

Obligate nutritional symbionts are particularly critical in insects feeding on restricted diets. In aphids and other phloem-feeding insects, intracellular symbionts such as *Buchnera* synthesise essential amino acids that are scarce or absent in plant sap. Experimental removal of these symbionts results in stunted growth, reduced fecundity, and high mortality, underscoring their indispensable role in host nutrition (Bennett & Moran, 2015; Liu et al., 2023). Similarly, *Wigglesworthia* in tsetse flies provides B vitamins required for larval development and adult fertility (Gupta & Nair, 2020).

Insects consuming structurally complex plant material rely on gut-associated microbial consortia for digestion. Termites and wood-feeding beetles host dense hindgut communities that degrade cellulose and hemicellulose through microbial fermentation, producing short-chain fatty acids that serve as major energy sources for the host (Tokuda et al., 2018; Wen et al., 2023). In pollinators, bacterial taxa such as *Gilliamella* and *Bifidobacterium* contribute to pollen wall degradation and carbohydrate fermentation, improving nutrient assimilation and feeding efficiency (De Beer et al., 2023; Zheng et al., 2019). Through these nutritional functions, microbiomes directly influence growth rates, body size, and developmental success, thereby shaping overall fitness.

Across insects, microbiome composition often affects growth rate, timing of metamorphosis, and stage transitions, but there are notable exceptions where development proceeds normally without a gut community. In holometabolous insects, where larval and adult stages occupy distinct ecological niches, microbial contributions to development are especially evident. In the wax moth *Galleria mellonella*, *Enterococcus innesii* dominates larval guts and delays larva-to-pupa transition; host AMPs later reduce *Enterococcus*, triggering pupation (Gi et al., 2023). In axenic *Drosophila melanogaster*, development from egg to adult is slower; adding live yeasts (e.g., *Saccharomyces cerevisiae*) restores faster development, partly via increased feeding during late larval stages, implicating a gut–brain axis effect (Jiménez-Padilla et al., 2024). Reviews highlight that gut symbionts can alter hormonal control, nutrient acquisition, and tissue differentiation, with carry-over effects on body size and timing of molts and metamorphosis (Girard et al., 2023). In the stinkbug *Plautia stali*, experimental manipulation of metamorphosis shows that remodelling of a specialised symbiotic organ changes symbiont function towards provisioning sulphur-rich amino acids for reproduction, illustrating stage-specific reprogramming of symbiosis across metamorphosis (Oishi et al., 2023). Metamorphosis often causes major microbiota turnover, potentially facilitating ecological niche shifts between larvae and adults (Manthey et al. 2023).

In social insects, microbiome-mediated nutritional effects can influence caste differentiation and colony organisation. Although host genetics

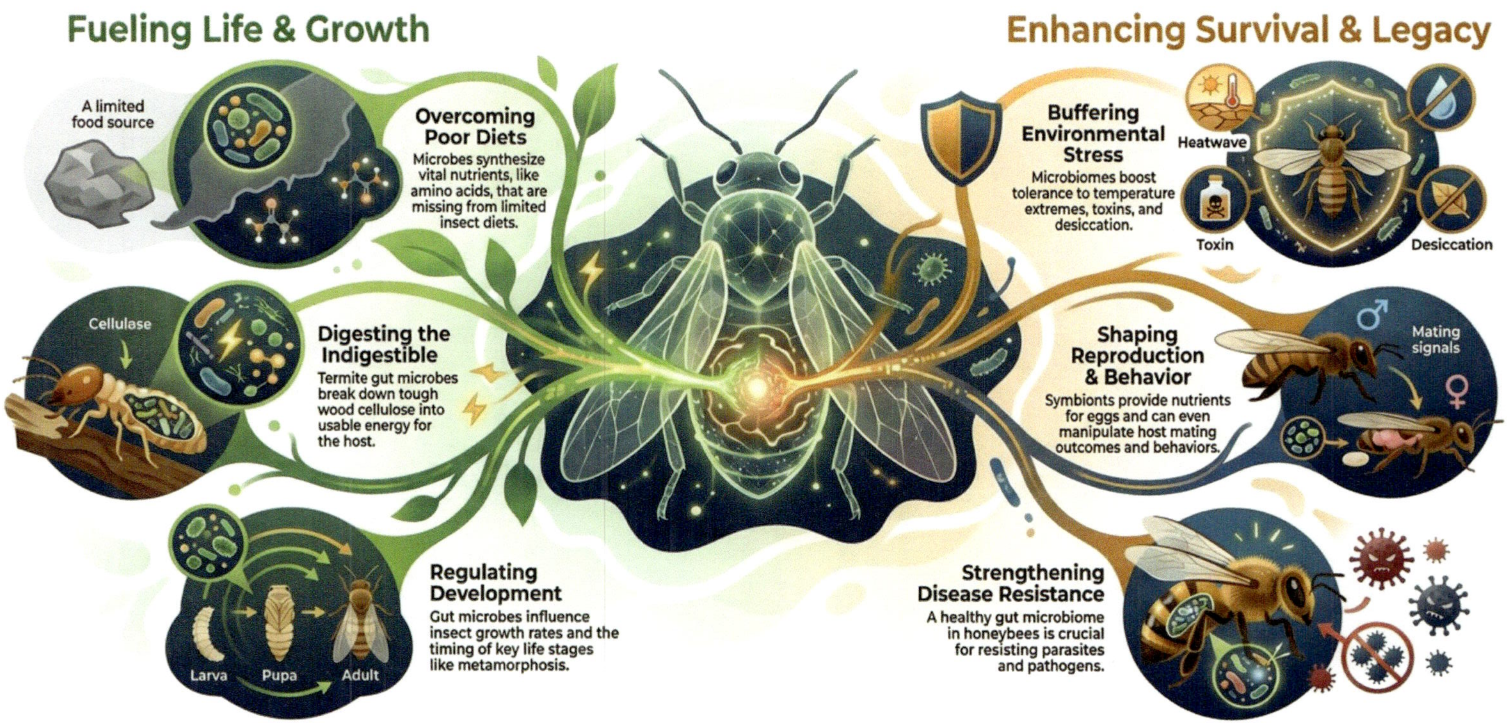

Figure 1.5 Microbiome contributions to insect host fitness and adaptation. This schematic summarises the major pathways through which insect-associated microbiomes enhance host fitness across the life cycle. Microbial symbionts support nutrition and growth by compensating for dietary limitations, synthesising essential nutrients, and facilitating digestion of recalcitrant substrates. During development, microbiomes influence growth rate, developmental timing, and metamorphic transitions through interactions with host metabolism and hormonal signalling. Microbial partners also shape reproductive success and transgenerational outcomes by supporting oogenesis, fertility, and, in some systems, by manipulating reproductive processes. Behavioural effects arise from microbial metabolites that modulate feeding, foraging, oviposition, and social interactions. In addition, microbiomes improve host resilience by improving tolerance to environmental stressors, strengthening immune defences, and contributing to detoxification of plant secondary metabolites and anthropogenic chemicals. Together, these organised functions form an integrated microbiome-driven fitness framework that supports insect survival, ecological adaptation, and evolutionary success.

and environmental cues play dominant roles, microbial metabolites may contribute indirectly by modulating larval nutrition and metabolic signalling. Developmental stability supported by a functional microbiome therefore enhances both individual performance and colony-level fitness (Su et al., 2024).

Microbial symbionts influence multiple aspects of insect reproduction, ranging from nutrient provisioning to direct manipulation of reproductive systems. Nutritional symbionts increase reproductive output by ensuring sufficient availability of amino acids, vitamins, and cofactors required for oogenesis and embryonic development (Douglas, 2015). Reproductive manipulators such as *Wolbachia* exert profound effects on host fertility and population dynamics. By inducing cytoplasmic incompatibility, parthenogenesis, feminisation, or male killing, these bacteria alter mating outcomes and sex ratios in host populations (Al Darwish et al., 2025; Fricke et al., 2025). In some cases, such manipulations increase the reproductive success of infected females, promoting rapid spread of the symbiont through host populations.

Beyond intracellular symbionts, gut microbiota may also influence reproductive behaviour and mate choice. In *Drosophila*, microbial metabolites affect cuticular hydrocarbon profiles and mating preferences, linking microbiome composition to sexual selection and assortative mating. These effects demonstrate that microbial contributions to reproduction extend beyond physiology to behavioural and evolutionary dimensions (Everaerts et al., 2010).

Insect behaviour is shaped not only by host neural circuits but also by microbial metabolic activity. Microorganisms associated with insects produce a wide range of volatile and non-volatile compounds that influence feeding, foraging, oviposition, and social interactions.

In fruit flies, yeasts and bacteria associated with fermenting substrates release volatiles that attract adults to optimal oviposition sites and nutrient-rich environments (Becher et al., 2012). These microbial cues guide host behaviour in ways that enhance larval survival and resource exploitation. In pollinators, microbial modification of nectar chemistry can influence foraging preferences and learning behaviour, indirectly affecting plant–pollinator interactions.

In social insects, microbiome-mediated modulation of cuticular hydrocarbons may contribute to nestmate recognition and colony cohesion (Lorenzi & d'Ettorre, 2020). By shaping chemical communication systems, microbes influence social organisation and cooperative behaviour, which are key determinants of fitness in eusocial species.

Insects frequently encounter environmental stressors such as temperature extremes, desiccation, toxins, and pathogens (Walton et al., 2024). The microbiome often enhances host resilience under these conditions, facilitating adaptation to challenging environments. Gut-associated microbes can improve thermal tolerance by modulating metabolic pathways involved in energy balance and stress responses (Jyoti & Dey, 2025). Certain bacterial taxa increase survival during heat or cold stress by influencing host physiology and metabolic flexibility. Similarly, microbial detoxification of plant secondary metabolites and anthropogenic chemicals allows insects to exploit chemically defended resources and survive in polluted habitats (Lemoine et al., 2020).

Microbiomes also enhance immune resilience, priming host defences and reducing pathogen susceptibility under stress (Jyoti & Dey, 2025). In honeybees, disruption of core gut microbiota by pesticides or antibiotics reduces stress tolerance and increases mortality, whereas restoration of microbial balance improves survival (El Khoury et al., 2024). Through these mechanisms, microbiomes function as buffers that mitigate environmental stress and expand the ecological niches available to insect hosts.

A stable and functional microbiome is closely associated with increased longevity and improved disease resistance in insects. Multiple studies show that disrupting the core gut microbiota increases parasite susceptibility, mortality, and physiological stress in honeybees. Diet-induced dysbiosis (aged pollen/substitutes) reduces the core symbiont *S. alvi*, increases opportunists (*Frischella perrara*, *Parasaccharibacter apium*), and correlates with impaired development, higher mortality, and higher *Nosema* prevalence (Maes et al., 2016). Experimental skewing of early colonisation with *S. alvi* or nutritional/social stress produces dysbiosis and increases susceptibility to the trypanosomatid parasite *Lotmaria passim*, with broad changes in immunity and detoxification gene expression (Schwarz et al., 2016). Antibiotic treatment causes long-lasting gut community disruption and elevated mortality, especially when bees are challenged with opportunistic bacteria such as *Serratia* (Raymann et al., 2017). Conversely, maintenance or restoration of healthy microbiomes can reverse negative outcomes, highlighting the causal role of microbial communities in sustaining insect health across the lifespan. By supplementing nutrition, supporting development, enhancing reproductive success, modulating behaviour, and increasing tolerance to environmental stress, microbial communities play a central role in insect adaptation and ecological success. These contributions are not isolated effects but interconnected processes that integrate metabolism, immunity, and behaviour into a coherent fitness strategy.

1.5 MICROBIOME-MEDIATED DETOXIFICATION AND VECTOR BIOLOGY

Insects are continuously exposed to a wide range of chemical stressors originating from both natural and anthropogenic sources. These include plant secondary metabolites, microbial toxins, heavy metals, and synthetic pesticides. Survival in chemically complex environments requires effective detoxification strategies, many of which are mediated not only by host-encoded enzymes but also by associated microbial communities (Lemoine et al., 2020).

Plant-feeding insects, in particular, encounter diverse allelochemicals that function as anti-herbivore defences. Similarly, agricultural pests and disease vectors are subjected to intense selection pressure from insecticides used in crop protection and vector control. The insect microbiome has emerged as a critical component of detoxification capacity, expanding the biochemical repertoire available to the host and shaping ecological success under chemical stress.

Herbivorous insects frequently exploit plants that produce toxic compounds such as alkaloids, terpenoids, phenolics, and glucosinolates. Microbial symbionts contribute to detoxification through several mechanisms, including enzymatic degradation, chemical modification, and sequestration of toxic compounds. Gut-associated bacteria can metabolise plant toxins before they are absorbed into host tissues, thereby reducing physiological damage. In some leaf-feeding insects, microbial degradation of tannins and phenolic compounds improves nutrient availability and feeding efficiency, allowing hosts to exploit chemically defended plants. These microbial detoxification processes not only enhance host survival but also influence host plant range and feeding specialisation, contributing to insect diversification and ecological expansion.

Exposure to synthetic insecticides represents a strong selective force on both insects and their microbiomes. Increasing evidence demonstrates that microbial symbionts can directly contribute to insecticide resistance by degrading or modifying toxic compounds before they exert lethal effects on the host. In several agricultural pests, gut-associated bacteria possess enzymes capable of hydrolysing organophosphate and carbamate insecticides, thereby reducing toxicity. Pyrethroid resistance in mosquitoes and cockroaches has also been linked to microbial degradation pathways that complement host detoxification enzymes. Although evidence for neonicotinoid degradation by insect-associated microbes is still emerging, microbial modulation of absorption and metabolic intermediates may reduce effective toxicity. Microbiome-mediated resistance can operate independently of host genetic resistance mechanisms, highlighting microbes as autonomous drivers of resistance phenotypes.

Microbiome-mediated detoxification has important evolutionary consequences. By reducing the effective dose of toxic compounds, microbial symbionts can buffer insects against chemical stress, allowing survival of individuals that might otherwise be eliminated. This buffering effect facilitates the persistence and spread of host populations under pesticide pressure.

In addition, microbial communities are hotspots for horizontal gene transfer. Genes encoding detoxification enzymes can move among bacterial taxa within the insect gut, accelerating the evolution of community-level resistance. Repeated exposure to pesticides selects for microbial taxa with detoxification capacity, leading to stable, resistance-enhancing microbiomes that persist across generations. These processes complicate pest management by creating multilayered resistance systems that involve both host and microbial components. The dual role of insect-associated microbiomes as both protective agents against chemical stress and modulators of disease transmission is summarised schematically in Figure 1.6.

In disease vectors, microbiomes influence not only survival and resistance to insecticides but also pathogen transmission dynamics. Vector competence, defined as the ability of an insect to acquire, maintain, and transmit pathogens, is strongly modulated by gut-associated microbial

Figure 1.6 Microbiome-mediated detoxification and vector competence in insects. This schematic highlights the dual role of insect-associated microbiomes in host adaptation and disease transmission. Microbial symbionts enhance survival by detoxifying plant secondary metabolites and synthetic insecticides, thereby contributing to resistance development under chemical stress. At the same time, these microbial communities modulate vector competence by influencing pathogen establishment and immune responses within the insect gut. Together, these interconnected functions illustrate how microbiomes act as a key interface between detoxification, resistance evolution, and pathogen transmission in insect hosts.

communities (Dennison et al., 2014; Weiss & Aksoy, 2011).

Mosquitoes: Mosquito gut microbiomes typically include bacteria such as *Asaia*, *Enterobacter*, *Serratia*, and *Pseudomonas*. Blood feeding induces dramatic shifts in microbial composition due to changes in pH, temperature, iron availability, and oxidative stress (Weiss & Aksoy, 2011). Certain bacterial taxa inhibit *Plasmodium* development through the production of ROS or the activation of host immune pathways. For example, *Enterobacter* species can suppress malaria parasites in the midgut, reducing transmission potential. Other microbes influence arbovirus replication, either enhancing or suppressing infections such as dengue and Zika (Dennison et al., 2014; Weiss & Aksoy, 2011).

Tsetse Flies: Tsetse flies harbour a relatively simple but highly specialised microbiome dominated by *Wigglesworthia*, *Sodalis*, and *Wolbachia*. *Wigglesworthia* is essential for immune system maturation, and its absence increases susceptibility to *Trypanosoma* infections. *Sodalis* has been explored as a vehicle for paratransgenesis, in which engineered symbionts express antiparasitic molecules to reduce disease transmission (Gupta & Nair, 2020).

Other Vectors: Sand flies, kissing bugs, fleas, and other ectoparasites also rely on microbial symbionts for nutrition, immune regulation, and pathogen interaction. Although less extensively studied, these systems demonstrate that microbiome–pathogen interactions are a general feature of vector biology rather than an exception.

Understanding microbiome-mediated detoxification and pathogen modulation has enabled the development of innovative control strategies that target microbial partners rather than insects alone (Branda et al., 2024). One prominent example is the introduction of *Wolbachia* into *Aedes* mosquito populations. *Wolbachia* reduces viral replication and shortens mosquito lifespan, significantly decreasing dengue transmission without the use of chemical insecticides.

Paratransgenesis represents another promising approach. By engineering symbiotic bacteria to express antimicrobial or antiparasitic molecules, researchers can interfere with pathogen development inside vectors such as tsetse flies and kissing bugs.

Manipulation of environmental microbial exposure, probiotic supplementation, and disruption of microbial detoxification pathways are also being explored as tools to restore pesticide sensitivity or reduce vector competence. Microbial symbionts expand host detoxification capacity, accelerate resistance evolution, and modulate vector competence, positioning microbiomes as critical determinants of both agricultural outcomes and public health risk. By integrating microbial metabolism with applied entomology, these findings underscore the need to consider insect–microbe interactions when designing sustainable pest and vector management strategies.

1.6 ENVIRONMENTAL MODULATION AND MODEL SYSTEMS

Insect-associated microbiomes are shaped not only by host genetics, diet, and immune regulation but also by the broader environmental context in which insects live. Habitat characteristics, climate variables, microbial reservoirs, and anthropogenic pressures collectively determine microbial exposure and influence which taxa can colonise and persist within insect hosts (Schmidt & Engel, 2021).

Environmental microbes serve as the primary source pool for many insect microbiomes. Soil, plant surfaces, decaying organic matter, and water bodies harbour diverse microbial communities that insects encounter during feeding, development, and reproduction. Consequently, variation in habitat type and quality can lead to pronounced differences in microbiome composition, even among closely related insect populations (Schmidt & Engel, 2021).

Honeybees represent one of the best-characterised and most informative model systems for studying insect-associated microbiomes. Unlike many insects with highly variable or environmentally derived gut communities, honeybees maintain a compact, host-specific, and socially transmitted microbiome that is remarkably conserved across individuals and geographic regions (De Beer et al., 2023; Zheng et al., 2019). The adult worker honeybee gut harbours a core community dominated by a small number of bacterial lineages, including *S. alvi*, *G. apicola*, *Bifidobacterium* spp., and *Lactobacillus* Firm-4 and Firm-5 clades. These taxa colonise distinct gut compartments and engage in cooperative metabolic interactions that support digestion, immune regulation, and barrier integrity (De Beer et al., 2023). Newly emerged adult bees possess nearly sterile guts and acquire their microbiome through social interactions within the hive, such as trophallaxis, grooming, and contact with nestmates and hive materials. This mode of social transmission ensures colony-level microbiome consistency and highlights the importance of behavioural ecology in shaping microbial inheritance (Zheng et al., 2019) (Figure 1.7).

The honeybee microbiome plays central roles in nutrition, immunity, and colony health. Gut bacteria contribute enzymes that degrade pollen walls and complex carbohydrates, enhancing nutrient availability from pollen and nectar. Fermentative metabolism produces organic acids that regulate gut pH and inhibit opportunistic pathogens (De Beer et al., 2023). Microbial interactions with the gut epithelium modulate immune signalling pathways, maintaining tolerance towards beneficial taxa while preserving responsiveness to pathogens. Stable microbiome composition has been linked to resistance against parasites such as *Nosema* and *Crithidia*, which are associated with reduced worker lifespan and colony decline. Because of its simplicity, functional specialisation, and experimental accessibility, the honeybee microbiome has become a powerful system for dissecting host–microbe interactions and for assessing the impacts of environmental stressors.

Climatic factors such as temperature and humidity strongly influence microbial physiology and, by extension, insect-associated microbiomes. Heat stress can reduce microbial diversity or favour thermotolerant taxa, leading to altered metabolic output and immune interactions. Seasonal variation in resource availability further shapes microbiome composition. Pollinators experience seasonal shifts in gut communities corresponding to changes in floral diversity and nectar chemistry. Similarly, termites and other detritivores exhibit seasonal microbiome restructuring linked to moisture availability and substrate composition.

Habitat type also plays a critical role. Insects inhabiting agricultural landscapes, forests, or urban environments encounter distinct microbial reservoirs, leading to environment-specific microbiome signatures. These differences can affect digestion, immunity, and stress tolerance.

Human activities impose strong selective pressures on insect microbiomes. Exposure to pesticides, heavy metals, and industrial pollutants can directly alter microbial community structure by selecting for tolerant taxa or by eliminating sensitive symbionts (Lemoine et al., 2020).

Urbanisation often reduces plant diversity and alters food resources, particularly for pollinators. Reduced dietary diversity has been associated with decreased microbial diversity and impaired immune stability in bees, increasing susceptibility to pathogens (De Beer et al., 2023). Artificial food sources, such as sugar feeders, may further modify microbiome composition by providing nutritionally simplified substrates. These anthropogenic effects highlight the vulnerability of insect microbiomes to environmental disturbance and underscore the need for microbiome-aware management practices.

In herbivorous insects, host plants exert strong indirect effects on microbiome composition. Plant secondary metabolites shape gut microbial communities by selecting for taxa capable of tolerating or degrading specific compounds. Insects harbouring microbial detoxification partners can expand their host range, feeding on plants that are inaccessible to other species. Coevolution among insects, plants, and microbes can therefore drive ecological specialisation and diversification. Microbiome-mediated plant–insect interactions represent an important but often underappreciated dimension of coevolutionary dynamics.

Invasive insects frequently undergo rapid microbiome restructuring upon entering new environments. Changes in diet, habitat, and microbial exposure can lead to altered community composition that enhances adaptability to novel conditions. Microbiome plasticity may contribute to invasion success by facilitating detoxification, nutrient acquisition, and stress tolerance in unfamiliar ecosystems. Comparative studies between native and invasive populations reveal that microbiome shifts often accompany range expansion and ecological dominance.

Insects exist within complex ecological networks involving predators, parasites, mutualists, and competitors. These interactions influence microbiome composition through changes in behaviour, diet, and microbial exposure. Predation pressure can alter feeding patterns and stress physiology, indirectly shaping gut communities. Parasitoids and pathogens introduce additional microbial challenges that may disrupt resident microbiota. Mutualistic interactions, such as ant–aphid associations, facilitate microbial exchange among partners, further integrating microbiomes into broader ecological systems. Climate, habitat, anthropogenic stress, host plant interactions, and community dynamics collectively shape microbial

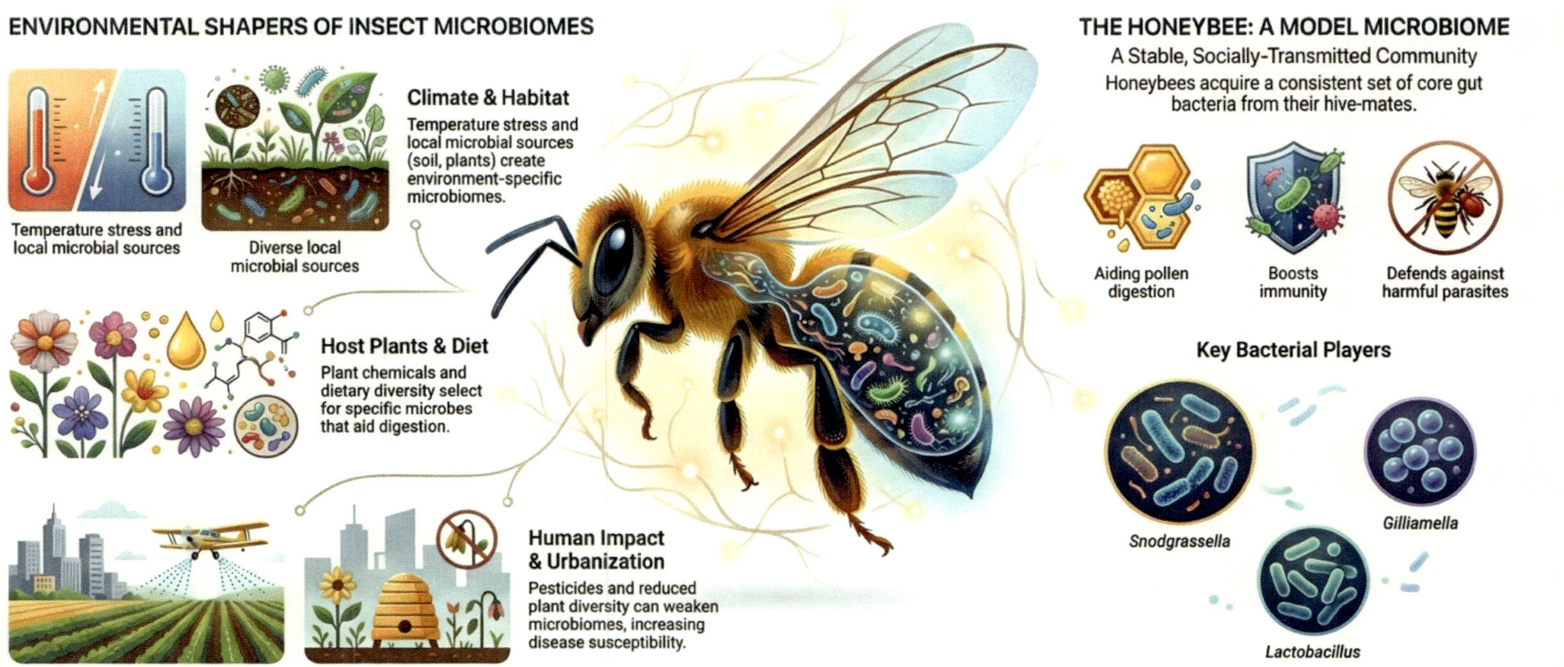

Figure 1.7 Environmental modulation of insect-associated microbiomes illustrated through the honeybee model system. The figure highlights how habitat, climate, diet, and anthropogenic stressors shape microbial acquisition, community structure, and functional outputs within insect hosts. Social transmission in honeybees exemplifies how stable, host-specific microbiomes can be maintained while remaining sensitive to environmental disturbance, linking ecological context to insect health and resilience.

exposure and selection. The honeybee microbiome exemplifies how a stable, socially transmitted microbial community can support host health while remaining sensitive to environmental disturbance. By situating insect microbiomes within their ecological landscape, this chapter provides essential context for interpreting variability and resilience across systems.

1.7 METHODS, APPLICATIONS, AND ONE HEALTH INTEGRATION

Robust investigation of insect-associated microbiomes requires careful consideration of experimental design, sampling strategies, molecular techniques, and analytical workflows. Insects often harbour low-biomass and highly compartmentalised microbial communities, making microbiome studies particularly sensitive to contamination and methodological bias (Siddiqui et al., 2022). Sampling strategies must account for gut compartmentalisation, life stage, diet, and environmental exposure. Dissecting foregut, midgut, and hindgut regions separately provides more accurate insights into spatial organisation and functional differentiation of microbial communities. Inclusion of negative controls and surface sterilisation protocols is essential, especially when working with small insects or intracellular symbionts (Siddiqui et al., 2022).

DNA extraction methods must effectively lyse both Gram-positive and Gram-negative bacteria while minimising host DNA contamination. Mechanical disruption, enzymatic digestion, and chemical lysis are often combined to maximise recovery of microbial DNA from insect tissues.

Marker gene amplicon sequencing, particularly targeting the bacterial 16S rRNA gene and fungal Internal Transcribed Spacer (ITS) regions, remains the most widely used approach for profiling insect microbiomes. This method is cost-effective and enables large-scale comparative analyses but is limited in taxonomic resolution and functional inference (Siddiqui et al., 2022). Shotgun metagenomic sequencing overcomes these limitations by providing species- or strain-level resolution and enabling direct identification of genes involved in digestion, detoxification, immune modulation, and pathogen inhibition (Dar et al., 2024). However, metagenomics requires greater sequencing depth and computational resources and is challenged by host DNA contamination in insect samples.

Functional insights are further enhanced through metatranscriptomics and metabolomics. Metatranscriptomics captures actively expressed microbial genes under specific conditions, while metabolomics identifies chemical products of host-microbe interactions, such as short-chain fatty acids, detoxification intermediates, and signalling molecules. Integration of these approaches provides a dynamic view of microbiome function beyond taxonomic composition (Dar et al., 2024).

Establishing causal relationships between microbiomes and host phenotypes requires experimental manipulation. Antibiotic treatment, gnotobiotic rearing, microbial supplementation, and targeted symbiont elimination are commonly used strategies to perturb microbial communities and assess functional outcomes (Zhang et al., 2021). Culturomics and isolation of insect-associated microbes enable direct functional testing, genome sequencing of symbionts, and development of engineered microbial strains (Sarhan et al., 2019). Cultured symbionts have been instrumental in elucidating metabolic interactions in systems such as the honeybee gut microbiome and in advancing paratransgenic approaches for vector control (De Beer et al., 2023). Imaging techniques, including fluorescence in situ hybridisation and confocal microscopy, complement sequencing-based methods by revealing spatial organisation and host–microbe interfaces at cellular resolution.

Insights into insect microbiomes have generated a wide range of applications in agriculture and environmental management. In pest insects, disruption of obligate symbionts can reduce survival, development, and reproduction, offering opportunities for precision pest control that minimise non-target effects. Microbiome-informed strategies also enhance pollinator health. Probiotic supplementation, improved dietary diversity, and reduced chemical exposure support stable gut microbiomes in bees, improving immunity, foraging efficiency, and colony resilience. Beyond pest and pollinator management, insects and their microbiomes contribute to nutrient cycling and waste processing. Termites, beetles, and fly larvae accelerate the decomposition of organic matter through microbial fermentation, while insect frass enriched with beneficial microbes is increasingly used as an organic fertiliser in sustainable agriculture.

In disease vectors, microbiome manipulation offers innovative alternatives to chemical insecticides. Introduction of *Wolbachia* into mosquito populations reduces viral replication and shortens lifespan, significantly lowering transmission of dengue and other arboviruses. Paratransgenesis represents another promising approach, in which engineered symbiotic bacteria express antiparasitic molecules within the vector gut. This strategy has been explored in tsetse flies and kissing bugs to reduce transmission of trypanosomes.

Environmental management of larval habitats and microbial exposure further influences vector microbiomes and competence, highlighting the importance of ecological context in vector control strategies. The integration of methodological advances applied microbiome interventions, and their broader implications for agriculture, ecosystem function, and public health is summarised in Figure 1.8.

The One Health concept recognises the interconnectedness of human, animal, and environmental health (Pitt & Gunn, 2024). Insects and their microbiomes occupy a central position within this framework, linking ecosystems, agriculture, and public health (Verma et al., 2023). Pollinators influence food security through crop fertilisation, while their microbiomes modulate resilience to environmental stress and disease. Disease vectors bridge wildlife, livestock, and human populations, with microbiomes shaping pathogen transmission dynamics (De Garine-Wichatitsky et al., 2021). Insects also act as reservoirs and dispersal agents for antimicrobial resistance genes, particularly in environments impacted by agriculture and urbanisation. Climate change further complicates these interactions by altering insect distributions, microbial communities, and disease risk (Harvey et al., 2020). Monitoring insect microbiomes can therefore provide early indicators of ecosystem disruption and emerging health threats. Advances in sequencing, experimental manipulation, and multi-omics have transformed understanding of host–microbe interactions in insects, enabling translation into agriculture, environmental management, and public health.

Translating insect microbiome research into practice requires interdisciplinary collaboration among entomologists, microbiologists, veterinarians, ecologists, and public health professionals. Policy-relevant implications include reducing pesticide reliance, promoting integrated pest management, conserving pollinator habitats, and incorporating microbiome considerations into environmental risk assessment. Microbiome-based interventions must be evaluated not only for efficacy but also for ecological safety, long-term stability, and potential non-target effects. A One Health perspective encourages holistic assessment of these interventions across biological and societal scales. Embedding insect microbiome research within a One Health framework highlights the potential of microbiome-informed strategies to promote sustainable agriculture, protect ecosystem services, and reduce the burden of vector-borne diseases in a rapidly changing world.

ACKNOWLEDGEMENT

The authors acknowledge the use of NotebookLM and Gemini (Google LLC) as AI-assisted tools for figure design support and conceptual organisation. All scientific content, data interpretation, and final conclusions remain the sole responsibility of the authors.

REFERENCES

Al Darwish, H., Tariq, M., Salama, S., Hart, T., & Sun, J. S. (2025). The microbiome as a driver of insect physiology, behavior, and control strategies. *Applied Microbiology*, 5(3), 90. https://doi.org/10.3390/applmicrobiol5030090

Bai, S., Yao, Z., Cai, Z., Ma, Q., Guo, Q., Zhang, P., Zhou, Q., Gu, J., Liu, S., Lemaitre, B., Li, X., & Zhang, H. (2025). Bacterial-induced Duox-ROS regulates the Imd immune pathway in the gut by modulating the peritrophic matrix. *Cell Reports*, 44(3), 115404. https://doi.org/10.1016/j.celrep.2025.115404

Bai, S., Yao, Z., Raza, M. F., Cai, Z., & Zhang, H. (2021). Regulatory mechanisms of microbial homeostasis in insect gut. *Insect Science*, 28(2), 286–301. https://doi.org/10.1111/1744-7917.12868

Becher, P. G., Flick, G., Rozpędowska, E., Schmidt, A., Hagman, A., Lebreton, S., Larsson, M. C., Hansson, B. S., Piškur, J., Witzgall, P., & Bengtsson, M. (2012). Yeast, not fruit volatiles mediate Drosophila melanogaster attraction, oviposition and development. *Functional Ecology*, 26(4), 822–828. https://doi.org/10.1111/j.1365-2435.2012.02006.x

Bennett, G. M., & Moran, N. A. (2015). Heritable symbiosis: The advantages and perils of an evolutionary rabbit hole. *Proceedings of the National Academy of Sciences of the United States of America*, 112(33), 10169–10176. https://doi.org/10.1073/pnas.1421388112

Birer, C., Moreau, C. S., Tysklind, N., Zinger, L., & Duplais, C. (2020). Disentangling the assembly mechanisms of ant cuticular bacterial communities

Figure 1.8 Insect microbiome research methods and application areas. The figure summarises advanced sequencing and imaging techniques used to study microbial diversity in insects and the functional roles of these microorganisms. It also highlights the critical importance of microbiome strategies within the framework of agricultural productivity, disease control, and the 'One Health' approach.

of two Amazonian ant species sharing a common arboreal nest. *Molecular Ecology*, 29(7), 1372–1385. https://doi.org/10.1111/mec.15400

Björk, J. R., Díez-Vives, C., Astudillo-García, C., Archie, E. A., & Montoya, J. M. (2019). Vertical transmission of sponge microbiota is inconsistent and unfaithful. *Nature Ecology & Evolution*, 3(8), 1172–1183. https://doi.org/10.1038/s41559-019-0935-x

Branda, F., Cella, E., Scarpa, F., Slavov, S. N., Bevivino, A., Moretti, R., Degafu, A. L., Pecchia, L., Rizzo, A., Defilippo, F., Moreno, A., Ceccarelli, G., Alcantara, L. C., Ferreira, A., Ciccozzi, M., & Giovanetti, M. (2024). Wolbachia-based approaches to controlling mosquito-borne viral threats: Innovations, AI integration, and future directions in the context of climate change. *Viruses*, 16(12), 1868. https://doi.org/10.3390/v16121868

Bright, M., & Bulgheresi, S. (2010). A complex journey: Transmission of microbial symbionts. *Nature Reviews. Microbiology*, 8(3), 218–230. https://doi.org/10.1038/nrmicro2262

Brune, A. (2014). Symbiotic digestion of lignocellulose in termite guts. *Nature Reviews Microbiology*, 12(3), 168–180. https://doi.org/10.1038/nrmicro3182

Ceja-Navarro, J. A., Karaoz, U., Bill, M., Hao, Z., White, R. A., Arellano, A., Ramanculova, L., Filley, T. R., Berry, T. D., Conrad, M. E., Blackwell, M., Nicora, C. D., Kim, Y.-M., Reardon, P. N., Lipton, M. S., Adkins, J. N., Pett-Ridge, J., & Brodie, E. L. (2019). Gut anatomical properties and microbial functional assembly promote lignocellulose deconstruction and colony subsistence of a wood-feeding beetle. *Nature Microbiology*, 4(5), 864–875. https://doi.org/10.1038/s41564-019-0384-y

Chen, B., Teh, B.-S., Sun, C., Hu, S., Lu, X., Boland, W., & Shao, Y. (2016). Biodiversity and activity of the gut microbiota across the life history of the insect herbivore *Spodoptera littoralis*. *Scientific Reports*, 6(1), 29505. https://doi.org/10.1038/srep29505

Dar, M. A., Xie, R., Jing, L., Qing, X., Ali, S., Pandit, R. S., Shaha, C. M., & Sun, J. (2024). Elucidating the structure, and composition of bacterial symbionts in the gut regions of wood-feeding termite, Coptotermes formosanus and their functional profile towards lignocellulolytic systems. *Frontiers in Microbiology*, 15, 1395568.

De Beer, J., Sole, C., Pirk, C., & Weldon, C. (2023). Nutritive and immunological benefits derived from the honey bee gut microbiome are threatened by anthropogenic stressors. *African Entomology*, 31, 1. https://doi.org/10.17159/2254-8854/2023/a14674

De Garine-Wichatitsky, M., Miguel, E., Kock, R., Valls-Fox, H., & Caron, A. (2021). The ecology of pathogens transmission at the wildlife-livestock interface: beyond disease ecology, towards socio-ecological system health. In Vicente Joaquín, Vercauteren Kurt C. , Gortázar Christian. *Diseases at the Wildlife-Livestock Interface: Research and Perspectives in a Changing World* (pp. 91–119). Springer.

Dennison, N. J., Jupatanakul, N., & Dimopoulos, G. (2014). The mosquito microbiota influences vector competence for human pathogens. *Current Opinion in Insect Science*, 3, 6–13. https://doi.org/10.1016/j.cois.2014.07.004

Dong, V. Q., Bui Thi, T. D., Pham, T. L., Nguyen, T. H., Nguyen, Q. H., Ha, T. T., & Pham, H. T. (2024). Profile of the gut microbial composition in Apis mellifera larvae collected in Ha Noi. *Academia Journal of Biology*, 46(2 SE-Articles), 63–70. https://doi.org/10.15625/2615-9023/20165

Douglas, A. E. (2009). The microbial dimension in insect nutritional ecology. *Functional Ecology*, 23, 38–47. https://doi.org/10.1111/j.1365-2435.2008.01442.x

Douglas, A. E. (2015). Multiorganismal insects: Diversity and function of resident microorganisms. *Annual Review of Entomology*, 60, 17–34. https://doi.org/10.1146/annurev-ento-010814-020822

El Khoury, S., Gauthier, J., Mercier, P. L., Moïse, S., Giovenazzo, P., & Derome, N. (2024). Honeybee gut bacterial strain improved survival and gut microbiota homeostasis in Apis mellifera exposed in vivo to clothianidin. *Microbiology Spectrum*, 12(10), e0057824. https://doi.org/10.1128/spectrum.00578-24

Engel, P., & Moran, N. A. (2013). The gut microbiota of insects–diversity in structure and function. *FEMS Microbiology Reviews*, 37(5), 699–735. https://doi.org/10.1111/1574-6976.12025

Everaerts, C., Farine, J.-P., Cobb, M., & Ferveur, J.-F. (2010). Drosophila cuticular hydrocarbons revisited: Mating status alters cuticular profiles. *PloS One*, 5(3), e9607. https://doi.org/10.1371/journal.pone.0009607

Fricke, L. C., Villalta, M. D., & Lindsey, A. R. I. (2025). Endosymbionts interacting with sex-determining genes and processes. *Current Opinion in Insect Science*, 72, 101410. https://doi.org/10.1016/j.cois.2025.101410

Gi, K. H., Jin-Soo, S., Joon-Hui, C., Soohyun, L., Jun-Seob, K., & Choong-Min, R. (2023). Population dynamics of intestinal enterococcus modulate *Galleria mellonella* metamorphosis. *Microbiology Spectrum*, 11(4), e02780–22. https://doi.org/10.1128/spectrum.02780-22

Girard, M., Luis, P., Valiente Moro, C., & Minard, G. (2023). Crosstalk between the microbiota and insect postembryonic development. *Trends in Microbiology*, 31(2), 181–196. https://doi.org/10.1016/j.tim.2022.08.013

Guilhot, R., Xuéreb, A., Lagmairi, A., Olazcuaga, L., & Fellous, S. (2023). Microbiota acquisition and transmission in Drosophila flies. *IScience*, *26*(9), 107656. https://doi.org/10.1016/j.isci.2023.107656

Gupta, A., & Nair, S. (2020). Dynamics of insect–Microbiome interaction influence host and microbial symbiont. *Frontiers in Microbiology*, *11*, 1357.

Gurung, K., Wertheim, B., & Falcao Salles, J. (2019). The microbiome of pest insects: It is not just bacteria. *Entomologia Experimentalis et Applicata*, *167*(3), 156–170. https://doi.org/10.1111/eea.12768

Harvey, J. A., Heinen, R., Gols, R., & Thakur, M. P. (2020). Climate change-mediated temperature extremes and insects: From outbreaks to breakdowns. *Global Change Biology*, *26*(12), 6685–6701.

Holt, J. R., Cavichiolli de Oliveira, N., Medina, R. F., Malacrinò, A., & Lindsey, A. R. I. (2024). Insect–microbe interactions and their influence on organisms and ecosystems. *Ecology and Evolution*, *14*(7), e11699. https://doi.org/10.1002/ece3.11699

Horak, R. D., Leonard, S. P., & Moran, N. A. (2020). Symbionts shape host innate immunity in honeybees. *Proceedings of the Royal Society B: Biological Sciences*, *287*(1933), 20201184. https://doi.org/10.1098/rspb.2020.1184

Hosokawa, T., Koga, R., Kikuchi, Y., Meng, X.-Y., & Fukatsu, T. (2010). Wolbachia as a bacteriocyte-associated nutritional mutualist. *Proceedings of the National Academy of Sciences*, *107*(2), 769–774. https://doi.org/10.1073/pnas.0911476107

Huang, K., Wang, J., Huang, J., Zhang, S., Vogler, A. P., Liu, Q., Li, Y., Yang, M., Li, Y., & Zhou, X. (2021). Host phylogeny and diet shape gut microbial communities within bamboo-feeding insects. *Frontiers in Microbiology*, *12*, 633075.

Janke, R. S., Kaftan, F., Niehs, S. P., Scherlach, K., Rodrigues, A., Svatoš, A., Hertweck, C., Kaltenpoth, M., & Flórez, L. V. (2022). Bacterial ectosymbionts in cuticular organs chemically protect a beetle during molting stages. *The ISME Journal*, *16*(12), 2691–2701. https://doi.org/10.1038/s41396-022-01311-x

Jiménez-Padilla, Y., Adewusi, B., Lachance, M.-A., & Sinclair, B. J. (2024). Live yeasts accelerate Drosophila melanogaster larval development. *Journal of Experimental Biology*, *227*(19), jeb247932. https://doi.org/10.1242/jeb.247932

Jing, T.-Z., Qi, F.-H., & Wang, Z.-Y. (2020). Most dominant roles of insect gut bacteria: Digestion, detoxification, or essential nutrient provision? *Microbiome*, *8*(1), 38. https://doi.org/10.1186/s40168-020-00823-y

Jones, A. G., Mason, C. J., Felton, G. W., & Hoover, K. (2019). Host plant and population source drive diversity of microbial gut communities in two polyphagous insects. *Scientific Reports*, *9*(1), 2792. https://doi.org/10.1038/s41598-019-39163-9

Jones, J. A., Garcia, N. I., & Moczek, A. P. (2024). Microbiome composition and turnover in the face of complex lifecycles and bottlenecks: Insights through the study of dung beetles. *Applied and Environmental Microbiology*, *91*(1), e01278–24. https://doi.org/10.1128/aem.01278-24

Ju, J.-F., Bing, X.-L., Zhao, D.-S., Guo, Y., Xi, Z., Hoffmann, A. A., Zhang, K.-J., Huang, H.-J., Gong, J.-T., Zhang, X., & Hong, X.-Y. (2020). Wolbachia supplement biotin and riboflavin to enhance reproduction in planthoppers. *The ISME Journal*, *14*(3), 676–687. https://doi.org/10.1038/s41396-019-0559-9

Jyoti, & Dey, P. (2025). Mechanisms and implications of the gut microbial modulation of intestinal metabolic processes. *Npj Metabolic Health and Disease*, *3*(1), 24. https://doi.org/10.1038/s44324-025-00066-1

Khan, S. A., Kojour, M. A. M., & Han, Y. S. (2023). Recent trends in insect gut immunity. *Frontiers in Immunology*, *14*, 1272143.

Lemoine, M. M., Engl, T., & Kaltenpoth, M. (2020). Microbial symbionts expanding or constraining abiotic niche space in insects. *Current Opinion in Insect Science*, *39*, 14–20. https://doi.org/10.1016/j.cois.2020.01.003

Lemoine, M. M., Wöhner, T., & Kaltenpoth, M. (2025). Microbial community dynamics in natural *Drosophila melanogaster* populations across seasons. *Environmental Microbiology*, *27*(6), e70104. https://doi.org/10.1111/1462-2920.70104

Liu, H., Yang, Q., Liu, J., Haq, I. U., Li, Y., Zhang, Q., Attia, K. A., Abushady, A. M., Liu, C., & Lv, N. (2023). Host plant-mediated effects on Buchnera symbiont: Implications for biological characteristics and nutritional metabolism of pea aphids (Acyrthosiphon pisum). *Frontiers in Plant Science*, *14*, 1288997.

Liu, N., Li, H., Chevrette, M. G., Zhang, L., Cao, L., Zhou, H., Zhou, X., Zhou, Z., Pope, P. B., Currie, C. R., Huang, Y., & Wang, Q. (2019). Functional metagenomics reveals abundant polysaccharide-degrading gene clusters and cellobiose utilization pathways within gut microbiota of a wood-feeding higher termite. *The ISME Journal*, *13*(1), 104–117. https://doi.org/10.1038/s41396-018-0255-1

Lorenzi, M. C., & d'Ettorre, P. (2020). Nestmate recognition in social insects: What does it mean to be chemically insignificant? *Frontiers in Ecology and Evolution*, *7*, 488.

Luo, J., Cheng, Y., Guo, L., Wang, A., Lu, M., & Xu, L. (2021). Variation of gut microbiota caused by an imbalance diet is detrimental to bugs' survival. *Science of The Total Environment*, *771*, 144880. https://doi.org/10.1016/j.scitotenv.2020.144880

Maes, P. W., Rodrigues, P. A. P., Oliver, R., Mott, B. M., & Anderson, K. E. (2016). Diet-related gut bacterial dysbiosis correlates with impaired development, increased mortality and Nosema disease in the honeybee (Apis mellifera). *Molecular Ecology*, 25(21), 5439–5450. https://doi.org/10.1111/mec.13862

Manthey, C., Johnston, P. R., Nakagawa, S., & Rolff, J. (2023). Complete metamorphosis and microbiota turnover in insects. *Molecular Ecology*, 32(23), 6543–6551. https://doi.org/10.1111/mec.16673

Mason, C. J., Auth, J., & Geib, S. M. (2023). Gut bacterial population and community dynamics following adult emergence in pest tephritid fruit flies. *Scientific Reports*, 13(1), 13723. https://doi.org/10.1038/s41598-023-40562-2

Mondal, S., Somani, J., Roy, S., Babu, A., & Pandey, A. K. (2023). Insect microbial symbionts: Ecology, interactions, and biological significance. *Microorganisms*, 11(11), 2665. https://doi.org/10.3390/microorganisms11112665

Mugo-Kamiri, L., Querejeta, M., Raymond, B., & Herniou, E. A. (2024). The effect of diet composition on the diversity of active gut bacteria and on the growth of Spodoptera exigua (Lepidoptera: Noctuidae). *Journal of Insect Science*, 24(2), 13. https://doi.org/10.1093/jisesa/ieae031

Oishi, S., Moriyama, M., Mizutani, M., Futahashi, R., & Fukatsu, T. (2023). Regulation and remodeling of microbial symbiosis in insect metamorphosis. *Proceedings of the National Academy of Sciences*, 120(40), e2304879120. https://doi.org/10.1073/pnas.2304879120

Pitt, S. J., & Gunn, A. (2024). The one health concept. *British Journal of Biomedical Science*, 81, 12366. https://doi.org/10.3389/bjbs.2024.12366

Powell, J. E., Leonard, S. P., Kwong, W. K., Engel, P., & Moran, N. A. (2016). Genome-wide screen identifies host colonization determinants in a bacterial gut symbiont. *Proceedings of the National Academy of Sciences*, 113(48), 13887–13892. https://doi.org/10.1073/pnas.1610856113

Ragini, G., Mani, M. K., Sharma, R., Bharadwaj, N., Subramanian, M., Nagarajan, S. A., & Rahi, M. (2025). Microbiome diversity in mosquitoes and sand flies: Implications for vector competence. *Parasites & Vectors*, 18(1), 388. https://doi.org/10.1186/s13071-025-06964-z

Raymann, K., Shaffer, Z., & Moran, N. A. (2017). Antibiotic exposure perturbs the gut microbiota and elevates mortality in honeybees. *PLoS Biology*, 15(3), e2001861. https://doi.org/10.1371/journal.pbio.2001861

Ricigliano, V. A., Fine, J. D., & Nicklisch, S. C. T. (2025). Harnessing biotechnology for bee pollinator health. *Trends in Biotechnology*. 44(1):111–127. https://doi.org/10.1016/j.tibtech.2025.05.027

Sai Charan, D., Vidya Madhuri, E., Rupali, J. S., Hariharan, S., Verma, S., Aravind, K., Paschapur, A., Singh, D. P., Kumar, A., & Prabha, R. (2025). Harnessing insect gut microbiota: Approaches and applications for next-generation crop management and agricultural sustainability. *Discover Agriculture*, 3(1), 281. https://doi.org/10.1007/s44279-025-00439-6

Sanaei, E., Chavez, J., Harris, E. V, Alcaide, T. Y., Baffour-Addo, K., Bugay, M. J., Adams, K. L., Zelaya, A., de Roode, J. C., & Gerardo, N. M. (2024). Microbiome analysis of monarch butterflies reveals effects of development and diet. *FEMS Microbiology Ecology*, 100(12), fiae143. https://doi.org/10.1093/femsec/fiae143

Sarhan, M. S., Hamza, M. A., Youssef, H. H., Patz, S., Becker, M., ElSawey, H., Nemr, R., Daanaa, H.-S. A., Mourad, E. F., Morsi, A. T., Abdelfadeel, M. R., Abbas, M. T., Fayez, M., Ruppel, S., & Hegazi, N. A. (2019). Culturomics of the plant prokaryotic microbiome and the dawn of plant-based culture media – A review. *Journal of Advanced Research*, 19, 15–27. https://doi.org/10.1016/j.jare.2019.04.002

Schmidt, K., & Engel, P. (2021). Mechanisms underlying gut microbiota–host interactions in insects. *Journal of Experimental Biology*, 224(2), jeb207696. https://doi.org/10.1242/jeb.207696

Schwarz, R. S., Moran, N. A., & Evans, J. D. (2016). Early gut colonizers shape parasite susceptibility and microbiota composition in honey bee workers. *Proceedings of the National Academy of Sciences*, 113(33), 9345–9350. https://doi.org/10.1073/pnas.1606631113

Sibinga, N. A., Werner, E., Tegtmeier, D., & De Smet, J. (2025). Animal board invited review: The need for, and the path towards, a functional understanding of the farmed insect microbiome. *Animal*, 19(8), 101575. https://doi.org/10.1016/j.animal.2025.101575

Siddiqui, J. A., Khan, M. M., Bamisile, B. S., Hafeez, M., Qasim, M., Rasheed, M. T., Rasheed, M. A., Ahmad, S., Shahid, M. I., & Xu, Y. (2022). Role of insect gut microbiota in pesticide degradation: A review. *Frontiers in Microbiology*, 13, 870462.

Smith, C. C., Srygley, R. B., Healy, F., Swaminath, K., & Mueller, U. G. (2017). Spatial structure of the mormon cricket gut microbiome and its predicted contribution to nutrition and immune function. *Frontiers in Microbiology*, 8, 801.

Soderholm, A. T., & Pedicord, V. A. (2019). Intestinal epithelial cells: At the interface of the microbiota and mucosal immunity. *Immunology*, 158(4), 267–280. https://doi.org/10.1111/imm.13117

Su, C., Xie, T., Jiang, L., Wang, Y., Wang, Y., Nie, R., Zhao, Y., He, B., Ma, J., Yang, Q., & Hao, J. (2024). Host genetics and larval host plant modulate microbiome structure and evolution underlying the intimate insect-microbe-plant interactions in Parnassius species on the Qinghai-Tibet Plateau. *Ecology and Evolution*, 14(4), e11218. https://doi.org/10.1002/ece3.11218

Sudakaran, S., Retz, F., Kikuchi, Y., Kost, C., & Kaltenpoth, M. (2015). Evolutionary transition in symbiotic syndromes enabled diversification of phytophagous insects on an imbalanced diet. *The ISME Journal*, 9(12), 2587–2604. https://doi.org/10.1038/ismej.2015.75

Suenami, S., Koto, A., & Miyazaki, R. (2023). Basic structures of gut bacterial communities in eusocial insects. *Insects*, 14(5), 444. https://doi.org/10.3390/insects14050444

Tokuda, G., Mikaelyan, A., Fukui, C., Matsuura, Y., Watanabe, H., Fujishima, M., & Brune, A. (2018). Fiber-associated spirochetes are major agents of hemicellulose degradation in the hindgut of wood-feeding higher termites. *Proceedings of the National Academy of Sciences of the United States of America*, 115(51), E11996–E12004. https://doi.org/10.1073/pnas.1810550115

Verma, R. C., Waseem, M. A., Sharma, N., Bharathi, K., Singh, S., Anto Rashwin, A., Pandey, S. K., & Singh, B.V. (2023). The role of insects in ecosystems, an in-depth review of entomological research. *International Journal of Environment and Climate Change*, 13(10), 4340–4348.

Vienna, K., & Mikheyev, A. M. (2021). Honey bee larval and adult microbiome life stages are effectively decoupled with vertical transmission overcoming early life perturbations. *MBio*, 12(6), e02966-21. https://doi.org/10.1128/mBio.02966-21

Walton, A., Herman, J. J., & Rueppell, O. (2024). Social life results in social stress protection: A novel concept to explain individual life-history patterns in social insects. *Biological Reviews*, 99(4), 1444–1457. https://doi.org/10.1111/brv.13074

Wang, K., Gao, P., Geng, L., Liu, C., Zhang, J., & Shu, C. (2022). Lignocellulose degradation in Protaetia brevitarsis larvae digestive tract: Refining on a tightly designed microbial fermentation production line. *Microbiome*, 10(1), 90. https://doi.org/10.1186/s40168-022-01291-2

Wang, Y., Tang, J., Chen, Y., Chen, S., Chen, S., Yu, X., Wan, C., Xiang, G., Chen, Y., & Li, Q. (2025). The ecological–Evolutionary game of the insect gut microbiome: Environmental drivers, host regulation, and prospects for cross-cutting applications. *Veterinary Sciences*, 12(9), 866. https://doi.org/10.3390/vetsci12090866

Wang, Y.-B., Ren, F.-R., Yao, Y.-L., Sun, X., Walling, L. L., Li, N.-N., Bai, B., Bao, X.-Y., Xu, X.-R., & Luan, J.-B. (2020). Intracellular symbionts drive sex ratio in the whitefly by facilitating fertilization and provisioning of B vitamins. *The ISME Journal*, 14(12), 2923–2935. https://doi.org/10.1038/s41396-020-0717-0

Weiss, B., & Aksoy, S. (2011). Microbiome influences on insect host vector competence. *Trends in Parasitology*, 27(11), 514–522. https://doi.org/10.1016/j.pt.2011.05.001

Wen, S.-Y., Ali, A., Huang, I.-C., Liu, J.-S., Chen, P.-Y., Padma Viswanadha, V., Huang, C.-Y., & Kuo, W.-W. (2023). Doxorubicin induced ROS-dependent HIF1α activation mediates blockage of IGF1R survival signaling by IGFBP3 promotes cardiac apoptosis. *Aging*, 15(1), 164–178. https://doi.org/10.18632/aging.204466

Xiao, X., Yang, L., Pang, X., Zhang, R., Zhu, Y., Wang, P., Gao, G., & Cheng, G. (2017). A Mesh-Duox pathway regulates homeostasis in the insect gut. *Nature Microbiology*, 2, 17020. https://doi.org/10.1038/nmicrobiol.2017.20

Yasika, Y., & Shivakumar, M. S. (2025). A comprehensive account of functional role of insect gut microbiome in insect orders. *Journal of Natural Pesticide Research*, 11, 100110. https://doi.org/10.1016/j.napere.2024.100110

Yueqing, A. L., Garcia, S., & Hambäck, A. P. (2025). Microbial transfer through fecal strings on eggs affects leaf beetle microbiome dynamics. *MSystems*, 10(6), e01723-24. https://doi.org/10.1128/msystems.01723-24

Yun, J.-H., Roh, S. W., Whon, T. W., Jung, M.-J., Kim, M.-S., Park, D.-S., Yoon, C., Nam, Y.-D., Kim, Y.-J., Choi, J.-H., Kim, J.-Y., Shin, N.-R., Kim, S.-H., Lee, W.-J., & Bae, J.-W. (2014). Insect gut bacterial diversity determined by environmental habitat, diet, developmental stage, and phylogeny of host. *Applied and Environmental Microbiology*, 80(17), 5254–5264. https://doi.org/10.1128/AEM.01226-14

Zhang, S., Wang, Z., Luo, Q., Zhou, L., Du, X., & Ren, Y. (2025). Effects of microbes on insect host physiology and behavior mediated by the host immune system. *Insects*, 16(1), 82. https://doi.org/10.3390/insects16010082

Zhang, W., Tettamanti, G., Bassal, T., Heryanto, C., Eleftherianos, I., & Mohamed, A. (2021). Regulators and signalling in insect antimicrobial innate immunity: Functional molecules and cellular pathways. *Cellular Signalling*, 83, 110003. https://doi.org/10.1016/j.cellsig.2021.110003

Zheng, H., Perreau, J., Powell, J. E., Han, B., Zhang, Z., Kwong, W. K., Tringe, S. G., & Moran, N. A. (2019). Division of labor in honey bee gut microbiota for plant polysaccharide digestion. *Proceedings of the National Academy of Sciences*, 116(51), 25909–25916. https://doi.org/10.1073/pnas.1916224116

CHAPTER TWO

Microbiomes in Agricultural Insect Pests

Zhangrong Song and Chun Nin Adam Wong

2.1 INTRODUCTION

Over 1 million insect species have been formally described, a small fraction of which (<10%; exact estimates vary among studies) are considered agricultural pests. These pest species are concentrated within six insect orders: Lepidoptera (caterpillars and moths), Hemiptera (aphids, whiteflies, hoppers, and true bugs), Coleoptera (beetles), Diptera (flies), Thysanoptera (thrips), and Orthoptera (locusts) (Royal Botanic Gardens, 2017). Despite their modest taxonomic representation, agricultural insect pests exert a disproportionate impact on food production, causing up to 38% agricultural losses globally (Junaid and Gokce, 2024). Historically, pest management research has been grounded in a host-centric framework. However, it is increasingly recognized that interactions between insects and their associated microorganisms can substantially influence key pest traits (Bordenstein and Theis, 2015; Zheng et al., 2023), including physiology, nutrition, immunity, reproduction, and ecological interactions (Douglas, 2015; Holt et al., 2024; Ntiri and Wong, 2025). These microbial associates, which include bacteria, fungi, and viruses, can also shape the pest niche of host plant species (Gurung et al., 2019; Lemoine et al., 2020). Recognition of the functional importance of insect-associated microbiomes has expanded opportunities for novel pest management strategies, such as symbiont disruption and targeted manipulation of microbial functions (Qadri et al., 2020; Yuxin Zhang et al., 2023). Thus, understanding microbiome structure and function is increasingly central to developing sustainable management strategies and predicting pest responses to changing environmental conditions and agricultural practices.

In this chapter, we synthesize current knowledge on the microbiomes of well-studied agricultural pests, focusing on four of the six major pest orders, while also highlighting key examples from other orders where microbiome research has provided important insights. The implications of microbiome research for pest management are also discussed.

2.2 LEPIDOPTERAN PEST MICROBIOMES

2.2.1 Microbiome Diversity in Lepidopteran Pests

Lepidopteran pests, including armyworms, cutworms, bollworms, borers, loopers, and a wide variety of moth larvae, are among the most destructive herbivores affecting global agricultural production. Their ecological success is often attributed to physiological plasticity (Woestmann and Saastamoinen, 2016) and an ability to overcome or even exploit plant chemical defenses during colonization (N. Zhang et al., 2024; S. Zhang et al., 2024). Studies of lepidopteran larval gut microbiomes generally report low-diversity communities that are highly heterogenous among individuals and species. Several physiological features of the lepidopteran larval gut are believed to underlie this characteristic variability, including rapid food passage, highly alkaline conditions, and a gut epithelium lined by a

DOI: 10.1201/9781003639176-2

peritrophic matrix, all of which limit stable bacterial colonization (Mason, 2020;Teh et al., 2016). In some species, such as *Manduca sexta*, larvae develop normally in the absence of a resident gut microbiome, and the caterpillar gut communities harbor few taxa closely resembling those associated with ingested plant material. These observations suggest that, in such systems, gut-associated bacteria are transient, environmentally acquired, and functionally dispensable (Hammer et al., 2017). However, studies of major agricultural pests such as the fall armyworm (*Spodoptera frugiperda*) and corn earworm (*Helicoverpa zea*) demonstrate that bacterial populations can establish and persist through larval development, although their abundance and community structure are strongly influenced by diet and host body size. In certain species like the diamondback moth (*Plutella xylostella* (L.)), experimental manipulation of the gut microbiome has been associated with measurable changes in host performance, such as altered responses to insecticides (Xia et al., 2018), indicating context-dependent functional effects (Mason, 2020). Additionally, studies across dozens of lepidopteran species have consistently detected recurrent bacterial taxa, suggesting that even dynamic microbiomes strongly shaped by environmental factors can exert significant functional influence (Mason et al., 2020; Shao et al., 2024; Wang et al., 2025).

At higher taxonomic levels, lepidopteran larval microbiomes are dominated by members of the phyla Proteobacteria, Firmicutes, Bacteroidota, and Actinobacteriota. Within Proteobacteria, both alpha- and Gamma-proteobacteria are commonly represented, including genera such as *Methylobacterium*, *Pseudomonas*, *Acinetobacter*, *Sphingomonas*, *Klebsiella*, and *Enterobacter* (Shao et al., 2024). Many of these taxa are well known as phyllosphere inhabitants, suggesting that their presence in the larval gut reflects repeated ingestion from host plants. A limited subset of bacterial families consistently dominates surveys of both laboratory and field populations, including Enterobacteriaceae, Bacillaceae, Pseudomonadaceae, Staphylococcaceae, and Enterococcaceae (Voirol et al., 2018; Wang et al., 2025). At the genus level, *Pseudomonas*, *Bacillus*, *Staphylococcus*, *Enterobacter*, and *Enterococcus* recur across different species, life stages, and geographic ranges (Deguenon et al., 2021; Govindharaj et al., 2025; Hu et al., 2022; Palacio et al., 2021; Wang et al., 2025; Xia et al., 2017; Xie et al., 2010).

Firmicutes, particularly members of the classes Bacilli and Clostridia, often dominate larval gut communities. Genera such as *Enterococcus*, *Staphylococcus*, *Lactococcus*, *Lactobacillus*, *Streptococcus*, and *Bacillus* are often abundant (Shao et al., 2024). These bacteria are less common on plant surfaces, implying that they may be selectively retained by the gut environment or enriched through ingestion of decaying material, frass-contaminated foliage, or conspecific interactions.

Collectively, these patterns are consistent with a system in which the gut continuously filters and redistributes environmental microbes, favoring lineages capable not only of tolerating the extreme physiochemical conditions of the caterpillar midgut, but also of persisting across complete metamorphosis and repeated transitions between plant-associated environments and the insect hosts.

Beyond bacteria, other microbial groups can be present (Douglas, 2015) but appear far less consistent. Fungal taxa, particularly yeasts, have been isolated from several lepidopteran species, but current evidence suggests they are more transient than bacterial associates (Šigut et al., 2022). Viral associations show even greater variability. Persistent viral infections have been reported in certain species, such as densoviruses (HaDN2) in *Helicoverpa armigera* (Xu et al., 2012) and dsRNA viruses in *Homona magnanima* (Nakanishi et al., 2008), but the ecological and physiological consequences of these associations remain poorly understood.

Unlike several other major pest groups, Lepidoptera generally lack obligatory intracellular bacterial symbionts besides the reproductive symbiont *Wolbachia*. *Wolbachia* infections are widespread in this insect order and are estimated to occur in up to 80% of lepidopteran species (Ahmed et al., 2015). *Wolbachia* can influence host reproduction by inducing cytoplasmic incompatibility, male killing, or altered fecundity, thereby shaping population dynamics. Other symbionts, including *Spiroplasma* and *Rickettsia*, occur at lower prevalence in lepidopterans, and their ecological roles remain less clearly defined (Araújo et al., 2021; Shao et al., 2024).

2.2.2 Drivers of Lepidopteran Microbiome Variability

A defining driver of microbiome variability in Lepidoptera is larval feeding habits. Lepidopteran larvae are adapted for rapid biomass accumulation, by consuming large quantities of foliage almost continuously and increasing body mass by several orders of magnitude before pupation.

For example, major agricultural pests such as *S. frugiperda* and *H. armigera* consume several times their body mass in plant tissue per day during larval development, processing large cumulative volumes of food prior to metamorphosis (Ajmal et al., 2024; P. Wang et al., 2019). This feeding strategy results in continuous influx and rapid passage of ingested material through the gut, limiting microbial residence time and opportunities for stable establishment (Mason, 2020). These constraints are reinforced by the physiological conditions within the larval midgut. The midguts of many lepidopteran larvae maintain one of the most alkaline environments known among terrestrial animals, frequently above ten in pH (X. Zhang et al., 2022). Such high alkaline conditions are disruptive to bacterial membrane integrity and metabolism, restricting persistence to alkaline-tolerant taxa. Although the peritrophic matrix is not unique to Lepidoptera, its interaction with high pH, enzymatic activity, and continuous feeding further confines microbes to the gut lumen or to ingested material rather than promoting stable association with epithelial surfaces (Teh et al., 2016).

Complete metamorphosis imposes an additional filter on microbial persistence. During pupation, larval gut tissues are dismantled and rebuilt. In many species like *Grapholita molesta*, metamorphosis results in a complete community turnover (Hammer et al., 2017; Manthey et al., 2023; X. Wang et al., 2020). The reorganization of gut tissues during pupation eliminates most larval microbes, and adult gut microbiomes are reassembled and shaped by environmental exposure (Hammer et al., 2017; Shao et al., 2024).

Diet also plays a significant role in shaping microbiome composition (Mason et al., 2020; Sanaei et al., 2024). Because most lepidopteran larvae feed on foliage that hosts its own phyllosphere microbiomes, microbial exposure varies with plant species, leaf age, surface chemistry, and local environmental conditions. Polyphagous pests such as *H. armigera* or *S. frugiperda* tend to harbor a wider range of bacteria (Rodrigues et al., 2023; Wu et al., 2024), corresponding to the diversity of plants they encounter. Larvae reared on different host plants retain distinct microbial signatures, and these differences persist even after the insects are transitioned to a common artificial diet (Mason et al., 2020). Field-collected larvae generally show greater microbial richness than laboratory colonies (Broderick et al., 2004; Gomes et al., 2020). In some species, soil-associated microbes also contribute substantially to gut communities, as observed in the cabbage moth *Mamestra brassicae* (Hannula et al., 2019).

Despite the strong host and environmental filters, persistent and repeatable microbial associations have been documented in certain lepidopteran species. In *Spodoptera littoralis*, strains of *Enterococcus mundtii* can persist in the gut throughout larval development and into adulthood (Govindharaj et al., 2025). In *M. sexta*, eggs may carry metabolically active *Enterococcus* on their surface, enabling early colonization of larvae (Brinkmann et al., 2008). In *Brithys crini*, a subset of 29 genera, mainly belonging to the Proteobacteria phylum, is conserved across developmental stages (González-Serrano et al., 2020). These examples indicate that microbial persistence occurs under specific physiological and ecological conditions.

2.2.3 Functional Roles of Lepidopteran Microbiomes

Despite their often transient and variable nature, lepidopteran gut microbiomes can greatly influence host performance, particularly through modulating plant–insect interactions. During feeding, lepidopteran larvae introduce oral secretions and regurgitant containing gut-associated bacteria and their metabolites into wounded plant tissues. These microbial components have been shown to suppress jasmonic acid (JA)-mediated plant defenses or alter crosstalk between JA and salicylic acid (SA) pathways that result in reduced plant resistance and enhanced herbivore performance (Gu et al., 2025; Yamasaki et al., 2021). For example, application of *Enterobacter ludwigii* isolated from *H. zea* regurgitant to wounded tomato plants suppressed JA-responsive defenses but elicited expression of SA-responsive genes (Wang et al., 2017).

Gut microbes also contribute to the detoxification of plant defensive compounds. Many host plants produce phenolics, terpenoids, and protease inhibitors that reduce herbivore performance. In lepidopteran pests, bacteria such as *Enterobacter* spp. produce enzymes that neutralize reactive oxygen species (Mazumdar et al., 2021; Xia et al., 2017), while other taxa provide glutathione-S-transferases capable of conjugating lipophilic plant toxins (Xia et al., 2017). In conifer-feeding species such as *Lymantria dispar*, microbes capable of degrading monoterpenes have been detected, suggesting contribution to coping with terpene-rich foliage (Broderick et al., 2004; Powell and

Raffa, 1999). Microbiome-dependent detoxification effects also extend to chemical insecticides (Malook et al., 2025). In *P. xylostella*, specific gut bacteria such as *Enterococcus* sp. can enhance resistance to chlorpyrifos, whereas other strains, including *Serratia* sp., reduce resistance (Xia et al., 2018). These effects are mediated largely through modulation of host immune responses, specifically the expression of antimicrobial peptides, rather than direct enzymatic degradation. In *S. frugiperda*, persistent insecticide exposure affects gut microbiota assembly, and disruption of the microbial community by antibiotic treatment slows the rate and heritability of resistance evolution (Y. Zhang et al., 2024). Together, these studies indicate that microbiome composition and community dynamics can influence insecticide susceptibility and the evolution of resistance.

Another microbial contribution lies in nutrient acquisition. Many lepidopteran larvae feed on foliage that is low in nitrogen and contains complex polysaccharides that are difficult to digest by endogenous insect enzymes alone. Gut bacteria can supplement host digestion by producing plant cell wall-degrading enzymes including cellulases, pectinases, and xylanases (Watanabe and Tokuda, 2010), as documented in *Bombyx mori* (Anand et al., 2010), *P. xylostella* (Gandotra et al., 2018), and several saturniid species (Munivenkatappa and Shivashankar, 2025). These activities may enhance digestion efficiency and expand the range of plant tissues that larvae can exploit. Microbial contributions to nitrogen metabolism have also been proposed, although evidence remains limited. Nevertheless, bacteria harboring genes associated with nitrogen fixation have been isolated from the gut of *P. xylostella* (Indiragandhi et al., 2007, 2008).

The microbiome also influences interactions between lepidopteran pests and entomopathogens. Some bacterial taxa enhance susceptibility to microbial control agents, while others reduce it. For example, the efficacy of *Bacillus thuringiensis* (Bt) against *L. dispar* depends on the presence of certain midgut bacteria (Broderick et al., 2006). Conversely, larvae reared under aseptic conditions show elevated vulnerability to pathogens, such as in *Galleria mellonella* (Upfold et al., 2023) and *P. xylostella* (Raymond et al., 2009), due to the absence of gut bacteria leading to reduced colonization resistance. Several gut bacteria produce antimicrobial compounds that selectively inhibit pathogens, thereby contributing to the insect's immune defense. For example, in *S. littoralis*, *E. mundtii* selectively clears gut pathogens from the gut lumen by secreting a stable antimicrobial peptide (mundticin KS) (Shao et al., 2017). These complex dynamics help explain why biological control agents sometimes perform inconsistently across field populations and highlight the need to consider microbiome context in control strategies.

2.3 HEMIPTERAN PEST MICROBIOMES

2.3.1 Microbiome Diversity in Hemipteran Pests

Hemipteran pests, including Sternorrhyncha such as aphids, whiteflies, and psyllids, and Auchenorrhyncha such as planthoppers, are among the most intensively studied agricultural insect pests. Some species are also significant vectors of plant pathogens (Mora et al., 2021). These sap-feeding pests feed on nutritionally imbalanced phloem sap, which is rich in carbohydrates but deficient in essential amino acids, vitamins, and other nutrients (Douglas, 2006). Their guts generally lack a chitin-based peritrophic matrix and instead have a specialized double-layered lipoprotein structure known as the perimicrovillar membrane (PMM) to manage high-volume fluid ingestion (Gutiérrez-Cabrera et al., 2016; Nardi et al., 2019). Across Hemiptera, microbial diversity is typically characterized by low taxonomic richness (Jing et al., 2014; Kuechler et al., 2012; Mao and Bennett, 2020; Michalik et al., 2023b). Often, hemipteran pests harbor heritable symbionts localized in specialized bacteriocytes (Douglas, 2015; Jing et al., 2014) that provide essential nutrients lacking in phloem sap (Douglas, 2003; Gündüz and Douglas, 2009; Moran, 1992). These obligate symbionts are maternally transmitted, exhibit substantial genome reduction, and have persisted in stable associations with their hosts over millions of years (Baumann, 2005). In contrast, facultative symbionts show variable prevalence yet can modulate immunity, stress tolerance, pathogen transmission, and interactions with natural enemies (Darby et al., 2003; Oliver et al., 2003; Tsuchida et al., 2004).

Aphids represent the classic model of insect–microbe symbiosis. Their microbiomes are typically dominated by the intracellular obligate symbiont *Buchnera aphidicola*, which supplies essential amino acids to support aphid growth and reproduction (Douglas, 1998). Many aphid pests also carry facultative symbionts, such as *Hamiltonella defensa*, *Regiella insecticola*, *Serratia symbiotica*, *Rickettsiella*, *Rickettsia*, *Arsenophonus*, *Wolbachia*, *Spiroplasma*, and

Candidatus Fukatsuia symbiotica (previously X-type or PAXS), whose prevalence varies among host lineages and can produce species-specific microbiome signatures that track host relatedness (Gauthier et al., 2015; Maeda et al., 2024; Qin et al., 2022).

Whiteflies likewise possess a heritable symbiotic core dominated by the obligate bacteriocyte symbiont *Candidatus Portiera aleyrodidarum*, frequently accompanied by facultative endosymbionts including *Arsenophonus*, *Rickettsia*, *Wolbachia*, *Hamiltonella*, and in some invasive lineages *Cardinium* (Andreason et al., 2020; El Hamss et al., 2022). Beyond these intracellular symbionts, high-throughput sequencing surveys reveal a low-abundance but variable assemblage of additional bacteria associated with different life stages and populations (El Hamss et al., 2024).

In Auchenorrhyncha pests, ancestral symbionts such as *Sulcia* and *Nasuia* dominate microbial communities and are present across individuals and populations in most lineages (Abbà et al., 2022; Bennett and Moran, 2013; Mulio et al., 2024; Phauk et al., 2025a). These symbionts also reside in bacteriomes and exhibit extreme genome reduction, reflecting long-term vertical transmission and metabolic specialization complementary with their hosts (Bennett and Moran, 2013). In leafhoppers, secondary symbionts such as *Arsenophonus*, *Sodalis*, *Wolbachia*, *Rickettsia*, and *Lariskella* are frequently detected, often in a species-specific manner (Phauk et al., 2025a; H. Wang et al., 2019). Comparative surveys across *Macrosteles* species reveal striking contrasts between highly conserved microbial assemblages dominated by obligate symbionts but great facultative symbiont diversity, despite limited host genetic differentiation and geographic separation (Mulio et al., 2024). In some subfamilies such as Typhlocybinae, obligate symbionts have been completely lost following shifts to mesophyll feeding, resulting in microbiomes dominated by facultative bacteria (Kobiałka et al., 2025).

True bugs (Heteroptera) exhibit a distinct symbiotic architecture centered on extracellular gut symbionts localized to midgut crypts (Gonella et al., 2020; Kikuchi et al., 2009; Ohbayashi et al., 2015; Oishi et al., 2019). These communities are typically dominated by a single gammaproteobacterial lineage, most commonly *Pantoea* or *Erwinia*, that is vertically transmitted and maintained at high densities throughout development (Fourie et al., 2023; Kashkouli et al., 2021). Some facultative symbionts also have been detected in stink bugs, such as *Spiroplasma* and *Burkholderia* (Kakizawa et al., 2022; Kikuchi et al., 2011; Takeshita and Kikuchi, 2017). Outside these crypt-associated symbionts, stink bug gut compartments generally harbor low-diversity and compositionally variable bacterial communities (Shan et al., 2021).

2.3.2 Drivers of Hemipteran Microbiome Variability

Host evolutionary history is a dominant driver of microbiome structure in hemipteran pests (Lei et al., 2023; Mizutani et al., 2025; Qin et al., 2022; Shigenobu et al., 2000; Z. Yang et al., 2023). Across aphid species, microbiome similarity is consistently higher within species than between species, even when hosts feed on different plants, and microbiome similarity often correlates with aphid phylogenetic distance (McLean et al., 2019; Qin et al., 2021). This strong host-symbiont fidelity is enforced by vertical transmission, resulting in remarkable stability of obligate symbionts across populations, seasons, and geographic regions. For example, a survey on *Myzus persicae* (Sulzer) from 16 hosts and 30 regions report that their associated symbionts do not change with respect to host plant or geography (Xu et al., 2021).

While host species identity is a strong driver of symbiont identity and composition (Malacrinò, 2022; Qin et al., 2022), environmental variables can modulate symbiont abundance and functional relevance (Li et al., 2022; Z. Yang et al., 2023). In aphids, *Buchnera* titer varies both within individuals and among different clonal lineages, which is affected by several environmental factors, such as the temperature and precipitation (Perreau et al., 2021; Smith and Moran, 2020). In *Aphis gossypii*, *Buchnera* abundance negatively correlates with precipitation and temperature, while facultative bacteria *Acinetobacter*, *Serratia*, and *Aeromonas* show positive correlations with precipitation (Alimu et al., 2025). In planthoppers, microbiome composition varies across geographic strains and correlates with abiotic conditions such as temperature and precipitation, with implications for insecticide susceptibility (Gupta et al., 2022; Liu et al., 2020).

Diet and host plant identity also shape hemipteran microbiomes, particularly in polyphagous species, where feeding on different plants induces reproducible shifts in bacterial and fungal taxa (Pawar et al., 2022; Yan et al., 2021). For example, in *A. gossypii*, changes in host plant and plant viral infection status restructure aphid microbiome diversity and the relative abundance of *Arsenophonus* and *Sphingomonas*. In *Bemisia tabaci*, elevated *Cardinium* titers due to high temperature or host-plant shifts were associated with changes in microbial community composition and predicted functional profiles, whereas stable *Cardinium* levels are associated with minimal microbiome change (De Moya-Ruiz et al., 2025; Yang et al., 2025).

Sex and developmental stage further contribute to microbiome heterogeneity. Females often harbor higher abundances of obligate symbionts than males, consistent with their roles in vertical transmission. For instance, female rice-associated leafhoppers harbor a higher abundance of *Karelsulcia* and *Nasuia* than males (Phauk et al., 2025b). Besides abundance, the frequency of symbionts can also be affected by sex. In *B. tabaci*, females exhibited higher frequencies of *Rickettsia*, *Hamiltonella*, and *Cardinium* (Pan et al., 2012). Compared with holometabolous insects, hemipterans do not undergo the same extent of gut remodeling, allowing symbionts to persist across life stages with limited turnover (Manthey et al., 2023). Therefore, developmental transitions tend to influence symbiont relative abundance rather than driving whole community replacement. For example, in *Monalonion velezangeli*, whole-body bacterial microbiome diversity and composition persist in both nymphs and adults, dominated by *Wolbachia* (~92%) (Navarro-Escalante et al., 2024).

2.3.3 Functional Roles of Hemipteran Microbiomes

Hemipteran microbiomes contribute to pest success through several well-documented functional roles, including nutritional supplementation, detoxification, immunity and defense, reproduction, and vector competence. Across sap-feeding hemipterans, the primary microbiome function is nutritional supplementation. Obligate symbionts synthesize essential amino acids, vitamins, and cofactors absent from plant sap diets, enabling hosts to exploit nutritionally imbalanced diet (Luan, 2024; Michalik et al., 2023a; Moriyama and Fukatsu, 2022; Sudakaran et al., 2017). Long-term host–symbiont associations are commonly accompanied by extreme genome reduction (Otero-Bravo and Sabree, 2021). As hosts have co-evolved with these symbionts, metabolic pathways (e.g., amino acid biosynthesis) are often partitioned between symbionts and host genomes (Hansen and Moran, 2011; Russell et al., 2013). Hosts have also acquired genes horizontally that support symbiont maintenance and integration, as seen in *B. aphidicola* in aphids (Shigenobu et al., 2000), *C. P. aleyrodidarum* in whiteflies (Lei et al., 2023), *Candidatus Carsonella ruddii* in psyllids (Mizutani et al., 2025), and *Candidatus Pantoea* persica in specific stink bugs (Kashkouli et al., 2021).

Facultative symbionts can expand the nutritional repertoire of hemipteran pests (Oliver et al., 2006). For example, *Candidatus Liberibacter psyllaurous* in *Bactericera cockerelli* enhances host fitness by provisioning arginine and other nutrients (Kwak et al., 2025). In stink bugs, gut symbionts contribute to amino acid and vitamin biosynthesis, and experimental symbiont removal leads to severe developmental defects or mortality (Prieto et al., 2023).

Facultative symbionts strongly influence hemipteran interactions with natural enemies, pathogens, and parasitoids. In aphids, *H. defensa* protects against parasitoid wasps via a phage-encoded toxin that kills developing parasitoid larvae (Oliver et al., 2003, 2009). *Wolbachia* and *R. insecticola* confer resistance to a specialized fungal pathogen, *Pandora neoaphidis* (Heyworth and Ferrari, 2015; Higashi et al., 2024; Scarborough et al., 2005), although the underlying protective mechanisms remain largely unresolved. Besides fungal pathogens, *R. insecticola* also protects *Acyrthosiphon pisum* against *Acyrthosiphon pisum virus* (APV) (Higashi et al., 2023). In *Diaphorina citri*, the specialized symbiont *Profftella armatura* produces a defensive polyketide (Tanabe et al., 2022), and newly identified symbionts such as *Fukatsuia* and *S. symbiotica* may play similar roles in other psyllid species (Nakabachi et al., 2022a, 2022b).

Hemipteran microbiomes also contribute to host detoxification metabolism. In planthoppers, bacterial taxa such as *Staphylococcus sciuri* and *Acinetobacter soli* modulate expression of host detoxification genes like the key P450 gene *NlCYP6ER1*, glutathione S-transferase (GST) genes, and esterase (EST) genes, thereby influencing resistance to multiple classes of insecticides (Vasquez et al., 2024; Zhang et al., 2023). In whiteflies, microbiome shifts and the presence of specific bacterial genera *Rhodococcus* and *Exiguobacterium* and fungal genus *Wallemia* have been linked to tolerance to the heavy metal cadmium (Guo et al., 2023).

Symbionts further shape host population dynamics through effects on host reproduction, thermal tolerance, host plant specialization, and vector competence (Gauthier et al., 2015). In aphids, *S. symbiotica* improves host survival under heat stress (Burke et al., 2010; Tougeron et al., 2023), while psyllids infected with *Liberibacter* display altered physiology and in some cases increased fecundity (Pelz-Stelinski and Killiny, 2016; Ren et al., 2016). In the small brown planthopper *Laodelphax striatellus*, high microbial diversity has been linked to improved tolerance of extreme environments, potentially reducing the cost of migration (Liu et al., 2020). Symbiont density can be environmentally sensitive. Under heat stress, shifts in the relative titers of co-infecting endosymbionts, for example, increases in *Cardinium* with concomitant decreases in *Rickettsia* in *B. tabaci*, are accompanied by changes in predicted community functions. This highlights the importance of symbiont regulation for host performance under agroecosystem stress (K. Yang et al., 2023; Yang et al., 2025).

In whiteflies, symbiont-derived chaperonins proposed to protect circulative plant viruses during transit in the hemolymph (Kliot and Ghanim, 2013). Metagenomic profiling across seasons and regions of *B. tabaci* revealed that endosymbiotic communities (*Portiera*, *Rickettsia*, *Hamiltonella*, and *Arsenophonus*) vary with temperature and weather conditions, and greater symbiont diversity coincides with higher incidence of tomato leaf curl disease, suggesting a possible link between microbiome composition and vector competence (Sujatha et al., 2025). In cassava whitefly (*B. tabaci* SSA1), outbreaking status and mitotype are associated with distinct microbiome profiles, implying that microbiome variation may be associated with outbreak dynamics (El Hamss et al., 2025).

2.4 COLEOPTERAN PEST MICROBIOMES

2.4.1 Microbiome Diversity in Coleopteran Pests

Coleoptera comprises approximately 40% of described insect species and represents the most taxonomically and ecologically diverse insect order (Stork et al., 2015; Usnaddinovich and Abatovna, 2025). Coleopteran pests occur across agricultural and ecosystem contexts, including crop pests such as the Colorado potato beetle (*Leptinotarsa decemlineata*), corn rootworms (*Diabrotica* spp.), cereal leaf beetle (*Oulema melanopus*, Chrysomelidae), Japanese beetle (*Popillia japonica*, Scarabaeidae), weevils (Curculionidae), and wireworms (Elateridae), wood-boring pests such as the emerald ash borer (*Agrilus planipennis*, Buprestidae) and the Asian longhorned beetle (*Anoplophora glabripennis*, and stored-product pests exploiting dry substrates including the flour beetles (*Tribolium* spp., Tenebrionidae), drugstore beetles (*Stegobium paniceum*, Ptinidae), and carpet beetles (Dermestidae). As holometabolous insects, the diets and feeding strategies of Coleopteran pests markedly differ between adults and larvae (Hammer and Moran, 2019). These fundamentally different pest lifestyles along with life-stage differences form distinct digestive physiologies and symbiotic architectures.

Across Coleopteran lineages, gut organization ranges from relatively simple to highly compartmentalized, often featuring region-specific enlargements and modifications linked to dietary specialization, though the midgut is typically protected by a peritrophic matrix (Mason, 2020). Unlike hemipterans, which commonly rely on tightly integrated nutritional symbionts with highly reduced genomes, Coleoptera-associated microbiome spans a continuum of microbial relationships that range from environmentally acquired, flexible communities to highly specialized intracellular endosymbionts (Fukumori et al., 2022; Kudo et al., 2019; Salem and Kaltenpoth, 2022). Their gut microbial communities are dominated by Proteobacteria and Firmicutes, with variable contributions from Actinobacteria, Bacteroidetes, and fungi depending on diet, habitat, and developmental stage (Chouaia et al., 2019; Drahun et al., 2025; Ebert et al., 2021; Kudo et al., 2019; Mohammed et al., 2018).

The microbiome of Coleopteran crop pests shows distinct structures according to feeding niche and the plant tissue consumed. Root-feeding scarab larvae such as *P. japonica* harbor particularly diverse gut communities during soil-dwelling stages, reflecting continuous exposure to the soil microbial reservoir (Avila-Arias et al., 2022). These larval microbiomes are often enriched in anaerobic and fermentative taxa, including *Ruminococcaceae*, *Lachnospiraceae*, and *Clostridiaceae*, which decline sharply following pupation and adult emergence (Chouaia et al., 2019). In contrast, foliar-feeding beetles such as *L. decemlineata* and *O. melanopus* tend to exhibit low microbial diversity but greater compositional consistency across individuals and host plants (Tatiana A et al., 2022; Wielkopolan et al., 2021, 2022). These communities are frequently enriched in Enterobacteriaceae, Lactobacillaceae, and facultative symbionts such as *Wolbachia* and *Rickettsia*, which persist across life stages and geographic locations (Wielkopolan et al., 2021, 2024; Yu et al., 2021). Similar to Hemiptera, some endosymbionts, represented by genera such as *Nardonella* in weevils (Curculionidae), *Stammera* in leaf beetles (Chrysomelidae), are observed in dedicated bacteriomes and possess highly reduced genomes (Fukumori et al., 2022; Mason, 2020; Salem et al., 2017; Weiss and Kaltenpoth, 2016).

Stored-product pests within this order occupy relatively uniform grain- or warehouse-associated substrates. Their microbiomes are shaped primarily by life stage and diet. For example, the microbiome of red flour beetle *Tribolium castaneum* varies with diet and developmental stage (Agarwal and Agashe, 2020). Larvae and adults reared on wheat flour exhibit marked differences in gut bacterial diversity and composition (Bi et al., 2024), whereas beetles reared on sauce-flavored Daqu maintain similar microbiomes across life stages, dominated by Bacteroidota and Proteobacteria alongside Firmicutes and Actinobacteriota, and fungi *Aspergillus* within the phylum Ascomycota (Lü et al., 2025).

Wood-boring Coleopteran pests, such as passalids, cerambycids, and some scarabs, show highly complex microbiome configurations. These insects feed on lignocellulose-rich substrates that are recalcitrant to enzymatic breakdown, characterized by severe nutritional limitation, and high concentrations of plant secondary metabolites (Wang et al., 2022). Their guts harbor diverse microbial consortia including bacteria, yeasts, and filamentous fungi capable of fermenting cellulose, hemicellulose, and pectin (Douglas, 2015; Scully et al., 2013, 2014; Tokuda, 2019). These communities often show spatial structuring across enlarged gut chambers, with anaerobic or micro-aerobic niches harboring specialized lineages (Ebert et al., 2021; Mohammed et al., 2018; Suárez-Moo et al., 2020). Such microbial architecture enables beetles to exploit nutrient-poor substrates that would otherwise be inaccessible. For example, in cerambycids such as *A. glabripennis*, larval gut communities comprise diverse bacterial and fungal taxa with predicted lignocellulolytic capacities, including Enterobacteriaceae, Actinobacteria, and consistently associated fungi such as *Fusarium solani* (Schloss et al., 2006; Scully et al., 2013, 2014; Wang et al., 2022).

2.4.2 Drivers of Coleopteran Microbiome Variability

In Coleoptera, microbiome assembly is shaped by developmental transitions, feeding ecology, and environmental exposure (Rodriguez-Fernandez et al., 2025; Yun et al., 2014), with host phylogeny exerting more context-dependent influence.

As in lepidopterans, holometabolous development imposes a major constraint on microbial persistence in Coleopteran species. During metamorphosis, much of the larval gut community is lost, leading to extensive reassembly in the adult stage (Manthey et al., 2023). Consequently, larval microbiomes, particularly in soil- and wood-feeding species, are strongly influenced by environmental acquisition. For example, larvae of scarab beetles such as the green June beetle *Cotinis nitida* and Japanese beetle *P. japonica* in continuous contact with soil and rhizosphere microbes harbor highly diverse and variable microbial communities. In contrast, adults exhibit reduced richness and greater host regulation, consistent with a transition toward a more selective gut environment (Avila-Arias et al., 2022; Kucuk et al., 2023). Within individuals, gut compartmentalization also contributes to microbiome structuring: physicochemical gradients in pH, oxygen availability, and redox potential between midgut and hindgut regions create distinct microhabitats that favor aerobic or facultatively anaerobic lignocellulose-degrading taxa (Avila-Arias et al., 2022). In Japanese beetle *P. japonica*, their midgut microbiome contains transient communities that vary with the surrounding soil environment, whereas hindgut microbiomes are more conserved across individuals (Avila-Arias et al., 2022).

Feeding ecology is a central determinant of microbiome diversity in Coleopteran pests. Dietary chemistry acts as a powerful selective force shaping both the composition and functional profile of gut microbial communities. Root-feeding larvae consume carbohydrate-rich but nitrogen-poor substrates, favoring fermentative microbes capable of degrading complex polysaccharides and recycling nitrogen, such as *Clostridiales* and *Bacteroidales* in *P. japonica* (Avila-Arias et al., 2022), and *Citrobacter* and related Enterobacteriaceae in weevils (Meng et al., 2019). Foliar-feeding and wood-boring pests encounter a wide range of plant secondary metabolites, including alkaloids, glycoalkaloids, and phenolics. These compounds select for microbial taxa with detoxification and stress-response capabilities. For example, bark beetles and weevils that attack conifers depend on microbial associates capable of metabolizing terpenes (Zulak and Bohlmann, 2010) and resin compounds (Trapp and Croteau, 2001). Host plant identity further modulates microbiome composition and function. For example, wood-boring pests larvae developing on preferred host trees harbor more diverse microbial communities and exhibit higher lignocellulolytic enzyme activities than those feeding on resistant hosts (Wang et al., 2022).

Stored-product pests highlight how environmental exposure imposes additional filters on Coleopteran microbiomes. These pests live in anthropogenic environments, characterized by low moisture and exposure to uniform, processed substrates (Hagstrum, 2006). Interestingly, in *T. castaneum* inhabiting dry Daqu warehouses, only a small proportion of gut-associated bacteria overlapped with the substrate (~5.1% in adults and ~7.3% in larvae), whereas fungal overlap exceeded 89%, indicating fundamentally different acquisition and retention mechanisms for bacteria versus fungi (Lü et al., 2025).

Finally, geographic origin and invasion history can influence microbiome composition, but these factors generally exert secondary effects relative to developmental stage, feeding ecology, and environmental exposure (Paddock et al., 2022).

2.4.3 Functional Roles of Coleopteran Microbiomes

Coleopteran microbiomes are highly diverse and functionally versatile, reflecting the ecological breadth of the order. Across species, symbionts provide essential digestive enzymes, supplement limiting nutrients, detoxify plant and microbial toxins, and contribute to host defense and tolerance to environmental stressors (García-Lozano and Salem, 2025; Jing et al., 2020; Salem and Kaltenpoth, 2022).

A major role of Coleopteran microbiomes is enabling the digestion of structurally complex plant materials. Many beetles specialize on plant tissues that their endogenous enzymes alone cannot efficiently process. Coleopteran pests such as passalids, cerambycids, and scarabs harbor yeast-like symbionts and bacteria that break down lignocellulose into short-chain fatty acids by producing cellulases, hemicellulases, xylanases, pectinases, glycoside hydrolases, laccases, and peroxidases (Andert et al., 2010; Ebert et al., 2021; Geib et al., 2009, 2010; Lee et al., 2004; Mohammed et al., 2018; Schwarz et al., 2023; Scully et al., 2013; Suh et al., 2008). Larvae of soil-dwelling and root-feeding *P. japonica* and *Phyllophaga* spp. harbor dense hindgut communities dominated by Firmicutes (e.g., Clostridia) and Bacteroidota, involved in anaerobic fermentation and polysaccharide degradation (Frias et al., 2023; Sheng et al., 2015). In wood-boring pest *A. glabripennis*, larval gut microbiome is enriched in genes associated with lignocellulose degradation, the fermentation of wood sugars, and the biosynthesis of essential amino acids, vitamins, and sterols (Scully et al., 2013, 2014). Some Coleopteran pests rely on intracellular symbionts for nutritional supplementation. An example is the weevil symbiont *Nardonella*, synthesizing precursors for tyrosine, a key metabolic requirement for cuticle formation and sclerotization (Anbutsu et al., 2017). Weevils lacking these symbionts develop thin, weak cuticles, thus suffer reduced survival and desiccation resistance, as seen in *Candidatus Shikimatogenerans silvanidophilus–Oryzaephilus surinamensis* symbiosis (Engl et al., 2018; Kanyile et al., 2022). In *T. castaneum*, gut-associated Firmicutes, Bacteroidota, and Proteobacteria support the degradation of complex carbohydrates and plant polysaccharides and nitrogen metabolism (Agarwal and Agashe, 2020; Korša et al., 2022; Lü et al., 2025). Similarly, the tortoise beetle *Cassida rubiginosa* relies on the bacterium *Stammera*, whose genome is almost entirely devoted to producing pectinases (McKenna, 2020; Salem et al., 2017, 2020). Loss of this symbiont severely impairs larval development (Salem et al., 2017).

Microbial associates also play key roles in detoxifying plant defense compounds or manipulating plant defense pathways. Foliar-feeding beetles, in particular, exploit chemically defended host plants through microbial mediation. For example, the cabbage stem flea beetle (*Psylliodes chrysocephala*), a major pest of oilseed rape, relies on *Pantoea* symbionts to detoxify toxic isothiocyanates (Shukla and Beran, 2020). Antibiotic-fed beetles accumulated over 11-fold higher levels of unmetabolized isothiocyanates. In addition, *Pantoea* was detected in undamaged Brassica tissues, suggesting that the pest may acquire the symbiont from its diet. The coffee berry borer *Hypothenemus hampei* harbors bacteria capable of degrading caffeine, an otherwise toxic alkaloid (Ceja-Navarro et al., 2015). In the corn rootworm *Diabrotica virgifera virgiferaverifier* LeConte, gut microbiome facilitates physiological adaptation to transient feeding on soybean through elevating tolerance to cysteine protease inhibitors in soybean foliage (Chu et al., 2013). Similar to *H. zea* described previously, the Colorado potato beetle (*L. decemlineata*) larvae also deploy symbiotic bacteria via oral secretions to manipulate plant signaling, specifically inducing the SA signaling pathway in tomato plants, while suppressing JA-mediated defenses (Chung et al., 2013).

Microbiome can induce both positive and negative responses to insecticides (Malook et al., 2025). In Colorado potato beetle (*L. decemlineata*), gut microbiome disruption by antibiotics reduced host susceptibility to Bt and avermectins, while reintroduction of core taxa restores or enhances susceptibility (Polenogova et al., 2021). Yet, larvae experimentally supplemented with *Lactococcus lactis* also had five-fold lower mortality rates than untreated larvae upon exposure to the insecticide (Stara and Hubert, 2023). In *O. melanopus*, gut bacteria *Acinetobacter*, *Stenotrophomonas*, *Lactococcus* and *Pseudomonas* enhance their larval survival after pesticide treatments of cypermethrin and chlorpyrifos (Wielkopolan et al., 2024), and microbial diversity shifts with insecticide type and dose.

Defense is another key function of Coleopteran pest microbiomes. For example, *Tribolium* spp. carry bacteria that can inhibit fungal pathogens, offering protection at the egg and larval stage (Janke et al., 2022). By promoting cuticle formation, symbiotic bacteria also enhance resistance to fungal pathogen in *O. surinamensis* (Kanyile et al., 2022).

2.5 DIPTERAN PEST MICROBIOMES

Diptera (true flies) comprise over 150,000 described species worldwide and includes numerous economically important agricultural pests that affect diverse crop systems (Pape et al., 2011; Stork, 2018). Agricultural dipteran pests damage crops through feeding, oviposition, larval tunneling, and gall formation. Among these, tephritid fruit flies (Tephritidae) represent the most economically significant dipteran pests globally, including species such as the Mediterranean fruit fly (*Ceratitis capitata*), oriental fruit fly (*Bactrocera dorsalis*), and Mexican fruit fly (*Anastrepha ludens*), alongside other key agricultural pests such as the spotted wing drosophila (*Drosophila suzukii*), leaf-miner flies (Agromyzidae), and gall midges (Cecidomyiidae).

Unlike several hemipteran pests that rely on vertically transmitted intracellular obligate symbionts, dipteran pest microbiomes are established through early-life microbial exposure from multiple sources, including maternal deposition on eggs and larval developmental habitat (Sontowski and Van Dam, 2020). These microbiomes are functionally integral to host biology, contributing to nutrition, detoxification, immune regulation, reproduction, and plant–insect interactions (Hamby and Becher, 2016; Siden-Kiamos et al., 2025; Sontowski and Van Dam, 2020). Across dipteran taxa, microbiome composition is often shaped by environmental acquisition, with host-mediated filtering playing a context-dependent role, resulting in communities that are often variable in taxonomic composition yet support more conserved functional outcomes (Hendrycks et al., 2025; Nowak et al., 2025).

2.5.1 Microbiome Diversity in Dipteran Pest

Dipteran agricultural pest microbiomes are typically characterized by low taxonomic richness but high variability across species, populations, and ecological contexts. Among dipteran pests, the gut microbiomes of tephritid fruit flies (Tephritidae) have been the most intensively studied. They are dominated by a limited subset of bacterial lineages, such as the genera *Klebsiella*, *Enterobacter*, *Providencia*, and *Raoultella*, most frequently within the Proteobacteria (especially Enterobacteriaceae) that persist across developmental stages and environments (Deutscher et al., 2019; Mason et al., 2023; Raza et al., 2020).

The larvae of many tephritid pests can develop in diverse plant tissues, including fruits, vegetables, and flowers that are nutritionally rich but chemically defended (Hafsi et al., 2016). Adult tephritids also rely on microbial cues, such as volatile organic compounds (VOCs), for feeding, aggregation, and oviposition site selection (Biasazin et al., 2018; He et al., 2022). This ecological context has likely favored microbiomes that are functionally conserved but taxonomically flexible (De Cock et al., 2020; Malacrinò et al., 2018; Morrow et al., 2015; Tithi, 2022). For example, strains of *Klebsiella pneumoniae*, *Enterobacter cloacae*, and *Staphylococcus aureus* and their associated volatile compounds have been confirmed to attract *A. ludens*, although their relative abundance is context-dependent (Kuzina et al., 2001; Raza et al., 2020; Robacker and Flath, 1995). These bacteria often enter fruit tissues during oviposition and proliferate within the fruit, where they enhance larval development by contributing to carbohydrate, amino acid, and vitamin/cofactor metabolism (Tairo et al., 2025; Ventura et al., 2018).

The gut microbiome of spotted wing drosophila (*D. suzukii*, Drosophilidae) contrasts with that of tephritids. It is consistently enriched in Proteobacteria, particularly acetic acid bacteria (Acetobacteraceae) such as *Acetobacter* and *Gluconobacter*, as well as Enterobacteriaceae including *Klebsiella*, *Kluyvera*, and *Providencia* (Bhandari et al., 2025; Chandler et al., 2014; Martinez-Sañudo et al., 2018; Vacchini et al., 2017). In addition to bacteria, yeasts are frequently associated with *D. suzukii* larvae and may persist through metamorphosis (Guilhot et al., 2023; Hamby et al., 2012), although fungal community composition is largely determined by the breeding substrate.

2.5.2 Drivers of Dipteran Microbiome Variability

Dipteran microbiomes are influenced by host identity, diet, rearing environment, and developmental stage. Sex and seasonal effects have also been documented in several pest species (Raza et al., 2020).

The impacts of host identity on dipteran pest microbiome varies by species. In some cases, microbial communities share similar dominate bacteria but differ markedly among genera at diversity (e.g., alpha diversity is higher in Dacinae than in Ceratitinae) (Ravigné et al., 2022). However, studies on *Bactrocera latifrons*, *B. dorsalis*, *C. capitata*, and *Zeugodacus cucurbitae*, demonstrates that different fruits fly species exhibit distinct microbiomes even under controlled uniformed laboratory conditions (Aoki et al., 2025).

Laboratory or mass-rearing conditions typically reduce microbial diversity and alter dominant taxa relative to field populations (Ben-Yosef et al., 2008; De Cock et al., 2020; Malacrinò et al., 2018). For instance, when reared under controlled laboratory conditions, several tepheritids, such as *B. dorsalis* and *C. capitata*, exhibit distinct microbiomes as compared to field populations (Aoki et al., 2025; De Cock et al., 2020; Hendrycks et al., 2025; Majumder et al., 2022; Mullens et al., 2024). Mass rearing and irradiation have been found to alter the fly gut microbiome, including proliferation of pathogens but reduction in specific symbionts like *Klebsiella* (Ben-Yosef et al., 2008). These shifts are especially relevant for sterile insect technique (SIT) as they may contribute to reduced host mating competitiveness and overall fitness.

Many dipterans smear bacteria onto egg surfaces during oviposition, which are subsequently acquired by hatching larvae (Marchini et al., 2002; Mazzon et al., 2008; Wang et al., 2014), This has been reported in tephritids such as *C. capitata* and *Bactrocera oleae* (Estes et al., 2009; Marchini et al., 2002; Mazzon et al., 2008),

Dipteran microbiome assembly is also strongly structured by developmental stage. For example, *A. ludens* larval microbiome closely reflect the microbial composition of the host fruit, but much of it is filtered out during metamorphosis, leading to adult microbiomes that are less diverse and variable (Aluja et al., 2021). In the melon fruit fly *Bactrocera cucurbitae* (Coquillett), bacterial alpha diversity was significantly lower in the ovaries than in the eggs, but not between the primary and mature ovaries. Pseudomonadota, Deinococcota, Bacteroidota, Bacillota, and Actinomycetota were the dominant phyla in all three stages (Chen et al., 2023; Tian et al., 2023).

Diet, host plant identity, and environmental microbial pools are also major determinants of microbiome composition. In *B. dorsalis*, *Musca domestica*, and *Delia radicum*, microbiomes are assembled primarily from microbial pools associated with host plants, decaying organic matter, and soil (Aluja et al., 2021; Aoki et al., 2025; Kempraj et al., 2024; Park et al., 2019; Van Den Bosch and Welte, 2020). Host plants differ in nutritional profiles, microbial communities, and chemical defenses, all of which shape microbial acquisition and persistence. For instance, the gut microbiota of *B. dorsalis* and *C. capitata* larvae varies across two different host plants from four different locations (De Cock et al., 2020). As a result, polyphagous species, such as *C. capitata*, *Ceratitis quilicii*, and *B. dorsalis* exploiting a broad or seasonally variable host range, often harbor more flexible microbiomes, whereas oligophagous like *Z. cucurbitae* and monophagous like *B. oleae* tend to show greater microbial stability (De Cock et al., 2020; Yamlahi et al., 2025).

Seasonal and climatic factors also influence microbiome structure. During winter, spotted wing drosophila gut communities transition to microbes associated with cold conditions and decomposition, such as bacteria *Pseudomonas*, fungi *Guehomyces*, and *Cryptococcus* (Fountain et al., 2018). Such seasonal turnover highlights the responsiveness of dipteran microbiomes to environmental variation.

2.5.3 Functional Roles of Dipteran Microbiomes

Dipteran microbiomes influence host physiology and agricultural impact through various nutritional and metabolic mechanisms (Al Darwish et al., 2025; Gupta and Nair, 2020; Raza et al., 2020). Additionally, certain microbes have been shown to produce VOCs that influence host–plant interactions and dipteran pest behavior.

A key function of dipteran microbiomes is nutritional supplementation, particularly during larval development in nutritionally imbalanced substrates. Fruit fly gut bacteria enhance digestion of plant cell wall polysaccharides, such as pectins, hemicellulose, and provide essential amino acids and B vitamins during larval development within fruits (Akami et al., 2019; Consuegra et al., 2020; McMullen et al., 2020). For example, *Enterobacter* and *Klebsiella* species in *B. dorsalis* and *C. capitata* provide pectin-degrading enzymes including pectinases, cellulases, and amylases, while provisioning B vitamins and nitrogenous compounds to support larval development on nitrogen-poor fruits and enhance adult reproductive output (Gu et al., 2024; Ren et al., 2022; Zaada et al., 2019).

Microbiomes also facilitate pest colonization of chemically defended plant hosts and contribute to insecticide resistance. In *Rhagoletis pomonella*, gut bacteria contribute to detoxification of the phenolic compound phloridzin (Lauzon et al., 2003), while genomic analyses of olive fruit fly *B. oleae* symbionts, such as *Erwinia dacicola*, have identified genes encoding detoxification and digestive enzymes essential for survival in unripe olives rich in oleuropein (Ben-Yosef et al., 2015; Blow et al., 2020). In *B. dorsalis*, gut bacterium *Citrobacter* sp. enhances resistance to the organophosphate insecticide trichlorphon via symbiont-mediated degradation (Cheng et al., 2017), and lactic acid bacteria (*Enterococcus casseliflavus* and *L. lactis*) increase resistance to pyrethroid β-cypermethrin by upregulating detoxification pathways (Zeng et al., 2024).

Microbial metabolites have been shown to mediate plant–insect–microbe interactions and can shape dipteran pest behavior. For example, certain *Gluconobacter* strains produce volatiles that attract *D. suzukii* (Mazzetto et al., 2016). *B. dorsalis* avoids oviposition on egg-infested fruits 48 h after initial oviposition through recognizing bacterially induced increases in β-caryophyllene emissions (Li et al., 2020) but is attracted to volatiles such as 3-hexenyl acetate produced by the gut bacterium *Citrobacter freundii* (He et al., 2022).

Microbiomes also influence pest reproduction and mating. In *B. dorsalis*, gut bacteria alter mate choice by enhancing female attractiveness and increasing male sperm allocation (Damodaram et al., 2016). Such effects highlight the role of symbionts in sexual selection and reproductive success.

Beyond reproduction and nutrition, symbiotic bacteria can modulate host immune function. Resident microbes can prime immune responses, affecting pathogen or parasite susceptibility. For example, in *B. dorsalis*, resident gut bacteria *L. lactis* can increase the emergence success of parasitoids, *Fopius arisanus* and *Diachasmimorpha longicaudata*, while other bacteria can increase parasitoid body size and developmental time or reduce parasitoid fecundity depending on the symbiont and wasp species (Gwokyalya et al., 2023).

2.6 OTHER IMPORTANT AGRICULTURAL PEST EXAMPLES

Beyond Lepidoptera, Hemiptera, Coleoptera, and Diptera, microbiomes have been shown to play important roles in digestion, detoxification, immunity, and ecological interactions in other notable agricultural pests. As commonly seen in orthopterans and isopterans polyphagous insects feeding on leaves, stems, detritus, or decaying matter (Arora et al., 2022; Hance et al., 2025; Hirata et al., 2023), microbiome diversity tends to be highest in these insects with frequent environmental exposure and diets rich in structurally complex or microbe-rich substrates (Wang et al., 2025).

2.6.1 Orthopteran Pests

Orthopteran pests, including grasshoppers, crickets, and locusts, harbor gut communities dominated by fermentative bacteria capable of digesting plant fibers and metabolizing dietary phenolics. These insects often have expanded hindgut regions (W. Li et al., 2023; Ling et al., 2022; Ng et al., 2018) for microbial fermentation to convert plant carbohydrates into short-chain fatty acids. Their digestive systems lack the extreme alkalinity seen in lepidopteran caterpillars, allowing a broader range of taxa to persist. Together with their wide dietary breath, orthopteran microbiomes are often more diverse and less specialized than those of sap-feeding insects (Hance et al., 2025). Overall, Proteobacteria and Firmicutes are commonly associated with these pests, but the dominant genera vary by species (Ling et al., 2022; Mead et al., 1988; Muratore et al., 2020; J. Wang et al., 2020; Yun et al., 2014). A detailed summary can be found in Table 1 of a recent review (Hance et al., 2025).

Functionally, orthopteran microbiomes resemble those of other chewing herbivorous feeding on chemically defended plants (Douglas, 2015; Zhao et al., 2022). In locusts, hindgut bacteria produce volatile metabolites that act as aggregation pheromones. These bacteria are enriched in gregarious locusts compared to solitary ones and are thought to play a role in the phase transition and swarm formation (Shi et al., 2014).

2.6.2 Agricultural Pest Termites

Termites are known for their reliance on gut microbial symbioses involving complex cooperation between bacteria, archaea, and protists for digestion and nutrient assimilation. Based on feeding habits, termites can be divided into wood-feeding termites, as well as soil-feeding, grass-feeding, and fungus-growing termites. Some members are considered agricultural pests.

Agricultural pest termites generally exhibit greater gut microbial richness and diversity than wood-feeding termites. Some soil-feeders have several folds greater microbial richness and diversity indices compared to wood feeders. This increased diversity is linked to the more complex and variable substrates encountered in soil and litter environments. The gut microbiome of wood-feeding termites is typically dominated by flagellated protozoa and bacteria including Spirochaetes, Fibrobacteres, and other nitrogen fixation microorganisms from Firmicutes, Actinobacteria, Proteobacteria, and Bacteroidetes (Diouf et al., 2015; Su et al., 2016; Victorica et al., 2020; Vikram et al., 2021). These bacterial and protist communities cooperate in lignocellulose degradation, nitrogen fixation, and fermentation, reflecting adaptation to nitrogen-poor wood diets (Ahmad et al., 2021; Arora et al., 2022; Soukup et al., 2021). In contrast, grass-feeding termites gut microbiomes are enriched in Firmicutes and other fermentative bacteria adapted for hemicellulose-rich diets (Menezes et al., 2018; Mikaelyan et al., 2015). Soil-feeding termites harbor an even more diverse microbiome (He et al., 2013; Marynowska et al., 2020), dominated by Firmicutes and Planctomycetes, as well as Proteobacteria and Actinobacteria (Köhler et al., 2008; Korsa et al., 2023).

Fungus-growing termites attack a wide range of crops including sugarcane, maize, rice, wheat by attacking the roots and stems. Their guts are free of protists but abundant in archaea and bacteria. They have a complex tripartite symbiosis involving the host, cultivated *Termitomyces* fungi, and bacteria dominated by Firmicutes and Bacteroidetes. These symbiotic systems contribute carbohydrate-active enzymes that ferment cellulose and fungal biomass to produce short-chain fatty acids (Benndorf et al., 2020; Hu et al., 2019; Otani et al., 2019).

2.7 IMPLICATIONS FOR AGRICULTURAL PEST MANAGEMENT

Across major agricultural pest orders, insect microbiomes are increasingly recognized as integral components of pest biology. Microbial contributions to development, nutrition, detoxification, resistance to plant defenses, behavior, and immunity collectively affect the overall pest fitness and responses to management interventions, such as chemical insecticides and biological control agents.

From a management perspective, microbiome research highlights several complementary opportunities. First, the dependence of pests on microbial symbionts creates vulnerabilities that can be exploited. Disruption of symbiont transmission or function can delay host development or impair reproduction (Hassan et al., 2020; Lv et al., 2018; Xin Yee Tan and Shigenobu, 2025), demonstrating that symbionts are viable targets for non-chemical control strategies.

Second, microbiome-mediated detoxification and insecticide resistance create both risks and opportunities for pest management (Al Darwish et al., 2025). In several pest systems, symbionts contribute directly to insecticide degradation or enhance host tolerance, and resistant populations often show consistent shifts in microbiome composition (Malook et al., 2025). These findings suggest that microbiomes can serve as potential markers for resistance, while targeting detoxifying symbionts may restore susceptibility or mitigate resistance development (Haider et al., 2025).

In pest species whose microbiomes are environmentally acquired, manipulation of soil or phyllosphere microbial communities may affect pest performance (Das et al., 2023; S. Zhang et al., 2022). Such approach may be especially relevant for managing invasive pests, which often depend on symbionts to establish in new environments or novel food sources (Gupta et al., 2023).

For insect vectors of plant pathogens, understanding how symbionts modulate pathogen acquisition or transmission can inform early detection programs through identifying vector populations most likely to spread disease and can guide targeted vector suppression strategies (Perilla-Henao and Casteel, 2016). Advances in microbiome profiling provide new opportunities to identify previously unrecognized symbionts that influence disease dynamics, while symbiont-based interventions, such as manipulating *Wolbachia* infections, hold promises for reducing vector competence or altering population structure (Roldán et al., 2024).

Furthermore, VOCs produced by microbes can be leveraged to influence pest behavior in management. Microbial volatiles capable of manipulating mating or reproductive behavior can be used in mating disruption. For example, rectal bacteria (*Bacillus* sp.) associated with male *B. dorsalis* produce sex pheromones that attract mature virgin females (Ren et al., 2021). Attract-and-kill and push-and-pull strategies can exploit these microbial-based volatiles, as baits to attract pests. Yeasts such as *Hanseniaspora uvarum*, *Saccharomyces cerevisiae*, and related fermentation compounds have been widely used for *D. suzukii* trapping systems (Bueno et al., 2020; Harmon et al., 2019). Some microbial volatiles have also been identified as repellents for *D. suzukii* in laboratory assays, such as ethyl butanoate and ethyl (E)-but-2-enoate (Quadrel et al., 2025).

Finally, microbiome manipulation can enhance autocidal programs. In dipteran pests like Mediterranean fruit flies *C. capitata*, microbiome manipulation (e.g., probiotic treatments) has shown promise in improving mating competitiveness and enhanced SIT outcomes (Ami et al., 2010; Haytham et al., 2024), which demonstrate how microbial approach can strengthen autocidal programs.

REFERENCES

Abbà, S., Rossi, M., Vallino, M., Galetto, L., Marzachì, C., Turina, M., 2022. Metatranscriptomic assessment of the microbial community associated with the flavescence dorée phytoplasma insect vector *Scaphoideus titanus*. *Front. Microbiol.* 13, 866523. https://doi.org/10.3389/fmicb.2022.866523

Agarwal, A., Agashe, D., 2020. The red flour beetle Tribolium castaneum: A model for host-microbiome interactions. *PLOS ONE* 15, e0239051. https://doi.org/10.1371/journal.pone.0239051

Ahmad, F., Yang, G., Liang, S., Zhou, Q., Gaal, H.A., Mo, J., 2021. Multipartite symbioses in fungus-growing termites (Blattodea: Termitidae, Macrotermitinae) for the degradation of lignocellulose. *Insect Sci.* 28, 1512–1529. https://doi.org/10.1111/1744-7917.12890

Ahmed, M.Z., Araujo-Jnr, E.V., Welch, J.J., Kawahara, A.Y., 2015. Wolbachia in butterflies and moths: Geographic structure in infection frequency. *Front. Zool.* 12, 16. https://doi.org/10.1186/s12983-015-0107-z

Ajmal, M.S., Ali, S., Jamal, A., Saeed, M.F., Radicetti, E., Civolani, S., 2024. Feeding and growth response of fall armyworm *Spodoptera frugiperda* (Lepidoptera: Noctuidae) towards different host plants. *Insects* 15, 789. https://doi.org/10.3390/insects15100789

Akami, M., Andongma, A.A., Zhengzhong, C., Nan, J., Khaeso, K., Jurkevitch, E., Niu, C.-Y., Yuval, B., 2019. Intestinal bacteria modulate the foraging behavior of the oriental fruit fly *Bactrocera dorsalis* (Diptera: Tephritidae). *PLOS ONE* 14, e0210109. https://doi.org/10.1371/journal.pone.0210109

Al Darwish, H., Tariq, M., Salama, S., Hart, T., Sun, J.S., 2025. The microbiome as a driver of insect physiology, behavior, and control strategies. *Appl. Microbiol.* 5, 90. https://doi.org/10.3390/applmicrobiol5030090

Alimu, A., Gao, Y., Liu, J., Lu, Y., 2025. Geographic factors influence communities of symbiotic bacterial communities in Aphis gossypii across China's major cotton regions. *Front. Microbiol.* 16, 1569543. https://doi.org/10.3389/fmicb.2025.1569543

Aluja, M., Zamora-Briseño, J.A., Pérez-Brocal, V., Altúzar-Molina, A., Guillén, L., Desgarennes, D., Vázquez-Rosas-Landa, M., Ibarra-Laclette, E.,

Alonso-Sánchez, A.G., Moya, A., 2021. Metagenomic survey of the highly polyphagous *Anastrepha ludens* developing in ancestral and exotic hosts reveals the lack of a stable microbiota in larvae and the strong influence of metamorphosis on adult gut microbiota. *Front. Microbiol.* 12, 685937. https://doi.org/10.3389/fmicb.2021.685937

Ami, E.B., Yuval, B., Jurkevitch, E., 2010. Manipulation of the microbiota of mass-reared Mediterranean fruit flies *Ceratitis capitata* (Diptera: Tephritidae) improves sterile male sexual performance. *ISME J.* 4, 28–37. https://doi.org/10.1038/ismej.2009.82

Anand, A.A.P., Vennison, S.J., Sankar, S.G., Prabhu, D.I.G., Vasan, P.T., Raghuraman, T., Geoffrey, C.J., Vendan, S.E., 2010. Isolation and characterization of bacteria from the gut of *Bombyx mori* that degrade cellulose, xylan, pectin and starch and their impact on digestion. *J. Insect Sci.* 10, 1–20. https://doi.org/10.1673/031.010.10701

Anbutsu, H., Moriyama, M., Nikoh, N., Hosokawa, T., Futahashi, R., Tanahashi, M., Meng, X.-Y., Kuriwada, T., Mori, N., Oshima, K., Hattori, M., Fujie, M., Satoh, N., Maeda, T., Shigenobu, S., Koga, R., Fukatsu, T., 2017. Small genome symbiont underlies cuticle hardness in beetles. *Proc. Natl. Acad. Sci.* 114. https://doi.org/10.1073/pnas.1712857114

Andert, J., Marten, A., Brandl, R., Brune, A., 2010. Inter-and intraspecific comparison of the bacterial assemblages in the hindgut of humivorous scarab beetle larvae (Pachnoda spp.): Differences in the gut microbiota of Pachnoda spp. *FEMS Microbiol. Ecol.* 74, 439–449. https://doi.org/10.1111/j.1574-6941.2010.00950.x

Andreason, S.A., Shelby, E.A., Moss, J.B., Moore, P.J., Moore, A.J., Simmons, A.M., 2020. Whitefly endosymbionts: Biology, evolution, and plant virus interactions. *Insects* 11, 775. https://doi.org/10.3390/insects11110775

Aoki, S., Weaver, M., Simmonds, T.J., Shikano, I., Geib, S.M., Mason, C.J., 2025. Effects of species and sex on the gut microbiome of four laboratory-reared fruit fly lines (Diptera: Tephritidae) using full-length 16S rRNA PacBio Kinnex sequencing. *BMC Microbiol.* 25, 455. https://doi.org/10.1186/s12866-025-04025-0

Araújo, F.S.D., Barcelos, R.M., Mendes, T.A.D.O., Mafra, C., 2021. Molecular evidence of Rickettsia felis in Phereoeca sp. *Rev. Bras. Parasitol. Veterinária* 30, e015620. https://doi.org/10.1590/s1984-29612021017

Arora, J., Kinjo, Y., Šobotník, J., Buček, A., Clitheroe, C., Stiblik, P., Roisin, Y., Žifčáková, L., Park, Y.C., Kim, K.Y., Sillam-Dussès, D., Hervé, V., Lo, N., Tokuda, G., Brune, A., Bourguignon, T., 2022. The functional evolution of termite gut microbiota. *Microbiome* 10, 78. https://doi.org/10.1186/s40168-022-01258-3

Avila-Arias, H., Scharf, M.E., Turco, R.F., Richmond, D.S., 2022. Soil environments influence gut prokaryotic communities in the larvae of the invasive Japanese beetle Popillia japonica newman. *Front. Microbiol.* 13, 854513. https://doi.org/10.3389/fmicb.2022.854513

Baumann, P., 2005. Biology of bacteriocyte-associated endosymbionts of plant sap-sucking insects. *Annu. Rev. Microbiol.* 59, 155–189. https://doi.org/10.1146/annurev.micro.59.030804.121041

Benndorf, R., Schwitalla, J.W., Martin, K., De Beer, Z.W., Vollmers, J., Kaster, A.-K., Poulsen, M., Beemelmanns, C., 2020. Nocardia macrotermitis sp. nov. and Nocardia aurantia sp. nov., isolated from the gut of the fungus-growing termite Macrotermes natalensis. *Int. J. Syst. Evol. Microbiol.* 70, 5226–5234. https://doi.org/10.1099/ijsem.0.004398

Bennett, G.M., Moran, N.A., 2013. Small, smaller, smallest: The origins and evolution of ancient dual symbioses in a phloem-feeding insect. *Genome Biol. Evol.* 5, 1675–1688. https://doi.org/10.1093/gbe/evt118

Ben-Yosef, M., Jurkevitch, E., Yuval, B., 2008. Effect of bacteria on nutritional status and reproductive success of the Mediterranean fruit fly *Ceratitis capitata*. *Physiol. Entomol.* 33, 145–154. https://doi.org/10.1111/j.1365-3032.2008.00617.x

Ben-Yosef, M., Pasternak, Z., Jurkevitch, E., Yuval, B., 2015. Symbiotic bacteria enable olive fly larvae to overcome host defences. *R. Soc. Open Sci.* 2, 150170. https://doi.org/10.1098/rsos.150170

Bhandari, R., Wong, A.C.-N., Lee, J.C., Boyd, A., Shelby, K., Ringbauer, J., Kang, D.S., 2025. Microbiome composition and co-occurrence dynamics in wild *Drosophila suzukii* are influenced by host crop, fly sex, and sampling location. *Microbiol. Spectr.* 13, e02608–24. https://doi.org/10.1128/spectrum.02608-24

Bi, J., Zhang, L., He, Y., 2024. Metabolomics and microbiome provide new insights into mechanisms of the variations between the larva and adult stages of the model beetle Tribolium castaneum (Herbst). *J. Asia-Pac. Entomol.* 27, 102241. https://doi.org/10.1016/j.aspen.2024.102241

Biasazin, T.D., Chernet, H.T., Herrera, S.L., Bengtsson, M., Karlsson, M.F., Lemmen-Lechelt, J.K., Dekker, T., 2018. Detection of volatile constituents from food lures by tephritid fruit flies. *Insects* 9, 119. https://doi.org/10.3390/insects9030119

Blow, F., Gioti, A., Goodhead, I.B., Kalyva, M., Kampouraki, A., Vontas, J., Darby, A.C., 2020. Functional genomics of a symbiotic community: Shared traits in the olive fruit fly gut microbiota. *Genome Biol. Evol.* 12, 3778–3791. https://doi.org/10.1093/gbe/evz258

Bordenstein, S.R., Theis, K.R., 2015. Host biology in light of the microbiome: Ten principles of holobionts and hologenomes. *PLOS Biol.* 13, e1002226. https://doi.org/10.1371/journal.pbio.1002226

Brinkmann, N., Martens, R., Tebbe, C.C., 2008. Origin and diversity of metabolically active gut bacteria from laboratory-bred larvae of *Manduca sexta* (Sphingidae, Lepidoptera, Insecta). *Appl. Environ. Microbiol.* 74, 7189–7196. https://doi.org/10.1128/AEM.01464-08

Broderick, N.A., Raffa, K.F., Goodman, R.M., Handelsman, J., 2004. Census of the bacterial community of the gypsy moth larval midgut by using culturing and culture-independent methods. *Appl. Environ. Microbiol.* 70, 293–300. https://doi.org/10.1128/AEM.70.1.293-300.2004

Broderick, N.A., Raffa, K.F., Handelsman, J., 2006. Midgut bacteria required for *Bacillus thuringiensis* insecticidal activity. *Proc. Natl. Acad. Sci.* 103, 15196–15199. https://doi.org/10.1073/pnas.0604865103

Bueno, E., Martin, K.R., Raguso, R.A., Mcmullen, J.G., Hesler, S.P., Loeb, G.M., Douglas, A.E., 2020. Response of wild spotted wing drosophila (Drosophila suzukii) to microbial volatiles. *J. Chem. Ecol.* 46, 688–698. https://doi.org/10.1007/s10886-019-01139-4

Burke, G., Fiehn, O., Moran, N., 2010. Effects of facultative symbionts and heat stress on the metabolome of pea aphids. *ISME J.* 4, 242–252. https://doi.org/10.1038/ismej.2009.114

Ceja-Navarro, J.A., Vega, F.E., Karaoz, U., Hao, Z., Jenkins, S., Lim, H.C., Kosina, P., Infante, F., Northen, T.R., Brodie, E.L., 2015. Gut microbiota mediate caffeine detoxification in the primary insect pest of coffee. *Nat. Commun.* 6, 7618. https://doi.org/10.1038/ncomms8618

Chandler, J.A., James, P.M., Jospin, G., Lang, J.M., 2014. The bacterial communities of *Drosophila suzukii* collected from undamaged cherries. *PeerJ* 2, e474. https://doi.org/10.7717/peerj.474

Chen, L., Tian, Z., Ma, W., Wang, J., Huang, Q., Yongping, Z., Gao, X., Chen, H., Zhou, Z., 2023. Comparison of bacterial diversity in *Bactrocera cucurbitae* (Coquillett) ovaries and eggs based on 16S rRNA sequencing. *Sci. Rep.* 13, 11793. https://doi.org/10.1038/s41598-023-38992-z

Cheng, D., Guo, Z., Riegler, M., Xi, Z., Liang, G., Xu, Y., 2017. Gut symbiont enhances insecticide resistance in a significant pest, the oriental fruit fly *Bactrocera dorsalis* (Hendel). *Microbiome* 5, 13. https://doi.org/10.1186/s40168-017-0236-z

Chouaia, B., Goda, N., Mazza, G., Alali, S., Florian, F., Gionechetti, F., Callegari, M., Gonella, E., Magoga, G., Fusi, M., Crotti, E., Daffonchio, D., Alma, A., Paoli, F., Roversi, P.F., Marianelli, L., Montagna, M., 2019. Developmental stages and gut microenvironments influence gut microbiota dynamics in the invasive beetle *Popillia japonica* Newman (Coleoptera: Scarabaeidae). *Environ. Microbiol.* 21, 4343–4359. https://doi.org/10.1111/1462-2920.14797

Chu, C.-C., Spencer, J.L., Curzi, M.J., Zavala, J.A., Seufferheld, M.J., 2013. Gut bacteria facilitate adaptation to crop rotation in the western corn rootworm. *Proc. Natl. Acad. Sci.* 110, 11917–11922. https://doi.org/10.1073/pnas.1301886110

Chung, S.H., Rosa, C., Scully, E.D., Peiffer, M., Tooker, J.F., Hoover, K., Luthe, D.S., Felton, G.W., 2013. Herbivore exploits orally secreted bacteria to suppress plant defenses. *Proc. Natl. Acad. Sci.* 110, 15728–15733. https://doi.org/10.1073/pnas.1308867110

Consuegra, J., Grenier, T., Baa-Puyoulet, P., Rahioui, I., Akherraz, H., Gervais, H., Parisot, N., Da Silva, P., Charles, H., Calevro, F., Leulier, F., 2020. Drosophila-associated bacteria differentially shape the nutritional requirements of their host during juvenile growth. *PLOS Biol.* 18, e3000681. https://doi.org/10.1371/journal.pbio.3000681

Damodaram, K.J.P., Ayyasamy, A., Kempraj, V., 2016. Commensal bacteria aid mate-selection in the fruit fly, *Bactrocera dorsalis*. *Microb. Ecol.* 72, 725–729. https://doi.org/10.1007/s00248-016-0819-4

Darby, A.C., Tosh, C.R., Walters, K.F.A., Douglas, A.E., 2003. The significance of a facultative bacterium to natural populations of the pea aphid *Acyrthosiphon pisum*. *Ecol. Entomol.* 28, 145–150. https://doi.org/10.1046/j.1365-2311.2003.00492.x

Das, T., Bhattacharyya, A., Bhar, A., 2023. Linking phyllosphere and rhizosphere microbiome to the plant–insect interplay: The new dimension of tripartite interaction. *Physiologia* 3, 129–144. https://doi.org/10.3390/physiologia3010011

De Cock, M., Virgilio, M., Vandamme, P., Bourtzis, K., De Meyer, M., Willems, A., 2020. Comparative microbiomics of tephritid frugivorous pests (Diptera: Tephritidae) from the field: A tale of high variability across and within species. *Front. Microbiol.* 11, 1890. https://doi.org/10.3389/fmicb.2020.01890

De Moya-Ruiz, C., Ferriol, I., Juárez, M., Hurtado-Ruiz, M.A., Gómez, P., 2025. Host plant switching and viral infections reshape the microbiome of the aphid vector *Aphis gossypii*. *Phytobiomes J.* 9, 391–402. https://doi.org/10.1094/PBIOMES-03-25-0017-R

Deguenon, J.M., Dhammi, A., Ponnusamy, L., Travanty, N.V., Cave, G., Lawrie, R., Mott, D., Reisig, D., Kurtz, R., Roe, R.M., 2021. Bacterial microbiota of field-collected Helicoverpa zea (Lepidoptera: Noctuidae) from transgenic Bt and non-Bt cotton. *Microorganisms* 9, 878. https://doi.org/10.3390/microorganisms9040878

Deutscher, A.T., Chapman, T.A., Shuttleworth, L.A., Riegler, M., Reynolds, O.L., 2019. Tephritid-microbial interactions to enhance fruit fly performance in sterile insect technique programs. *BMC Microbiol.* 19, 287. https://doi.org/10.1186/s12866-019-1650-0

Diouf, M., Roy, V., Mora, P., Frechault, S., Lefebvre, T., Hervé, V., Rouland-Lefèvre, C., Miambi, E., 2015. Profiling the succession of bacterial communities throughout the life stages of a higher termite Nasutitermes arborum (Termitidae, Nasutitermitinae) using 16S rRNA gene pyrosequencing. *PLOS ONE* 10, e0140014. https://doi.org/10.1371/journal.pone.0140014

Douglas, A.E., 1998. Nutritional interactions in insect-microbial symbioses: Aphids and their symbiotic bacteria *Buchnera*. *Annu. Rev. Entomol.* 43, 17–37. https://doi.org/10.1146/annurev.ento.43.1.17

Douglas, A.E., 2003. The nutritional physiology of aphids, in: *Advances in Insect Physiology*. Elsevier, pp. 73–140. https://doi.org/10.1016/S0065-2806(03)31002-1

Douglas, A.E., 2006. Phloem-sap feeding by animals: Problems and solutions. *J. Exp. Bot.* 57, 747–754. https://doi.org/10.1093/jxb/erj067

Douglas, A.E., 2015. Multiorganismal insects: Diversity and function of resident microorganisms. *Annu. Rev. Entomol.* 60, 17–34. https://doi.org/10.1146/annurev-ento-010814-020822

Drahun, I., Morrison, K., Poole, E.A., Van Herk, W.G., Cassone, B.J., 2025. Characterisation of the bacteriomes harboured by major wireworm pest species in the Canadian Prairies. *Insect Mol. Biol.* 34, 203–217. https://doi.org/10.1111/imb.12962

Ebert, K.M., Arnold, W.G., Ebert, P.R., Merritt, D.J., 2021. Hindgut microbiota reflects different digestive strategies in dung beetles (Coleoptera: Scarabaeidae: Scarabaeinae). *Appl. Environ. Microbiol.* 87, e02100–20. https://doi.org/10.1128/AEM.02100-20

El Hamss, H., Ally, H.M., Delatte, H., Omongo, C.A., Colvin, J., Maruthi, M.N., 2025. Outbreak-driven differences in the microbiome composition and diversity of two cassava whitefly Bemisia tabaci mitotypes SSA1-SG1 and SSA1-SG2. *Front. Microbiol.* 16, 1597836. https://doi.org/10.3389/fmicb.2025.1597836

El Hamss, H., Maruthi, M.N., Ally, H.M., Omongo, C.A., Wang, H.-L., Van Brunschot, S., Colvin, J., Delatte, H., 2022. Spatio-temporal changes in endosymbiont diversity and composition in the African cassava whitefly, Bemisia tabaci SSA1. *Front. Microbiol.* 13, 986226. https://doi.org/10.3389/fmicb.2022.986226

El Hamss, H., Maruthi, M.N., Omongo, C.A., Wang, H.-L., Van Brunschot, S., Colvin, J., Delatte, H., 2024. Microbiome diversity and composition in Bemisia tabaci SSA1-SG1 whitefly are influenced by their host's life stage. *Microbiol. Res.* 278, 127538. https://doi.org/10.1016/j.micres.2023.127538

Engl, T., Eberl, N., Gorse, C., Krüger, T., Schmidt, T.H.P., Plarre, R., Adler, C., Kaltenpoth, M., 2018. Ancient symbiosis confers desiccation resistance to stored grain pest beetles. *Mol. Ecol.* 27, 2095–2108. https://doi.org/10.1111/mec.14418

Estes, A.M., Hearn, D.J., Bronstein, J.L., Pierson, E.A., 2009. The olive fly endosymbiont, "candidatus Erwinia dacicola," Switches from an intracellular existence to an extracellular existence during Host insect development. *Appl. Environ. Microbiol.* 75, 7097–7106. https://doi.org/10.1128/AEM.00778-09

Fountain, M.T., Bennett, J., Cobo-Medina, M., Conde Ruiz, R., Deakin, G., Delgado, A., Harrison, R., Harrison, N., 2018. Alimentary microbes of winter-form *Drosophila suzukii*. *Insect Mol. Biol.* 27, 383–392. https://doi.org/10.1111/imb.12377

Fourie, A., Venter, S.N., Slippers, B., Fourie, G., 2023. Pantoea bathycoeliae sp. nov and Sodalis sp. are core gut microbiome symbionts of the two-spotted stink bug. *Front. Microbiol.* 14, 1284397. https://doi.org/10.3389/fmicb.2023.1284397

Frias, J., Garriga, A., Peñalver, Á., Teixeira, M., Beltrí, R., Toubarro, D., Simões, N., 2023. Exploring gut microbiome variations between Popillia japonica populations of azores. *Microorganisms* 11, 1972. https://doi.org/10.3390/microorganisms11081972

Fukumori, K., Oguchi, K., Ikeda, H., Shinohara, T., Tanahashi, M., Moriyama, M., Koga, R., Fukatsu, T., 2022. Evolutionary dynamics of host organs for microbial symbiosis in tortoise leaf beetles (Coleoptera: Chrysomelidae: Cassidinae). *mBio* 13, e03691–21. https://doi.org/10.1128/mbio.03691-21

Gandotra, S., Bhuyan, P.M., Gogoi, D.K., Kumar, A., Subramanian, S., 2018. Screening of nutritionally important gut bacteria from the lepidopteran insects through qualitative enzyme assays. *Proc. Natl. Acad. Sci. India Sect. B Biol. Sci.* 88, 329–337. https://doi.org/10.1007/s40011-016-0762-7

García-Lozano, M., Salem, H., 2025. Microbial bases of herbivory in beetles. *Trends Microbiol.* 33, 151–163. https://doi.org/10.1016/j.tim.2024.08.004

Gauthier, J.-P., Outreman, Y., Mieuzet, L., Simon, J.-C., 2015. Bacterial communities associated with host-adapted populations of pea aphids revealed by deep sequencing of 16S ribosomal DNA. *PLOS ONE* 10, e0120664. https://doi.org/10.1371/journal.pone.0120664

Geib, S.M., Jimenez-Gasco, M.D.M., Carlson, J.E., Tien, M., Hoover, K., 2009. Effect of host tree species on cellulase activity and bacterial community composition in the gut of larval Asian longhorned beetle. *Environ. Entomol.* 38, 686–699. https://doi.org/10.1603/022.038.0320

Geib, S.M., Tien, M., Hoover, K., 2010. Identification of proteins involved in lignocellulose degradation using in gel zymogram analysis combined with mass spectroscopy-based peptide analysis of gut proteins from larval Asian longhorned beetles, *Anoplophora glabripennis*. *Insect Sci.* 17, 253–264. https://doi.org/10.1111/j.1744-7917.2010.01323.x

Gomes, A.F.F., Omoto, C., Cônsoli, F.L., 2020. Gut bacteria of field-collected larvae of *Spodoptera frugiperda* undergo selection and are more diverse and active in metabolizing multiple insecticides than laboratory-selected resistant strains. *J. Pest Sci.* 93, 833–851. https://doi.org/10.1007/s10340-020-01202-0

Gonella, E., Orrù, B., Marasco, R., Daffonchio, D., Alma, A., 2020. Disruption of host-symbiont associations for the symbiotic control and management of pentatomid agricultural pests—A review. *Front. Microbiol.* 11, 547031. https://doi.org/10.3389/fmicb.2020.547031

González-Serrano, F., Pérez-Cobas, A.E., Rosas, T., Baixeras, J., Latorre, A., Moya, A., 2020. The gut microbiota composition of the moth Brithys crini reflects insect metamorphosis. *Microb. Ecol.* 79, 960–970. https://doi.org/10.1007/s00248-019-01460-1

Govindharaj, G.-P.-P., Annamalai, M., Choudhary, J.S., Khan, R.M., Basana-Gowda, G., Patil, N., Panda, R.M., Srivastava, K., Mohapatra, S.D., 2025. Significant variations of bacterial communities among the developmental stages of Scirpophaga incertulas (Walker) (Lepidoptera: Crambidae). *Sci. Rep.* 15, 8552. https://doi.org/10.1038/s41598-025-93048-8

Gu, C., Yan, C., Wang, M., Lin, Y., 2025. Study on secondary metabolites of Spodoptera litura Gut-associated bacteria and their induction of tomato resistance to pests. *J. Phytopathol.* 173, e70062. https://doi.org/10.1111/jph.70062

Gu, J., Yao, Z., Lemaitre, B., Cai, Z., Zhang, H., Li, X., 2024. Intestinal commensal bacteria promote *Bactrocera dorsalis* larval development through the vitamin B6 synthesis pathway. *Microbiome* 12, 227. https://doi.org/10.1186/s40168-024-01931-9

Guilhot, R., Xuéreb, A., Lagmairi, A., Olazcuaga, L., Fellous, S., 2023. Microbiota acquisition and transmission in Drosophila flies. *iScience* 26, 107656. https://doi.org/10.1016/j.isci.2023.107656

Gündüz, E.A., Douglas, A.E., 2009. Symbiotic bacteria enable insect to use a nutritionally inadequate diet. *Proc. R. Soc. B Biol. Sci.* 276, 987–991. https://doi.org/10.1098/rspb.2008.1476

Guo, L., Li, Z., Xu, J., 2023. Effects of cadmium stress on bacterial and fungal communities in the whitefly Bemisia tabaci. *Int. J. Mol. Sci.* 24, 13588. https://doi.org/10.3390/ijms241713588

Gupta, A., Nair, S., 2020. Dynamics of insect–microbiome interaction influence host and microbial symbiont. *Front. Microbiol.* 11, 1357. https://doi.org/10.3389/fmicb.2020.01357

Gupta, A., Sinha, D.K., Nair, S., 2022. Shifts in Pseudomonas species diversity influence adaptation of brown planthopper to changing climates and geographical locations. *iScience* 25, 104550. https://doi.org/10.1016/j.isci.2022.104550

Gupta, S., Chakraborty, A., Roy, A., 2023. Prospects for deploying microbes against tree-killing beetles (Coleoptera) in Anthropocene. *Front. For. Glob. Change* 6, 1182834. https://doi.org/10.3389/ffgc.2023.1182834

Gurung, K., Wertheim, B., Falcao Salles, J., 2019. The microbiome of pest insects: It is not just bacteria. *Entomol. Exp. Appl.* 167, 156–170. https://doi.org/10.1111/eea.12768

Gutiérrez-Cabrera, A.E., Córdoba-Aguilar, A., Zenteno, E., Lowenberger, C., Espinoza, B., 2016. Origin, evolution and function of the hemipteran perimicrovillar membrane with emphasis on Reduviidae that transmit Chagas disease. *Bull. Entomol. Res.* 106, 279–291. https://doi.org/10.1017/S0007485315000929

Gwokyalya, R., Weldon, C.W., Herren, J.K., Gichuhi, J., Makhulu, E.E., Ndlela, S., Mohamed, S.A., 2023. Friend or foe: Symbiotic bacteria in *Bactrocera dorsalis*–Parasitoid associations. *Biology* 12, 274. https://doi.org/10.3390/biology12020274

Hafsi, A., Facon, B., Ravigné, V., Chiroleu, F., Quilici, S., Chermiti, B., Duyck, P.-F., 2016. Host plant range of a fruit fly community (Diptera: Tephritidae): Does fruit composition influence larval performance? *BMC Ecol.* 16, 40. https://doi.org/10.1186/s12898-016-0094-8

Hagstrum, D.W., 2006. *Fundamentals of Stored-Product Entomology, 1. Aufl. ed, American Associate of Cereal Chemists International.* Elsevier Reference Monographs, s.l.

Haider, K., Abbas, D., Galian, J., Ghafar, M.A., Kabir, K., Ijaz, M., Hussain, M., Khan, K.A., Ghramh, H.A., Raza, A., 2025. The multifaceted roles of gut microbiota in insect physiology, metabolism, and environmental adaptation: Implications for pest management strategies. *World J. Microbiol. Biotechnol.* 41, 75. https://doi.org/10.1007/s11274-025-04288-9

Hamby, K.A., Becher, P.G., 2016. Current knowledge of interactions between Drosophila suzukii and microbes, and their potential utility for pest management. *J. Pest Sci.* 89, 621–630. https://doi.org/10.1007/s10340-016-0768-1

Hamby, K.A., Hernández, A., Boundy-Mills, K., Zalom, F.G., 2012. Associations of yeasts with spotted-wing drosophila (Drosophila suzukii; Diptera: Drosophilidae) in cherries and raspberries. *Appl. Environ. Microbiol.* 78, 4869–4873. https://doi.org/10.1128/AEM.00841-12

Hammer, T.J., Janzen, D.H., Hallwachs, W., Jaffe, S.P., Fierer, N., 2017. Caterpillars lack a resident gut microbiome. *Proc. Natl. Acad. Sci.* 114, 9641–9646. https://doi.org/10.1073/pnas.1707186114

Hammer, T.J., Moran, N.A., 2019. Links between metamorphosis and symbiosis in holometabolous insects. *Philos. Trans. R. Soc. B Biol. Sci.* 374, 20190068. https://doi.org/10.1098/rstb.2019.0068

Hance, T., Hamidovic, A., Suraporn, S., 2025. Insight into the microbiota of orthopteran in relation to gut compartmentalisation. *Insects* 16, 555. https://doi.org/10.3390/insects16060555

Hannula, S.E., Zhu, F., Heinen, R., Bezemer, T.M., 2019. Foliar-feeding insects acquire microbiomes from the soil rather than the host plant. *Nat. Commun.* 10, 1254. https://doi.org/10.1038/s41467-019-09284-w

Hansen, A.K., Moran, N.A., 2011. Aphid genome expression reveals host–symbiont cooperation in the production of amino acids. *Proc. Natl. Acad. Sci.* 108, 2849–2854. https://doi.org/10.1073/pnas.1013465108

Harmon, D.S., Haseeb, M., Kanga, L.H.B., Liburd, O.E., 2019. Evaluation of monitoring traps and lures for Drosophila suzukii (Diptera: Drosophilidae) in berry plantings in Florida. *Insects* 10, 313. https://doi.org/10.3390/insects10100313

Hassan, B., Siddiqui, J.A., Xu, Y., 2020. Vertically transmitted gut bacteria and nutrition influence the immunity and fitness of *Bactrocera dorsalis* Larvae. *Front. Microbiol.* 11, 596352. https://doi.org/10.3389/fmicb.2020.596352

Haytham, H., Kamel, C., Wafa, D., Salma, F., Naima, B.M., George, T., Ameur, C., Msaad Guerfali, M., 2024. Probiotic consortium modulating the gut microbiota composition and function of sterile Mediterranean fruit flies. *Sci. Rep.* 14, 1058. https://doi.org/10.1038/s41598-023-50679-z

He, M., Chen, H., Yang, X., Gao, Y., Lu, Y., Cheng, D., 2022. Gut bacteria induce oviposition preference through ovipositor recognition in fruit fly. *Commun. Biol.* 5, 973. https://doi.org/10.1038/s42003-022-03947-z

He, S., Ivanova, N., Kirton, E., Allgaier, M., Bergin, C., Scheffrahn, R.H., Kyrpides, N.C., Warnecke, F., Tringe, S.G., Hugenholtz, P., 2013. Comparative metagenomic and metatranscriptomic analysis of hindgut paunch microbiota in wood- and dung-feeding higher termites. *PLoS ONE* 8, e61126. https://doi.org/10.1371/journal.pone.0061126

Hendrycks, W., Mullens, N., Bakengesa, J., Kabota, S., Tairo, J., Backeljau, T., Majubwa, R., Mwatawala, M., De Meyer, M., Virgilio, M., 2025. Deterministic and stochastic effects drive the gut microbial diversity in cucurbit-feeding fruit flies (Diptera, Tephritidae). *PLOS ONE* 20, e0313447. https://doi.org/10.1371/journal.pone.0313447

Heyworth, E.R., Ferrari, J., 2015. A facultative endosymbiont in aphids can provide diverse ecological benefits. *J. Evol. Biol.* 28, 1753–1760. https://doi.org/10.1111/jeb.12705

Higashi, C.H.V., Nichols, W.L., Chevignon, G., Patel, V., Allison, S.E., Kim, K.L., Strand, M.R., Oliver, K.M., 2023. An aphid symbiont confers protection against a specialized RNA virus, another increases vulnerability to the same pathogen. *Mol. Ecol.* 32, 936–950. https://doi.org/10.1111/mec.16801

Higashi, C.H.V., Patel, V., Kamalaker, B., Inaganti, R., Bressan, A., Russell, J.A., Oliver, K.M., 2024. Another tool in the toolbox: Aphid-specific Wolbachia protect against fungal pathogens. *Environ. Microbiol.* 26, e70005. https://doi.org/10.1111/1462-2920.70005

Hirata, K., Asahi, T., Kataoka, K., 2023. Spatial and sexual divergence of gut bacterial communities in field cricket teleogryllus occipitalis (Orthoptera: Gryllidae). *Microb. Ecol.* 86, 2627–2641. https://doi.org/10.1007/s00248-023-02265-z

Holt, J.R., Cavichiolli de Oliveira, N., Medina, R.F., Malacrinò, A., Lindsey, A.R.I., 2024. Insect–microbe interactions and their influence on organisms and ecosystems. *Ecol. Evol.* 14, e11699. https://doi.org/10.1002/ece3.11699

Hu, H., Da Costa, R.R., Pilgaard, B., Schiøtt, M., Lange, L., Poulsen, M., 2019. Fungiculture in termites is associated with a mycolytic gut bacterial community. *mSphere* 4, e00165–19. https://doi.org/10.1128/mSphere.00165-19

Hu, L., Sun, Z., Xu, C., Wang, J., Mallik, A.U., Gu, C., Chen, D., Lu, L., Zeng, R., Song, Y., 2022. High nitrogen in maize enriches gut microbiota conferring insecticide tolerance in lepidopteran pest Spodoptera litura. *iScience* 25, 103726. https://doi.org/10.1016/j.isci.2021.103726

Indiragandhi, P., Anandham, R., Madhaiyan, M., Poonguzhali, S., Kim, G.H., Saravanan, V.S., Sa, T., 2007. Cultivable bacteria associated with larval gut of prothiofos-resistant, prothiofos-susceptible and field-caught populations of diamondback moth, *Plutella xylostella* and their potential for, antagonism towards entomopathogenic fungi and host insect nutriti: Gut bacteria of *Plutella xylostella*. *J. Appl. Microbiol.* 103, 2664–2675. https://doi.org/10.1111/j.1365-2672.2007.03506.x

Indiragandhi, P., Anandham, R., Madhaiyan, M., Sa, T.M., 2008. Characterization of plant growth–promoting traits of bacteria isolated from larval guts of diamondback moth *Plutella xylostella* (Lepidoptera: Plutellidae). *Curr. Microbiol.* 56, 327–333. https://doi.org/10.1007/s00284-007-9086-4

Janke, R.S., Kaftan, F., Niehs, S.P., Scherlach, K., Rodrigues, A., Svatoš, A., Hertweck, C., Kaltenpoth, M., Flórez, L.V., 2022. Bacterial ectosymbionts in cuticular organs chemically protect a beetle during molting stages. *ISME J.* 16, 2691–2701. https://doi.org/10.1038/s41396-022-01311-x

Jing, T.-Z., Qi, F.-H., Wang, Z.-Y., 2020. Most dominant roles of insect gut bacteria: Digestion, detoxification, or essential nutrient provision? *Microbiome* 8, 38. https://doi.org/10.1186/s40168-020-00823-y

Jing, X., Wong, A.C., Chaston, J.M., Colvin, J., McKenzie, C.L., Douglas, A.E., 2014. The bacterial communities in plant phloem-sap-feeding insects. *Mol. Ecol.* 23, 1433–1444. https://doi.org/10.1111/mec.12637

Junaid, M., Gokce, A., 2024. Global agricultural losses and their causes. *Bull. Biol. Allied Sci. Res.* 2024, 66. https://doi.org/10.54112/bbasr.v2024i1.66

Kakizawa, S., Hosokawa, T., Oguchi, K., Miyakoshi, K., Fukatsu, T., 2022. Spiroplasma as facultative bacterial symbionts of stinkbugs. *Front. Microbiol.* 13, 1044771. https://doi.org/10.3389/fmicb.2022.1044771

Kanyile, S.N., Engl, T., Kaltenpoth, M., 2022. Nutritional symbionts enhance structural defence against predation and fungal infection in a grain pest beetle. *J. Exp. Biol.* 225, jeb243593. https://doi.org/10.1242/jeb.243593

Kashkouli, M., Castelli, M., Floriano, A.M., Bandi, C., Epis, S., Fathipour, Y., Mehrabadi, M., Sassera, D., 2021. Characterization of a novel *Pantoea* symbiont allows inference of a pattern of convergent genome reduction in bacteria associated with Pentatomidae. *Environ. Microbiol.* 23, 36–50. https://doi.org/10.1111/1462-2920.15169

Kempraj, V., Auth, J., Cha, D.H., Mason, C.J., 2024. Impact of larval food source on the stability of the *Bactrocera dorsalis* microbiome. *Microb. Ecol.* 87, 46. https://doi.org/10.1007/s00248-024-02352-9

Kikuchi, Y., Hosokawa, T., Fukatsu, T., 2011. An ancient but promiscuous host–symbiont association between *Burkholderia* gut symbionts and their heteropteran hosts. *ISME J.* 5, 446–460. https://doi.org/10.1038/ismej.2010.150

Kikuchi, Y., Hosokawa, T., Nikoh, N., Meng, X.-Y., Kamagata, Y., Fukatsu, T., 2009. Host-symbiont cospeciation and reductive genome evolution in gut symbiotic bacteria of acanthosomatid stinkbugs. *BMC Biol.* 7, 2. https://doi.org/10.1186/1741-7007-7-2

Kliot, A., Ghanim, M., 2013. The role of bacterial chaperones in the circulative transmission of plant viruses by insect vectors. *Viruses* 5, 1516–1535. https://doi.org/10.3390/v5061516

Kobiałka, M., Świerczewski, D., Walczak, M., Urbańczyk, W., 2025. Extremely distinct microbial communities in closely related leafhopper subfamilies: Typhlocybinae and Eurymelinae (Cicadellidae, Hemiptera). *mSystems* 10, e00603–25. https://doi.org/10.1128/msystems.00603-25

Köhler, T., Stingl, U., Meuser, K., Brune, A., 2008. Novel lineages of *Planctomycetes* densely colonize the alkaline gut of soil-feeding termites (*Cubitermes spp.*). *Environ. Microbiol.* 10, 1260–1270. https://doi.org/10.1111/j.1462-2920.2007.01540.x

Korša, A., Lo, L.K., Gandhi, S., Bang, C., Kurtz, J., 2022. Oral immune priming treatment alters microbiome composition in the red flour beetle Tribolium castaneum. *Front. Microbiol.* 13, 793143. https://doi.org/10.3389/fmicb.2022.793143

Korsa, G., Beyene, A., Ayele, A., 2023. Bacterial diversity from soil-feeding termite gut and their potential application. *Ann. Microbiol.* 73, 38. https://doi.org/10.1186/s13213-023-01741-8

Kucuk, R.A., Campbell, B.J., Lyon, N.J., Shelby, E.A., Caterino, M.S., 2023. Gut bacteria of adult and larval Cotinis nitida Linnaeus (Coleoptera: Scarabaeidae) demonstrate community differences according to respective life stage and gut region. *Front. Microbiol.* 14, 1185661. https://doi.org/10.3389/fmicb.2023.1185661

Kudo, R., Masuya, H., Endoh, R., Kikuchi, T., Ikeda, H., 2019. Gut bacterial and fungal communities in ground-dwelling beetles are associated with host food habit and habitat. *ISME J.* 13, 676–685. https://doi.org/10.1038/s41396-018-0298-3

Kuechler, S.M., Renz, P., Dettner, K., Kehl, S., 2012. Diversity of symbiotic organs and bacterial endosymbionts of lygaeoid bugs of the families blissidae and lygaeidae (Hemiptera: Heteroptera: Lygaeoidea). *Appl. Environ. Microbiol.* 78, 2648–2659. https://doi.org/10.1128/AEM.07191-11

Kuzina, L.V., Peloquin, J.J., Vacek, D.C., Miller, T.A., 2001. Isolation and identification of bacteria associated with adult laboratory Mexican fruit flies, *Anastrepha ludens* (Diptera: Tephritidae). *Curr. Microbiol.* 42, 290–294. https://doi.org/10.1007/s002840110219

Kwak, Y., Argandona, J.A., Miao, S., Son, T.J., Hansen, A.K., 2025. A dual insect symbiont and plant pathogen improves insect host fitness under arginine limitation. *mBio* 16, e03588–24. https://doi.org/10.1128/mbio.03588-24

Lauzon, C.R., Potter, S.E., Prokopy, R.J., 2003. Degradation and detoxification of the dihydrochalcone phloridzin by Enterobacter agglomerans, a bacterium associated with the apple pPest, Rhagoletis pomonella (Walsh) (Diptera: Tephritidae). *Environ. Entomol.* 32, 953–962. https://doi.org/10.1603/0046-225X-32.5.953

Lee, S.J., Kim, S.R., Yoon, H.J., Kim, I., Lee, K.S., Je, Y.H., Lee, S.M., Seo, S.J., Dae Sohn, H., Jin, B.R., 2004. cDNA cloning, expression, and enzymatic activity of a cellulase from the mulberry longicorn beetle, Apriona germari. *Comp. Biochem. Physiol. B Biochem. Mol. Biol.* 139, 107–116. https://doi.org/10.1016/j.cbpc.2004.06.015

Lei, T., Luo, N., Song, C., Yu, J., Zhou, Y., Qi, X., Liu, Y., 2023. Comparative genomics reveals three genetic groups of the whitefly obligate endosymbiont candidatus portiera aleyrodidarum. *Insects* 14, 888. https://doi.org/10.3390/insects14110888

Lemoine, M.M., Engl, T., Kaltenpoth, M., 2020. Microbial symbionts expanding or constraining abiotic niche space in insects. *Curr. Opin. Insect Sci.* 39, 14–20. https://doi.org/10.1016/j.cois.2020.01.003

Li, G., Sun, J., Meng, Y., Yang, C., Chen, Z., Wu, Y., Tian, L., Song, F., Cai, W., Zhang, X., Li, H., 2022. The impact of environmental habitats and diets on the gut microbiota diversity of true bugs (Hemiptera: Heteroptera). *Biology* 11, 1039. https://doi.org/10.3390/biology11071039

Li, H., Ren, L., Xie, M., Gao, Y., He, M., Hassan, B., Lu, Y., Cheng, D., 2020. Egg-surface bacteria are indirectly associated with oviposition aversion in *Bactrocera dorsalis*. *Curr. Biol.* 30, 4432–4440.e4. https://doi.org/10.1016/j.cub.2020.08.080

Li, M., Mao, C., Li, X., Jiang, L., Zhang, W., Li, M., Liu, H., Fang, Y., Liu, S., Yang, G., Hou, X., 2023. Edible insects: A new sustainable nutritional resource worth promoting. *Foods* 12, 4073. https://doi.org/10.3390/foods12224073

Ling, Y., Li, W.-J., Li, F.-F., Xue, X.-B., Gao, Y.-Y., Wang, L., Liang, K., Li, X.-J., 2022. Microbial gut diversity in four grasshopper species and its correlation with cellulose digestibility. *Front. Microbiol.* 13, 1002532. https://doi.org/10.3389/fmicb.2022.1002532

Liu, W., Zhang, X., Wu, N., Ren, Y., Wang, X., 2020. High diversity and functional complementation of alimentary canal microbiota ensure small brown planthopper to adapt different biogeographic environments. *Front. Microbiol.* 10, 2953. https://doi.org/10.3389/fmicb.2019.02953

Lü, J., Xu, S., Teng, C., Huang, R., Xiong, G., Cheng, Q., 2025. Comparative analysis of gut symbionts in Tribolium castaneum (Coleoptera: Tenebrionidae) and their dietary substrate, sauce-flavored Daqu. *Front. Insect Sci.* 5, 1614310. https://doi.org/10.3389/finsc.2025.1614310

Luan, J.-B., 2024. Insect bacteriocytes: Adaptation, development, and evolution. *Annu. Rev. Entomol.* 69, 81–98. https://doi.org/10.1146/annurev-ento-010323-124159

Lv, N., Wang, L., Sang, W., Liu, C.-Z., Qiu, B.-L., 2018. Effects of endosymbiont disruption on the nutritional dynamics of the pea aphid Acyrthosiphon pisum. *Insects* 9, 161. https://doi.org/10.3390/insects9040161

Maeda, G.P., Kelly, M.K., Sundar, A., Moran, N.A., 2024. Intracellular defensive symbiont is culturable and capable of transovarial, vertical transmission. *mBio* 15, e03253–23. https://doi.org/10.1128/mbio.03253-23

Majumder, R., Taylor, P.W., Chapman, T.A., 2022. Dynamics of the queensland fruit fly microbiome through the transition from nature to an established laboratory colony. *Microorganisms* 10, 291. https://doi.org/10.3390/microorganisms10020291

Malacrinò, A., 2022. Host species identity shapes the diversity and structure of insect microbiota. *Mol. Ecol.* 31, 723–735. https://doi.org/10.1111/mec.16285

Malacrinò, A., Campolo, O., Medina, R.F., Palmeri, V., 2018. Instar-and host-associated differentiation of bacterial communities in the Mediterranean fruit

fly Ceratitis capitata. *PLOS ONE* 13, e0194131. https://doi.org/10.1371/journal.pone.0194131
Malook, S.U., Arora, A.K., Wong, A.C.N., 2025. The role of microbiomes in shaping insecticide resistance: Current insights and emerging paradigms. *Curr. Opin. Insect Sci.* 69, 101346. https://doi.org/10.1016/j.cois.2025.101346
Manthey, C., Johnston, P.R., Nakagawa, S., Rolff, J., 2023. Complete metamorphosis and microbiota turnover in insects. *Mol. Ecol.* 32, 6543–6551. https://doi.org/10.1111/mec.16673
Mao, M., Bennett, G.M., 2020. Symbiont replacements reset the co-evolutionary relationship between insects and their heritable bacteria. *ISME J.* 14, 1384–1395. https://doi.org/10.1038/s41396-020-0616-4
Marchini, D., Rosetto, M., Dallai, R., Marri, L., 2002. Bacteria associated with the oesophageal bulb of the medfly *Ceratitis capitata* (Diptera:Tephritidae). *Curr. Microbiol.* 44, 120–124. https://doi.org/10.1007/s00284-001-0061-1
Martinez-Sañudo, I., Simonato, M., Squartini, A., Mori, N., Marri, L., Mazzon, L., 2018. Metagenomic analysis reveals changes of the *Drosophila suzukii* microbiota in the newly colonized regions. *Insect Sci.* 25, 833–846. https://doi.org/10.1111/1744-7917.12458
Marynowska, M., Goux, X., Sillam-Dussès, D., Rouland-Lefèvre, C., Halder, R., Wilmes, P., Gawron, P., Roisin, Y., Delfosse, P., Calusinska, M., 2020. Compositional and functional characterisation of biomass-degrading microbial communities in guts of plant fibre- and soil-feeding higher termites. *Microbiome* 8, 96. https://doi.org/10.1186/s40168-020-00872-3
Mason, C.J., 2020. Complex relationships at the intersection of insect gut microbiomes and plant defenses. *J. Chem. Ecol.* 46, 793–807. https://doi.org/10.1007/s10886-020-01187-1
Mason, C.J., Auth, J., Geib, S.M., 2023. Gut bacterial population and community dynamics following adult emergence in pest tephritid fruit flies. *Sci. Rep.* 13, 13723. https://doi.org/10.1038/s41598-023-40562-2
Mason, C.J., St. Clair, A., Peiffer, M., Gomez, E., Jones, A.G., Felton, G.W., Hoover, K., 2020. Diet influences proliferation and stability of gut bacterial populations in herbivorous lepidopteran larvae. *PLOS ONE* 15, e0229848. https://doi.org/10.1371/journal.pone.0229848
Mazumdar, T., Teh, B.S., Murali, A., Schmidt-Heck, W., Schlenker, Y., Vogel, H., Boland, W., 2021. Transcriptomics reveal the survival strategies of *Enterococcus mundtii* in the gut of *Spodoptera littoralis*. *J. Chem. Ecol.* 47, 227–241. https://doi.org/10.1007/s10886-021-01246-1
Mazzetto, F., Gonella, E., Crotti, E., Vacchini, V., Syrpas, M., Pontini, M., Mangelinckx, S., Daffonchio, D., Alma, A., 2016. Olfactory attraction of Drosophila suzukii by symbiotic acetic acid bacteria. *J. Pest Sci.* 89, 783–792. https://doi.org/10.1007/s10340-016-0754-7
Mazzon, L., Piscedda, A., Simonato, M., Martinez-Sanudo, I., Squartini, A., Girolami, V., 2008. Presence of specific symbiotic bacteria in flies of the subfamily Tephritinae (Diptera Tephritidae) and their phylogenetic relationships: Proposal of "Candidatus Stammerula tephritidis." *Int. J. Syst. Evol. Microbiol.* 58, 1277–1287. https://doi.org/10.1099/ijs.0.65287-0
McKenna, D., 2020. Evolution: Symbiotic microbes mediate host range of herbivorous beetles. *Curr. Biol.* 30, R893–R896. https://doi.org/10.1016/j.cub.2020.05.089
McLean, A.H.C., Godfray, H.C.J., Ellers, J., Henry, L.M., 2019. Host relatedness influences the composition of aphid microbiomes. *Environ. Microbiol. Rep.* 11, 808–816. https://doi.org/10.1111/1758-2229.12795
McMullen, J.G., Peters-Schulze, G., Cai, J., Patterson, A.D., Douglas, A.E., 2020. How gut microbiome interactions affect nutritional traits of *Drosophila melanogaster*. *J. Exp. Biol.* 223, jeb227843. https://doi.org/10.1242/jeb.227843
Mead, L.J., Khachatourians, G.G., Jones, G.A., 1988. Microbial ecology of the gut in laboratory stocks of the migratory grasshopper, *Melanoplus sanguinipes* (Fab.) (Orthoptera: Acrididae). *Appl. Environ. Microbiol.* 54, 1174–1181. https://doi.org/10.1128/aem.54.5.1174-1181.1988
Menezes, L., Alvarez, T.M., Persinoti, G.F., Franco, J.P., Squina, F., Moreira, E.A., Alvaredo Paixão, D.A., Costa-Leonardo, A.M., Da Silva, V.X., Clerici, M.T.P.S., Arab, A., 2018. Food storage by the savanna termite *Cornitermes cumulans* (Syntermitinae): A strategy to improve hemicellulose digestibility? *Microb. Ecol.* 76, 492–505. https://doi.org/10.1007/s00248-017-1128-2
Meng, F., Bar-Shmuel, N., Shavit, R., Behar, A., Segoli, M., 2019. Gut bacteria of weevils developing on plant roots under extreme desert conditions. *BMC Microbiol.* 19, 311. https://doi.org/10.1186/s12866-019-1690-5
Michalik, A., Bauer, E., Szklarzewicz, T., Kaltenpoth, M., 2023a. Nutrient supplementation by genome-eroded *Burkholderia* symbionts of scale insects. *ISME J.* 17, 2221–2231. https://doi.org/10.1038/s41396-023-01528-4

Michalik, A., Franco, D.C., Deng, J., Szklarzewicz, T., Stroiński, A., Kobiałka, M., Łukasik, P., 2023b. Variable organization of symbiont-containing tissue across planthoppers hosting different heritable endosymbionts. *Front. Physiol.* 14, 1135346. https://doi.org/10.3389/fphys.2023.1135346

Mikaelyan, A., Dietrich, C., Köhler, T., Poulsen, M., Sillam-Dussès, D., Brune, A., 2015. Diet is the primary determinant of bacterial community structure in the guts of higher termites. *Mol. Ecol.* 24, 5284–5295. https://doi.org/10.1111/mec.13376

Mizutani, M., Fujikawa, T., Fukatsu, T., Kakizawa, S., 2025. Complete genome of the mutualistic symbiont " *Candidatus* Carsonella ruddii" from a Japanese island strain of the Asian citrus psyllid *Diaphorina citri*. *Microbiol. Resour. Announc.* 14, e01082–24. https://doi.org/10.1128/mra.01082-24

Mohammed, W.S., Ziganshina, E.E., Shagimardanova, E.I., Gogoleva, N.E., Ziganshin, A.M., 2018. Comparison of intestinal bacterial and fungal communities across various xylophagous beetle larvae (Coleoptera: Cerambycidae). *Sci. Rep.* 8, 10073. https://doi.org/10.1038/s41598-018-27342-z

Mora, V., Ramasamy, M., Damaj, M.B., Irigoyen, S., Ancona, V., Ibanez, F., Avila, C.A., Mandadi, K.K., 2021. Potato zebra chip: An overview of the disease, control strategies, and prospects. *Front. Microbiol.* 12, 700663. https://doi.org/10.3389/fmicb.2021.700663

Moran, N.A., 1992. The evolution of aphid life cycles. *Annu. Rev. Entomol.* 37, 321–348. https://doi.org/10.1146/annurev.en.37.010192.001541

Moriyama, M., Fukatsu, T., 2022. Host's demand for essential amino acids is compensated by an extracellular bacterial symbiont in a hemipteran insect model. *Front. Physiol.* 13, 1028409. https://doi.org/10.3389/fphys.2022.1028409

Morrow, J.L., Frommer, M., Shearman, D.C.A., Riegler, M., 2015. The microbiome of field-caught and laboratory-adapted Australian tephritid fruit fly species with different host plant use and specialisation. *Microb. Ecol.* 70, 498–508. https://doi.org/10.1007/s00248-015-0571-1

Mulio, S.Å., Zwolińska, A., Klejdysz, T., Prus-Frankowska, M., Michalik, A., Kolasa, M., Łukasik, P., 2024. Limited variation in microbial communities across populations of *Macrosteles* leafhoppers (Hemiptera: Cicadellidae). *Environ. Microbiol. Rep.* 16, e13279. https://doi.org/10.1111/1758-2229.13279

Mullens, N., Hendrycks, W., Bakengesa, J., Kabota, S., Tairo, J., Svardal, H., Majubwa, R., Mwatawala, M., De Meyer, M., Virgilio, M., 2024. Anna Karenina as a promoter of microbial diversity in the cosmopolitan agricultural pest Zeugodacus cucurbitae (Diptera, Tephritidae). *PLOS ONE* 19, e0300875. https://doi.org/10.1371/journal.pone.0300875

Munivenkatappa, S., Shivashankar, M., 2025. Lignocellulose bioconversion by gut microbiota of saturniidae species: A sustainable path to biofuel production. *Int. J. Chem. Res.* 9, 1–10. https://doi.org/10.22159/ijcr.2025v9i4.306

Muratore, M., Prather, C., Sun, Y., 2020. The gut bacterial communities across six grasshopper species from a coastal tallgrass prairie. *PLOS ONE* 15, e0228406. https://doi.org/10.1371/journal.pone.0228406

Nakabachi, A., Inoue, H., Hirose, Y., 2022a. Microbiome analyses of 12 psyllid species of the family Psyllidae identified various bacteria including Fukatsuia and Serratia symbiotica, known as secondary symbionts of aphids. *BMC Microbiol.* 22, 15. https://doi.org/10.1186/s12866-021-02429-2

Nakabachi, A., Inoue, H., Hirose, Y., 2022b. High-resolution Microbiome Analyses of Nine Psyllid Species of the Family Triozidae Identified Previously Unrecognized but Major Bacterial Populations, including *Liberibacter* and *Wolbachia* of Supergroup O. *Microbes Environ.* 37, n/a. https://doi.org/10.1264/jsme2.ME22078

Nakanishi, K., Hoshino, M., Nakai, M., Kunimi, Y., 2008. Novel RNA sequences associated with late male killing in *Homona magnanima*. *Proc. R. Soc. B Biol. Sci.* 275, 1249–1254. https://doi.org/10.1098/rspb.2008.0013

Nardi, J.B., Miller, L.A., Bee, C.M., 2019. Luminal membranes in the midgut of the lace bug Corythucha ciliata. *Cell Tissue Res.* 375, 685–696. https://doi.org/10.1007/s00441-018-2943-6

Navarro-Escalante, L., Benavides, P., Acevedo, F.E., 2024. Diversity of bacterial symbionts associated with the tropical plant bug Monalonion velezangeli (Hemiptera: Miridae) revealed by high-throughput 16S-rRNA sequencing. *Peer Community J.* 4, e7. https://doi.org/10.24072/pcjournal.362

Ng, S.H., Stat, M., Bunce, M., Simmons, L.W., 2018. The influence of diet and environment on the gut microbial community of field crickets. *Ecol. Evol.* 8, 4704–4720. https://doi.org/10.1002/ece3.3977

Nowak, K.H., Hartop, E., Prus-Frankowska, M., Buczek, M., Kolasa, M.R., Roslin, T., Ovaskainen, O., Łukasik, P., 2025. What lurks in the dark? An innovative framework for studying diverse wild insect microbiota. *Microbiome* 13, 186. https://doi.org/10.1186/s40168-025-02169-9

Ntiri, E.S., Wong, A.C.N., 2025. Microbial metabolites as engines of behavioral variation across animals. *Gut Microbes* 17, 2501191. https://doi.org/10.1080/19490976.2025.2501191

Ohbayashi, T., Takeshita, K., Kitagawa, W., Nikoh, N., Koga, R., Meng, X.-Y., Tago, K., Hori, T., Hayatsu, M., Asano, K., Kamagata, Y., Lee, B.L., Fukatsu, T., Kikuchi, Y., 2015. Insect's intestinal organ for symbiont sorting. *Proc. Natl. Acad. Sci.* 112, E5179–E5188. https://doi.org/10.1073/pnas.1511454112

Oishi, S., Moriyama, M., Koga, R., Fukatsu, T., 2019. Morphogenesis and development of midgut symbiotic organ of the stinkbug Plautia stali (Hemiptera: Pentatomidae). *Zool. Lett.* 5, 16. https://doi.org/10.1186/s40851-019-0134-2

Oliver, K.M., Degnan, P.H., Hunter, M.S., Moran, N.A., 2009. Bacteriophages encode factors required for protection in a symbiotic mutualism. *Science* 325, 992–994. https://doi.org/10.1126/science.1174463

Oliver, K.M., Moran, N.A., Hunter, M.S., 2006. Costs and benefits of a superinfection of facultative symbionts in aphids. *Proc. R. Soc. B Biol. Sci.* 273, 1273–1280. https://doi.org/10.1098/rspb.2005.3436

Oliver, K.M., Russell, J.A., Moran, N.A., Hunter, M.S., 2003. Facultative bacterial symbionts in aphids confer resistance to parasitic wasps. *Proc. Natl. Acad. Sci.* 100, 1803–1807. https://doi.org/10.1073/pnas.0335320100

Otani, S., Zhukova, M., Koné, N.A., Da Costa, R.R., Mikaelyan, A., Sapountzis, P., Poulsen, M., 2019. Gut microbial compositions mirror caste-specific diets in a major lineage of social insects. *Environ. Microbiol. Rep.* 11, 196–205. https://doi.org/10.1111/1758-2229.12728

Otero-Bravo, A., Sabree, Z.L., 2021. Multiple concurrent and convergent stages of genome reduction in bacterial symbionts across a stink bug family. *Sci. Rep.* 11, 7731. https://doi.org/10.1038/s41598-021-86574-8

Paddock, K.J., Finke, D.L., Kim, K.S., Sappington, T.W., Hibbard, B.E., 2022. Patterns of microbiome composition vary across spatial scales in a specialist insect. *Front. Microbiol.* 13, 898744. https://doi.org/10.3389/fmicb.2022.898744

Palacio, M.F.H., Montoya, O.I., Saldamando, C.I., García-Bonilla, E., Junca, H., Cadavid-Restrepo, G.E., Moreno-Herrera, C.X., 2021. Dry and rainy seasons significantly alter the gut microbiome composition and reveal a key *Enterococcus* sp. (Lactobacillales: Enterococcaceae) core component in *Spodoptera frugiperda* (Lepidoptera: Noctuidae) corn strain from Northwestern Colombia. *J. Insect Sci.* 21, 10. https://doi.org/10.1093/jisesa/ieab076

Pan, H., Li, X., Zhang, Y., 2012. Sex affects the infection frequencies of symbionts in *Bemisia tabaci*. *Commun. Integr. Biol.* 5, 337–339. https://doi.org/10.4161/cib.20398

Pape, T., Blagoderov, V., Mostovski, M.B., 2011. Order diptera linnaeus, 1758. In: Zhang, Z.-Q. (Ed.) *Animal Biodiversity: An Outline of Higher-Level Classification and Survey of Taxonomic Richness*. Zootaxa 3148. https://doi.org/10.11646/zootaxa.3148.1.42

Park, R., Dzialo, M.C., Spaepen, S., Nsabimana, D., Gielens, K., Devriese, H., Crauwels, S., Tito, R.Y., Raes, J., Lievens, B., Verstrepen, K.J., 2019. Microbial communities of the house fly Musca domestica vary with geographical location and habitat. *Microbiome* 7, 147. https://doi.org/10.1186/s40168-019-0748-9

Pawar, M.M., Shivanna, B., Prasannakumar, M.K., Parivallal, P.B., Suresh, K., Meenakshi, N.H., 2022. Spatial distribution and community structure of microbiota associated with cowpea aphid (Aphis craccivora Koch). *3 Biotech* 12, 75. https://doi.org/10.1007/s13205-022-03142-1

Pelz-Stelinski, K.S., Killiny, N., 2016. Better together: Association with ' *Candidatus* liberibacter asiaticus' increases the reproductive fitness of its insect vector, *Diaphorina citri* (Hemiptera: Liviidae). *Ann. Entomol. Soc. Am.* 109, 371–376. https://doi.org/10.1093/aesa/saw007

Perilla-Henao, L.M., Casteel, C.L., 2016. Vector-borne bacterial plant pathogens: Interactions with hemipteran insects and plants. *Front. Plant Sci.* 7, 1163. https://doi.org/10.3389/fpls.2016.01163

Perreau, J., Zhang, B., Maeda, G.P., Kirkpatrick, M., Moran, N.A., 2021. Strong within-host selection in a maternally inherited obligate symbiont: *Buchnera* and aphids. *Proc. Natl. Acad. Sci.* 118, e2102467118. https://doi.org/10.1073/pnas.2102467118

Phauk, S., Assentato, L., Meas, S., Terenius, O., 2025a. Primary and secondary symbionts of cambodian cicadellidae and the role of parasitisation. *Environ. Microbiol. Rep.* 17, e70196. https://doi.org/10.1111/1758-2229.70196

Phauk, S., Assentato, L., Sin, S., Uk, O., Hap, S., Terenius, O., 2025b. Symbiont diversity of rice-associated leafhoppers (Cicadellidae) in the tropical floodplains of the tonle Sap Lake, Cambodia. *Microb. Ecol.* 88, 109. https://doi.org/10.1007/s00248-025-02619-9

Polenogova, O.V., Noskov, Y.A., Yaroslavtseva, O.N., Kryukova, N.A., Alikina, T., Klementeva, T.N., Andrejeva, J., Khodyrev, V.P., Kabilov, M.R., Kryukov, V.Y., Glupov, V.V., 2021. Influence of *Bacillus thuringiensis* and avermectins on gut physiology and microbiota in Colorado potato beetle: Impact of

enterobacteria on susceptibility to insecticides. *PLOS ONE* 16, e0248704. https://doi.org/10.1371/journal.pone.0248704
Powell, J.S., Raffa, K.F., 1999. Effects of selected *Larix laricina* terpenoids on *Lymantria dispar* (Lepidoptera: Lymantriidae) *Development and Behavior. Environ. Entomol.* 28, 148–154. https://doi.org/10.1093/ee/28.2.148
Prieto, S.V., Brunetti, M., Magoga, G., Orrù, B., Gonella, E., Montagna, M., Alma, A., 2023. The intrinsic diversity of Nezara viridula gut symbionts affects the host fitness decline promoted by primary symbiont elimination. *Entomol. Gen.* 43, 1151–1160. https://doi.org/10.1127/entomologia/2023/2224
Qadri, M., Short, S., Gast, K., Hernandez, J., Wong, A.C.-N., 2020. Microbiome innovation in agriculture: Development of microbial based tools for insect pest management. *Front. Sustain. Food Syst.* 4, 547751. https://doi.org/10.3389/fsufs.2020.547751
Qin, M., Chen, J., Jiang, L., Qiao, G., 2022. Insights into the species-specific microbiota of greenideinae (Hemiptera: Aphididae) with evidence of phylosymbiosis. *Front. Microbiol.* 13, 828170. https://doi.org/10.3389/fmicb.2022.828170
Qin, M., Chen, J., Xu, S., Jiang, L., Qiao, G., 2021. Microbiota associated with *Mollitrichosiphum* aphids (Hemiptera: Aphididae: Greenideinae): Diversity, host species specificity and phylosymbiosis. *Environ. Microbiol.* 23, 2184–2198. https://doi.org/10.1111/1462-2920.15391
Quadrel, A., Ferguson, B., Rering, C.C., Urbaneja-Bernat, P., Rodriguez-Saona, C., 2025. Two volatiles from anthracnose-infected blueberries trigger electrophysiological and aversive behavioral responses in *Drosophila suzukii* (Diptera: Drosophilidae). *J. Econ. Entomol.* 118, 2432–2442. https://doi.org/10.1093/jee/toaf032
Ravigné, V., Becker, N., Massol, F., Guichoux, E., Boury, C., Mahé, F., Facon, B., 2022. Fruit fly phylogeny imprints bacterial gut microbiota. *Evol. Appl.* 15, 1621–1638. https://doi.org/10.1111/eva.13352
Raymond, B., Johnston, P.R., Wright, D.J., Ellis, R.J., Crickmore, N., Bonsall, M.B., 2009. A mid-gut microbiota is not required for the pathogenicity of *Bacillus thuringiensis* to diamondback moth larvae. *Environ. Microbiol.* 11, 2556–2563. https://doi.org/10.1111/j.1462-2920.2009.01980.x
Raza, M.F., Yao, Z., Bai, S., Cai, Z., Zhang, H., 2020. Tephritidae fruit fly gut microbiome diversity, function and potential for applications. *Bull. Entomol. Res.* 110, 423–437. https://doi.org/10.1017/S0007485319000853
Ren, L., Ma, Y., Xie, M., Lu, Y., Cheng, D., 2021. Rectal bacteria produce sex pheromones in the male oriental fruit fly. *Curr. Biol.* 31, 2220–2226.e4. https://doi.org/10.1016/j.cub.2021.02.046
Ren, S.-L., Li, Y.-H., Zhou, Y.-T., Xu, W.-M., Cuthbertson, A.G.S., Guo, Y.-J., Qiu, B.-L., 2016. Effects of *Candidatus Liberibacter asiaticus* on the fitness of the vector *Diaphorina citri*. *J. Appl. Microbiol.* 121, 1718–1726. https://doi.org/10.1111/jam.13302
Ren, X., Cao, S., Akami, M., Mansour, A., Yang, Y., Jiang, N., Wang, H., Zhang, G., Qi, X., Xu, P., Guo, T., Niu, C., 2022. Gut symbiotic bacteria are involved in nitrogen recycling in the tephritid fruit fly *Bactrocera dorsalis*. *BMC Biol.* 20, 201. https://doi.org/10.1186/s12915-022-01399-9
Robacker, D.C., Flath, R.A., 1995. Attractants fromStaphylococcus aureus cultures for Mexican fruit fly, *Anastrepha ludens*. *J. Chem. Ecol.* 21, 1861–1874. https://doi.org/10.1007/BF02033682
Rodrigues, P.A.P., Cavichiolli De Oliveira, N., Omoto, C., Girke, T., Cônsoli, F.L., 2023. Host plant affects the larval gut microbial communities of the generalist herbivores *Helicoverpa armigera* and *Spodoptera frugiperda*. bioRxiv https://doi.org/10.1101/2023.03.08.531690
Rodriguez-Fernandez, I.A., Santiago-Rodriguez, T.M., Figueroa-Pratts, P.G., Cintrón-Berríos, K., Rodriguez-Cornier, N.D., Toranzos, G.A., 2025. Gut microbial community structure of the adult citrus root weevil Diaprepes abbreviatus. *Front. Insect Sci.* 5, 1676003. https://doi.org/10.3389/finsc.2025.1676003
Roldán, E.L., Stelinski, L.L., Pelz-Stelinski, K.S., 2024. Reduction of *Wolbachia* in *Diaphorina citri* (*Hemiptera: Liviidae*) increases phytopathogen acquisition and decreases fitness. *J. Econ. Entomol.* 117, 733–749. https://doi.org/10.1093/jee/toae089
Royal Botanic Gardens, K., 2017. *State of the World's Plants 2017*. Royal Botanic Gardens, Kew.
Russell, C.W., Bouvaine, S., Newell, P.D., Douglas, A.E., 2013. Shared metabolic pathways in a coevolved insect-bacterial symbiosis. *Appl. Environ. Microbiol.* 79, 6117–6123. https://doi.org/10.1128/AEM.01543-13
Salem, H., Bauer, E., Kirsch, R., Berasategui, A., Cripps, M., Weiss, B., Koga, R., Fukumori, K., Vogel, H., Fukatsu, T., Kaltenpoth, M., 2017. Drastic genome reduction in an herbivore's pectinolytic symbiont. *Cell* 171, 1520–1531.e13. https://doi.org/10.1016/j.cell.2017.10.029
Salem, H., Kaltenpoth, M., 2022. Beetle–bacterial symbioses: Endless forms most functional. *Annu. Rev. Entomol.* 67, 201–219. https://doi.org/10.1146/annurev-ento-061421-063433

Salem, H., Kirsch, R., Pauchet, Y., Berasategui, A., Fukumori, K., Moriyama, M., Cripps, M., Windsor, D., Fukatsu, T., Gerardo, N.M., 2020. Symbiont digestive range reflects host plant breadth in herbivorous beetles. *Curr. Biol.* 30, 2875–2886.e4. https://doi.org/10.1016/j.cub.2020.05.043

Sanaei, E., Chavez, J., Harris, E.V., Alcaide, T.Y., Baffour-Addo, K., Bugay, M.J., Adams, K.L., Zelaya, A., de Roode, J.C., Gerardo, N.M., 2024. Microbiome analysis of monarch butterflies reveals effects of development and diet. *FEMS Microbiol. Ecol.* 100, fiae143. https://doi.org/10.1093/femsec/fiae143

Scarborough, C.L., Ferrari, J., Godfray, H.C.J., 2005. Aphid protected from pathogen by endosymbiont. *Science* 310, 1781–1781. https://doi.org/10.1126/science.1120180

Schloss, P.D., Delalibera, I., Handelsman, J., Raffa, K.F., 2006. Bacteria associated with the guts of two wood-boring beetles: *Anoplophora glabripennis* and *Saperda vestita* (Cerambycidae). *Environ. Entomol.* 35, 625–629. https://doi.org/10.1603/0046-225X-35.3.625

Schwarz, M., Beza-Beza, C.F., Mikaelyan, A., 2023. Wood fibers are a crucial microhabitat for cellulose- and xylan- degrading bacteria in the hindgut of the wood-feeding beetle Odontotaenius disjunctus. *Front. Microbiol.* 14, 1173696. https://doi.org/10.3389/fmicb.2023.1173696

Scully, E.D., Geib, S.M., Carlson, J.E., Tien, M., McKenna, D., Hoover, K., 2014. Functional genomics and microbiome profiling of the Asian longhorned beetle (Anoplophora glabripennis) reveal insights into the digestive physiology and nutritional ecology of wood feeding beetles. *BMC Genomics* 15, 1096. https://doi.org/10.1186/1471-2164-15-1096

Scully, E.D., Geib, S.M., Hoover, K., Tien, M., Tringe, S.G., Barry, K.W., Glavina Del Rio, T., Chovatia, M., Herr, J.R., Carlson, J.E., 2013. Metagenomic profiling reveals lignocellulose degrading system in a microbial community associated with a wood-feeding beetle. *PLoS ONE* 8, e73827. https://doi.org/10.1371/journal.pone.0073827

Shan, H., Wu, W., Sun, Z., Chen, J., Li, H., 2021. The gut microbiota of the insect infraorder pentatomomorpha (Hemiptera: Heteroptera) *for the Light of Ecology and Evolution. Microorganisms* 9, 464. https://doi.org/10.3390/microorganisms9020464

Shao, Y., Chen, B., Sun, C., Ishida, K., Hertweck, C., Boland, W., 2017. Symbiont-derived antimicrobials contribute to the control of the lepidopteran gut microbiota. *Cell Chem. Biol.* 24, 66–75. https://doi.org/10.1016/j.chembiol.2016.11.015

Shao, Y., Mason, C.J., Felton, G.W., 2024. Toward an integrated understanding of the lepidoptera microbiome. *Annu. Rev. Entomol.* 69, 117–137. https://doi.org/10.1146/annurev-ento-020723-102548

Sheng, P., Li, Y., Marshall, S., Zhang, H., 2015. High genetic diversity of microbial cellulase and hemicellulase genes in the hindgut of *Holotrichia parallela* larvae. *Int. J. Mol. Sci.* 16, 16545–16559. https://doi.org/10.3390/ijms160716545

Shi, W., Guo, Y., Xu, C., Tan, S., Miao, J., Feng, Y., Zhao, H., St. Leger, R.J., Fang, W., 2014. Unveiling the mechanism by which microsporidian parasites prevent locust swarm behavior. *Proc. Natl. Acad. Sci.* 111, 1343–1348. https://doi.org/10.1073/pnas.1314009111

Shigenobu, S., Watanabe, H., Hattori, M., Sakaki, Y., Ishikawa, H., 2000. Genome sequence of the endocellular bacterial symbiont of aphids Buchnera sp. *APS. Nature* 407, 81–86. https://doi.org/10.1038/35024074

Shukla, S.P., Beran, F., 2020. Gut microbiota degrades toxic isothiocyanates in a flea beetle pest. *Mol. Ecol.* 29, 4692–4705. https://doi.org/10.1111/mec.15657

Siden-Kiamos, I., Pantidi, G., Vontas, J., 2025. The journey of the bacterial symbiont through the olive fruit fly: Lessons learned and open questions. *Insects* 16, 789. https://doi.org/10.3390/insects16080789

Šigut, M., Pyszko, P., Šigutová, H., Višňovská, D., Kostovčík, M., Kotásková, N., Dorňák, O., Kolařík, M., Drozd, P., 2022. Fungi are more transient than bacteria in caterpillar gut microbiomes. *Sci. Rep.* 12, 15552. https://doi.org/10.1038/s41598-022-19855-5

Smith, T.E., Moran, N.A., 2020. Coordination of host and symbiont gene expression reveals a metabolic tug-of-war between aphids and *Buchnera*. *Proc. Natl. Acad. Sci.* 117, 2113–2121. https://doi.org/10.1073/pnas.1916748117

Sontowski, R., Van Dam, N.M., 2020. Functional variation in dipteran gut bacterial communities in relation to their diet, life cycle stage and habitat. *Insects* 11, 543. https://doi.org/10.3390/insects11080543

Soukup, P., Větrovský, T., Stiblik, P., Votýpková, K., Chakraborty, A., Sillam-Dussès, D., Kolařík, M., Odriozola, I., Lo, N., Baldrian, P., Šobotník, J., Bourguignon, T., 2021. Termites are associated with external species-specific bacterial communities. *Appl. Environ. Microbiol.* 87, e02042–20. https://doi.org/10.1128/AEM.02042-20

Stara, J., Hubert, J., 2023. Does Leptinotarsa decemlineata larval survival after pesticide treatment depend on microbiome composition? Pest Manag. Sci. 79, 4921–4930. https://doi.org/10.1002/ps.7694

Stork, N.E., 2018. How many species of insects and other terrestrial arthropods are there on earth? *Annu. Rev. Entomol.* 63, 31–45. https://doi.org/10.1146/annurev-ento-020117-043348

Stork, N.E., McBroom, J., Gely, C., Hamilton, A.J., 2015. New approaches narrow global species estimates for beetles, insects, and terrestrial arthropods. *Proc. Natl. Acad. Sci.* 112, 7519–7523. https://doi.org/10.1073/pnas.1502408112

Su, L., Yang, L., Huang, S., Su, X., Li, Y., Wang, F., Wang, E., Kang, N., Xu, J., Song, A., 2016. Comparative gut microbiomes of four species representing the higher and the lower termites. *J. Insect Sci.* 16, 97. https://doi.org/10.1093/jisesa/iew081

Suárez-Moo, P., Cruz-Rosales, M., Ibarra-Laclette, E., Desgarennes, D., Huerta, C., Lamelas, A., 2020. Diversity and composition of the gut microbiota in the developmental stages of the dung beetle copris incertus say (Coleoptera, Scarabaeidae). *Front. Microbiol.* 11, 1698. https://doi.org/10.3389/fmicb.2020.01698

Sudakaran, S., Kost, C., Kaltenpoth, M., 2017. Symbiont acquisition and replacement as a source of ecological innovation. *Trends Microbiol.* 25, 375–390. https://doi.org/10.1016/j.tim.2017.02.014

Suh, S.-O., Nguyen, N.H., Blackwell, M., 2008. Yeasts isolated from plant-associated beetles and other insects: Seven novel *Candida* species near *Candida albicans*. *FEMS Yeast Res.* 8, 88–102. https://doi.org/10.1111/j.1567-1364.2007.00320.x

Sujatha, S., Sindhura, K.A.V., Koti, P.S., Hiremath, S., Muttappagol, M., Vinay Kumar, H.D., Shankarappa, K.S., Venkataravanappa, V., Reddy, K.M.S., Reddy, C.N.L., 2025. Influence of weather and seasonal factors on whitefly dynamics, associated endosymbiotic microbiomes, and Begomovirus transmission causing tomato leaf curl disease: Insights from a metagenomic perspective. *Front. Microbiol.* 16, 1555058. https://doi.org/10.3389/fmicb.2025.1555058

Tairo, J., Hendrycks, W., Kabota, S., Zinga, M., Majubwa, R., Kudra, A., Mwatawala, M., De Meyer, M., Virgilio, M., 2025. A first characterisation of the microbiome and bacterial metabolic functions of DACUS VERTEBRATUS (Diptera, Tephritidae) in agroecological and conventional farming. *J. Appl. Entomol.* 149, 1207–1216. https://doi.org/10.1111/jen.13450

Takeshita, K., Kikuchi, Y., 2017. Riptortus pedestris and Burkholderia symbiont: An ideal model system for insect–microbe symbiotic associations. *Res. Microbiol.* 168, 175–187. https://doi.org/10.1016/j.resmic.2016.11.005

Tanabe, N., Takasu, R., Hirose, Y., Kamei, Y., Kondo, M., Nakabachi, A., 2022. Diaphorin, a polyketide produced by a bacterial symbiont of the Asian citrus psyllid, inhibits the growth and cell division of *Bacillus subtilis* but promotes the growth and metabolic activity of *Escherichia coli*. *Microbiol. Spectr.* 10, e01757-22. https://doi.org/10.1128/spectrum.01757-22

Tatiana A, E., Andrey V, Y., Mariia V, D., Alla A, G., Tamara I, K., Vera S, S., Olga V, E., 2022. Antimicrobial activity of bacteria isolated from Leptinotarsa decemlineata and Solanum tuberosum. *Ann. Environ. Sci. Toxicol.* 6, 105–119. https://doi.org/10.17352/aest.000061

Teh, B.-S., Apel, J., Shao, Y., Boland, W., 2016. Colonization of the intestinal tract of the polyphagous pest *Spodoptera littoralis* with the GFP-tagged indigenous gut bacterium *Enterococcus mundtii*. *Front. Microbiol.* 7, 928. https://doi.org/10.3389/fmicb.2016.00928

Tian, Z., Chen, L., Chen, G., Wang, J., Ma, C., Zhang, Y., Gao, X., Chen, H., Zhou, Z., 2023. Effect of host shift on the gut microbes of *Bactrocera cucurbitae* (Coquillett) (Diptera: Tephritidae). *Front. Microbiol.* 14, 1264788. https://doi.org/10.3389/fmicb.2023.1264788

Tithi, N.A., 2022. Tephritids Gut Microbionts: Diversity, Volatile Emissions and Their Impact on Fly Behaviour, in: *The 2nd International Electronic Conference on Diversity (IECD 2022)—New Insights into the Biodiversity of Plants, Animals and Microbes. Presented at the International Electronic Conference on Diversity*, MDPI, p. 26. https://doi.org/10.3390/IECD2022-12418

Tokuda, G., 2019. Plant cell wall degradation in insects: Recent progress on endogenous enzymes revealed by multi-omics technologies, in: *Advances in Insect Physiology*. Elsevier, pp. 97–136. https://doi.org/10.1016/bs.aiip.2019.08.001

Tougeron, K., Iltis, C., Rampnoux, E., Goerlinger, A., Dhondt, L., Hance, T., 2023. Still standing: The heat protection delivered by a facultative symbiont to its aphid host is resilient to repeated thermal stress. *Curr. Res. Insect Sci.* 3, 100061. https://doi.org/10.1016/j.cris.2023.100061

Trapp, S., Croteau, R., 2001. Defensive resin biosynthesis in conifers. *Annu. Rev. Plant Physiol. Plant Mol. Biol.* 52, 689–724. https://doi.org/10.1146/annurev.arplant.52.1.689

Tsuchida, T., Koga, R., Fukatsu, T., 2004. Host plant specialization governed by facultative symbiont. *Science* 303, 1989–1989. https://doi.org/10.1126/science.1094611

Upfold, J., Rejasse, A., Nielsen-Leroux, C., Jensen, A.B., Sanchis-Borja, V., 2023. The immunostimulatory role of an Enterococcus-dominated gut microbiota in host protection against bacterial and fungal pathogens in Galleria mellonella larvae. *Front. Insect Sci.* 3, 1260333. https://doi.org/10.3389/finsc.2023.1260333

Usnaddinovich, T.J., Abatovna, J.A., 2025. Insecta: Ecology of coleoptera. *Am. J. Agric. Hortic. Innov.* 5, 14–16. https://doi.org/10.37547/ajahi/Volume05Issue03-04

Vacchini, V., Gonella, E., Crotti, E., Prosdocimi, E.M., Mazzetto, F., Chouaia, B., Callegari, M., Mapelli, F., Mandrioli, M., Alma, A., Daffonchio, D., 2017. Bacterial diversity shift determined by different diets in the gut of the spotted wing fly *Drosophila suzukii* is primarily reflected on acetic acid bacteria. *Environ. Microbiol. Rep.* 9, 91–103. https://doi.org/10.1111/1758-2229.12505

Van Den Bosch, T.J.M., Welte, C.U., 2020. The microbial diversity of cabbage pest *Delia radicum* across multiple life stages. *Front. Microbiol.* 11, 315. https://doi.org/10.3389/fmicb.2020.00315

Vasquez, Y.M., Li, Z., Xue, A.Z., Bennett, G.M., 2024. Chromosome-level genome assembly of the aster leafhopper (*Macrosteles quadrilineatus*) reveals the role of environment and microbial symbiosis in shaping pest insect genome evolution. *Mol. Ecol. Resour.* 24, e13919. https://doi.org/10.1111/1755-0998.13919

Ventura, C., Briones-Roblero, C.I., Hernández, E., Rivera-Orduña, F.N., Zúñiga, G., 2018. Comparative analysis of the gut bacterial community of four anastrepha fruit flies (Diptera: Tephritidae) *Based on Pyrosequencing. Curr. Microbiol.* 75, 966–976. https://doi.org/10.1007/s00284-018-1473-5

Victorica, M.R., Soria, M.A., Batista-García, R.A., Ceja-Navarro, J.A., Vikram, S., Ortiz, M., Ontañon, O., Ghio, S., Martínez-Ávila, L., Quintero García, O.J., Etcheverry, C., Campos, E., Cowan, D., Arneodo, J., Talia, P.M., 2020. Neotropical termite microbiomes as sources of novel plant cell wall degrading enzymes. *Sci. Rep.* 10, 3864. https://doi.org/10.1038/s41598-020-60850-5

Vikram, S., Arneodo, J.D., Calcagno, J., Ortiz, M., Mon, M.L., Etcheverry, C., Cowan, D.A., Talia, P., 2021. Diversity structure of the microbial communities in the guts of four neotropical termite species. *PeerJ* 9, e10959. https://doi.org/10.7717/peerj.10959

Voirol, L.R.P., Frago, E., Kaltenpoth, M., Hilker, M., Fatouros, N.E., 2018. Bacterial symbionts in lepidoptera: Their diversity, *Transmission, and Impact on the Host. Front. Microbiol.* 9, 556. https://doi.org/10.3389/fmicb.2018.00556

Wang, A., Yao, Z., Zheng, W., Zhang, H., 2014. Bacterial communities in the gut and reproductive organs of *Bactrocera minax* (Diptera: Tephritidae) based on 454 pyrosequencing. *PLoS ONE* 9, e106988. https://doi.org/10.1371/journal.pone.0106988

Wang, H., Wu, N., Liu, Y., Kundu, J.K., Liu, W., Wang, X., 2019. Higher bacterial diversity of gut microbiota in different natural populations of leafhopper vector does not influence WDV transmission. *Front. Microbiol.* 10, 1144. https://doi.org/10.3389/fmicb.2019.01144

Wang, J., Bai, J., Zheng, F., Ling, Y., Li, X., Wang, Jing, Zhi, Y.-C., Li, X.-J., 2020. Diversity of the gut microbiome in three grasshopper species using 16S rRNA and determination of cellulose digestibility. *PeerJ* 8, e10194. https://doi.org/10.7717/peerj.10194

Wang, J., Peiffer, M., Hoover, K., Rosa, C., Zeng, R., Felton, G.W., 2017. Helicoverpa zea gut-associated bacteria indirectly induce defenses in tomato by triggering a salivary elicitor(s). *New Phytol.* 214, 1294–1306. https://doi.org/10.1111/nph.14429

Wang, L., Li, C., Wang, X., Wang, G., Shang, S., Dou, Z., Luo, Y., 2022. Gut Lignocellulose activity and microbiota in Asian longhorned beetle and their predicted contribution to larval nutrition. *Front. Microbiol.* 13, 899865. https://doi.org/10.3389/fmicb.2022.899865

Wang, P., Furlong, M.J., Walsh, T.K., Zalucki, M.P., 2019. Moving to keep fit: Feeding behavior and movement of *Helicoverpa armigera* (Lepidoptera: Noctuidae) on artificial diet with different protein: Carbohydrate ratios. *J. Insect Sci.* 19, 20. https://doi.org/10.1093/jisesa/iez098

Wang, X., Sun, S., Yang, X., Cheng, J., Wei, H., Li, Z., Michaud, J.P., Liu, X., 2020. Variability of gut microbiota across the life cycle of *Grapholita molesta* (Lepidoptera: Tortricidae). *Front. Microbiol.* 11, 1366. https://doi.org/10.3389/fmicb.2020.01366

Wang, Y., Tang, J., Chen, Yao, Chen, Shuyi, Chen, Sumin, Yu, X., Wan, C., Xiang, G., Chen, Yaping, Li, Q., 2025. The ecological–Evolutionary game of the insect gut microbiome: Environmental drivers, host regulation, and prospects for cross-cutting applications. *Vet. Sci.* 12, 866. https://doi.org/10.3390/ijms18112450

Watanabe, H., Tokuda, G., 2010. Cellulolytic systems in insects. *Annu. Rev. Entomol.* 55, 609–632. https://doi.org/10.1146/annurev-ento-112408-085319

Weiss, B., Kaltenpoth, M., 2016. Bacteriome-localized intracellular symbionts in pollen-feeding beetles of the genus dasytes (Coleoptera, Dasytidae). *Front. Microbiol.* 7. https://doi.org/10.3389/fmicb.2016.01486

Wielkopolan, B., Frąckowiak, P., Wieczorek, P., Obrępalska-Stęplowska, A., 2022. The impact of *Oulema melanopus*—Associated bacteria on the wheat defense response to the feeding of their insect hosts. *Cells* 11, 2342. https://doi.org/10.3390/cells11152342

Wielkopolan, B., Krawczyk, K., Szabelska-Beręsewicz, A., Obrępalska-Stęplowska, A., 2021. The structure of the cereal leaf beetle (*Oulema melanopus*) microbiome depends on the insect's developmental stage, host plant, and origin. *Sci. Rep.* 11, 20496. https://doi.org/10.1038/s41598-021-99411-9

Wielkopolan, B., Szabelska-Beręsewicz, A., Gawor, J., Obrępalska-Stęplowska, A., 2024. Cereal leaf beetle-associated bacteria enhance the survival of their host upon insecticide treatments and respond differently to insecticides with different modes of action. *Environ. Microbiol. Rep.* 16, e13247. https://doi.org/10.1111/1758-2229.13247

Woestmann, L., Saastamoinen, M., 2016. The importance of trans-generational effects in Lepidoptera. *Curr. Zool.* 62, 489–499. https://doi.org/10.1093/cz/zow029

Wu, L., Hu, C., Liu, T., 2024. Metagenomic profiling of gut microbiota in Fall Armyworm (*Spodoptera frugiperda*) larvae fed on different host plants. *BMC Microbiol.* 24, 337. https://doi.org/10.1186/s12866-024-03481-4

Xia, X., Gurr, G.M., Vasseur, L., Zheng, D., Zhong, H., Qin, B., Lin, J., Wang, Y., Song, F., Li, Y., Lin, H., You, M., 2017. Metagenomic sequencing of diamondback moth gut microbiome unveils key holobiont adaptations for herbivory. *Front. Microbiol.* 8, 663. https://doi.org/10.3389/fmicb.2017.00663

Xia, X., Sun, B., Gurr, G.M., Vasseur, L., Xue, M., You, M., 2018. Gut microbiota mediate insecticide resistance in the diamondback moth, *Plutella xylostella* (L.). *Front. Microbiol.* 9, 25. https://doi.org/10.3389/fmicb.2018.00025

Xie, J., Vilchez, I., Mateos, M., 2010. Spiroplasma bacteria enhance survival of *Drosophila hydei* Attacked by the Parasitic Wasp Leptopilina heterotoma. *PLoS ONE* 5, e12149. https://doi.org/10.1371/journal.pone.0012149

Xin Yee Tan, K., Shigenobu, S., 2025. Targeted disruption of the cls gene in Buchnera aphidicola impairs membrane integrity and host symbiont dynamics. *iScience* 28, 113178. https://doi.org/10.1016/j.isci.2025.113178

Xu, P., Cheng, P., Liu, Z., Li, Y., Murphy, R.W., Wu, K., 2012. Complete genome sequence of a monosense densovirus infecting the cotton bollworm, *Helicoverpa armigera*. *J. Virol.* 86, 10909–10909. https://doi.org/10.1128/JVI.01912-12

Xu, S., Jiang, L., Qiao, G., Chen, J., 2021. Diversity of bacterial symbionts associated with Myzus persicae (Sulzer) (Hemiptera: Aphididae: Aphidinae) revealed by 16S rRNA Illumina sequencing. *Microb. Ecol.* 81, 784–794. https://doi.org/10.1007/s00248-020-01622-6

Yamasaki, Y., Sumioka, H., Takiguchi, M., Uemura, T., Kihara, Y., Shinya, T., Galis, I., Arimura, G., 2021. Phytohormone-dependent plant defense signaling orchestrated by oral bacteria of the herbivore *Spodoptera litura*. *New Phytol.* 231, 2029–2038. https://doi.org/10.1111/nph.17444

Yamlahi, Y.E., Remmal, I., Maurady, A., Britel, M.R., Bakali, A.H., Mokhtar, N.B., Galiatsatos, I., Stathopoulou, P., Tsiamis, G., 2025. Characterization of the olive fly (*Bactrocera oleae*) microbiome across diverse geographic regions of Morocco. *Insect Sci.* 32, 1669–1680. https://doi.org/10.1111/1744-7917.70126

Yan, X. -T., Ye, Z. -X., Wang, X., Zhang, C. -X., Chen, J. -P., Li, J. -M., Huang, H. -J., 2021. Insight into different host range of three planthoppers by transcriptomic and microbiomic analysis. *Insect Mol. Biol.* 30, 287–296. https://doi.org/10.1111/imb.12695

Yang, K., Li, C.-R., Qin, P.-H., Yuan, M.-Y., Chu, D., 2025. Conditional effects of *Cardinium* on microbiota in an invasive whitefly under different ecological factors. *Microbiol. Spectr.* 13, e02240–24. https://doi.org/10.1128/spectrum.02240-24

Yang, K., Qin, P., Yuan, M., Chen, L., Zhang, Y., Chu, D., 2023. Infection density pattern of *Cardinium* affects the responses of bacterial communities in an invasive whitefly under heat conditions. *Insect Sci.* 30, 1149–1164. https://doi.org/10.1111/1744-7917.13141

Yang, Z., Luo, J., Men, Y., Liu, Z., Zheng, Z., Wang, Y., Xie, Q., 2023. Different roles of host and habitat in determining the microbial communities of plant-feeding true bugs. *Microbiome* 11, 244. https://doi.org/10.1186/s40168-023-01702-y

Yu, Y., Wang, Y., Li, H., Yu, X., Shi, W., Zhai, J., 2021. Comparison of microbial communities in colorado potato beetles (Leptinotarsa decemlineata Say) collected from different sources in China. *Front. Microbiol.* 12, 639913. https://doi.org/10.3389/fmicb.2021.639913

Yun, J.-H., Roh, S.W., Whon, T.W., Jung, M.-J., Kim, M.-S., Park, D.-S., Yoon, C., Nam, Y.-D., Kim, Y.-J., Choi, J.-H., Kim, J.-Y., Shin, N.-R., Kim, S.-H., Lee, W.-J., Bae, J.-W., 2014. Insect gut bacterial diversity determined by environmental habitat, diet, developmental stage, and phylogeny of host. *Appl. Environ. Microbiol.* 80, 5254–5264. https://doi.org/10.1128/AEM.01226-14

Zaada, D.S.Y., Ben-Yosef, M., Yuval, B., Jurkevitch, E., 2019. The host fruit amplifies mutualistic interaction between Ceratitis capitata larvae and associated bacteria. *BMC Biotechnol.* 19, 92. https://doi.org/10.1186/s12896-019-0581-z

Zeng, T., Fu, Q., Luo, F., Dai, J., Fu, R., Qi, Y., Deng, X., Lu, Y., Xu, Y., 2024. Lactic acid bacteria modulate the CncC pathway to enhance resistance to β-cypermethrin in the oriental fruit fly. *ISME J.* 18, wrae058. https://doi.org/10.1093/ismejo/wrae058

Zhang, N., Qian, Z., He, J., Shen, X., Lei, X., Sun, C., Fan, J., Felton, G.W., Shao, Y., 2024. Gut bacteria of lepidopteran herbivores facilitate digestion of plant toxins. *Proc. Natl. Acad. Sci.* 121, e2412165121. https://doi.org/10.1073/pnas.2412165121

Zhang, S., Li, Z., Shu, J., Xue, H., Guo, K., Zhou, X., 2022. Soil-derived bacteria endow Camellia weevil with more ability to resist plant chemical defense. *Microbiome* 10, 97. https://doi.org/10.1186/s40168-022-01290-3

Zhang, S., Song, F., Wang, J., Li, X., Zhang, Y., Zhou, W., Xu, L., 2024. Gut microbiota facilitate adaptation of invasive moths to new host plants. *ISME J.* 18, wrae031. https://doi.org/10.1093/ismejo/wrae031

Zhang, X., Zhang, F., Lu, X., 2022. Diversity and functional roles of the gut microbiota in lepidopteran insects. *Microorganisms* 10, 1234. https://doi.org/10.3390/microorganisms10061234

Zhang, Y., Mao, K., Chen, K., Zhao, Z., Ju, F., 2024. Symbiont community assembly shaped by insecticide exposure and feedback on insecticide resistance of *Spodoptera frugiperda*. *Commun. Biol.* 7, 1194. https://doi.org/10.1038/s42003-024-06892-1

Zhang, Yunhua, Cai, T., Yuan, M., Li, Z., Jin, R., Ren, Z., Qin, Y., Yu, C., Cai, Y., Shu, R., He, S., Li, J., Wong, A.C.N., Wan, H., 2023. Microbiome variation correlates with the insecticide susceptibility in different geographic strains of a significant agricultural pest, Nilaparvata lugens. *Npj Biofilms Microbiomes* 9, 2. https://doi.org/10.1038/s41522-023-00369-5

Zhang, Yuxin, Zhang, S., Xu, L., 2023. The pivotal roles of gut microbiota in insect plant interactions for sustainable pest management. *Npj Biofilms Microbiomes* 9, 66. https://doi.org/10.1038/s41522-023-00435-y

Zhao, L., Wang, W.-Q., Xu, S.-Q., Guan, D.-L., 2022. The comparison of gut bacteria communities and the functions among the sympatric grasshopper species from the loess plateau. *Front. Microbiol.* 13, 806927. https://doi.org/10.3389/fmicb.2022.806927

Zheng, R., Wang, Q., Wu, R., Paradkar, P.N., Hoffmann, A.A., Wang, G.-H., 2023. Holobiont perspectives on tripartite interactions among microbiota, mosquitoes, and pathogens. *ISME J.* 17, 1143–1152. https://doi.org/10.1038/s41396-023-01436-7

Zulak, K.G., Bohlmann, J., 2010. Terpenoid biosynthesis and specialized vascular cells of conifer defense. *J. Integr. Plant Biol.* 52, 86–97. https://doi.org/10.1111/j.1744-7909.2010.00910.x

CHAPTER THREE

Adaptation through Association

THE ROLE OF INSECT MICROBIOME IN INSECT RESILIENCE, PLASTICITY, AND SURVIVAL

Ayushi Gupta, Ramcharan Bhattacharya, and Suresh Nair

3.1 INTRODUCTION

Insect pests cause substantial damage to the agricultural produce, thereby threatening the food security and livelihood of millions of people worldwide. Among the several biotic and abiotic factors that inflict crop damage, insect pests alone cause 30–40% loss in yield annually (Sharma et al. 2017). To combat this, modern agricultural practices involve extensive use of insecticides and resistant cultivars. However, insects continue to evolve and adapt, and in the process, insects often overcome these control measures in a very short frame of time. This rapid adaptability poses a significant threat to global food production and also affects the long-term sustainability of current pest management practices. Studies have shown genome plasticity has allowed insects to quickly adapt to changing environmental conditions and colonize novel ecological niches. However, the molecular mechanisms underpinning genome plasticity and rapid insect adaptations are yet to be fully understood (Ashra and Nair 2022).

Usually, insect adaptation is governed by both genetic and epigenetic factors and their interaction with the external environment (Gupta and Nair 2025). This interaction initiates a cascade of genetic and random molecular events within insects, facilitating their adaptation. Recently, some studies have hinted at the role of epigenetic processes such as DNA methylation in conferring adaptive advantages to insects (Gupta and Nair 2021, 2022). Besides, it has been demonstrated that in some insect species transpositional activity profoundly influences the genomic organization and structure, resulting in altered insect phenotypes under changing environmental conditions (González et al. 2010; Mateo et al. 2014; Hof et al. 2016). The ability of transposons to move and replicate within the genome (Kazazian 2004) makes them the perfect candidate for providing the insect's genome with its dynamic nature. Moreover, conventional efforts to understand insect resilience have attributed rapid insect adaptations to genetic mutations, high reproductive rates, migratory nature, and complex insect-host interactions as important components. However, lately several studies have provided valuable insights into a crucial and often underappreciated factor that can potentially contribute to insect adaptability – the insect microbiome. Comprising complex communities of bacteria, fungi, and other microorganisms, the insect microbiome, especially the gut microbiota, plays an essential role in mediating a wide range of physiological and ecological functions that enhance insect survival under wide range of biotic and abiotic stresses.

DOI: 10.1201/9781003639176-3

The microbial symbionts provide several advantages to their insect hosts, including nutritional benefits, immunity and defence against predators, pathogens, and parasites, synthesis of essential amino acids and vitamins, and detoxifying xenobiotic compounds (Janson et al. 2008; Gupta and Nair 2020). For instance, symbiotic bacteria compensate for the nutritional deficiencies in sap-feeding insects, such as aphids, whiteflies, and planthoppers (Russell et al. 2014; Santos-Garcia et al. 2020). In contrast, in others, microbial partners contribute to increased tolerance against heat, cold, desiccation, pathogens, and resistance host plant varieties (Horgan et al. 2019). Besides, bacterial symbionts are also known to influence developmental timing, and in some cases, alter host behaviour to improve fitness and survivability (Douglas et al. 2015; Wielkopolan and Obrępalska-Stęplowska 2016).

Interestingly, the gut microbiome of insects is not static – it is highly dynamic and responsive to environmental conditions. Data from various studies have revealed rapid microbial shifts within an insect body with changing geographic and climatic conditions (Gupta and Nair 2022). Additionally, variations in host plant type, pesticide exposure, environmental fluctuations, and other ecological factors can cause significant changes in microbial composition (Nguyen et al. 2021). This plasticity enables insects to rapidly adjust to novel environments and withstand stressors without immediately undergoing irreversible genetic changes. Microbial flexibility has been associated with enhanced virulence, host plant switching, ability to detoxify pesticides, and even host range expansion in some insects (Thia et al. 2025). Studies have shown that microbial symbionts can contribute to the insect's survival and fitness under harsh environments by interacting with various molecular processes within an insect's body. For instance, bacterial symbionts influence the gene expression patterns of their insect host via epigenetic modifications. These changes not only have transgenerational effects but also determine the long-term adaptive potential in insects (Mogilicherla and Roy 2023). Further, some bacterial species interact with transposable elements and influence the expression of stress-responsive genes, further expanding the mechanisms insects use to adapt to changing environmental conditions (Lai and Wang 2025).

Considering the role and importance of microbiome for insect survival, fitness, and longevity, it is crucial to integrate microbiome into our understanding of insect ecology and evolution to gain new insights on sustainable pest management. Targeting or manipulating the insect microbiome may provide a handle to disrupt essential physiological functions, reduce pest fitness, and/or enhance the effectiveness of biological control agents. Hence, the insect microbiome offers a promising avenue for developing more sustainable and ecologically sound pest management strategies. Moreover, as ecosystems face escalating pressures from climate shifts to pollution and habitat disruption, the microbial communities within hosts and environments play a pivotal role in modulating stress responses. Therefore, understanding microbiome–insect interactions in the context of environmental change is crucial. This chapter explores the central role of the insect microbiome in promoting resilience, plasticity, and survival. Here, focus has been given on the current knowledge on microbiome composition and function, its influence on insect physiology and behaviour, and the environmental factors that shape microbial communities. By highlighting the complex interplay between insects and their microbial partners, this chapter aims to provide a comprehensive understanding of how these relationships contribute to the evolutionary success of insect pests in agricultural ecosystems.

3.2 THE INSECT MICROBIOME: COMPOSITION AND FUNCTIONS

A diverse array of microbial taxa, including bacteria, archaea, fungi, and protists, resides within an insect's body. Of these, the most widely studied are the gut-associated bacteria (dominated by phyla such as Proteobacteria, Firmicutes, Bacteroidetes, and Actinobacteria) (Douglas 2015). The gut microflora shapes insect physiology, development, immunity, behaviour, and ecological adaptability (Engel and Moran 2013). Insects are highly dependent on their gut microflora for survival and normal functions related to their life cycle (Gupta and Nair 2020). And based on the extent of dependence, their association with microbes can be classified as obligate (or primary) and facultative (or secondary). Whichever the case, both obligate and facultative microbes are known to form intricate symbiotic relationships with their insect hosts, drastically influencing their biology and evolution (Gupta and Nair 2020).

Further, based on the nature of association, the insect microbiome is broadly categorized into core and transient communities. The core microbiota refers to those microbial species that are uniformly present in a population and often coevolve with the host. These symbionts typically reside in specialized organs such as bacteriocytes and are generally associated with various essential metabolic and defensive functions in an insect (Moran et al. 2008). For example, *Buchnera aphidicola* in aphids compensates for its nutrient-deficient diet (i.e., phloem sap) by synthesizing various amino acids, vitamins, and sterols necessary for normal development and reproduction (Moran 2021). Likewise, *Wigglesworthia glossinidia* found in tsetse flies provides essential B-vitamins for the fly's nutrition and supports its reproductive success; honeybees maintain a core gut microbiota consisting of *Snodgrassella alvi*, *Gilliamella apicola*, and *Lactobacillus* spp., critical for nutrient metabolism and immune regulation (Weiss et al. 2022; Zheng et al. 2018). Being crucial for insect survival and fitness, such bacterial species form part of the core microbiota. In contrast, the transient microbiota comprises microbes acquired by an insect from the environment during its lifetime. Therefore, it can exhibit significant variation between individuals and populations. These microbes may not always establish long-term associations but can influence crucial processes such as insect digestion, immunity, or detoxification in response to environmental fluctuations (Holt et al. 2024). For instance, *Drosophila melanogaster* acquires most of its microbiota from its environment, particularly from the fruit on which it feeds and lays eggs (Guilhot et al. 2023). Similarly, mosquitoes harbour highly variable and environmentally influenced microbial communities (Coon et al. 2016).

Although insect genetics and host–plant interactions have been the prime focus of several studies to understand insect survival and resilience, with molecular advancements, the microbial community, particularly those species that exhibit changes in response to the changing ecological and environmental conditions, is gaining attention. These microbial species are generally acquired and maintained in the population depending on the requirements of the insect host. The microbial symbionts in insects can be acquired and transmitted (1) vertically (i.e., the direct transfer of microbes from parent to offspring to ensure the stable inheritance of obligate symbionts as in the case of *Buchnera* in aphids), (2) horizontally (i.e., the acquisition of microbes through social interactions or mating), or (3) environmentally (i.e., when insects acquire microbes from soil, host plants, or other substrates during feeding or development). The environmental acquisition of microbes is common in generalist herbivores and insects with broad ecological ranges, such as the fall armyworm (*Spodoptera frugiperda*) (Villa et al. 2023; Chu et al. 2025). However, the microbiome composition in most insects is determined by combining these strategies, depending on their life stage, diet, and exposure to environmental conditions. For example, in the black soldier fly (*Hermetia illucens*), the microbiome composition changes across developmental stages and is primarily influenced by larval substrate rather than vertical inheritance (Auger et al. 2023); the brown planthopper (*Nilaparvata lugens*) undergoes rapid shifts in its microbiome composition in response to changing climates and geographical locations (Gupta and Nair 2022). Therefore, it is reasonable to conclude that the insect microbiome is highly dynamic, shaped predominantly by its functional contribution to insect survival, fitness, and longevity.

The symbiotic bacterial species perform a wide array of functional roles in an insect's body, including the synthesis of essential nutrients such as amino acids, vitamins, and sterols; regulation of insect development and reproduction, energy metabolism; and modulation of the immune system and defence response against pathogens (Yasika and Shivakumar 2025). For instance, in *Bactrocera dorsalis* larvae, gut bacteria such as *Enterobacter cloacae* contribute to vitamin B6 biosynthesis under nutrient-poor diets (Gu et al. 2024); while in *Spodoptera litura* and *Tuta absoluta*, gut-associated *Enterococcus mundtii* aids in lysine production and detoxification of host plant compounds (Chen Y et al. 2024). Further, in *S. frugiperda*, disruption of gut microbiota through antibiotics has been linked to reduced larval growth and reproductive output, thereby demonstrating the microbiome's role in fitness and fecundity (Chen et al. 2021). Similarly, microbial symbionts also affect mating behaviour, fecundity, and even sex ratio distortion in a few insect species (Werren et al. 2008). Furthermore, microbes can prime the host's immune responses, compete with pathogens for resources, or produce antimicrobial compounds. In bumble bees, for example, the gut microbiota confers resistance against *Crithidia bombi* and *Nosema* spp., reducing infection severity (Mockler et al. 2018). Likewise, *Wolbachia* is also known to reduce mosquito susceptibility to viral infections, thereby impacting vector competence (Branda et al. 2024). Taken together, these studies indicate that the symbiotic microbial species are a crucial part of an insect's life cycle, playing diverse roles and determining their evolutionary success.

In addition, one of the most essential functions of the microbiome is enhancing insect adaptability to environmental and host-related stresses. Studies have shown that microbial symbionts aid in the detoxification of plant secondary metabolites and pesticides, enabling insects to feed on resistant plant varieties and survive chemical exposure (Moran 2021; Yasika and Shivakumar 2025). Besides, alterations in microbial community composition have been associated with host plant switching and pesticide resistance, as seen in the case of planthoppers and aphids (Horgan et al. 2019). Further, the microbiome is also known to facilitate insect adaptation to various abiotic stressors such as temperature and humidity, expanding their ecological range (Darwish et al. 2025). Collectively, it is reasonable to state that the insect microbiome is a key determinant of host resilience, offering metabolic, immunological, and ecological benefits that facilitate insect survival in diverse and changing environments. We also believe that understanding these microbial communities' composition, transmission, and functional dynamics would open new avenues for pest control strategies, particularly those aiming to disrupt essential microbial functions or exploit symbiont-mediated vulnerabilities.

3.3 TYPES OF ENVIRONMENTAL STRESSORS AFFECTING INSECTS

Insects are routinely exposed to multiple environmental stressors, including abiotic and biotic factors, significantly challenging insect survival, development, reproduction, and evolutionary trajectories. Abiotic stressors are physical or chemical factors in the environment that do not have a biological origin, such as temperature extremes (e.g., heat stress, cold stress, rapid temperature fluctuations), desiccation/humidity fluctuations (e.g., loss of water, low relative humidity), toxins and pollutants (e.g., heavy metals, pesticides, and xenobiotics), UV radiation, and oxygen limitation (or hypoxia particularly in the case of aquatic or high-altitude insects). In contrast, biotic stressors for insects are derived from their interaction with other organisms such as pathogens (e.g., viruses, fungi, bacteria), parasites, predators, and host plant changes. Further, it can also result from competition (inter- or intraspecific) for resources between organisms thriving in the same ecosystem. Furthermore, host plant defence mechanisms (e.g., plant-derived allelochemicals and morphological traits) that inhibit insect feeding also contribute to insect biotic stress.

Adaptation to these stressors involves physiological, morphological, genetic, and ecological processes. Insects deploy multiple, sometimes overlapping mechanisms to resist or tolerate environmental stressors. These mechanisms include behavioural responses (e.g., seeking shelter and nocturnal shifts, choosing microclimates to avoid extremes) and physiological or biochemical adaptations (e.g., recruitment of heat shock proteins (HSPs), antifreeze proteins, detoxification enzymes such as cytochrome P450s and glutathione-S-transferases and modulation of membrane lipid composition; Zhang et al. 2019; Ramniwas et al. 2023).

In addition, over the past few years, we have increasing evidence pointing towards the central role of insect microbiome in mediating adaptation. For instance, Ren et al. (2023) demonstrated that continuous exposure to heat stress across the developmental stages of *Sarcophaga peregrina* larvae altered its life-history traits (i.e., shorter development time and higher larval growth rate) and induced significant shifts in its gut bacterial community structure, implicating the role of microbiome in contributing towards the generation of a heat-tolerant phenotype. Similarly, it has been shown that in *Zeugodacus cucurbitae*, temperature stress and diet quality significantly alter the microbial composition, affecting its fecundity and longevity. Additionally, this study also indicated the role of microbiome in acting as a buffer against multiple abiotic stressors (i.e., poor diet and heat in this case) in phytophagous insects (Hafsi et al. 2024).

Further, the microbiome is also suggested to provide insects with the capacity to withstand various biotic stressors. For instance, in *Spodoptera litura*, exposure to the fungus, *Metarhizium flavoviride*, triggers multiple metabolic pathways and induces shifts in the gut microbiome composition. Besides, this study also demonstrated how the microbiome and host metabolome interact under biotic stress (Juefeng et al. 2024). Likewise, in some

Lepidopteran pests, the gut bacteria are known to facilitate the digestion of plant toxins and secondary metabolites (~~Zang~~Zhang et al. 2024), thereby ensuring their survival on toxic plants. The scientific literature abounds with examples derived from diverse insect models wherein the microbiome is shown to be involved in mediating various processes crucial to their survival, such as osmoregulation and water retention mechanisms and developmental and life-history adjustments (including diapause, quiescence, modulation of developmental rate, changes in reproductive timing, and fecundity; Mondal et al. 2023). Moreover, we now have information indicating that the microbiome actively drives genetic evolutionary adaptations and epigenetic plasticity in insects (Kolodny and Schuleburg 2020; Lai and Wang 2025). Collectively, we now have sufficient evidence to state that the microbiome helps shape the evolutionary landscape of insect populations by mediating their survival under extreme and/or changing environmental conditions, be it biotic or abiotic. From the above, it is clear that these microbial partners can aid in detoxification, bolster immune responses, and even influence gene expression related to chemosensation and metabolism (Siddiqui et al. 2022; Khan et al. 2023). As such, the microbiome acts as a dynamic buffer, enabling insects to navigate and survive under rapidly changing ecological conditions. And therefore, it has become pertinent to understand the mechanisms underlying these complex insect–microbiome interactions to better comprehend insect ecology and insect adaptation to climate change. Subsequently, the information thus obtained can be used to devise efficient and sustainable pest control strategies. With this view, the following section focuses on mechanisms underlying microbiome-mediated stress adaptations in insects.

3.4 MICROBIOME-MEDIATED MECHANISMS OF STRESS ADAPTATION

Recent work has highlighted that microbial symbionts serve as essential mediators of adaptation via multiple routes. (1) Detoxification: microbial symbionts aid in the degradation of plant allelochemicals, secondary metabolites, and xenobiotics (Kline and Joshi 2024). For instance, the gut bacteria of several Lepidopteran insects facilitate the digestion of plant toxins (Zhang et al. 2024). Further, in some insects gut microbes metabolize or detoxify pesticides by regulating the expression of enzymes associated with host detoxification pathways (e.g., cytochrome P450 enzymes; Jaffar et al. 2022). (2) Metabolic assistance: gut microbes provide essential nutrients (such as amino acids, vitamins, osmolytes, etc.) to their insect host to enable it to meet its energy requirements under stress (Jing et al. 2020; Zhang et al. 2023). Besides, in some cases, it has been observed that microbes are associated with the production and/or regulation of the synthesis of metabolites that act as cryoprotectants and heat protectants or help buffer nutritional deficits under stress (Iltis et al. 2022; Chabanol and Gendrin 2024). For example, in cave-dwelling beetles, the gut microbiota ferments the organic matter to allow the extraction of nutrients from scarce or detrital sources (Moldovan et al. 2023). (3) Modulation of the immune system: microbiome can help insects combat pathogen infection, which often increases under stress by competing with pathogens, producing antimicrobial compounds, or initiating immune responses (Engel and Moran 2013). Data from studies have indicated that under biotic stress (pathogens), symbionts often prime or enhance immune responses in their insect host (Singh et al. 2021). In fact, in some cases, perturbing the microbiome (e.g., via antibiotics) reduced the capacity of insects to tolerate stress (Zhang et al. 2022; Rupawate et al. 2023; Archer et al. 2025). (4) Thermal tolerance: microbiome is capable of influencing the expression of HSPs, oxidative stress response genes, and genes involved in maintaining cellular integrity under stress (Chen et al. 2024). Besides, changes in microbiome composition have been observed when insects are exposed to heat/cold stress. And these changes have been correlated with improved survival, faster development, and altered metabolism in various insect models (e.g., in *S. peregrina*, Yang et al. 2024; *Z. cucurbitae*, Asimakis et al. 2021; Hafsi et al. 2024).

Data from various studies have linked the presence or absence of certain symbionts with modulation of the host transcriptomic response under stress (Zhao et al. 2023; Fan et al. 2025). Stress conditions impact gut microbes' diversity, relative abundance, or functionality (Lange et al. 2023; Charaabi et al. 2025). And these shifts are often directed towards the enrichment of stress-tolerant taxa. While most of these shifts are reversible, some may have long-term or even transgenerational effects. The insect microbiome is suggested to play a vital role in driving genetic adaptation and epigenetic regulation, particularly under environmental pressures

(Kim et al. 2016). These changes often result in enhanced survival in fluctuating or stressful environments, such as extreme temperatures, through natural selection mechanisms, epigenetic modifications, and transgenerational inheritance. For example, under thermal stress – a major selective pressure on insects – the microbiome can impact the expression and selection of host genes related to thermal tolerance, by producing metabolites that enhance thermal resistance or inducing immune and stress response pathways; e.g., a study on *D. melanogaster* indicated that insects raised with specific bacterial taxa like *Acetobacter* and *Lactobacillus* sp. exhibited greater heat tolerance (Moghadam et al. 2018). These bacteria alter gut physiology and have been shown to induce the expression of HSPs in the host. In fact, studies have shown that over generations, insects harbouring beneficial microbes may show selection for alleles that work synergistically with the microbiome, conferring higher fitness under thermal stress (Henry et al. 2021). For instance, the facultative symbiont *Serratia symbiotica* in aphids enhances host survival at high temperatures by protecting the essential symbiont *B. aphidicola*. This mutualistic relationship favours aphid genotypes that maintain and regulate the abundance of these symbionts, thereby influencing host allele frequencies over time (Corbin et al. 2017; Tougeron et al. 2023).

Another mechanism by which the microbiome mediates rapid insect adaptations is by modulating epigenetic processes, such as DNA methylation, histone modification, and expression of non-coding RNA. These modifications can affect developmental timing, immunity, and stress tolerance. For instance, in mosquitoes (e.g., *Anopheles* spp., *Aedes aegypti*), gut microbes have been shown to influence histone acetylation and DNA methylation patterns of various immune-related genes (Zhang et al. 2013; Kim et al. 2016). And these epigenetic changes are believed to prime mosquitoes and promote their survival upon pathogen attack. To bring about this effect, usually microbes produce metabolites like short-chain fatty acids (SCFAs) and vitamins, which act as epigenetic modulators of host gene expression (Watt et al. 2020).

Further, the microbiome also contributes to transgenerational plasticity in insects, i.e., the ability of parental environments to influence the phenotype of the offspring through vertical transmission of microbes and epigenetic memory. Such a phenomenon enables the offspring to be pre-adapted to environmental challenges that the parental generation faced. For example, *Drosophila* exposed to heat stress can pass on microbial communities to offspring that enhance thermal tolerance (Moghadam et al. 2018; Jaramillo and Castañeda 2021). This inheritance of microbes contributes to a phenotypic memory of stress, thereby facilitating adaptive plasticity in insects. Furthermore, in some insects such as *Tribolium castaneum* and *N. lugens*, it has been observed that exposure to environmental stress leads to DNA methylation changes that persist across generations (Feliciello et al. 2013; Gupta and Nair 2022). Further, these changes can sometimes be microbiome-mediated, i.e., through microbial induction of stress response pathways, which eventually affect the evolutionary outcomes for insects.

It is evident from the above-referred studies that the gut microbiome of insects is an integral part of an insect's life cycle. Therefore, it is plausible to consider the insect and its microbiome as 'holobionts', where selective pressures act on both to influence their evolution. And while we have made remarkable progress in deciphering and understanding this relationship, many microbiome-mediated mechanisms operating in an insect's body are yet to be fully unravelled. Besides, more information on microbial taxa or strains that are most important under given stresses (especially when an insect is exposed to multiple stresses, simultaneously) is critical to develop control strategies aimed at manipulating/targeting microbes to combat pest outbreaks. Besides, while we have some leads from a few insect models, the molecular mechanisms by which the microbiome regulates host gene expression (e.g., HSPs, antioxidant enzymes) are still unclear.

Nevertheless, we know that the microbiome of an insect is highly dynamic and responsive to environmental stress. This dynamic nature of the insect microbiome makes it an ideal candidate for facilitating rapid insect adaptations. Therefore, a comprehensive knowledge of how microbes respond to external environmental changes is essential for fully understanding their role in insect adaptation. With this view, the next section of this chapter highlights some recent studies that show how host–microbe associations change in response to environmental challenges. Further, we discuss symbiont turnover, transmission, and co-evolutionary dynamics of the insect microbiome under stress conditions.

3.5 INSECT–MICROBIOME DYNAMICS UNDER STRESS

As is evident, insect–microbe associations represent vast and diverse relationships that influence nearly all aspects of their life cycle. However, these associations are not static but highly dynamic and responsive to ecological pressures and evolutionary shifts. In the face of rapid environmental fluctuations – such as climate change, habitat degradation, pollution, and the introduction of novel pathogens – these symbiotic relationships are constantly under intense selection pressure (Gupta and Nair 2020). While the insect host retains the resilient symbionts, others get replaced to maintain homeostasis and insect survival. Environmental variability drastically influences the structure and function of insect-associated microbiomes. Stressors cause rapid adaptive shifts in microbial community composition, allowing insects to survive in new or stressful environments (Gupta and Nair 2020). Such responsive symbiotic shifts highlight the importance of functional flexibility in symbiosis, allowing hosts to cope with unpredictable and challenging environments. In fact, this phenomenon is generally referred to as 'symbiont turnover', i.e., the replacement or acquisition of new symbiotic partners over time. It can occur across an organism's lifespan, across generations, or in response to specific environmental cues (Manthey et al. 2023).

Symbiont turnover can involve complete replacement of symbiont taxa or shifts in relative abundance within a complex microbial community. It can be driven by processes such as environmental filtering (i.e., external conditions select for symbionts with traits suited to new conditions), immune system modulation of the insect host (i.e., when host actively selects or rejects specific symbionts), existing competitive dynamics within the microbial community (i.e., dominant taxa may get replaced by others under new conditions), or symbiont loss and acquisition (i.e., hosts loses/gains symbionts) under stress conditions. In either case, this phenomenon can have functional implications for the insect host, which can be both beneficial and deleterious. For instance, symbiont turnover can sometimes lead to functional redundancy (i.e., a situation when multiple species perform similar functions) and reduced host performance (i.e., if the new symbionts are less beneficial or even pathogenic). However , it can also result in functional innovation, wherein new symbionts can introduce novel traits or capabilities (Manthey et al. 2023). A classic example is of aphids harbouring *B. aphidicola* (an obligate symbiont) that provide essential amino acids. However, aphids also harbour additional symbionts (e.g., *S. symbiotica* that confers thermal tolerance under heat stress, *Hamiltonella defensa* that provides protection against parasitoid wasps, and *Regiella insecticola* that defends against fungal pathogens) that undergoes shifts in response to environmental conditions. This context-dependent symbiont turnover allows aphid populations to adapt rapidly to changing conditions, without requiring for the immediate changes in the host genome (Tougeron et al. 2023; Oliver and Higashi 2019).

Further, microbiome dynamics and its stability within insects also depends on the mode of their transmission – whether symbionts are passed from parent to offspring (vertical transmission) or acquired afresh from the environment (horizontal transmission). Vertical transmission is often observed for species that show high fidelity and inter-dependence on their insect host (e.g., *Buchnera* in aphids, *Wigglesworthia* in tsetse flies). In fact, they are so tightly linked to their insect host that they possess a reduced genome (Nikoh et al. 2011), often as a result of relaxed selection on redundant genes. And generally, due to narrow genetic diversity, these species are less flexible in terms of their adaptability under fluctuating environments (references). However, vertically transmitted symbioses are highly stable and crucial to insect survival. For instance, cicadas carry *Hodgkinia* and *Sulcia* which have a highly reduced genome size (compared to their free-living counterparts), and together these bacterial species supply essential amino acids essential for insect survival (McCutcheon et al. 2009).

In contrast, horizontally transmitted symbionts offer greater ecological flexibility, allowing hosts to acquire symbionts already adapted to local environments. In addition, acquisition of new symbionts can result in novel interactions and initiation of new mutualistic or parasitic relationships. For example, the bean bug (*Riptortus pedestris*) acquires *Burkholderia* from soil each generation. Here, the insect actively selects the most beneficial strain from environmental pools, conferring pesticide resistance (Kikuchi and Yumoto 2013). This environmental acquisition strategy allows adaptation to locally prevalent conditions, such as pesticide-laden soils or altered climates, demonstrating the ecological flexibility of horizontal transmission.

Furthermore, insect and these symbiotic microbes often co-evolve under the influence of selective pressures that shape their evolutionary trajectories. In the case of obligate symbiosis, as observed for aphids, *Buchnera* has co-evolved for ~150 million years, and during the process, has lost many genes, while relying on the host for its metabolic functions. Likewise, in cicadas, *Hodgkinia's* genome has fragmented into multiple lineages, each retaining part of the original genome, reflecting extreme co-evolutionary specialization (Gupta and Nair 2020). And it's not just the symbiont but in many cases the symbionts too affect the evolution of their insect hosts. For instance, *Wolbachia* – a widespread symbiont in arthropods is known to manipulate host reproduction, e.g., induce cytoplasmic incompatibility, parthenogenesis (Turelli et al. 2022). In many cases, host genomes have evolved complementary adaptations, such as specialized tissues (e.g., bacteriocytes) or regulatory mechanisms to ensure symbiont sustenance and activity (Mao et al. 2018; Alarcón et al. 2022; Luan 2024). In fact, long-term co-evolution is believed to be a major player in governing major evolutionary transitions, such as the evolution of herbivory in insects via microbial digestion of plant polymers (Zhang et al. 2023; García-Lozano and Salem 2025) and expanded thermal or nutritional niches in insects (Lemoine et al. 2020; Gupta and Nair 2020). Such innovations underscore the creative power of co-evolution (of the insect and the microbiome) in generating new biological capabilities in insects.

Moreover, the nature of host-microbe interaction can range along a mutualism–parasitism continuum, influenced by genetic, ecological factors, environmental conditions, and evolutionary history. And this can be context-dependent; i.e., mutualists can also become parasitic if environmental conditions shift (e.g., nutrient availability/scarcity). Although, in most cases, the insect and its microbiome act in synergy to ensure their survival, at times, they may also experience evolutionary conflict at times. This happens especially if the reproductive interests of the host and symbiont diverge, as in the case of 'selfish' symbionts like *Wolbachia* (discussed above). Therefore, it is plausible to state that microbiome shifts under stress can either be detrimental or adaptive. While in some cases, microbial shifts under stress can help insects detoxify pollutants, digest novel food sources, or resist pathogens, in others, stress-induced dysbiosis could lead to reduced reproductive success, increased susceptibility to disease, or altered host development (Darwish et al. 2025). Thus, the microbiome can act both as a buffer and a bottleneck for insect resilience and adaptation under changing environmental conditions.

Understanding drivers, mechanisms, and consequences of symbiont turnover, transmission modes, and co-evolutionary dynamics is crucial to comprehend insect–microbiome relationships in a rapidly changing world fully. As environmental perturbations continue to disrupt ecosystems globally, the capacity of organisms to manage their symbiotic relationships may determine not only their survival, but the evolutionary trajectories of entire lineages. Hence, we believe future research focusing on developing predictive models for symbiotic shifts under climate change, exploring the potential for symbiont engineering or microbiome transplantation in conservation and agriculture, and deciphering the genomic signatures of co-evolution and functional integration is critical for effective pest management.

3.6 EXPLORING INSECT–MICROBIOME DYNAMICS FOR INSECT MANAGEMENT AND CONSERVATION

The microbiome is an important, yet often overlooked, component of insect ecology and resilience. Information on how microbial communities respond to environmental stress is a requisite to understanding insect adaptations and population dynamics fully. As climate change accelerates, it is essential to integrate information on microbiome dynamics for insect research and management to devise intelligent strategies towards pest control, conservation biology, and environmental monitoring. This is because studies have shown that microbes can be engineered/targeted to disrupt essential physiological functions or interfere with the reproduction of notorious pests. Symbiont-based strategies, such as the introduction of *Wolbachia*, have been successfully used to reduce vector-borne disease transmission and limit the population size of mosquitoes (Minwuyelet et al. 2023). Likewise, the microbiome can be exploited to enhance the efficacy of biocontrol agents. For example, entomopathogenic fungi or bacteria may be more effective when deployed under conditions that support favourable insect–microbe interactions or suppress defensive symbionts in pest species. Further, monitoring shifts in insect microbiomes can be used to predict environmental stress. Microbiome profiling can indicate the presence of pollutants, toxins, and other contaminants contributing to ecosystem degradation, thereby acting as a valuable tool for ecological monitoring. Furthermore, the microbiome can also be exploited to restore or maintain the population of beneficial insects such as bees and butterflies. In addition, with the observed influence of the microbiome on various aspects of insect physiology and development, it would be logical to consider its manipulation as a viable strategy for enhanced stress tolerance in pollinators and endangered insect species and ultimately ensure their survival in degraded habitats. Besides, the microbiome can be engineered to produce stress-resistant insect strains for controlled release in an agricultural setting. Hence, studying insect–microbiome dynamics under stress can have profound implications for ecology, agriculture, public health, and conservation. Therefore, by unravelling how stress alters these intricate relationships, we can better predict insect responses, mitigate the impacts of environmental change, and design innovative, microbiome-centric strategies to protect both beneficial and threatened insect species.

3.7 FUTURE DIRECTIONS AND CHALLENGES

Although studies into insect–microbiome interactions have opened several promising avenues in insect research, they suffer from several challenges and limitations. Understanding the multifaceted insect–microbe relationship not only requires novel methodologies but also a forward-thinking approach to responsibly harness microbial potential in supporting insect resilience and survival. A major frontier in insect microbiome research lies in integrating multi-omics approaches (including genomics, transcriptomics, proteomics, and metabolomics) to better understand how microbes mediate rapid insect adaptations to stress and govern host physiology, function, and survival under environmental pressures. However, as stated earlier, significant challenges remain, especially about data standardization and analysis pipelines, and computational modelling capable of handling the vast and complex datasets generated. Nevertheless, new technologies such as spatial metagenomics and single-cell sequencing allow precise mapping of microbe–host interactions at the cellular level. And these innovations will uncover previously hidden layers of symbiotic complexity and further our understanding of the intricacies of insect–microbiome interactions.

Besides, insects in natural and agricultural systems are often subjected to multiple, concurrent stressors, such as temperature fluctuations, pesticide exposure, and nutrient scarcity. Therefore, understanding how insect microbiomes respond to these multifactorial stresses simultaneously is critical to predicting their resilience and adaptability. Further, it is essential to state that as most current studies are primarily limited to single-stressor models, future research must move towards experimental designs that reflect real-world complexity, including interactive and synergistic stress effects. Furthermore, as discussed in previous sections, microbiome engineering can be a powerful and promising way to enhance insect tolerance to pathogens, pollutants, or climate extremes or modulate insect behaviour for biocontrol purposes. For instance, genetically modified gut bacteria could be programmed to express antifungal compounds or biosensors (Mondal et al. 2023). However, studies related to these areas are still in their infancy and require a deeper understanding of microbe–host compatibility, gene stability, and environmental survivability to be commercially exploited in the immediate future. Moreover, despite being a potential pest control or conservation strategy, it raises ethical and ecological concerns (Lange et al. 2022; Wang et al. 2025). Also, before introducing engineered or foreign microbial strains into natural insect populations, thorough analyses have to be made, as they can have unforeseen impacts on non-target species, and ecosystem stability. Therefore, rigorous risk assessment rules and suitable regulatory guidelines that evaluate the benefits of microbiome interventions against potential bioethical dilemmas must be drafted. Besides, public engagement and transparent communication are also necessary to build societal trust in microbiome-based technologies.

Although we have made remarkable progress in deciphering several complex insect–microbiome interactions, transgenerational transmission of microbiomes, and their evolutionary implications, many fundamental questions about microbiomes' role in insect speciation and adaptation still remain unanswered. For instance, the exact molecular mechanisms underlying microbiome-mediated stress tolerance, in many insect species, are unclear and the information is still restricted mainly to standardized model systems. Further, the evolutionary dynamics of many insect–microbiome associations across several species are yet to be unravelled. This is necessary for performing cross-species comparisons to identify universal versus host-specific microbial traits. Altogether, we believe that insect-microbiome research stands at a transformative juncture today. And by leveraging interdisciplinary tools and embracing ethical innovation, we can unlock new strategies for enhancing insect resilience and ecosystem stability in a rapidly changing world.

3.8 CONCLUSION

The intricate relationships between insects and their microbiomes are pivotal in shaping insect resilience, adaptability, and survival under environmental change. Studies have indicated that far from being passive passengers, microbial partners actively influence host physiology, development, immunity, and stress responses – highlighting their integral role in insect plasticity and ecological success of both partners. As stressors, generated as a consequence of climate change, habitat loss, and anthropogenic chemicals, intensify, understanding the microbiome's contributions becomes not just scientifically compelling but ecologically urgent. This chapter underscores the growing body of evidence that indicates insect–microbe associations are dynamic, context-dependent, and evolutionarily significant. Emerging technologies and interdisciplinary approaches allow us to probe these associations with unprecedented resolution and speed. Therefore, to harness the promise of these studies – whether for conservation, pest management, or climate resilience – we must immediately address various associated challenges currently faced, from unravelling complex multi-stressor interactions to navigating the ethical terrain of microbial manipulation, before their ultimate introduction into the environment. Further, while it is evident that this newfound information on microbial symbioses offers promise in enhancing insect health and ecosystem stability, we need to find avenues to leverage it to our advantage quickly. As we look ahead, a collaborative, cautious, and curiosity-driven approach will be key to unlocking the full potential of these microscopic allies in supporting insect life on a changing planet.

ACKNOWLEDGMENTS

AG thanks the Department of Science and Technology (DST), Government of India, for the INSPIRE Faculty Fellowship. AG and RB acknowledge the support provided by the ICAR and ICAR-NIPB. SN thanks ICGEB for core grants and the award of the Arturo Falaschi Emeritus Scientist position (2021–24).

REFERENCES

Alarcón ME, Polo PG, Akyüz SN, Rafiqi AM (2022) Evolution and ontogeny of bacteriocytes in insects. *Front Physiol* 13:1034066.

Archer CR, Dobelmann J, Doublet V, Lindeboom S, Linke L, Wilfert L (2025) Is antibiotic (mis)use an emerging ecological threat to wild insects? *J Insect Physiol* 164:104837.

Ashra H, Nair S (2022) Trait plasticity during plant-insect interactions: from molecular mechanisms to impact on community dynamics. *Plant Sci* 317:111188.

Asimakis E, Stathopoulou P, Sapounas A, Khaeso K, Batargias C, Khan M, Tsiamis G (2021) New insights on the *Zeugodacus cucurbitae* (Coquillett) bacteriome. *Microorganisms* 9:659.

Auger L, Deschamps MH, Vandenberg G, Derome N (2023) Microbiota is structured by gut regions, life stage, and diet in the Black Soldier Fly (*Hermetia illucens*). *Front Microbiol* 14:1221728.

Branda F, Cella E, Scarpa F, Slavov SN, Bevivino A, Moretti R, Degafu AL, Pecchia L, Rizzo A, Defilippo F, Moreno A, Ceccarelli G, Alcantara LCJ, Ferreira A, Ciccozzi M, Giovanetti M. (2024) *Wolbachia*-based approaches to controlling mosquito-borne viral threats: innovations, AI integration, and future directions in the context of climate change. *Viruses* 16:1868.

Chabanol E, Gendrin M (2024) Insects and microbes: best friends from the nursery. *Curr Opin Insect Sci* 66:101270.

Charaabi K, Hamdene H, Djobbi W, Fadhel S, Tanfouri N, Saidi M, Meriem, Guerfali M (2025) Assessing gut microbiota diversity and functional potential in resistant and susceptible strains of the mediterranean fruit fly. *Sci Rep* 15:33456.

Chen L-J, Li Z-Z, Liu W, Lyu B (2024) Impact of high temperature and drought stress on the microbial community in wolf spiders. *Ecotoxicol Environ Saf* 283:116801.

Chen Y, Chen Y, Zhang Y, Sun Z, Li Y, Ding J, Zhang G, Du E, Zi X, Tian C, Zhao W, Gui F (2024) Role of *Enterococcus mundtii* in gut of the tomato leaf miner (*Tuta absoluta*) to detoxification of Chlorantraniliprole. *Pestic Biochem Physiol* 204:106060.

Chen Y, Zhou H, Lai Y, Chen Q, Yu XQ, Wang X (2021) Gut microbiota dysbiosis influences metabolic homeostasis in *Spodoptera frugiperda*. *Front Microbiol* 12:727434.

Chu B, Ge S, He W, Sun X, Ma J, Yang X, Lv C, Xu P, Zhao X, Wu K (2025) Gut symbiotic bacteria enhance reproduction in *Spodoptera frugiperda* by regulating juvenile hormone III and 20-hydroxyecdysone pathways. *Microbiome* 13:1–17.

Coon KL, Brown MR, Strand MR (2016) Mosquitoes host communities of bacteria that are essential for development but vary greatly between local habitats. *Mol Ecol* 25:5806.

Corbin C, Heyworth ER, Ferrari J, Hurst GDD (2017) Heritable symbionts in a world of varying temperature. *Heredity (Edinb)* 118:10–20.

Darwish H Al, Tariq M, Salama S, Hart T, Sun JS, Al Darwish H, Tariq M, Salama S, Hart T, Sun JS (2025) The microbiome as a driver of insect physiology, behavior, and control strategies. *Appl Microbiol* 5:90.

Douglas AE (2015) Multiorganismal insects: diversity and function of resident microorganisms. *Annu Rev Entomol* 60:17–34.

Engel P, Moran NA (2013) The gut microbiota of insects – diversity in structure and function. *FEMS Microbiol Rev* 37:699–735.

Fan Q, Sun H, Liang P (2025) The influence of microorganism on insect-related pesticide resistance. *Agriculture* 15:1519.

Feliciello I, Parazajder J, Akrap I, Ugarković D (2013) First evidence of DNA methylation in insect *Tribolium castaneum*. *Epigenetics* 8:534–541.

García-Lozano M, Salem H (2025) Microbial bases of herbivory in beetles. *Trends Microbiol* 33:151–163.

González J, Karasov TL, Messer PW, Petrov DA (2010) Genome-wide patterns of adaptation to temperate environments associated with transposable elements in Drosophila. *PLoS Genet* 6:e1000905.

Gu J, Yao Z, Lemaitre B, Cai Z, Zhang H, Li X (2024) Intestinal commensal bacteria promote *Bactrocera dorsalis* larval development through the vitamin B6 synthesis pathway. *Microbiome* 12:227.

Guilhot R, Xuéreb A, Lagmairi A, Olazcuaga L, Fellous S (2023) Microbiota acquisition and transmission in *Drosophila* flies. *iScience* 26:107656.

Gupta A, Nair S (2020) Dynamics of insect–microbiome interaction influence host and microbial symbiont. *Front Microbiol* 11:1357.

Gupta A, Nair S (2021) Methylation patterns of Tf2 retrotransposons linked to rapid adaptive stress response in the brown planthopper (*Nilaparvata lugens*). *Genomics* 113:4214–4226.

Gupta A, Nair S (2022) Heritable epigenomic modifications influence stress resilience and rapid adaptations in the brown planthopper (*Nilaparvata lugens*). *Int J Mol Sci* 23:8728.

Gupta A, Nair S (2025) Epigenetic processes in insect adaptation to environmental stress. *Curr Opin Insect Sci* 67:101294.

Hafsi A, Moquet L, Hendrycks W, De Meyer M, Virgilio M, Delatte H (2024) Evidence for a gut microbial community conferring adaptability to diet quality and temperature stressors in phytophagous insects: the melon fruit fly *Zeugodacus cucurbitae* (Diptera: Tephritidae) as a case study. *BMC Microbiol* 24:514.

Henry LP, Bruijning M, Forsberg SKG, Ayroles JF (2021) The microbiome extends host evolutionary potential. *Nat Commun* 12:1–13.

Hof van't AE, Campagne P, Rigden DJ, Yung CJ, Lingley J, Quail MA, Hall N, Darby AC, Saccheri IJ (2016) The industrial melanism mutation in British peppered moths is a transposable element. *Nature* 534:102–105.

Holt JR, Cavichiolli de Oliveira N, Medina RF, Malacrinò A, Lindsey ARI (2024) Insect–microbe interactions and their influence on organisms and ecosystems. *Ecol Evol* 14:e11699.

Horgan FG, Srinivasan TS, Crisol-Martínez E, Almazan MLP, Ramal AF, Oliva R, Quibod IL, Bernal CC (2019) Microbiome responses during virulence adaptation by a phloem-feeding insect to resistant near-isogenic rice lines. *Ecol Evol* 9:11911–11929.

Iltis C, Tougeron K, Hance T, Louâpre P, Foray V (2022) A perspective on insect–microbe holobionts facing thermal fluctuations in a climate-change context. *Environ Microbiol* 24:18–29.

Jaffar S, Ahmad S, Lu Y (2022) Contribution of insect gut microbiota and their associated enzymes in insect physiology and biodegradation of pesticides. *Front Microbiol* 13:979383.

Janson EM, Stireman JO, Singer MS, Abbot P (2008) Phytophagous insect-microbe mutualisms and adaptive evolutionary diversification. *Evolution (N Y)* 62:997–1012.

Jaramillo A, Castañeda LE (2021) Gut microbiota of *Drosophila subobscura* contributes to its heat tolerance and is sensitive to transient thermal stress. *Front Microbiol* 12:654108.

Jing TZ, Qi FH, Wang ZY (2020) Most dominant roles of insect gut bacteria: digestion, detoxification, or essential nutrient provision? *Microbiome* 8:1–20.

Juefeng Z, Fang L, Haiying Z, Liwei L, Jianming C (2024) Integrated microbiome and metabolomic analysis of *Spodoptera litura* under *Metarhizium flavoviride* qc1401 stress. *Int Microbiol* 28:721.

Kazazian HH (2004) Mobile elements: drivers of genome evolution. *Science* 303:1626–1632.

Khan SA, Kojour MAM, Han YS (2023) Recent trends in insect gut immunity. *Front Immunol* 14:1272143.

Kikuchi Y, Yumoto I (2013) Efficient colonization of the bean bug *Riptortus pedestris* by an environmentally transmitted *Burkholderia* symbiont. *Appl Environ Microbiol* 79:2088–2091.

Kim D, Thairu MW, Hansen AK (2016) Novel insights into insect-microbe interactions—role of epigenomics and small RNAs. *Front Plant Sci* 7:206558.

Kline O, Joshi NK (2024) Microbial symbiont-based detoxification of different phytotoxins and synthetic toxic chemicals in insect pests and pollinators. *J Xenobiotics* 14:753–771.

Kolodny O, Schulenburg H (2020) Microbiome-mediated plasticity directs host evolution along several distinct time scales. *Philos Trans R Soc B Biol Sci* 375:20190589.

Lai Y, Wang S (2025) Epigenetic regulation in insect-microbe interactions. *Annu Rev Entomol* 70:293–311.

Lange C, Boyer S, Bezemer TM, Lefort MC, Dhami MK, Biggs E, Groenteman R, Fowler SV, Paynter Q, Verdecia Mogena AM, Kaltenpoth M. (2023) Impact of intraspecific variation in insect microbiomes on host phenotype and evolution. *ISME J* 17:1798.

Lange L, Berg G, Cernava T, Champomier-Vergès MC, Charles T, Cocolin L, Cotter P, D'Hondt K, Kostic T, Maguin E, Makhalanyane T, Meisner A, Ryan M, Kiran GS, de Souza RS, Sanz Y, Schloter M, Smidt H, Wakelin S, Sessitsch A. (2022) Microbiome ethics, guiding principles for microbiome research, use and knowledge management. *Environ Microbiome* 17:1–8.

Lemoine MM, Engl T, Kaltenpoth M (2020) Microbial symbionts expanding or constraining abiotic niche space in insects. *Curr Opin Insect Sci* 39:14–20.

Luan JB (2024) Insect bacteriocytes: adaptation, development, and evolution. *Annu Rev Entomol* 69:81–98.

Manthey C, Johnston PR, Nakagawa S, Rolff J (2023) Complete metamorphosis and microbiota turnover in insects. *Mol Ecol* 32:6543–6551.

Mao M, Yang X, Bennett GM (2018) Evolution of host support for two ancient bacterial symbionts with differentially degraded genomes in a leafhopper host. *Proc Natl Acad Sci USA* 115:E11691–E11700.

Mateo L, Ullastres A, González J (2014) A transposable element insertion confers xenobiotic resistance in *Drosophila*. *PLoS Genet* 10:e1004560.

McCutcheon JP, McDonald BR, Moran NA (2009) Convergent evolution of metabolic roles in bacterial co-symbionts of insects. *Proc Natl Acad Sci USA* 106:15394.

Minwuyelet A, Petronio GP, Yewhalaw D, Sciarretta A, Magnifico I, Nicolosi D, Di Marco R, Atenafu G (2023) Symbiotic *Wolbachia* in mosquitoes and its role in reducing the transmission of mosquito-borne diseases: updates and prospects. *Front Microbiol* 14:1267832.

Mockler BK, Kwong WK, Moran NA, Koch H (2018) Microbiome structure influences infection by the parasite *Crithidia bombi* in bumble bees. *Appl Environ Microbiol* 84:e02335–17.

Moghadam NN, Thorshauge PM, Kristensen TN, de Jonge N, Bahrndorff S, Kjeldal H, Nielsen JL (2018) Strong responses of *Drosophila melanogaster* microbiota to developmental temperature. *Fly* 12:1–12.

Mogilicherla K, Roy A (2023) Epigenetic regulations as drivers of insecticide resistance and resilience to climate change in arthropod pests. *Front Genet* 13:1044980.

Moldovan OT, Carrell AA, Bulzu PA, et al (2023) The gut microbiome mediates adaptation to scarce food in Coleoptera. *Environ Microbiome* 18:80.

Mondal S, Somani J, Roy S, Babu A, Pandey AK (2023) Insect microbial symbionts: ecology, interactions, and biological significance. *Microorganisms* 11:2665.

Moran NA (2021) Microbe profile: *Buchnera aphidicola*: ancient aphid accomplice and endosymbiont exemplar. *Microbiology* 167:001127.

Moran NA, McCutcheon JP, Nakabachi A (2008) Genomics and evolution of heritable bacterial symbionts. *Annu Rev Genet* 42:165–190.

Nguyen J, Lara-Gutiérrez J, Stocker R (2021) Environmental fluctuations and their effects on microbial communities, populations and individuals. *FEMS Microbiol Rev* 45:fuaa068.

Nikoh N, Hosokawa T, Oshima K, Hattori M, Fukatsu T (2011) Reductive evolution of bacterial genome in insect gut environment. *Genome Biol Evol* 3:702.

Oliver KM, Higashi CH (2019) Variations on a protective theme: *Hamiltonella defensa* infections in aphids variably impact parasitoid success. *Curr Opin Insect Sci* 32:1–7.

Ramniwas S, Tyagi PK, Sharma A, Kumar G (2023) Editorial: abiotic stress and physiological adaptive strategies of insects. *Front Physiol* 14:1210052.

Ren L, Zhang X, Yang F, Jocelin NF, Shang Y, Wang Q, Liu Z, Guo Y (2023) Effects of heat tolerance on the gut microbiota of *Sarcophaga peregrina* (Diptera: Sarcophagidae) and impacts on the life history traits. *Parasit Vectors* 16:364.

Rupawate PS, Roylawar P, Khandagale K, Gawande S, Ade AB, Jaiswal DK, Borgave S (2023) Role of gut symbionts of insect pests: a novel target for insect-pest control. *Front Microbiol* 14:1146390.

Russell CW, Poliakov A, Haribal M, Jander G, van Wijk KJ, Douglas AE (2014) Matching the supply of bacterial nutrients to the nutritional demand of the animal host. *Proc R Soc B Biol Sci* 281:20141163.

Santos-Garcia D, Mestre-Rincon N, Zchori-Fein E, Morin S (2020) Inside out: microbiota dynamics during host-plant adaptation of whiteflies. *ISME J* 14:847–856.

Sharma S, Kooner R, Arora R. (2017) Insect pests and crop losses. In Breeding Insect Resistant Crops for Sustainable Agriculture, ed. R Arora and S Sandhu, (pp 45-66). Springer, Singapore.

Siddiqui JA, Khan MM, Bamisile BS, Hafeez M, Qasim M, Rasheed MT, Rasheed MA, Ahmad S, Shahid MI, Xu Y (2022) Role of insect gut microbiota in pesticide degradation: a review. *Front Microbiol* 13:870462.

Singh S, Singh A, Baweja V, Roy A, Chakraborty A, Singh IK (2021) Molecular rationale of insect-microbes symbiosis—from insect behaviour to mechanism. *Microorganisms* 9:2422.

Thia JA, Dorai APS, Hoffmann AA (2025) Symbiotic bacteria and pest control: plant toxins, chemical pesticides, and fungal entomopathogens. *Trends Microbiol* 33:976–988.

Tougeron K, Iltis C, Rampnoux E, Goerlinger A, Dhondt L, Hance T (2023) Still standing: the heat protection delivered by a facultative symbiont to its aphid host is resilient to repeated thermal stress. *Curr Res Insect Sci* 3:100061.

Turelli M, Katznelson A, Ginsberg PS (2022) Why *Wolbachia*-induced cytoplasmic incompatibility is so common. *Proc Natl Acad Sci USA* 119:e2211637119.

Villa SM, Chen JZ, Kwong Z, Acosta A, Vega NM, Gerardo NM (2023) Specialized acquisition behaviors maintain reliable environmental transmission in an insect-microbial mutualism. *Curr Biol* 33:2830–2838.e4.

Wang Y, Tang J, Chen Y, Chen S, Chen S, Yu X, Wan C, Xiang G, Chen Y, Li Q (2025) The ecological–evolutionary game of the insect gut microbiome: environmental drivers, host regulation, and prospects for cross-cutting applications. *Vet Sci* 12:866.

Watt R, Parkin K, Martino D (2020) The potential effects of short-chain fatty acids on the epigenetic regulation of innate immune memory. *Challenges* 11:25.

Weiss BL, Rio RVM, Aksoy S (2022) Microbe profile: *Wigglesworthia glossinidia*: the tsetse fly's significant other. *Microbiology* 168:001242.

Werren JH, Baldo L, Clark ME (2008) *Wolbachia*: master manipulators of invertebrate biology. *Nat Rev Microbiol* 6:741–751.

Wielkopolan B, Obrępalska-Stęplowska A (2016) Three-way interaction among plants, bacteria, and Coleopteran insects. *Planta* 244:313–332.

Yang F, Ma Q, Zhang X, Shang Y, Ngando FJ, Ren L, Cai J (2024) The gut bacterial composition across life stages of *Sarcophaga peregrina* (Diptera: Sarcophagidae) and the effects of amikacin on their development. *J Med Entomol* 61:1093–1104.

Yasika Y, Shivakumar MS (2025) A comprehensive account of functional role of insect gut microbiome in insect orders. *J Nat Pestic Res* 11:100110.

Zhang DW, Xiao ZJ, Zeng BP, Li K, Tang YL (2019) Insect behavior and physiological adaptation mechanisms under starvation stress. *Front Physiol* 10:163.

Zhang G, Hussain M, O'Neill SL, Asgari S (2013) *Wolbachia* uses a host microRNA to regulate transcripts of a methyltransferase, contributing to dengue virus inhibition in *Aedes aegypti*. *Proc Natl Acad Sci USA* 110:10276–10281.

Zhang N, Qian Z, He J, Shen X, Lei X, Sun C, Fan J, Felton GW, Shao Y (2024) Gut bacteria of Lepidopteran herbivores facilitate digestion of plant toxins. *Proc Natl Acad Sci* 121:e2412165121.

Zhang X, Wang X, Guo Z, Liu X, Wang P, Yuan X, Li Y (2022) Antibiotic treatment reduced the gut microbiota diversity, prolonged the larval development period and lessened adult fecundity of *Grapholita molesta* (Lepidoptera: Tortricidae). *Insects* 13:838.

Zhang Y, Zhang S, Xu L (2023) The pivotal roles of gut microbiota in insect plant interactions for sustainable pest management. *NPJ Biofilms Microbiomes* 9:66.

Zhao P, Rensing C, Wang D (2023) Symbiotic bacteria modulate *Lymantria dispar* immunity by altering community proportions after infection with LdMNPV. *Int J Mol Sci* 24:9694.

Zheng H, Steele MI, Leonard SP, Motta EVS, Moran NA (2018) Honey bees as models for gut microbiota research. *Lab Anim (NY)* 47:317.

CHAPTER FOUR

The Gut Microbiome and Insect Sociality

Yehuda Ben-Shahar

4.1 INTRODUCTION

The study of host–microbe interactions in insects traces its origins to the early nineteenth century, exemplified by the work of Agostino Bassi in 1835, who demonstrated that muscardine disease of silkworms was caused by a transmissible fungal pathogen, providing the first experimental evidence that microorganisms could directly influence animal biology (Bassi 1835). This discovery established insects as tractable systems for studying biological interactions between hosts and microbes and predated similar demonstrations in vertebrates. Subsequent work in the latter half of the nineteenth century, most notably by Louis Pasteur, further embedded insects within the developing germ theory of disease through studies of silkworm pathologies (Pasteur 1926). Although primarily focused on pathogenic relationships, these fundamental studies laid the conceptual and methodological groundwork for recognizing that insects routinely harbor diverse microbial communities. By the early twentieth century, research began to shift from disease-centric views toward the recognition of stable, beneficial, and often obligate host–microbial associations, particularly in the gut (Cleveland 1923; Keeble 1912; Nuttall 1923).

Early studies of insect–bacteria mutualism focused primarily on obligate nutritional symbioses, particularly in herbivorous and detritivorous insects whose diets are otherwise insufficient to support host metabolism (Akman Gündüz and Douglas 2009; Gullan and Kosztarab 1997; Engel and Moran 2013). Studies of termites, cockroaches, aphids, and other insects established that microbial partners can profoundly impact host fitness by extending host ecological niches by enabling cellulose degradation, nitrogen recycling, or essential amino acid provisioning (Dietrich, Köhler, and Brune 2014). These studies primarily framed microbes as indirect physiological enablers of host fitness.

However, research on reproductive parasites and inherited symbionts further expanded the potential conceptual depth of insect–microbe interactions. For example, studies have indicated that some microbes can manipulate host reproduction by altering sex ratios, inducing incompatibilities, or biasing mating outcomes (Kageyama, Hoshizaki, and Ishikawa 1998; Majerus 2003; Hurst and Jiggins 2000; Giordano, Jackson, and Robertson 1997; Breeuwer and Werren 1990; Turelli 1994). These findings have signified that microbial–host interactions potentially extend far beyond the indirect support of host nutrition. Instead, they uncovered intricate interdependencies across hosts and their associated microbiomes that are driven by shared gene networks with deep evolutionary roots that mutually affect population structures and evolutionary trajectories of bacteria and their animal hosts. Although not without some emerging controversies (Douglas and Werren 2016), these advancements set the stage for the recent rise of the modern holobiont perspective, which views host behavior, physiology, and social traits as systems-levels emergent properties of integrated host–microbe systems (Rosenberg and

DOI: 10.1201/9781003639176-4

Zilber-Rosenberg 2016; Guerrero, Margulis, and Berlanga 2013; Koide 2023; Richardson 2017; Bordenstein and Network 2024). Here we provide an extended view of this perspective by discussing the emergent role of host–microbiome interactions in insect sociality and related behaviors.

4.2 THE CONTRIBUTIONS OF MICROBIOMES TO SPECIES IDENTITY, GROUP MEMBERSHIP, AND KIN RECOGNITION

Insects rely primarily on chemical communication to recognize conspecifics, sex identity, and kin recognition and to determine group or colony membership. These recognition cues are mostly mediated by cuticular hydrocarbons (CHCs), a complex blend of long-chain hydrocarbons coating the insect cuticle, which act as pheromones that encode information about species identity, genetic relatedness, developmental stage, reproductive status, and social role (Drijfhout, Kather, and Martin 2009; Blomquist 2003).

A growing body of evidence indicates that the gut and cuticle-associated microbiomes contribute to the production, modification, and stabilization of chemical recognition cues, thereby influencing species and group membership. As we describe below, commensal microbes can affect chemical recognition systems through several non-mutually exclusive mechanisms, which include direct biosynthesis of pheromone-like hydrocarbons, metabolic modification of host-produced hydrocarbons, or indirect impact on host lipid metabolism to alter CHC composition. As discussed in detail below, the close association between core microbiomes and their insect hosts, which often follow very similar bacterial–host phylogenetic trees, indicates that the association between insects and their microbiomes has played an important role in arthropod evolution, including the emergence of extreme forms of animal sociality exhibited by eusocial species.

4.3 INSECT SOCIALITY FROM THE HOLOBIONT PERSPECTIVE

The currently prevailing view is that insect behaviors are closely linked to their gut microbiomes through reciprocal feedback loops. Studies have shown that social interactions across conspecifics drive microbial exposure, transmission, and persistence within host populations, while microbial communities, especially those in the gut, can influence host physiology and behavior through chemical, metabolic, neurophysiological, and immune pathways (Lewis and Lizé 2015; Johnson and Foster 2018; Schmickl and Karsai 2018; Biedermann and Rohlfs 2017; Griffiths et al. 2026; Zhang, Chen, et al. 2024; Engel and Moran 2013). These reciprocal interactions seem to mutually benefit the fitness of hosts and associated microbiomes via networks whose complexity seems to be directly associated with the level of social complexity of the host species (Week et al. 2025; Mazel, Prasad, and Engel 2025; Maccaro and Whyte 2026; Griffiths et al. 2026).

4.4 MICROBIOMES CONTRIBUTE TO AGGREGATION AND GROUP BEHAVIORS

Aggregation and group behaviors represent some of the most basal forms of insect sociality. Because aggregation increases the host local population density and encounter rates, microbiome-dependent host aggregation has direct consequences for the fitness of the host and the microbiome by affecting mating, competition, pathogen transmission, and microbial dispersal (Ferreira and Moita 2019; Nehring and Steiger 2018; McKinney, Valdez, and Ben-Shahar 2021). Strong experimental evidence indicates that gut microbes play an important causal role in insect aggregation and group behaviors by directly producing or modifying the host chemical signals that function as aggregation pheromones via their action as attractants (chemicals that attract multiple conspecifics to a specific location) or arrestants (chemicals that increase the likelihood that an individual will stay in one specific location) (Engl and Kaltenpoth 2018; Guo et al. 2023; Zhang, Zhu, et al. 2024; Weiss 2006; Sarkar et al. 2020).

Specific examples of the direct and indirect roles of bacteria in the synthesis and/or release of aggregation pheromones include the following: **(1)** The secretion of short-chain fatty acids/volatile carboxylic acids and phenolic compounds, produced by gut bacteria, in feces recruits conspecifics in German cockroach (Wada-Katsumata et al. 2015); **(2)** The conversion of plant- or host-derived compounds into aggregation chemical signals, effectively outsourcing steps of pheromone biosynthesis to the microbiome. For example, in locusts, the plant metabolite vanillic acid is converted by gut bacteria in secreted feces into guaiacol/phenol, which are key components of aggregation pheromone in this animal group (Dillon, Vennard, and Charnley 2002; Lavy et al. 2020); **(3)** In some insects, gut bacteria are required for the release of aggregation pheromones synthesized by the host, which often include long-chain alkanes and methyl-branched compounds (Dong et al. 2025).

Remarkably, although not as well understood at the molecular and cellular levels, data indicate that bacteria can also affect host aggregation by directly modulating how the nervous system perceives and processes aggregation signals. Recent studies indicate that these effects are likely to be mediated by diverse mechanisms that drive neuronal plasticity via direct modulation of olfactory sensory neurons by bacterial metabolites (Fischer et al. 2017), the direct regulation of signaling by inhibitory and excitatory neurotransmitters, as well as via neuronal modulation by biogenic amines (Mangold and Hughes 2021; Belelli et al. 2025; Liberti and Engel 2020; Hu et al. 2025; Nagpal and Cryan 2021; Sherwin et al. 2019). The gut microbiome can also interact with the nervous system indirectly via immune-related modulation of the relevant neural circuits (Zhang et al. 2025; Mangold and Hughes 2021).

4.5 THE ROLE OF MICROBIOMES IN MATE CHOICE AND REPRODUCTIVE BEHAVIORS

Empirical research indicates that gut microbiomes can influence insect mate choice, which can directly impact host fitness (Lewis and Lize 2015; Griffiths et al. 2026; Sharon et al. 2010; Qiao et al. 2019; He et al. 2022). As in aggregation behavior, these effects are mostly associated with bacteria-driven regulation of chemical mating signals, particularly CHCs and volatile pheromones, which are essential for mating success (Lu et al. 2012; Lu et al. 2014; Vernier et al. 2023). Experimental manipulations of microbiomes have been shown to alter chemical profiles associated with attractiveness and, in some cases, lead to assortative mating (Wang et al. 2023).

Because of its genetic tractability, many of the studies of the role of microbiomes in assortative mating have focused on the vinegar fly, *Drosophila melanogaster*, and related species. Initially, studies reported that manipulating diet generated rapid assortative mating, that disrupting the microbiota (*e.g.*, antibiotics) abolished the preference, and that re-association with particular bacteria (notably *Lactobacillus*) could restore the phenotype, consistent with a microbiome-mediated shift in sexual cues such as CHCs (Sharon et al. 2010; Ringo, Sharon, and Segal 2011; Sharon et al. 2011). However, subsequent follow-up studies have indicated that the robustness of the effect of the microbiome on courtship and mate choice is context-dependent and can vary across experimental designs, genotypes, and husbandry conditions (Leftwich et al. 2017). Nevertheless, as studies of the role of microbiome in assortative mating have expanded to other insect species, the emerging consensus suggests that microbial communities can contribute to prezygotic isolation when microbiomes covary with diet, habitat, and populations (Shropshire and Bordenstein 2016).

In addition to sexual attraction, mating success of insects can be influenced by antagonistic social interactions such as competition and aggression (Bath, Morimoto, and Wigby 2018; Toyoshima and Matsuo 2023; Gaspar, Dias, and Vasconcelos 2022; Saltz 2013). Although most studies on the role of the microbiome on insect mate choice and mating success are focused on pheromonal communication, emerging evidence suggests that gut microbiomes can also modulate mating success via central neuronal regulation of aggression. A noted example of such an effect is the recent discovery that the gut microbiome can impact male aggression levels in *Drosophila* via the central modulation of octopaminergic signaling pathways (Jia et al. 2021).

Together, these findings suggest that microbial communities can contribute to prezygotic isolation, which can covary with diet, habitat, and population structure. Furthermore, the above studies indicate that these effects can be driven by either the direct microbial release of chemical cues (Engl and Kaltenpoth 2018; Ren et al. 2021), or the indirect impact of bacterial metabolism on the biochemical pathways underlying chemical signal synthesis by the host (Leftwich, Hutchings, and Chapman 2018; Leftwich et al. 2017). These mechanisms provide a plausible route by which ecological factors influencing microbiomes may translate into behavioral divergence and point to the intriguing hypothesis that microbiome–host interactions can impact insect speciation on the evolutionary timescale (Shropshire and Bordenstein 2016; Lange et al. 2023; Sharon et al. 2011).

4.6 MICROBIOMES AND EUSOCIALITY

Eusociality represents the most complex form of social organization in animals and is classically defined in species that form colonies and exhibit reproductive division of labor between queens and non-reproductive workers, cooperative brood care, and overlapping generations within a colony (Wilson 1971). In insects, eusociality has evolved independently multiple times, most prominently in the orders Hymenoptera (ants, bees, and wasps), Blattodea (cockroaches and termites), and to a more limited extent in certain thrips and aphids (Bourke 2011; Korb and Heinze 2016). These lineages span deep evolutionary time and diverse ecological niches yet converge on remarkably similar social architectures, suggesting strong selective pressures favoring coordinated group living. While genetic, developmental, and ecological mechanisms underlying eusociality have been extensively studied (Robinson, Grozinger, and Whitfield 2005; Ben-Shahar et al. 2002; Wilson and Holldobler 2005), growing evidence indicates that microbial symbionts represent an additional layer influencing the emergence, maintenance, and elaboration of eusocial traits.

4.6.1 The Gut Microbiome and Eusocial Complexity

Eusocial insects differ from solitary and gregarious species not only in behavior but also in the scale, stability, and robustness of their social interactions. Colonies create densely connected social networks characterized by frequent physical contact, trophallaxis, shared nesting environments, and long-term cohabitation (Wilson 1971). Theoretical and empirical studies suggest that these features fundamentally alter the ecological landscape experienced by microbes, directly impacting bacterial fitness. Consequently, bacterial–host coevolution facilitates reliable horizontal transmission while favoring specialization and host fidelity (Debray, Tung, and Archie 2024; Archie and Tung 2015).

As social complexity increases, insect microbiomes tend to shift from transient, environmentally derived communities toward conserved and host-specific assemblages. This pattern is particularly well documented in honeybees and bumblebees, which, relative to mammalian gut microbiomes, harbor a simpler, highly conserved core gut microbiome that is transmitted socially between nestmates (Neu, Allen, and Roy 2021; Moran 2015). Similar microbiome population architectures have been observed in ants (Sanders et al. 2014; Moreau 2020; Hu and Moreau 2025) and termites (Brune and Dietrich 2015), where gut symbionts form highly structured and evolutionarily stable communities tightly linked to host phylogeny and caste (Bourguignon et al. 2018).

4.6.2 Microbiome-Dependent Regulation of Caste, Division of Labor, and Collective Physiology in Colonies of Eusocial Insects

A defining feature of eusociality is reproductive and behavioral caste differentiation. While the molecular, genetic, and developmental mechanisms that regulate case determination and behavioral division of labor within colonies are variable across eusocial species (Okwaro and Korb 2023; Kocher and Kingwell 2024; Robinson 2009), microbial communities can modulate metabolic and endocrine pathways underlying these processes.

In termites, obligate gut symbionts are essential for lignocellulose digestion, nitrogen fixation, and nitrogen recycling, functions that enable the sterile worker and soldier castes to sustain the colony despite severe dietary limitations (Breznak 1982; Hongoh et al. 2008; Brune 2014). Similarly, in social bees, gut bacteria influence nutrient assimilation, insulin signaling, juvenile hormone synthesis, and immune functions, which are all directly linked to the regulation of age-related division of labor in honey bees (Wang et al. 2024; Sinotte et al. 2020; Vernier et al. 2024; Jones et al. 2018).

At the colony level, socially transmitted microbiomes may contribute to physiological homogenization among workers, reducing inter-individual variance in metabolic state or immune responsiveness. Such homogenization has been proposed to stabilize task allocation and response thresholds in large colonies, supporting efficient collective behavior (Wang et al. 2024; Baud et al. 2023; Vernier et al. 2024). In this view, the microbiome functions as a shared physiological substrate that complements neural and endocrine regulation of social organization, which is coined here as the "Shared Gut Hypothesis".

4.6.3 The Gut Microbiome and Social Immunity

Despite the disease risks imposed by dense social living, eusocial insects often exhibit remarkable resistance to pathogens. This paradox is partly explained by "social immunity", a suite of collective behaviors and physiological mechanisms that reduce disease transmission and its impact on colony-level fitness (Cremer, Armitage, and Schmid-Hempel 2007). Microbiomes are integral to the robustness of social immunity, and its long-term effects on colony health. While the mechanisms by which the shared gut microbiome increases colony-level health, data from studies in diverse social bees indicate that the core gut bacteria can competitively exclude pathogens, stimulate basal immune activity, and produce antimicrobial compounds (Koch and Schmid-Hempel 2011; Kwong, Mancenido, and Moran 2017).

In ants and social bees, microbial communities associated with the gut, cuticle, and nest environment contribute to colony-level defense, sometimes in combination with socially exchanged antibiotics and other antimicrobial secretions (Currie et al. 1999; Schultz 1999; Mattoso, Moreira, and Samuels 2012). These interactions suggest that host social behaviors and microbial symbionts co-construct a multilayered immune system, which operates at both the individual and colony scales.

4.6.4 Gut Microbiomes and the Evolution of Eusocial Traits

The interaction between insect sociality and the microbiome is unlikely to conform to a single level of selection. Instead, selection may favor traits that benefit hosts, microbes, or colonies under different ecological conditions. Chemical communication systems are especially likely to mediate eco-evolutionary feedback, as microbiome-driven shifts in signaling can alter social structure, mating patterns, and population connectivity.

Collectively, these findings raise the possibility that microbial symbionts have not merely adapted to eusocial hosts but may have facilitated their evolution. Stable, socially transmitted microbiomes could reduce the physiological costs of group living by improving nutrition, buffering environmental variability, and mitigating pathogen pressure (Engel and Moran 2013; Douglas 2022). During the early stages of social evolution, such benefits may have reinforced selection for prolonged and stable associations between solitary hosts, followed by cooperative brood care and reproductive specialization.

From a coevolutionary perspective, eusocial colonies represent exceptional ecological niches for microbes: long-lived, multi-host systems with reliable transmission routes and reduced dependence on environmental dispersal. These conditions may favor symbionts whose fitness is aligned with colony success rather than individual reproduction (Johnson and Foster 2018; Foster et al. 2017). Over evolutionary time, this alignment could select for microbes capable of subtly influencing host development, behavior, or social cohesion, further entrenching host–microbe interdependence.

In this framework, eusocial insects and their microbiomes can be viewed as deeply integrated evolutionary units, where increasing social complexity and microbial dependency reinforce one another. Understanding the origins and diversification of eusociality may therefore require extending classical genetic and behavioral models to incorporate holobiont-level analyses, in which microbial partners contribute to one of the most profound evolutionary transitions in animal life.

4.7 HOST SOCIALITY FROM THE BACTERIAL PERSPECTIVE

Thus far, most published studies about the roles gut microbiomes play in insect sociality are primarily framed in the context of host biology and fitness, and evolution. This perspective has been intellectually driven, at least in part, by the early finding that in *Drosophila*, the host association with specific bacterial clades is mostly transient and plastic (Adair et al. 2018; Wong, Chaston, and Douglas 2013; Obadia et al. 2017). However, as analyses of gut microbiomes are broadening to include host species across the insect phylogeny, it seems that the somewhat weak host–microbiome association in *Drosophila* is an exception rather than the rule. This is primarily supported by finding that many insect species harbor specific obligatory core bacterial taxa that often exhibit clear genomic signatures of host–bacterial coevolution (Groussin, Mazel, and Alm 2020; Shapira 2016; Maritan et al. 2024).

These exciting findings indicate that from a microbial perspective, insect sociality represents an opportunity for enhanced transmission, and consequently, the expansion of highly specialized ecological niches that provide unique fitness advantages to the underlying microbial taxa. Under this hypothetical framework, it is not surprising that microbes have evolved the power to promote aggregation, social tolerance, and mating success of their insect hosts. As a result, by manipulating their hosts' contact rates or reproductive outputs, gut bacteria increase their own fitness. In this sense, social interaction of the host can be viewed as transmission networks that microbes may evolve to exploit.

Conversely, hosts may evolve social behaviors that stabilize beneficial microbial communities. Social transmission, grooming, and collective defense can reduce pathogen load while maintaining core symbionts. These behaviors effectively externalize aspects of immune function and microbial regulation to the social level. In addition, our recent findings that gut bacteria play an important role in defining the chemical signatures of hive membership in honeybees, independently of host genetics, indicate that at least some of the key aspects of eusociality have been exclusively selected for in the bacterial components of the social hologenome (Vernier et al. 2020). This discovery signifies the direct role of the evolution of insect sociality in bacterial fitness and points to the provocative hypothesis that the selective drive for the emergence of eusociality, at least in social bees, may have been initially driven by bacterial rather than host fitness.

REFERENCES

Adair, Karen L, Marita Wilson, Alyssa Bost, and Angela E Douglas. 2018. Microbial community assembly in wild populations of the fruit fly Drosophila melanogaster. *The ISME Journal* 12 (4):959–972.

Akman Gündüz, E, and AE Douglas. 2009. Symbiotic bacteria enable insect to use a nutritionally inadequate diet. *Proceedings of the Royal Society B: Biological Sciences* 276 (1658):987–991.

Archie, Elizabeth A, and Jenny Tung. 2015. Social behavior and the microbiome. *Current Opinion in Behavioral Sciences* 6:28–34.

Bassi, Agostino. 1835. *Del mal del segno calcinaccio o moscardino*: Instituto Geografico de Agostini.

Bath, Eleanor, Juliano Morimoto, and Stuart Wigby. 2018. The developmental environment modulates mating-induced aggression and fighting success in adult female Drosophila. *Functional Ecology* 32 (11):2542–2552.

Baud, Gilles LC, Aiswarya Prasad, Kirsten M Ellegaard, and Philipp Engel. 2023. Turnover of strain-level diversity modulates functional traits in the honeybee gut microbiome between nurses and foragers. *Genome Biology* 24 (1):283.

Belelli, Delia, Jeremy J Lambert, Murphy Lam Yim Wan, Ana Rita Monteiro, David J Nutt, and Jerome D Swinny. 2025. From bugs to brain: unravelling the GABA signalling networks in the brain–gut–microbiome axis. *Brain* 148 (5):1479–1506.

Ben-Shahar, Y, A Robichon, MB Sokolowski, and GE Robinson. 2002. Influence of gene action across different time scales on behavior. *Science* 296 (5568):741–4.

Biedermann, Peter HW, and Marko Rohlfs. 2017. Evolutionary feedbacks between insect sociality and microbial management. *Current Opinion in Insect Science* 22:92–100.

Blomquist, Gary. 2003. *Insect pheromone biochemistry and molecular biology: The biosynthesis and detection of pheromones and plant volatiles*: Elsevier.

Bordenstein, Seth R, and The Holobiont Biology Network. 2024. The disciplinary matrix of holobiont biology. *Science* 386 (6723):731–732.

Bourguignon, Thomas, Nathan Lo, Carsten Dietrich, et al. 2018. Rampant host switching shaped the termite gut microbiome. *Current Biology* 28 (4):649–654. e2.

Bourke, Andrew FG. 2011. *Principles of social evolution*: Oxford University Press.

Breeuwer, Johannes AJ, and John H Werren. 1990. Microorganisms associated with chromosome destruction and reproductive isolation between two insect species. *Nature* 346 (6284):558–560.

Breznak, John A. 1982. Intestinal microbiota of termites and other xylophagous insects. *Annual Reviews in Microbiology* 36 (1):323–323.

Brune, Andreas, and Carsten Dietrich. 2015. The gut microbiota of termites: digesting the diversity in the light of ecology and evolution. *Annual Review of Microbiology* 69 (1):145–166.

Brune, Andreas. 2014. Symbiotic digestion of lignocellulose in termite guts. *Nature Reviews Microbiology* 12 (3):168–180.

Cleveland, Lemuel Roscoe. 1923. Symbiosis between termites and their intestinal protozoa. *Proceedings of the National Academy of Sciences* 9 (12):424–428.

Cremer, Sylvia, Sophie AO Armitage, and Paul Schmid-Hempel. 2007. Social immunity. *Current Biology* 17 (16):R693–R702.

Currie, Cameron R, James A Scott, Richard C Summerbell, and David Malloch. 1999. Fungus-growing ants use antibiotic-producing bacteria to control garden parasites. *Nature* 398 (6729):701–704.

Debray, Reena, Jenny Tung, and Elizabeth A Archie. 2024. Ecology and evolution of the social microbiome. *Annual Review of Ecology, Evolution, and Systematics* 55:89–114.

Dietrich, Carsten, Tim Köhler, and Andreas Brune. 2014. The cockroach origin of the termite gut microbiota: patterns in bacterial community structure reflect major evolutionary events. *Applied and Environmental Microbiology* 80 (7):2261–2269.

Dillon, RJ, CT Vennard, and AK Charnley. 2002. A note: gut bacteria produce components of a locust cohesion pheromone. *Journal of Applied Microbiology* 92 (4):759–763.

Dong, Jinyang, Xiang Yao, Yanru Zhang, et al. 2025. Gut bacteria mediate aggregation pheromone release in the borer beetle Trigonorhinus sp. *Insects* 16 (10):999.

Douglas, Angela E, and John H Werren. 2016. Holes in the hologenome: why host-microbe symbioses are not holobionts. *MBio* 7 (2):10–128. 10.1128/mbio.02099-15.

Douglas, Angela E. 2022. *Insects and their beneficial* microbes: Princeton University Press.

Drijfhout, F.P., Kather, R., and Martin, S.J. (2009) The role of cuticular hydrocarbons in insects. In: *Behavioral and Chemical Ecology* Eds. W. Zhang and H. Liu. Nova Science Publishers, Inc. 91–114.

Engel, Philipp, and Nancy A Moran. 2013. The gut microbiota of insects–diversity in structure and function. *FEMS Microbiology Reviews* 37 (5):699–735.

Engl, Tobias, and Martin Kaltenpoth. 2018. Influence of microbial symbionts on insect pheromones. *Natural Product Reports* 35 (5):386–397.

Ferreira, Clara H, and Marta A Moita. 2019. What can a non-eusocial insect tell us about the neural basis of group behaviour? *Current Opinion in Insect Science* 36:118–124.

Fischer, Caleb N, Eric P Trautman, Jason M Crawford, Eric V Stabb, Jo Handelsman, and Nichole A Broderick. 2017. Metabolite exchange between microbiome members produces compounds that influence Drosophila behavior. *Elife* 6:e18855.

Foster, Kevin R, Jonas Schluter, Katharine Z Coyte, and Seth Rakoff-Nahoum. 2017. The evolution of the host microbiome as an ecosystem on a leash. *Nature* 548 (7665):43–51.

Gaspar, Miguel, Sophie Dias, and Maria Luísa Vasconcelos. 2022. Mating pair drives aggressive behavior in female Drosophila. *Current Biology* 32 (21):4734–4742. e4.

Giordano, Rosanna, Jan J Jackson, and Hugh M Robertson. 1997. The role of Wolbachia bacteria in reproductive incompatibilities and hybrid zones of Diabrotica beetles and Gryllus crickets. *Proceedings of the National Academy of Sciences* 94 (21):11439–11444.

Griffiths, Jessica A, Khemlal Nirmalkar, Wei-Li Wu, Rosa Krajmalnik-Brown, and Sarkis K Mazmanian. 2026. The gut microbiome shapes social behaviour across animal species. *Nature Reviews Microbiology*: 24 (5):328–343.

Groussin, Mathieu, Florent Mazel, and Eric J Alm. 2020. Co-evolution and co-speciation of host-gut bacteria systems. *Cell Host & Microbe* 28 (1):12–22.

Guerrero, Ricardo, Lynn Margulis, and Mercedes Berlanga. 2013. Symbiogenesis: the holobiont as a unit of evolution. *International Microbiology* 16 (3):133–143.

Gullan, Penny J, and Michael Kosztarab. 1997. Adaptations in scale insects. *Annual Review of Entomology* 42 (1):23–50.

Guo, Xiaojiao, Helen He, Jianghua Sun, and Le Kang. 2023. Plasticity of aggregation pheromones in insects. *Current Opinion in Insect Science* 59: 101098.

He, Muyang, Huimin Chen, Xiaorui Yang, Yang Gao, Yongyue Lu, and Daifeng Cheng. 2022. Gut bacteria induce oviposition preference through ovipositor recognition in fruit fly. *Communications Biology* 5 (1):973.

Hongoh, Yuichi, Vineet K Sharma, Tulika Prakash, et al. 2008. Genome of an endosymbiont coupling N2 fixation to cellulolysis within protist cells in termite gut. *Science* 322 (5904):1108–1109.

Hu, Juncheng, Ran Bi, Yuxuan Luo, et al. 2025. The gut microbiome promotes locomotion of Drosophila larvae via octopamine signaling. *Insect Science* 32 (1):277–289.

Hu, Yi, and Corrie S Moreau. 2025. Nutritional symbiosis between ants and their symbiotic microbes. *Annual Review of Entomology* 71, 35–49.

Hurst, GD, and Francis M Jiggins. 2000. Male-killing bacteria in insects: mechanisms, incidence, and implications. *Emerging Infectious Diseases* 6 (4):329.

Jia, Yicong, Shan Jin, Kunkun Hu, et al. 2021. Gut microbiome modulates Drosophila aggression through octopamine signaling. *Nature Communications* 12 (1):2698.

Johnson, Katerina V-A, and Kevin R Foster. 2018. Why does the microbiome affect behaviour? *Nature Reviews Microbiology* 16 (10):647–655.

Jones, Julia C, Carmelo Fruciano, Jo Marchant, et al. 2018. The gut microbiome is associated with behavioural task in honey bees. *Insectes Sociaux* 65 (3):419–429.

Kageyama, Daisuke, Sugihiko Hoshizaki, and Yukio Ishikawa. 1998. Female-biased sex ratio in the Asian corn borer, Ostrinia furnacalis: evidence for the occurrence of feminizing bacteria in an insect. *Heredity* 81 (3):311–316.

Keeble, Frederick. 1912. *Plant-animals: a study in symbiosis*: University Press.

Koch, Hauke, and Paul Schmid-Hempel. 2011. Socially transmitted gut microbiota protect bumble bees against an intestinal parasite. *Proceedings of the National Academy of Sciences* 108 (48):19288–19292.

Kocher, Sarah, and Callum Kingwell. 2024. The molecular substrates of insect eusociality. *Annual Review of Genetics* 58 (1):273–295.

Koide, Roger T. 2023. On holobionts, holospecies, and holoniches: the role of microbial symbioses in ecology and evolution. *Microbial Ecology* 85 (4):1143–1149.

Korb, Judith, and Jürgen Heinze. 2016. Major hurdles for the evolution of sociality. *Annual Review of Entomology* 61 (1):297–316.

Kwong, Waldan K, Amanda L Mancenido, and Nancy A Moran. 2017. Immune system stimulation by the native gut microbiota of honey bees. *Royal Society Open Science* 4 (2):170003.

Lange, Claudia, Stéphane Boyer, T Martijn Bezemer, et al. 2023. Impact of intraspecific variation in insect microbiomes on host phenotype and evolution. *The ISME Journal* 17 (11):1798–1807.

Lavy, Omer, Uri Gophna, Eran Gefen, and Amir Ayali. 2020. Locust bacterial symbionts: an update. *Insects* 11 (10):655.

Leftwich, Philip T, Matthew I Hutchings, and Tracey Chapman. 2018. Diet, gut microbes and host mate choice: understanding the significance of microbiome effects on host mate choice requires a case by case evaluation. *Bioessays* 40 (12):1800053.

Leftwich, PhilipT, NaomiVE Clarke, Matthew I Hutchings, and Tracey Chapman. 2017. Gut microbiomes and reproductive isolation in Drosophila. *Proceedings of the National Academy of Sciences* 114 (48):12767–12772.

Lewis, Zenobia, and Anne Lizé. 2015. Insect behaviour and the microbiome. *Current Opinion in Insect Science* 9:86–90.

Liberti, Joanito, and Philipp Engel. 2020. The gut microbiota—brain axis of insects. *Current Opinion in Insect Science* 39:6–13.

Lu, B, A LaMora, Y Sun, MJ Welsh, and Y Ben-Shahar. 2012. ppk23-Dependent chemosensory functions contribute to courtship behavior in Drosophila melanogaster. *PLoS Genet* 8 (3):e1002587.

Lu, B, KM Zelle, R Seltzer, A Hefetz, and Y Ben-Shahar. 2014. Feminization of pheromone-sensing neurons affects mating decisions in Drosophila males. *Biol Open* 3 (2):152–60.

Maccaro, Jessica, and Brian Whyte. 2026. Sociality and the Necrobiome. M. Eric Benbow, Jeffery K. Tomberlin, Aaron M. Tarone. In *Carrion Ecology, Evolution, and Their Applications*: CRC Press, 518–546.

Majerus, Michael EN. 2003. *Sex wars: genes, bacteria, and biased sex ratios*: Princeton University Press.

Mangold, Colleen A, and David P Hughes. 2021. Insect behavioral change and the potential contributions of neuroinflammation—a call for future research. *Genes* 12 (4):465.

Maritan, Elisa, Andrea Quagliariello, Enric Frago, Tomaso Patarnello, and Maria Elena Martino. 2024. The role of animal hosts in shaping gut microbiome variation. *Philosophical Transactions of the Royal Society B* 379 (1901):20230071.

Mattoso, Thalles C, Denise Moreira, and Richard I Samuels. 2012. Symbiotic bacteria on the cuticle of the leaf-cutting ant Acromyrmex subterraneus subterraneus protect workers from attack by entomopathogenic fungi. *Biology Letters* 8 (3):461–464.

Mazel, Florent, Aiswarya Prasad, and Philipp Engel. 2025. Host specificity of gut microbiota associated with social bees: patterns and processes. *Microbiology and Molecular Biology Reviews* 89:e00080–23.

McKinney, Ross M, Ryan Valdez, and Yehuda Ben-Shahar. 2021. The genetic architecture of larval aggregation behavior in Drosophila. *Journal of Neurogenetics* 35 (3):274–284.

Moran, Nancy A. 2015. Genomics of the honey bee microbiome. *Current Opinion in Insect Science* 10:22–28.

Moreau, Corrie S. 2020. Symbioses among ants and microbes. *Current Opinion in Insect Science* 39:1–5.

Nagpal, Jatin, and John F Cryan. 2021. Microbiota-brain interactions: moving toward mechanisms in model organisms. *Neuron* 109 (24):3930–3953.

Nehring, Volker, and Sandra Steiger. 2018. Sociality and communicative complexity: insights from the other insect societies. *Current Opinion in Insect Science* 28:19–25.

Neu, AT, EE Allen, and K Roy. 2021. Defining and quantifying the core microbiome: challenges and prospects. *Proceedings of the National Academy of Sciences of the United States of America* 118 (51): e2104429118.

Nuttall, George HF. 1923. Symbiosis in animals and plants. *The American Naturalist* 57 (652):449–475.

Obadia, Benjamin, Zehra Tüzün Güvener, Vivian Zhang, et al. 2017. Probabilistic invasion underlies natural gut microbiome stability. *Current Biology* 27 (13):1999–2006. e8.

Okwaro, Louis A, and Judith Korb. 2023. Epigenetic regulation and division of labor in social insects. *Current Opinion in Insect Science* 58:101051.

Pasteur, Louis. 1926. *Etudes sur la maladie des vers a soie*. Vol. 4: Masson.

Qiao, Huili, Ian W Keesey, Bill S Hansson, and Markus Knaden. 2019. Gut microbiota affects development and olfactory behavior in Drosophila melanogaster. *Journal of Experimental Biology* 222 (5):jeb192500.

Ren, Lu, Yingao Ma, Mingxue Xie, Yongyue Lu, and Daifeng Cheng. 2021. Rectal bacteria produce sex pheromones in the male oriental fruit fly. *Current Biology* 31 (10):2220–2226. e4.

Richardson, Lauren A. 2017. Evolving as a holobiont. *PLoS Biology* 15 (2):e2002168.

Ringo, John, Gil Sharon, and Daniel Segal. 2011. Bacteria-induced sexual isolation in Drosophila. *Fly* 5 (4):310–315.

Robinson, GE, CM Grozinger, and CW Whitfield. 2005. Sociogenomics: social life in molecular terms. *Nat Rev Genet* 6 (4):257–70.

Robinson, Gene E. 2009. Division of labor in insect societies. Vincent H. Resh, Ring T. Cardé, Academic Press. *Encyclopedia of insects*:297–299.

Rosenberg, Eugene, and Ilana Zilber-Rosenberg. 2016. Microbes drive evolution of animals and plants: the hologenome concept. *MBio* 7 (2):10–128. 10.1128/mbio. 01395-15.

Saltz, Julia B. 2013. Genetic composition of social groups influences male aggressive behaviour and fitness in natural genotypes of Drosophila melanogaster. *Proceedings of the Royal Society B: Biological Sciences* 280 (1771):20131926.

Sanders, Jon G, Scott Powell, Daniel JC Kronauer, Heraldo L Vasconcelos, Megan E Frederickson, and Naomi E Pierce. 2014. Stability and phylogenetic correlation in gut microbiota: lessons from ants and apes. *Molecular Ecology* 23 (6):1268–1283.

Sarkar, Amar, Siobhán Harty, Katerina V-A Johnson, et al. 2020. The role of the microbiome in the neurobiology of social behaviour. *Biological Reviews* 95 (5):1131–1166.

Schmickl, Thomas, and Istvan Karsai. 2018. Integral feedback control is at the core of task allocation and resilience of insect societies. *Proceedings of the National Academy of Sciences* 115 (52):13180–13185.

Schultz, Ted R. 1999. Ants, plants and antibiotics. *Nature* 398 (6730):747–748.

Shapira, Michael. 2016. Gut microbiotas and host evolution: scaling up symbiosis. *Trends in Ecology & Evolution* 31 (7):539–549.

Sharon, Gil, Daniel Segal, JM Ringo, A Hefetz, Ilana Zilber-Rosenberg, and Eugene Rosenberg. 2010. Commensal bacteria play a role in mating preference of Drosophila melanogaster. *Proceedings of the National Academy of Sciences U S A* 107 (46): 20051–6.

Sharon, Gil, Daniel Segal, Ilana Zilber-Rosenberg, and Eugene Rosenberg. 2011. Symbiotic bacteria are responsible for diet-induced mating preference in Drosophila melanogaster, providing support for the hologenome concept of evolution. *Gut Microbes* 2 (3):190–192.

Sherwin, Eoin, Seth R Bordenstein, John L Quinn, Timothy G Dinan, and John F Cryan. 2019. Microbiota and the social brain. *Science* 366 (6465):eaar2016.

Shropshire, J Dylan, and Seth R Bordenstein. 2016. Speciation by symbiosis: the microbiome and behavior. *MBio* 7 (2): e01785. 10.1128/mbio. 01785-15.

Sinotte, Veronica M, Justinn Renelies-Hamilton, Benjamin A Taylor, et al. 2020. Synergies between division of labor and gut microbiomes of social insects. *Frontiers in Ecology and Evolution* 7:503.

Toyoshima, Naoki, and Takashi Matsuo. 2023. Fight outcome influences male mating success in Drosophila prolongata. *Journal of Ethology* 41 (2):119–127.

Turelli, Michael. 1994. Evolution of incompatibility-inducing microbes and their hosts. *Evolution* 48 (5):1500–1513.

Vernier, Cassondra L, Iris M Chin, Boahemaa Adu-Oppong, et al. 2020. The gut microbiome defines social group membership in honey bee colonies. *Science Advances* 6 (42):eabd3431.

Vernier, Cassondra L, Lan Anh Nguyen, Tim Gernat, Amy Cash Ahmed, Zhenqing Chen, and Gene E Robinson. 2024. Gut microbiota contribute to variations in honey bee foraging intensity. *The ISME Journal* 18 (1):wrae030.

Vernier, CL, N Leitner, KM Zelle, et al. 2023. A pleiotropic chemoreceptor facilitates the production and perception of mating pheromones. *iScience* 26 (1):105882.

Wada-Katsumata, Ayako, Ludek Zurek, Godfrey Nalyanya, Wendell L Roelofs, Aijun Zhang, and Coby Schal. 2015. Gut bacteria mediate aggregation in the German cockroach. *Proceedings of the National Academy of Sciences* 112 (51):15678–15683.

Wang, Kang, Ming Zheng, Minqi Cai, et al. 2024. Possible interactions between gut microbiome and division of labor in honey bees. *Ecology and Evolution* 14 (8):e11707.

Wang, Zhengyan, Zhenzhen Chang, Zhiyuan Liu, and Shan Zhang. 2023. Influences of microbial symbionts on chemoreception of their insect hosts. *Insects* 14 (7):638.

Week, Bob, Shelbi L Russell, Hinrich Schulenburg, Brendan JM Bohannan, and Marjolein Bruijning. 2025. Applying evolutionary theory to understand host–microbiome evolution. *Nature Ecology & Evolution* 9 (10):1769–1780.

Weiss, Martha R. 2006. Defecation behavior and ecology of insects. *Annual Review of Entomology* 51 (1):635–661.

Wilson, Edward O. 1971. *The insect societies*: Belknap Press of Harvard University Press.

Wilson, EO, and B Holldobler. 2005. Eusociality: origin and consequences. *Proceedings of the National Academy of Sciences of the United States of America* 102 (38):13367–13371.

Wong, Adam CN, John M Chaston, and Angela E Douglas. 2013. The inconstant gut microbiota of Drosophila species revealed by 16S rRNA gene analysis. *The ISME journal* 7 (10):1922–1932.

Zhang, Hong, Yibin Zhu, Yibaina Wang, Liping Jiang, Xiaolu Shi, and Gong Cheng. 2024. Microbial interactions shaping host attractiveness: insights into dynamic behavioral relationships. *Current Opinion in Insect Science* 66:101275.

Zhang, Shan, Zhengyan Wang, Qiong Luo, Lizhen Zhou, Xin Du, and Yonglin Ren. 2025. Effects of microbes on insect host physiology and behavior mediated by the host immune system. *Insects* 16 (1):82.

Zhang, Wei, Xuanyu Chen, Ioannis Eleftherianos, Amr Mohamed, Ashley Bastin, and Nemat O Keyhani. 2024. Cross-talk between immunity and behavior: insights from entomopathogenic fungi and their insect hosts. *FEMS Microbiology Reviews* 48 (1):fuae003.

CHAPTER FIVE

Lignocellulose Degradation by Termite Gut Microorganisms

Alex Muppasani and Sunil Thomas

5.1 INTRODUCTION

Termites (order Isoptera) are a distinct clade of eusocial insects within the broader group of cockroaches, forming a sister group to the subsocial *Cryptocercus* wood roaches (Inward et al., 2007). Although over 3,000 termite species have been formally described, their overall taxonomic diversity remains relatively modest when compared to many other insect groups. Despite this, termites play disproportionately significant ecological roles, particularly in tropical and subtropical terrestrial ecosystems where they are often among the most abundant macroinvertebrates (Hellemans et al., 2024).

Functionally, termites contribute essential ecosystem services. They are primary decomposers of lignocellulosic plant material, breaking down dead wood and leaf litter and thus facilitating nutrient cycling. Their tunneling activity contributes to soil bioturbation, improving soil structure, aeration, and the infiltration of water (Jouquet et al., 2011). These roles make termites keystone species in many of the ecosystems they inhabit (Hellemans et al., 2024).

However, termites are also widely recognized for their economic impact as structural pests. Several species, particularly within a few genera, are notorious for damaging wooden infrastructure, including buildings, utility poles, fences, and other timber-based constructions. In many regions where they are present, termites account for more than half of total expenditures on urban pest control, not including the costs associated with insect-borne diseases. (Rust and Su, 2012) As a result, public perception of termites is often shaped more by their role as pests than by their ecological importance (Hellemans et al., 2024).

This dual identity—ecologically indispensable yet economically problematic—has shaped the focus of termite research. Much of the scientific attention has been concentrated on the relatively small subset of pest species, leaving many non-pest species comparatively understudied. Nonetheless, the combination of moderate species richness, high population densities, and significant ecological and economic relevance has positioned termites as a subject of continued interest among entomologists and taxonomists alike (Hellemans et al., 2024).

The digestive tract of termites is home to a remarkably dense and diverse community of microorganisms, forming one of the most specialized and functionally integrated symbiotic microbiomes found in insects. This complex assemblage of microbes plays a critical role in the host's ability to break down recalcitrant plant materials, particularly lignocellulose, enabling termites to exploit otherwise inaccessible dietary niches (Brune and Dietrich, 2015).

In lower termites, the symbiotic community is dominated by unique, co-evolved lineages of cellulolytic, flagellated protists. These protists, many of which are specific to termites and their close relatives, carry out the bulk of lignocellulose digestion. In turn, these protists harbor their own bacterial endo- and ectosymbionts, creating

DOI: 10.1201/9781003639176-5

a multi-tiered system of symbiosis (Ohkuma, 2008). In contrast, higher termites (family Termitidae) have lost these flagellates during the course of evolution and rely instead on dense and diverse communities of prokaryotes—bacteria and archaea—that occupy specialized compartments of the gut (Brune and Dietrich, 2015).

Although many of these microbial lineages remain uncultivated and poorly characterized, advances in molecular and genomic techniques are gradually revealing their functional roles and ecological dynamics. Comparative studies across different termite species with varied diets—ranging from wood and leaf litter to soil and fungus—are uncovering patterns in microbiome composition that correlate with both host phylogeny and dietary specialization. These patterns reflect not only the types of substrates ingested but also the evolving structure of the termite gut itself (Brune and Dietrich, 2015).

As the gut morphology becomes increasingly complex—featuring multiple compartments with distinct physicochemical conditions—it provides new ecological niches for microbial colonization. These microhabitats are shaped by factors such as the presence or absence of flagellates, the physical characteristics of digested wood fibers, and the gradient of oxygen and pH along the gut axis (Brune, 2014). Together, these conditions drive the differentiation and specialization of microbial communities within the gut (Brune and Dietrich, 2015).

This chapter explores the structure, diversity, and functional significance of the termite gut microbiome, highlighting how co-evolution between host and symbionts has led to one of the most intricate examples of mutualism in the animal kingdom.

5.2 COMPOSITION OF WOOD

Wood can be described as a three-dimensional biopolymer composite composed of an interconnected network of cellulose, hemicelluloses, and lignin, along with small quantities of extractives and inorganic materials (Cosgrove, 2005). While water is the major chemical constituent of a living tree, the dry mass of wood is primarily made up of sugar-based polymers (carbohydrates, 65–75%) bound with lignin (18–35%). On an elemental basis, dry wood typically consists of about 50% carbon, 6% hydrogen, 44% oxygen, and trace amounts of inorganic elements. Although chemical analyses can differentiate hardwoods (angiosperms) from softwoods (gymnosperms), they are not reliable for identifying individual species due to chemical variability within species and similarities among them. Generally, softwoods (conifers) contain higher levels of cellulose (40–45%) and lignin (26–34%), but lower amounts of pentosans (7–14%), whereas hardwoods (deciduous trees) have cellulose contents ranging from 38 to 49%, lignin from 23 to 30%, and pentosans from 19 to 26% (Rowell et al., 2013).

5.2.1 Lignin

Lignin is a complex class of aromatic polymers that plays a crucial role in providing rigidity, hydrophobicity, and resistance to microbial attack in vascular plants. It is predominantly found in the secondary cell walls of woody tissues and bark. Lignin is synthesized through the oxidative polymerization of three primary monolignols: p-coumaryl, coniferyl, and sinapyl alcohols (Boerjan et al., 2003; Ralph et al., 2019). The relative proportions and linkages of these monomers vary among plant species, resulting in structurally diverse lignin types.

This variability influences the physical properties of the wood and its resistance to biodegradation. For example, lignin polymers with a higher proportion of sinapyl alcohol units are generally more amenable to enzymatic degradation by fungi and insects due to their lower degree of cross-linking (Vane et al., 2003). Lignin's primary function is to protect plant tissues from pathogens, environmental stress, and herbivory, acting as a physical and chemical barrier.

5.2.2 Cellulose

Cellulose is the most abundant organic polymer on Earth and a fundamental component of plant cell walls. It is composed of linear chains of β-1,4-linked glucose units, forming highly ordered crystalline microfibrils that provide tensile strength to the cell wall (McNamara et al., 2015). These microfibrils are extremely stable, with estimates suggesting cellulose can have a half-life of several million years under certain conditions.

In woody plants, cellulose is intimately associated with lignin within the cell wall, and together, they make up the bulk of the biomass investment in trees (Alberts et al., 2002). This structural pairing forms a highly resilient matrix that is essential for supporting plant growth and form.

5.2.3 Hemicellulose and Extractives

Hemicellulose is a heterogeneous group of branched polysaccharides that, unlike cellulose, have a lower degree of polymerization and are more easily hydrolyzed. It accounts for approximately 20–30% of the biomass in many plant species. Hemicelluloses include xylans and mannans and are major constituents of wood. The wood hemicelluloses are structurally similar to hemicelluloses from agricultural crops. Hemicellulose binds with both cellulose microfibrils and lignin, playing a key role in cell wall architecture and flexibility (Abik et al., 2023).

In addition to structural polymers, wood contains water and a range of organic compounds known as extractives. These substances, although not critical to the structural integrity of the plant, contribute to the wood's physical and chemical properties, such as color, odor, and natural resistance to rot, fungal decay, and insect predation. Extractives include resins, tannins, oils, and other secondary metabolites that vary widely between species (War et al., 2012).

5.3 TERMITES

Often referred to as "white ants" because of their pale coloration and superficial resemblance to ants, termites are highly social insects that thrive predominantly in tropical and subtropical environments, though they are also distributed throughout much of the United States. Termite colonies exhibit a complex hierarchical structure centered around a reproductive queen. Her offspring, initially known as nymphs, are unspecialized juvenile termites capable of developing into different castes. The mechanisms behind nymph specialization remain an active area of study (Raina et al., 2004), with strong evidence suggesting that pheromonal or hormonal cues regulate their differentiation. Notably, when the queen is removed or dies, some nymphs can develop into secondary reproductives to sustain the colony's population.

Mature termite colonies can contain hundreds of millions of individuals organized in a caste-like system, consisting of reproductives (queen, king, and secondary reproductives), workers, and soldiers (Halim et al., 2025). Each caste performs specific roles vital to colony survival—workers handle foraging and nest maintenance, soldiers defend the colony, and reproductives ensure the continuation of the lineage.

In a broader biological context, termite species are often classified as "lower" or "higher" based on their symbiotic relationships with microorganisms in their digestive systems. Lower termites depend on flagellated protozoa and symbiotic bacteria in their hindguts to digest cellulose, the primary structural component of plant cell walls (Peterson and Scharf, 2016). While these termites produce some cellulase enzymes themselves, the protozoa are essential for efficient cellulose breakdown into glucose. In contrast, higher termites (family Termitidae), which account for roughly 70% of the more than 3,000 described termite species (Chouvenc et al., 2021), have lost these protozoan symbionts over the course of evolution. Instead, they rely on specialized gut bacteria that secrete lignocellulolytic enzymes—such as cellulases and hemicellulases—capable of decomposing lignocellulose, a complex matrix of cellulose, hemicellulose, and lignin. This remarkable adaptation allows higher termites to exploit a wider variety of plant materials and play an essential ecological role in nutrient cycling and organic matter decomposition in terrestrial ecosystems.

5.4 BIOGEOCHEMICAL CYCLE

The biogeochemical cycle refers to the natural processes through which essential elements—such as carbon, nitrogen, and phosphorus—continuously move between the Earth's surface, atmosphere, oceans, and living organisms (Cole et al., 2021). It encompasses a series of interconnected chemical cycles, each describing the movement of a particular nutrient or substance through the biotic (living) and abiotic (non-living) components of an ecosystem.

For example, in the carbon cycle, carbon primarily exists in the atmosphere as carbon dioxide (CO_2). Through the process of photosynthesis, plants, algae, and certain bacteria absorb CO_2 and convert it into glucose and oxygen, storing carbon within living tissues (Hohmann-Marriott and Blankenship, 2011). When animals consume these plants, carbon is transferred through the food web. Both plants and animals release CO_2 back into the atmosphere through respiration as they break down glucose for energy. When organisms die, decomposers such as bacteria and fungi further recycle carbon by breaking down organic material, releasing CO_2 via microbial respiration. Although this explanation covers the biological aspects of the carbon cycle, the complete cycle also includes geological and chemical processes such as oceanic carbon exchange and the combustion of fossil fuels. However, for the scope of this chapter, the focus will remain on biogeochemical processes involving living organisms.

Termites play a particularly important role in the biogeochemical cycling of carbon. In some ecosystems, they are responsible for decomposing nearly 50% of dead wood through the combined action of their own cellulase enzymes and symbiotic gut microbes (Wijas et al., 2024). It is estimated that about 42% of the carbon derived from wood consumed by termites is released into the atmosphere as gases (primarily CO_2), 40% is deposited into the environment as organic matter, and the remaining 18% is retained within termite biomass (Myer et al., 2021).

In addition to their role in decomposition, termite mounds act as significant reservoirs of both organic and inorganic carbon. These mounds often contain substantial amounts of pedogenic calcium carbonate ($CaCO_3$)—a mineral formed through the interaction of CO_2 and calcium ions in the soil. Termites in semi-arid regions are especially important for carbon sequestration, as their activities promote $CaCO_3$ accumulation (Clarke et al., 2023). This process, known as carbonation, captures atmospheric CO_2 in a stable, mineralized form, effectively locking away carbon from the environment.

During cellulose digestion, termites and their symbiotic microbes produce acetate, CO_2, and water. The CO_2 released during this process can react with calcium ions (Ca^{2+}) present in the soil to form calcium carbonate. Research shows that approximately 4% of the outer layers and 2% of the inner core of termite mounds are composed of carbonates, making these structures temporary carbon sinks (Mujinya et al., 2011). Over time, if mound soils are exposed to water or acidic conditions, the calcium carbonate may dissolve, releasing carbonate and bicarbonate ions into the soil or groundwater, and eventually some carbon may return to the atmosphere as CO_2.

When examining the nanoparticle composition of termite mounds, significant differences can be observed compared to surrounding soils. For instance, soil-feeding termites such as *Patawatermes turricola* consume the humic (organic) fraction of soil and construct their mounds using a mixture of fecal material and inorganic soil particles. As a result, the chemical and structural composition of their mounds differs markedly from that of the nearby soil environment (Duran-Bautista et al., 2025). This distinct composition further highlights the complex and dynamic role termites play in shaping soil chemistry and influencing global biogeochemical cycles.

5.5 SPECIFIC BACTERIA IN TERMITE GUTS

Higher termites lack the flagellated protozoa in their hindguts that lower termites depend on to ferment and digest cellulose. As a result, it is reasonable to predict—before even conducting microbial analysis—that the gut microbiomes of higher and lower termites differ significantly, and research has indeed confirmed this distinction.

In lower termites, flagellate protozoa play a key role in breaking down cellulose, classifying them as cellulolytic organisms. These protozoa, in partnership with gut bacteria, enable the efficient digestion of plant material. In contrast, higher termites, which are free of flagellates, achieve cellulose degradation through specialized fiber-associated bacteria that have evolved to occupy the same ecological niche once filled by protozoa. Genomic studies have shown that bacteria belonging to the phyla Fibrobacterota and Spirochaetota possess a comprehensive set of genes encoding cellulose-degrading enzymes. This genetic arsenal allows them to independently digest cellulose fibers from start to finish, demonstrating a remarkable example of evolutionary adaptation and microbial role replacement (Salgado et al., 2024).

In addition to these cellulolytic bacteria, the termite gut microbiome contains members of several other bacterial groups, including the Enterobacteriaceae family and genera such as *Bacteroides*, *Staphylococcus*, *Streptococcus*, and *Bacillus*. Together, these diverse microbial communities form a highly specialized symbiotic system that enables termites to efficiently decompose lignocellulosic materials, contributing significantly to nutrient cycling and energy flow in terrestrial ecosystems.

5.5.1 Inheritance of Termite Gut Microbiota

In termites and their close relatives, the wood-feeding cockroaches (*Cryptocercus*), the inheritance of beneficial gut microbes relies on a specialized social behavior known as proctodeal trophallaxis. This process involves the transfer of fecal material from one individual to another, effectively passing along the symbiotic microorganisms contained within it (Chouvenc et al., 2021). Though it may seem unappealing, the fecal matter exchanged is far from mere waste—it is a nutrient-rich mixture containing symbiotic bacteria, protozoa, and partially digested lignocellulosic material. These components are essential for the proper establishment and maintenance of the termite's hindgut microbiome, which plays a central role in cellulose digestion and nutrient recycling.

Proctodeal trophallaxis serves multiple critical functions within termite colonies. Firstly, it ensures that newly hatched or recently molted individuals, which have sterile or incomplete gut microbiota, are inoculated with the necessary cellulolytic and lignocellulolytic microorganisms. In lower termites, this includes flagellated protozoa that ferment cellulose, while in higher termites, the transmission primarily involves bacteria that produce lignocellulolytic enzymes capable of degrading cellulose and hemicellulose. Without this microbial inheritance, young termites would be unable to digest wood and would quickly perish (Mirabito and Rosengaus, 2016).

Secondly, this process helps maintain microbial uniformity across the entire colony. By continually sharing and redistributing gut microbes, termites sustain a stable and balanced microbiome throughout their social group. This consistency is vital not only for efficient digestion but also for colony-wide immunity. Many of the resident microbes inhibit or outcompete pathogenic bacteria and fungi, creating a biological defense system that protects the colony as a whole. Thus, proctodeal trophallaxis functions as both a nutritional and protective mechanism, reinforcing termite social cohesion and survival (Chouvenc et al., 2021).

From an evolutionary standpoint, proctodeal trophallaxis marks a pivotal shift in the way termites transmit symbiotic partners—from horizontal transmission (microbes acquired from the environment or peers) to vertical transmission, where symbionts are passed directly from parent to offspring (Nalepa, 2020). This transition has made termite–microbe relationships more stable and reliable across generations, strengthening the co-evolutionary bond between host and symbiont. Over millions of years, this intimate microbial inheritance has become indispensable to termite biology, transforming them into one of nature's most efficient lignocellulose recyclers (Chouvenc et al., 2021).

5.5.2 Digestion of Lignocellulose in Termite Guts

Lignocellulosic biomass, composed primarily of cellulose, hemicellulose, and lignin, represents one of the most abundant and recalcitrant forms of organic carbon on Earth (Sethupathy et al., 2021). Termites possess a remarkable ability to degrade this complex material, positioning them as critical players in the global carbon cycle. This capability relies heavily on the symbiotic microbiome within their hindguts, which includes bacteria, archaea, and in lower termites, flagellated protozoa. These symbionts work in concert to break down and ferment wood polysaccharides such as cellulose and hemicellulose, converting them into simpler sugars and subsequently into energy-rich metabolites.

Lignocellulose fermentation occurs under anaerobic conditions within the termite gut. Simple sugars released during cellulose and hemicellulose degradation are further processed into short-chain fatty acids (SCFAs)—primarily acetate, propionate, and butyrate—which serve as a major energy source for the host termite (Lazuka et al., 2018). Beyond providing nutrition, this microbial activity recycles carbon from plant biomass back into the environment in forms usable by other organisms, including CO_2, acetate, and methane, thereby linking termite digestion directly to ecosystem-level carbon cycling.

An important byproduct of this process is methane (CH_4), generated by methanogenic archaea residing in the termite hindgut. These archaea are often located near the hindgut epithelium or attached to filamentous prokaryotes associated with the gut wall, forming a stable microhabitat for methane production (Leadbetter and Breznak, 1996). Methane produced by termites globally is estimated to contribute 2–4% of the total atmospheric methane budget (Pester et al., 2007), highlighting the ecological significance of termite lignocellulose digestion beyond local ecosystems.

By breaking down lignocellulosic material into SCFAs, CO_2, and methane, termites effectively recycle carbon locked within wood and plant matter, returning it to the environment and sustaining a dynamic flow of nutrients. This intricate interplay between host and microbiome underscores termites' unique role as ecosystem engineers, facilitating both nutrient cycling and greenhouse gas emissions on a global scale.

5.5.3 Lignocellulose and Specialized Bacteria

Lignocellulose plays a critical role in the global carbon cycle, serving as one of the Earth's most abundant and carbon-rich resources. Each year, terrestrial plants produce over 200 billion tons of lignocellulosic biomass, much of which is not naturally decomposed (Chandel and Singh, 2011). As a product of photosynthesis, lignocellulose represents a renewable form of carbon, capturing atmospheric CO_2 and storing it in plant cell walls in the form of cellulose, hemicellulose, and lignin. Its breakdown and recycling are therefore essential for maintaining carbon flow through ecosystems (Gayathri et al., 2021).

Specialized microorganisms, including fermentative bacteria and protozoa, as well as acetogenic bacteria, play central roles in lignocellulose degradation (Hsin et al., 2025). These microbes work synergistically to deconstruct complex polysaccharides into simpler sugars and subsequently convert them into acetate, the primary product of lignocellulose fermentation. Acetate serves as a crucial energy source for the host organism, such as termites, and also contributes carbon to surrounding ecosystems.

In the termite gut, these microbial processes are finely tuned. Fermentative symbionts first depolymerize cellulose and hemicellulose into monosaccharides, which are then converted by acetogenic bacteria into acetate through reductive pathways. This microbial partnership allows termites to extract maximum energy from lignocellulosic material while simultaneously recycling carbon from otherwise indigestible plant biomass (Xie et al., 2023). By converting atmospheric carbon captured in plant material into bioavailable forms like acetate, CO_2, and methane, specialized gut microbes not only sustain termite nutrition but also contribute significantly to biogeochemical cycling and the broader flow of carbon in terrestrial ecosystems (Dar et al., 2024).

This intricate interplay between lignocellulose and specialized bacteria highlights how microbial communities have evolved to exploit one of nature's most abundant carbon reserves, transforming it into forms that sustain both the host and the ecosystem at large.

5.5.4 Fermentation of Lignocellulose

Acetogenic bacteria are key players in the production of acetate, a central energy source for termites. Acetogens are typically anaerobic bacteria that convert simple compounds such as CO_2 and H_2 into acetate using the Wood–Ljungdahl pathway (Ragsdale and Pierce, 2008):

2CO2 + 4H2 → CH3COOH + 2H2O2CO2 + 4H2 → CH3COOH + 2H2O

To understand this process, it is important to first consider the fermentation of lignocellulose in the termite hindgut. Lignocellulose is broken down into its component polysaccharides, cellulose and hemicellulose, by gut microbes (Hsin et al., 2025). Cellulose is a linear homopolymer of glucose units, while hemicellulose is a heteropolymer containing multiple sugars, including glucose, xylose, and mannose (Ahmad et al., 2022). Once hydrolyzed into monosaccharides, these sugars are fermented by anaerobic microbes through glycolysis, producing pyruvate, ATP, and NADH:

Glucose → Pyruvate + ATP + NADH

Pyruvate is then further oxidized in a process known as pyruvate oxidation, generating acetyl-CoA, CO_2, H_2, and sometimes other SCFAs such as ethanol, formate, propionate, butyrate, lactate, or succinate (Breznak and Switzer, 1986). Acetyl-CoA serves as a central intermediate in fermentation and can be converted into acetate through substrate-level phosphorylation, a reaction that also produces energy for the microbe:

Pyruvate (C3) → Acetyl-CoA (C2) + CO2 + H2
Acetyl-CoA + H2O → Acetate + Energy

While this describes the generic fermentation process, the Wood–Ljungdahl pathway is particularly important in environments where free sugars are scarce, such as the termite hindgut. In these conditions, simple sugars like glucose and xylose are immediately consumed by general fermentative microbes. CO_2 and H_2, which are products of sugar fermentation, can instead be directly converted into acetate without relying on abundant organic carbon sources (Ragsdale, 2008).

The metabolic strategies of termites differ depending on their caste of symbionts. In lower termites, cellulose and hemicellulose fermentation depends on protozoa and general bacteria, which convert sugars into acetate and other SCFAs. In higher termites, acetate production is largely mediated by autotrophic acetogenic bacteria, which convert CO_2 and H_2 directly into acetate, bypassing the need for sugars (Lee et al., 2022). This metabolic efficiency allows higher termites to thrive in wood with low sugar availability, giving them an evolutionary advantage. By maximizing energy extraction from lignocellulose, higher termites can sustain their colonies even in nutrient-limited environments (Peterson and Scharf, 2016), highlighting the critical role of specialized gut bacteria in termite ecology and the global carbon cycle.

5.6 SYMBIOTIC RELATIONSHIP BETWEEN FUNGI AND TERMITES

Higher termites exhibit not only distinct hindgut microbial compositions but also engage in external symbioses with fungi, particularly members of the Basidiomycota. Among these, termites of the subfamily Macrotermitinae exemplify a sophisticated mutualistic relationship with fungi that enhances their ability to decompose lignocellulosic biomass. Their hindgut microbiome remains exclusively prokaryotic (Auer et al., 2017), but the incorporation of fungi allows them to achieve near-complete lignocellulose breakdown.

The fungal partners are always species of the genus *Termitomyces*, which are critical for the decomposition process. Genomic studies reveal that *Termitomyces* reproduces both sexually through basidiospores and vegetatively via mycelial fragments, enabling flexible propagation strategies (Vreeberg et al., 2020). When new termite colonies are founded, the fungus must often be acquired from the surrounding environment, a process termed horizontal transmission. This reliance highlights the ecological interdependence between termites and local fungal diversity: the survival of a nascent colony depends on successfully encountering and incorporating the correct *Termitomyces* strain (Vreeburg et al., 2023).

In some cases, however, termites ensure vertical transmission of fungal symbionts. Reproductive alates—winged termites that disperse to establish new colonies—may carry fungal spores or mycelial fragments, directly inoculating the fungus into the new colony (Wisselink et al., 2020). Thus, both horizontal and vertical transmission strategies coexist, balancing environmental acquisition with lineage fidelity (Rosenberg and Rosenberg, 2022). This dual mode of transmission allows termite societies to selectively maintain their fungus gardens, likely as a means of defending against parasitic fungi or competitive microbial species (Makonde et al., 2013).

The fungus–termite partnership is highly cooperative: termites provide the fungi with partially digested plant material, creating an ideal substrate for mycelial growth (Schalk et al., 2021), while the fungi break down lignocellulose into nutrients more accessible to the termites. This mutualistic system exemplifies a complex ecological and evolutionary adaptation, where behavioral, microbial, and fungal controls converge to optimize nutrient cycling, colony nutrition, and survival. Such interactions underscore the integral role of termite–fungus symbioses in terrestrial ecosystems, particularly in the recycling of carbon and other plant-derived resources (Katoh et al., 2002).

5.7 INDUSTRIAL IMPORTANCE OF TERMITES

5.7.1 The Termite Gut as a Natural Bioreactor

The termite gut functions as an exceptionally efficient natural bioreactor, drawing significant attention in biofuel and bioprocessing research. This specialized microenvironment hosts a highly diverse community of symbiotic microorganisms capable of rapidly degrading lignocellulosic biomass—a resource traditionally considered recalcitrant. Studies have shown that termite gut microbiomes can depolymerize 74–99% of cellulose and 65–87% of hemicellulose within just 24 hours, surpassing the digestive efficiency of other natural systems, including ruminant animals. For instance, the gut microbiome of *Nasutitermes ephratae* achieved up to 45% degradation of wheat straw by weight, demonstrating the remarkable potential of these insects as models for industrial lignocellulose processing (Auer et al., 2017).

The digestive process begins in the termite foregut, where mechanical grinding breaks biomass into fine particles, vastly increasing the surface area available for microbial action. The partially digested material then moves to the hindgut, where a synergistic community of microbes produces an array of lignocellulolytic enzymes. These include cellulases and hemicellulases, which cleave cellulose and hemicellulose into simple sugars, and lignin-degrading enzymes such as laccases and peroxidases, which disrupt the protective lignin matrix, granting other enzymes access to polysaccharides. Remarkably, research has shown that over 60% of xylanase and cellulase activity in termite gut inocula is cell-bound, highlighting the efficiency of the microbe–enzyme association (Auer et al., 2017).

The coordinated action of these enzymes results in exceptionally high saccharification efficiency. Crude enzyme extracts from termite gut microbes can release up to 96.44% of sugars from substrates like xylan, demonstrating their industrial potential for converting plant biomass into fermentable sugars. This capacity has direct applications in biofuel production, biorefineries, and biomass valorization, where efficient and cost-effective breakdown of lignocellulose remains a major bottleneck.

Beyond biofuels, the termite gut serves as a model for designing biocatalytic systems, enzyme cocktails, and synthetic microbial consortia aimed at sustainable bioprocessing of renewable carbon sources. By understanding and harnessing the termite gut's enzymatic and microbial synergy, researchers are developing innovative strategies for transforming abundant plant biomass into energy, chemicals, and value-added products, bridging natural evolutionary efficiency with industrial technology.

5.7.2 Turning Plant Sugars into Fuel

Beyond breaking down lignocellulosic biomass, many microorganisms within the termite gut are capable of fermenting the resulting sugars into valuable biofuels and biochemicals, offering a compelling model for consolidated bioprocessing. One promising approach is the carboxylate platform, which provides an alternative to traditional biomass-to-ethanol conversion and has the potential to diversify lignocellulose-based biorefineries. In termite-inspired bioreactors, the main products of lignocellulose bioconversion are carboxylates, predominantly acetate (>85%), along with smaller amounts of propionate and butyrate (Auer et al., 2017). Propionate and butyrate are typically present as minor co-products, although some termite inocula, such as that from *Microcerotermes parvus*, can produce higher proportions of propionate (up to 20%). The total VFA accumulation in these systems can reach concentrations as high as 5.8 g/L, with the *N. ephratae* microbiome being the most productive.

Beyond liquid platform chemicals, the termite gut is also a potent producer of gaseous energy carriers, namely biohydrogen (H2) and biomethane (CH4). The strictly anaerobic conditions of the hindgut provide an ideal environment for hydrogenesis and methanogenesis, making the termite a compelling biomimetic model for producing these clean fuels (Ali et al., 2024).

Biohydrogen is a central intermediate in the gut's metabolic network. In lower termites, it is produced in large quantities by the cellulolytic flagellated protists as a byproduct of anaerobic fermentation in specialized organelles called hydrogenosomes. These protists are equipped with highly active iron-only hydrogenase enzymes that efficiently catalyze the reduction of protons to molecular hydrogen.

Certain gut yeasts can produce bioethanol, while the anaerobic environment of the termite hindgut also supports microbes that generate gaseous fuels such as biohydrogen and biomethane. This diversity highlights the termite gut as a natural prototype for industrial-scale biofuel production. Additionally, the gut harbors oleaginous (oil-producing) yeasts capable of accumulating substantial intracellular lipids. These microbial oils represent a promising feedstock for biodiesel production, providing an additional route for renewable energy generation (Sero et al., 2025).

The gut harbors a community of "oleaginous" yeasts, which are microorganisms capable of accumulating lipids, primarily in the form of triacylglycerols (TAGs), to levels exceeding 20% of their dry cell weight—the benchmark for industrial feasibility (Patel et al., 2020).

Bioprospecting efforts have successfully isolated and characterized several potent oleaginous yeast strains from the guts of wood-feeding termites. A particularly remarkable example is *Meyerozyma caribbica* strain SSA1654, which was found to accumulate lipids to an exceptional level of 47.25% of its dry weight (Al-Tohamy et al., 2021). Other promising oleaginous species identified from termite guts include *Meyerozyma guilliermondii* (44.7% lipid), *Vanrija humicola* (43.5% lipid), and *Debaryomyces hansenii* (34.0% lipid) (Ali et al., 2022).

5.7.3 Waste Management

The termite gut microbiome also demonstrates remarkable potential in bioremediation and waste management, particularly in degrading synthetic and industrial pollutants. Certain gut-associated bacteria, such as *Bacillus cereus* and *Lysinibacillus*, have been reported to degrade approximately 24.7% and 27.8% of polyethylene bags, respectively, within 30 days (Fitriana et al., 2022). Other microbial species, like *Acinetobacter tandoii*, can completely break down toxic industrial compounds such as phenol (Vane et al., 2003). Additionally, specific yeast species from termite-associated microbiomes have been successfully applied to the bioremediation of textile wastewater and azo dyes, illustrating the broad applicability of these microbes in environmental cleanup (Enagbonma et al., 2025).

This natural adaptation has direct applications in bioremediation. A strain of *A. tandoii* isolated from the gut of a wood-feeding termite was shown to be capable of utilizing phenol as its sole source of carbon and energy. This bacterium was able to completely degrade phenol at a concentration of 280 mg/L (Dexter and Boopathy, 2019).

The key to the yeasts' efficacy lies in their secretion of powerful extracellular oxidative enzymes, particularly manganese-dependent peroxidase (MnP) (Ali et al., 2021). This study constructed a novel yeast consortium, NYC-1, comprising four MnP-producing oleaginous strains isolated from wood-feeding termites, including *M. caribbica* and *D. hansenii*. This consortium demonstrated the ability to efficiently decolorize a range of different azo dyes at concentrations up to 250 mg/L, often achieving near-complete color removal within 24 hours. The MnP enzyme attacks the chromophore of the dye molecule, breaking the azo bond and initiating the degradation process, which can lead to complete mineralization into non-toxic products like CO_2 and H_2O.

Another study focused on a strain of *Pseudomonas aeruginosa* (LDPE-DB26) isolated from the gut of the Formosan subterranean termite, *Coptotermes formosanus*. This strain was able to achieve an 11.6% weight loss of low-density polyethylene (LDPE) film after 45 days of incubation (Ali et al., 2025). This degradation is not merely a physical fragmentation of the plastic into microplastics but involves true chemical alteration of the polymer structure. Molecular weight analysis of the LDPE film treated with *P. aeruginosa* revealed a 5.8% decrease in weight-averaged molecular weight and a substantial 32.7% reduction in number-averaged molecular weight. The significant drop in averaged molecular weight is a clear indicator of polymer chain scission, confirming that the bacteria are chemically breaking the long polyethylene chains into smaller fragments.

The combination of biofuel production and pollutant degradation underscores the termite gut microbiome's dual value as a biotechnological resource. By mimicking the metabolic capabilities and synergistic microbial networks found in termites, researchers can develop sustainable strategies for converting plant biomass into energy and for mitigating environmental pollutants, bridging natural microbial efficiency with industrial applications.

5.8 CONCLUSION

The evolutionary history of the termite holobiont—the termite and its associated gut microbiota—highlights the profound impact of symbiosis on ecological success and evolutionary innovation. The initial acquisition of cellulolytic protozoa allowed early termites to exploit the wood-feeding niche, driving the development of their complex social behaviors. A later, pivotal transition occurred in the ancestor of the Termitidae family, where the gut microbiome shifted from protozoan-dominated to prokaryote-dominated. This reorganization provided remarkable metabolic flexibility, enabling higher termites to access a broader range of ecological niches and establishing them as the most species-rich and ecologically dominant termite lineage. These findings underscore that evolution acts not only on the host but on the entire holobiont, and that shifts in symbiotic partnerships can profoundly reshape a host's evolutionary trajectory.

From a biotechnological perspective, the termite gut offers a natural blueprint for sustainable biomass conversion. Over millions of years, termites and their microbial partners have perfected a system capable of efficiently degrading lignocellulose and converting it into energy rich compounds, biofuels, and other valuable chemicals. Modern genomic and metagenomic tools allow researchers to unravel this intricate system, providing insights that can be translated into innovative biotechnologies. By emulating the termite gut's microbial efficiency and metabolic synergy, humanity can advance toward a sustainable bioeconomy, harnessing nature's evolutionary solutions to meet global energy and environmental challenges.

ACKNOWLEDGMENTS

The author (ST) acknowledges the Abraham Thomas Foundation for providing the resources for this work.

REFERENCES

Abik F, Palasingh C, Bhattarai M, Leivers S, Ström A, Westereng B, Mikkonen KS, Nypelö T. Potential of wood hemicelluloses and their derivates as food ingredients. *J Agric Food Chem*. 2023;71(6):2667–2683.

Ahmad F, Yang G, Zhu Y, Poulsen M, Li W, Yu T, Mo J. Tripartite symbiotic digestion of lignocellulose in the digestive system of a fungus-growing termite. *Microbiol Spectr*. 2022;10(6):e0123422.

Al-Tohamy R, Sun J, Khalil MA, Kornaros M, Ali SS. Wood-feeding termite gut symbionts as a promising source of novel manganese peroxidase-producing oleaginous yeasts for azo dye decolorization and biodiesel production. *Biotechnol Biofuels*. 2021;14(1):229.

Alberts B, Johnson A, Lewis J, Raff M, Roberts K, Walter P. The plant cell wall. In: *Molecular Biology of the Cell*. 4th ed. Garland Science; 2002.

Ali SS, Al-Tohamy R, Elsamahy T, Sun J. Harnessing recalcitrant lignocellulosic biomass for enhanced biohydrogen production: recent advances, challenges, and future perspective. *Biotechnol Adv*. 2024;72:108344.

Ali SS, Al-Tohamy R, Koutra E, Kornaros M, Khalil M, Elsamahy T, El-Shetehy M, Sun J. Coupling azo dye degradation and biodiesel production by manganese-dependent peroxidase-producing oleaginous yeasts isolated from wood-feeding termite gut symbionts. *Biotechnol Biofuels*. 2021;14(1):61.

Ali SS, Al-Tohamy R, Mohamed TM, Mahmoud YA, Ruiz HA, Sun L, Sun J. Could termites be hiding a goldmine of obscure yet promising yeasts for energy crisis solutions based on aromatic wastes? *Biotechnol Biofuels Bioprod*. 2022;15(1):35.

Ali SS, Sun J, Al-Tohamy R, Khalil MA, Elsamahy T, Schagerl M, Zhu D, El-Sapagh S. Harnessing and degradation mechanism of persistent polyethylene waste by newly isolated bacteria from waxworm and termite gut symbionts. *Microorganisms*. 2025;13(8):1929.

Auer L, Lazuka A, Sillam-Dussès D, Miambi E, O'Donohue M, Hernandez-Raquet G. Uncovering the potential of termite gut microbiome for lignocellulose bioconversion in anaerobic batch bioreactors. *Front Microbiol*. 2017;8:1–14.

Boerjan W, Ralph J, Baucher M. Lignin biosynthesis. *Annu Rev Plant Biol*. 2003;54:519–546.

Breznak JA, Switzer JM. Acetate synthesis from H(2) plus CO(2) by termite gut microbes. *Appl Environ Microbiol*. 1986;52(4):623–630.

Brune A. Symbiotic digestion of lignocellulose in termite guts. *Nat Rev Microbiol*. 2014;12(3):168–180.

Brune A, Dietrich C. The gut microbiota of termites: digesting the diversity in the light of ecology and evolution. *Annu Rev Microbiol*. 2015;69:145–166.

Chandel AK, Singh OV. Weedy lignocellulosic feedstock and microbial metabolic engineering: advancing the generation of biofuel. *Appl Microbiol Biotechnol*. 2011;89(5):1289–1303.

Chouvenc T, Šobotník J, Engel MS, Bourguignon T. Termite evolution: mutualistic associations, key innovations, and the rise of Termitidae. *Cell Mol Life Sci*. 2021;78(6):2749–2769.

Clarke CE, Francis ML, Sakala BJ, Hattingh M, Miller JA. Enhanced carbon storage in semi-arid soils through termite activity. *Catena*. 2023;232:107373.

Cole JJ, Hararuk O, Solomon CT. Chapter 7 –The carbon cycle: with a brief introduction to global biogeochemistry. In: Weathers KC, Strayer DL, Likens GE, eds. *Fundamentals of Ecosystem Science*. 2nd ed. Academic Press; 2021:131–160.

Cosgrove DJ. Growth of the plant cell wall. *Nat Rev Mol Cell Biol*. 2005;6(11):850–861.

Dar MA, Xie R, Zabed HM, Ali S, Zhu D, Sun J. Termite microbial symbiosis as a model for innovative design of lignocellulosic future biorefinery. *Biomass*. 2024;4(1):180–201.

Duran-Bautista EH, Silva-Olaya AM, Llanos-Cabrera MP, Yalanda-Sepúlveda K, Suárez JC. Soil-feeding termites build islands of soil fertility in pastures in the Colombian Amazon. *PLoS One*. 2025;20(4):e0322186.

Enagbonma BJ, Mmushi R, Babalola OO. Biotechnological utilization: the potential role of the termite gut symbiotic microbiome. *Symbiosis*. 2025;95(3):307–316.

Fitriana Y, Tampubolon DAT, Suharjo R, Lestari P, Swibawa IG. Lysinibacillus fusiformis and Paenibacillus alvei obtained from termites as antagonists of plant pathogenic fungi. *Plant Pathol J*. 2022;38(5):449–460.

Gayathri R, Mahboob S, Govindarajan M, Al-Ghanim KA, Ahmed Z, Al-Mulhm N, Vodovnik M, Vijayalakshmi S. A review on biological carbon sequestration: a sustainable solution for a cleaner air environment. *J King Saud Univ Sci*. 2021;33(2):101282.

Halim MA, Wong C, Zarkasi KZ, Rahman H, Abdullah NA, Majid AHA, Amin Z, Hussin NA. Exploring termite–microbe symbiosis: mechanisms, evolution and biotechnological potential. *Asia Pac J Mol Biol Biotechnol*. 2025;33(2), 41–53.

Hellemans S, Rocha MM, Wang M, et al. Genomic data provide insights into the classification of extant termites. *Nat Commun*. 2024;15:6724.

Hohmann-Marriott MF, Blankenship RE. Evolution of photosynthesis. *Annu Rev Plant Biol*. 2011;62:515–548.

Hsin KT, Lee H, Huang YC, Lin GJ, Lin PY, Lin YCJ, Chen PY. Lignocellulose degradation in bacteria and fungi: cellulosomes and industrial relevance. *Front Microbiol*. 2025;16:1–12.

Inward D, Beccaloni G, Eggleton P. Death of an order: a comprehensive molecular phylogenetic study confirms that termites are eusocial cockroaches. *Biol Lett*. 2007;3(3):331–335.

Jouquet P, Traoré S, Choosai C, Hartmann C, Bignell D. Influence of termites on ecosystem functioning: ecosystem services provided by termites. *Eur J Soil Biol*. 2011;47(4):215–222.

Katoh H, Miura T, Maekawa K, Shinzato N, Matsumoto T. Genetic variation of symbiotic fungi cultivated by the termite Odontotermes formosanus. *Mol Ecol*. 2002;11(8):1565–1572.

Lazuka A, Auer L, O'Donohue M, Hernandez-Raquet G. Anaerobic lignocellulolytic microbial consortium derived from termite gut: enrichment and community dynamics. *Biotechnol Biofuels*. 2018;11(1):284.

Leadbetter JR, Breznak JA. Physiological ecology of Methanobrevibacter species from termite hindguts. *Appl Environ Microbiol*. 1996;62(10):3620–3631.

Lee H, Bae J, Jin S, Kang S, Cho BK. Engineering acetogenic bacteria for efficient one-carbon utilization. *Front Microbiol*. 2022;13:865168.

Makonde HM, Boga HI, Osiemo Z, Mwirichia R, Stielow JB, Göker M, Klenk HP. Diversity of Termitomyces associated with fungus-farming termites. *PLoS One*. 2013;8(2):e56464.

McNamara JT, Morgan JLW, Zimmer J. A molecular description of cellulose biosynthesis. *Annu Rev Biochem*. 2015;84:895–921.

Mirabito D, Rosengaus RB. A double-edged sword? The cost of proctodeal trophallaxis in termites. *Insect Soc*. 2016;63(1):135–141.

Mujinya BB, Mees F, Boeckx P, Bodé S, Baert G, Erens H, Delefortrie S, Verdoodt A, Ngongo M, Van Ranst E. The origin of carbonates in termite mounds of the Lubumbashi area, D.R. Congo. *Geoderma*. 2011;165(1):95–105.

Myer A, Myer MH, Trettin CC, Forschler BT. The fate of carbon utilized by the subterranean termite Reticulitermes flavipes. *Ecosphere*. 2021;12(12):e03783.

Nalepa CA. Origin of mutualism between termites and flagellated gut protists: transition from horizontal to vertical transmission. *Front Ecol Evol*. 2020;8:1–10.

Ohkuma M. Symbioses of flagellates and prokaryotes in the gut of lower termites. *Trends Microbiol*. 2008;16(7):345–352.

Patel A, Karageorgou D, Rova E, Katapodis P, Rova U, Christakopoulos P, Matsakas L. An overview of potential oleaginous microorganisms and their role in biodiesel and omega-3 fatty acid-based industries. *Microorganisms*. 2020;8(3):434.

Pester M, Tholen A, Friedrich MW, Brune A. Methane oxidation in termite hindguts: absence of evidence and evidence of absence. *Appl Environ Microbiol*. 2007;73(6):2024–2028.

Peterson BF, Scharf ME. Lower termite associations with microbes: synergy, protection, and interplay. *Front Microbiol*. 2016;7:422.

Ragsdale SW. Enzymology of the Wood-Ljungdahl pathway of acetogenesis. *Ann N Y Acad Sci*. 2008;1125:129–136.

Ragsdale SW, Pierce E. Acetogenesis and the Wood-Ljungdahl pathway of CO_2 fixation. *Biochim Biophys Acta*. 2008;1784(12):1873–1898.

Raina A, Osbrink W, Park YI. Nymphs of the Formosan subterranean termite: aspects of formation and transformation. *Ann Entomol Soc Am*. 2004;97(4):757–764.

Ralph J, Lapierre C, Boerjan W. Lignin structure and its engineering. *Curr Opin Biotechnol*. 2019;56:240–249.

Rosenberg E, Zilber-Rosenberg I. Reconstitution and transmission of gut microbiomes between generations. *Microorganisms*. 2022;10(1):70.

Rowell RM, Pettersen R, Tshabalala MA. Cell wall chemistry. In: Rowell R, ed. *Handbook of Wood Chemistry and Wood Composites*. 2nd ed. CRC Press; 2013:33–72.

Rust MK, Su NY. Managing social insects of urban importance. *Annu Rev Entomol*. 2012;57:355–375.

Salgado JFM, Hervé V, Vera MAG, Tokuda G, Brune A. Unveiling lignocellulolytic potential: a genomic exploration of bacterial lineages within the termite gut. *Microbiome*. 2024;12(1):201.

Schalk F, Gostinčar C, Kreuzenbeck NB, Conlon BH, Sommerwerk E, Rabe P, Burkhardt I, Krüger T, Kniemeyer O, Brakhage AA, Gunde-Cimerman N, de Beer ZW, Dickschat JS, Poulsen M, Beemelmanns C. The termite fungal cultivar Termitomyces combines diverse enzymes and oxidative reactions for plant biomass conversion. *mBio*. 2021;12(3):e03551–20.

Sero ET, Amobonye A, Bhagwat P, Singh NK, de Carvalho JC, Permaul K, Pillai S. Exploring the termite gut as a hub of industrially important microbes and enzymes for biofuel production. *Biomass Bioenergy*. 2025;199:107899.

Sethupathy S, Morales GM, Li Y, Wang Y, Jiang J, Sun J, Zhu D. Harnessing microbial wealth for lignocellulose biomass valorization through secretomics: a review. *Biotechnol Biofuels*. 2021;14(1):154.

Van Dexter S, Boopathy R. Biodegradation of phenol by Acinetobacter tandoii isolated from the gut of the termite. *Environ Sci Pollut Res Int.* 2019;26(33):34067–34072.

Vane CH. The molecular composition of lignin in spruce decayed by white-rot fungi (Phanerochaete chrysosporium and Trametes versicolor) using pyrolysis-GC–MS and thermochemolysis with tetramethylammonium hydroxide. *International Biodeterioration & Biodegradation* 2003;51: 67–75.

Vreeburg SME, Auxier B, Jacobs B, Bourke PM, van den Heuvel J, Zwaan BJ, Aanen DK. A genetic linkage map and improved genome assembly of the termite symbiont Termitomyces cryptogamus. *BMC Genomics.* 2023;24(1):123.

Vreeburg SME, de Ruijter NCA, Zwaan BJ, da Costa RR, Poulsen M, Aanen DK. Asexual and sexual reproduction are two separate developmental pathways in a Termitomyces species. *Biol Lett.* 2020;16(8):20200394.

War AR, Paulraj MG, Ahmad T, Buhroo AA, Hussain B, Ignacimuthu S, Sharma HC. Mechanisms of plant defense against insect herbivores. *Plant Signal Behav.* 2012;7(10):1306–1320.

Wijas BJ, Flores-Moreno H, Allison SD, Rodriguez LC, Cheesman AW, Cernusak LA, Clement R, Cornwell WK, Duan ES, Eggleton P, Rosenfield MV, Yatsko AR, Zanne AE. Drivers of wood decay in tropical ecosystems: termites versus microbes. *Funct Ecol.* 2024;38(3):546–559.

Wisselink M, Aanen DK, van 't Padje A. The longevity of colonies of fungus-growing termites and the stability of the symbiosis. *Insects.* 2020;11(8):527.

Xie R, Dong C, Wang S, Danso B, Dar MA, Pandit RS, Pawar KD, Geng A, Zhu D, Li X, Xu Q, Sun J. Host-specific diversity of culturable bacteria in fungus-growing termites and their potential functions in lignocellulose bioconversion. *Insects.* 2023;14(4):403.

CHAPTER SIX

The Tick Microbiome

COMPLEX INTERPLAY BETWEEN TICK VECTORS, PATHOGENS, AND ENDOSYMBIONTS

Pitambar Dhakal, Yin Fu, and Longxian Zhang

6.1 BACKGROUND

6.1.1 General Account on Tick Diversity and Biology

Ticks (superfamily: Ixodoidea) are the obligatory nonpermanent sanguivorous ectoparasitic arthropods having a long history of existence since the Mid-late Cretaceous era with substantial evolutionary adaptations (Estrada-Peña and de la Fuente, 2018). They can parasitize and feed on all kinds of terrestrial vertebrate hosts, including amphibians, reptiles, birds, mammals (Anderson and Magnarelli, 2008), and even extinct reptiles like dinosaurs (Peñalver et al., 2017). Ticks are extensively distributed in diverse habitats worldwide, even in extremely cold environments such as Antarctica, with the highest species diversity in the tropics and subtropics (Pettersson et al., 2020). Considering species diversity, Ixodoidea is divided into three families – Ixodidae, Argasidae, and Nuttalliellidae – on the basis of morphological features (Damian et al., 2020). There are 911 tick species distributed worldwide, of which Ixodidae (hard ticks) is the most diverse and widespread with 14 genera and 709 species, Argasidae (soft ticks) includes 5 genera and 201 species, and Nuttalliellidae (monotypic) includes only a single genus and species (*Nuttalliella namaqua*) found in Europe (Barker and Murrell, 2004; Nava et al., 2009). Ticks cause serious public health concerns due to their global impact on humans and animals, and about 28 species are considered to be the vectors for human pathogens (Anderson and Magnarelli, 2008).

The tick life cycle completes through four life stages – eggs, larvae (six-legged), nymph, and adult. The two later stages are eight-legged. The duration of their life cycle varies but is relatively long. Ixodid ticks (members of the family Ixodidae) can survive up to 2–6 years while Argasid ticks can even live as long as 20 years (Anderson and Magnarelli, 2008). Thousands of eggs can be deposited by individual females. After hatching, they must feed on blood, the sole source of nutrition; hard ticks take one blood meal at each developmental stage while soft ticks require repeated blood meal at each stage (Anderson and Magnarelli, 2008). Molting and reproduction are regulated by temperature, day-length, and the amount and frequency of ingested blood. Based on their host preference and duration, life cycle of ticks can be divided into one-host, two-host, three-host, or multi-host life cycles (Figure 6.1). Ticks of one-host cycle (e.g., *Rhipicephalus annulatus*) remain on the same host for entire feeding stages (larval to adult stages) and only leave the host before laying eggs. The ticks using two hosts (e.g.,

 DOI: 10.1201/9781003639176-6

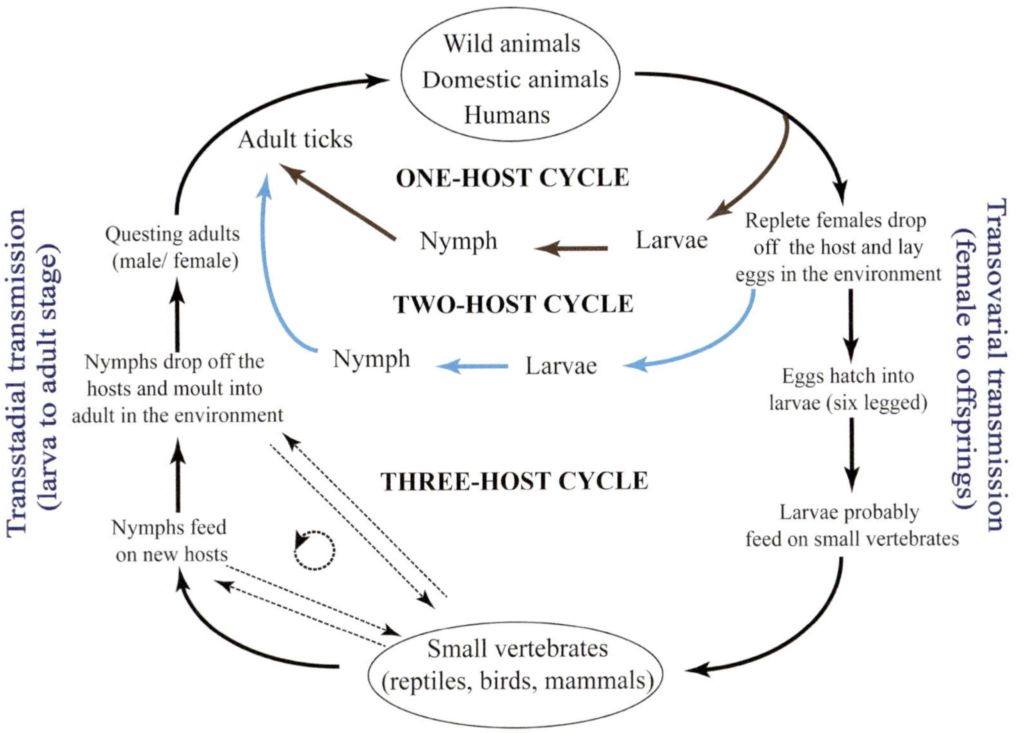

Figure 6.1 A generalized life cycle pattern of ticks. One-host cycle (brown arrows): larvae, nymph, and adult all feed on the same host; two-host cycle (blue arrows): larvae feed and molt in the first host, engorged nymphs leave the first host, molt into adults, adults attach to the second host, feed, and mate; three/multiple-host cycle (black arrows): larval feeding in the first host, larvae leave the first host, molt into nymph, and undergo overwintering, nymphal feeding on the second host, nymphs leave the second host and molt into adult, adults attach on the third host, feed, and mate. The dotted arrows (straight and circular) indicate repeated feeding of ticks through three/multiple-host cycle where sharing of microbes has higher possibilities. Illustration created on Adobe Illustrator CC 2019.

Hyalomma marginatum) undergo molting from larva to nymph on their first host and then leave the first host. The second host could be the same previous individual host of the same species, or another species. Likewise, ticks with three host cycles leave the host after larval and nymphal stages. The three hosts may or may not be the same species or even may be the same individual, which depends on the availability of the hosts. Ticks of the genera *Ixodes*, *Amblyomma*, *Dermacentor*, and *Rhipicephalus* usually lead three-host cycle. Argasid ticks (members of the family Argasidae) due to their two or more nymphal stages (maybe as many as eight instars) (Anderson and Magnarelli, 2008) undergo multi-host life cycle. Each nymphal instar requires a blood meal and may use the same or different host. Therefore, the tick life cycle has 'on-host' and 'off-host' phases during which the latter phases are exposed to harsh environmental conditions, including extreme temperature fluctuations and humidity, and both the phases can potentially acquire symbionts and/or pathogens.

6.1.2 Chapter Objective

This chapter is mainly aimed at summarizing the existing data on tick microbial communities, including tick-borne pathogens and endosymbionts, and their tripartite interactions and dependence for existence. Moreover, tick immune system and its role in interaction with pathogens has been highlighted since ticks provide a microhabitat for both pathogens and endosymbionts.

6.2 TICKS AS POTENTIAL DISEASE VECTORS

Arthropods, including ticks, vector a plethora of pathogenic and non-pathogenic microbes within them. The entire conglomerate of symbionts, commensals, and pathogens which inhabit a particular niche or multiple loci within the host's body are considered to be microbiome (Hooper and Gordon, 2001) (Figure 6.2). Among potential arthropod vectors, ticks stand the second position in terms of their vectorial potential (Xue et al., 2023). They harbor a wide array of severely pathogenic viruses, bacteria, protozoa, and helminths,; therefore, they serve as competent vectors as they have been known to transmit these pathogens to humans, domestic animals, and wild animals (Moming et al., 2024). Approximately 10% of tick species are known to serve as vectors of pathogens (Gondard et al., 2017). Ticks are provided with transovarially transmitting bacterial endosymbionts (Kloc et al., 2024) as well as maternal microbiome which serves as the initial inoculum in their eggs and subsequent larvae (Narasimhan et al., 2014). Besides eubacteria, the microbiome harbored by both hard and soft ticks comprises viruses (Shi et al., 2018; Antúnez et al., 2024), archaea (Zhang et al., 2021), and eukaryotes including protozoa (Hornok et al., 2022a,b), fungi (Bruley et al., 2025), and helminths, particularly filarial nematodes (Ajileye et al., 2025) which have been briefly discussed in this chapter. However, Ixodidae stands as the most prominent carrier of zoonotic diseases (Antúnez et al., 2024). Ticks generally acquire pathogens in their early life stage (larval or nymphal stages) while sucking blood and later transmit them to the other hosts. Acquisition of pathogens or symbionts within the ticks occurs during blood meal from a host which contains single or multiple pathogens, or sequence of feeding during their different life stages on various infected hosts, generally the reservoirs (Halos et al., 2005). Even a single tick can vector multiple pathogens at the same time since a tick switches to various hosts during life stages and several Tick-borne pathogens (TBPs) coexist in the same field, leading to the possibility of transmitting coinfection to the animals and humans as tick-borne disease (TBD) epidemics (Moutailler et al., 2016).

6.2.1 Tick-Borne Viruses (TBVs)

The recognition of ticks as carriers and vectors of pathogenic viruses dates back over a century, marked by the identification of the flavivirus Louping ill virus (LIV) (Stockman, 1918) as the cause of severe encephalitis in sheep (Shi et al., 2018). Since then, a wide array of viruses has been recognized as being circulated between ticks and their hosts and constitutes tick-borne viruses (TBVs) (Table 6.1). Their prevalence is on the rise, with the most frequently reported cases often linked to widespread epidemics (Shi et al., 2018). Until recently, 105 TBVs have been recognized to be capable of successful host-to-host transmission, of which 59 are found to be human pathogens (Tian et al., 2025). Due to anthropogenic changes, TBVs are being identified with increasing prevalence and new TBVs are being reported. Approximately, 10% of all tick species (including Argasidae and Ixodidae) are known to serve as vectors of arboviruses (arthropod-borne viruses – viruses transmitted by blood-sucking arthropods). There are highly specific virus–tick–vertebrate host associations with regard to the dissemination of TBVs. About 25% of these are associated with diseases, and notably, all are zoonotic (Kazimirova et al., 2017).

Considering genomic configuration, except the African swine fever virus (ASFV) (family: Asfarviridae) which has a DNA genome, other TBVs possess RNA genomes including single-stranded or double-stranded, positive-sense or negative-sense, monopartite or segmented viruses (Vandegrift and Kapoor, 2019; Madison-Antenucci et al., 2020). The spectrum of TBVs taxonomically belongs to two orders (Bunyavirales, Mononegavirales), nine families (Nairoviridae, Peribunyaviridae, Phenuiviridae, Nyamiviridae, Rhabdoviridae, Asfarviridae, Flaviviridae, Orthomyxoviridae, and Reoviridae), and 12 genera, along with other unassigned members (Shi et al., 2018). The tick-borne flaviviruses, categorized into seabird and mammalian groups (International Committee on Taxonomy of Viruses (ICTV), 2024), belong to the genus *Orthoflavivirus* provided with a positive-sense single-stranded RNA genome. Seabird group comprises Tyuleniy virus, Meaban virus, and Saumarez Reef virus, while the mammalian group consists of human-pathogenic viruses, the tick-borne encephalitis (TBE) serocomplex, and other viruses which have more than 70% nucleotide homology in their envelope gene (Yu and Park, 2024). The family Nairoviridae includes the tick-borne *Orthonairovirus* genus, which possesses three segments of a negative-sense single-strand RNA genome (Kuhn et al., 2024) and consists of 53 viruses being assigned to seven valid serogroups as species (Aslam et al., 2023). It includes the Crimean–Congo hemorrhagic fever virus (CCHFV), Dugbe virus, Nairobi sheep disease virus, Hazara orthonairovirus, Kasokero virus, and Erve virus. Most human Nairoviridae infections are associated with CCHFV, with high mortality

TABLE 6.1

Globally reported common tick-borne viruses and their vector ticks

Tick-borne viruses	Location	Tick species	Sources of ticks	References
African swine fever virus (ASFV)	Senegal, Malawi, Madagascar, Portugal, Mozambique	*Ornithodoros sonrai*, *Ornithodoros moubata*, *Ornithodoros porcinus*, *Ornithodoros erraticus*, *O. porcinus domesticus*, *O. porcinus*	Pig pens/farms, warthog burrows, rodent burrows in wild condition	Boinas et al. (2004), Vial et al. (2007), Ravaomanana et al. (2010), Quembo et al. (2018), and Anholt et al. (2023)
African swine fever virus (ASFV) (genotype I)	Zambia, South Africa	*Ornithodoros Moubata*	Loose soil and litter of national park, warthog, private farms, nature reserves	Chambaro et al. (2020)
African swine fever virus (ASFV) genotype III	South Africa	*Ornithodoros compactus*	Warthog, private farms, nature reserves	Craig et al. (2022)
African swine fever virus (ASFV) (genotype IX)	Kenya	*O. porcinus*	Warthog burrows	Obanda et al. (2024)
African swine fever virus (ASFV) (genotype XV)	Tanzania	*O. moubata* (complex group)	Warthog burrows	Peter et al. (2021)
African swine fever virus (ASFV) (genotype XX, XXI, XXII)	South Africa	*Ornithodoros waterbergensis*, *Ornithodoros phacochoerus*	Warthog, private farms, nature reserves	Craig et al. (2022)
Alkhurma hemorrhagic fever virus (AHFV)	Turkey, Saudi Arabia, Greece	*Hyalomma rufipes*, *Ornithodoros savignyi*, *H. rufipes*	Migratory birds, camels, camel resting places	Charrel et al. (2007) and Hoffman et al. (2018)
Alongshan virus (ALSV)	Switzerland, Germany, Finland, Russia, China	*Ixodes ricinus*, *Ixodes persulcatus*, *Dermacentor reticulatus*, *Haemaphysalis longicornis*	Cattle, vegetation, hunted game animals (roe deer, red deer, fallow deer, wild boar), vegetation (flagging)	Kuivanen et al. (2019), Kholodilov et al. (2020), Kholodilov et al. (2022), Stegmuller et al. (2023), Ebert et al. (2023), Cai et al. (2023), Su et al. (2024), and Xiao et al. (2024)
Alphadintovirus mayetiola	China	*Haemaphysalis qinghaiensis*	Yak, Tibetan sheep	Zhang et al. (2024)

(*Continued*)

TABLE 6.1
(Continued)

Tick-borne viruses	Location	Tick species	Sources of ticks	References
Amblyomma dissimile mivirus (ADM)	Trinidad and Tobago	*A. dissimile*	Wildlife (*Iguana mi virus* and *Mazama americana*) (postmortem collection)	Sameroff et al. (2021)
Baishan forest tick virus (BSFTV)	China (China-Korea border	*I. persulcatus*	Cattle, sheep, grassland, forest	Wang et al. (2024)
Balambala tick virus (BMTV)	Romania	*Hyalomma marginatum*	Sheep, goat, environment	Bratuleanu et al. (2024)
Beiji nairovirus (BJNV)	China, China (China-Korea border	*I. persulcatus*	Cattle, sheep, vegetation	Cai et al. (2023), Wang et al. (2023), and Su et al. (2024)
Bole tick virus 3 (BLTV3)	China	*H. marginatum*, *H. longicornis*	Goat, other animals, vegetation	Su et al. (2024) and Xiao et al. (2024)
Bole tick virus 4 (BLTV4)	China, Romania	*Dermacentor nuttalli*, *H. marginatum*	Sheep, goat, vegetation	Su et al. (2024) and Bratuleanu et al. (2024)
Bourbon virus	USA	*Amblyomma americanum*	Vegetation	Aziati et al. (2023)
Bunya virus (soft tick bunyavirus; STBV)	Japan	*Argas vespertilionis*	Feces beneath bat colonies, human habitats	Oba et al. (2016)
Bunyamwera orthobunyavirus	Brazil	*Amblyomma* spp.	Vegetation	Binder et al. (2019)
Butlers creek tick virus (Chuviridae)	Australia	*H. bancrofti*	Vegetation	Gofton et al. (2022)
Changping tick virus 1 (*Phlebovirus*)	China	*Dermacentor* sp.	Cattle, vegetation	Cai et al. (2023)
Changping tick virus 2 (Chuviridae)	China	*H. longicornis*	Animals, vegetation	Xiao et al. (2024)

Colorado tick fever virus (CTFV)	USA	*Dermacentor occidentalis*, *Dermacentor andersoni*	Vegetation: national park vegetation	Geissler et al. (2014), Williamson et al. (2019), and Padgett et al. (2022)
Crimean–Congo hemorrhagic fever virus (CCHFV)	Spain	*I. simplex*	Bats (*Hypsugo savii*, *Myotis emarginatus*)	Moraga-Fernández et al. (2024)
	West Africa	*Hyalomma impeltatum*	Cattle, camels	Schulz et al. (2021)
	Albania	*Hyalomma* spp.	Cattle	Papa et al. (2009)
	Bulgaria	*H. marginatum*, *Rhipicephalus sanguineus*, *I. ricinus*	Cattle	Gergova et al. (2012)
	Cameroon	*Hyalomma truncatum*	Cattle	Simo Tchetgna et al. (2023)
	China	*Hyalomma asiaticum*, *H. asiaticum*, *D. nuttalli*, *Hyalomma detritum*	Vegetation, goat, cattle	Xia et al. (2011), Guo et al. (2017), and Moming et al. (2018)
	Ghana	*H. truncatum*, *H. rufipes*, *Amblyomma variegatum*	Cattle, horse	Bentil et al. (2023)
	Guinea	*A. variegatum*, *Rhipicephalus geigyi*, *R. sanguineus* (*s.l.*), *Rhipicephalus decoloratus*, *R. geigyi*, *Rhipicephalus annulatus*	Cattle, domestic and roving dogs, small mammals	Naidenova et al. (2020)
	India	*Hyalomma anatolicum*	Household	Yadav et al. (2014)
	Iran	*H. marginatum*, *H. anatolicum*, *Hyalomma dromedarii*, *H. detritum*, *H. asiaticum*, *Heliocybe sulcata*, *R. sanguineus*	Sheep, goats, cows, chickens, camel, rodents, turtles, hedgehogs	Yaser et al. (2011), Kayedi et al. (2015), Biglari et al. (2016), Farhadpour et al. (2016), and Champour et al. (2016a, 2016b), and Saghafipour et al. (2019)

(*Continued*)

TABLE 6.1
(Continued)

Tick-borne viruses	Location	Tick species	Sources of ticks	References
	Ivory Coast	*A. variegatum*, *Hyalomma impressum*, *R. (Boophilus) geigyi*, *Rhipicephalus (Boophilus) microplus*, *R. sanguineus*	Cattle	Adjogoua et al. (2021)
	Kazakhstan	*H. asiaticum*, *I. ricinus*	Vegetation	Sultankulova et al. (2022)
	Mongolia	*H. asiaticum*	Livestock, vegetation	Voorhees et al. (2018)
	Russia	*H. marginatum*, *Hyalomma scupense*, *Rhipicephalus rossicus*, *Rhipicephalus turanicus*, *Rhipicephalus bursa*, *R. annulatus*, *Dermacentor marginatus*, *Haemaphysalis punctata*, *I. ricinus*	Cattle, sheep, horse, *Lepus europaeus*, man, vegetation, *Corvus frugilegus*, *Galerida cristata*	Tsapko et al. (2022)
	Senegal	*Rhipicephalus evertsi*, *Rhipicephalus guilhoni*	Sheep	Mhamadi et al. (2022)
	Spain	*Hyalomma lusitanicum*, *H. marginatum*	Domestic and wild ungulates, red deer (*Cervus elaphus*)	Negredo et al. (2019)
	Sudan	*H. impeltatum*	Sheep	Chitimia-Dobler et al. (2019)
	Turkey	*Hyalomma aegyptium*, *R. sanguineus* complex, *R. bursa*, *H. marginatum*, *R. annulatus*, *Ixodes* sp., *H. anatolicum*, *H. detritum*, *H. marginatum*, *Hyalomma turanicum*, *R. turanicus*, *Haemaphysalis concinna*	Vegetation, dogs, sheep, goat, cattle, cat, human, tortoises (*Testudo hermanni*, *Testudo graeca*), great reed warblers (*Acrocephalus arundinaceus*), European robins (*Erithacus rubecula*)	Tonbak et al. (2006), Gargili et al. (2011), Tekin et al. (2012), Leblebicioglu et al. (2014), Dincer et al. (2017), Kar et al. (2020), and Dincer et al. (2022)
	United Arab Emirates	*H. dromedarii*	Camel	Camp et al. (2020)

	West Africa	*H. dromedarii*	Cattle, camels	Schulz et al. (2021)
Crimean–Congo hemorrhagic fever virus (CCHFV) (African genotype II (Africa 2) clade)	Uganda	*R. decoloratus*	Cattle	Wampande et al. (2021)
Crimean–Congo hemorrhagic fever virus (CCHFV) (lineages Africa I, Africa III)	Mauritania (West Africa)	*H. dromedarii*, *H. rufipes*	Cattle, camel	Schulz et al. (2021)
Crimean–Congo hemorrhagic fever virus (CCHFV); AP92-like Kastoria strain	Greece	*R. bursa*	Sheep	Papa et al. (2014)
Crimean–Congo hemorrhagic fever virus (CCHFV) (lineages Europe 1)	Greece	*R. sanguineus* (*s.l.*)	Sheep, goats	Papa et al. (2017)
Crimean–Congo hemorrhagic fever virus (CCHFV) (lineages Europe 2)	Greece	*R. bursa*	Sheep, goats	Papa et al. (2017)
Crimean–Congo hemorrhagic fever virus (CCHFV) (*Orthonairovirus*)	West Africa	*H. rufipes*	Cattle, camels	Schulz et al. (2021)
Crimean–Congo hemorrhagic fever virus (CCHFV) (strain HANM-18)	China	*H. asiaticum*	Sheep	Kong et al. (2022)
Crimean–Congo hemorrhagic fever virus (CCHFV) (strain HANM-19)	China	*H. asiaticum*	Camel	Kong et al. (2022)

(*Continued*)

TABLE 6.1
(Continued)

Tick-borne viruses	Location	Tick species	Sources of ticks	References
Crimean–Congo hemorrhagic fever virus (CCHFV) (strain HANM-20)	China	*H. dromedarii*	Camel	Kong et al. (2022)
Dabie bandavirus	China	*H. longicornis*	Vegetation	Lei et al. (2025)
Dabieshan tick virus (DBTV)	China	*H. longicornis*	Vegetation, animals	Shao et al. (2020), Bai et al. (2023), and Xiao et al. (2024)
Deer tick virus (DTV)	USA	*Ixodes scapularis*	*Odocoileus virginianus* (white-tailed deer)	Smalley et al. (2022)
Dugbe virus (DUGV)	Cameroon	*A. variegatum*	Cattle	Simo Tchetgna et al. (2023)
	Ghana	*A. variegatum*	Cattle	Kobayashi et al. (2017)
	Senegal	*A. variegatum*	Cattle, sheep, goat	Ndiaye et al. (2024)
Genoa virus (Chuviridae)	Australia	*Ixodes holocyclus*	Vegetation	Gofton et al. (2022)
Granville quaranjavirus (Orthomyxoviridae)	China	*H. longicornis*	Animals, vegetation	Xiao et al. (2024)
Heartland virus (HRTV)	USA	*A. americanum*	Vegetation	Savage et al. (2018), Newman et al. (2020), and Tuten et al. (2020)
Heilongjiang Tick Virus 1 HLJ/SL-3	China	*Dermacentor silvarum*	Cattle, vegetation	Cai et al. (2023)
Hepelivirales sp.	China	*H. longicornis*	Animals, vegetation	Xiao et al. (2024)
Huangpi tick virus 3 (Rhabdoviridae)	China	*H. longicornis*	Animals, vegetation	Xiao et al. (2024)

Hubei toti-like virus 24	China	*H. longicornis*	Animals, vegetation	Xiao et al. (2024)
H. marginatum rhabdovirus (HMRV)	Romania	*H. marginatum*	Sheep, goat, environment	Bratuleanu et al. (2024)
Ingleside virus (Virgaviridae)	Australia	*I. holocyclus*	Vegetation	Gofton et al. (2022)
Iwanai Valley virus (IWVV)	Japan	*Ixodes ovatus*	Vegetation	Matsumura et al. (2024)
I. holocyclus Iflavirus (Iflaviridae)	Australia	*I. holocyclus*	Vegetation	Gofton et al. (2022)
I. scapularis-associated virus 2	China	*H. longicornis*	Animals, vegetation	Xiao et al. (2024)
I. scapularis-associated virus 1 (ISAV1)	China	*I. persulcatus*	Forest shrubs	Su et al. (2024)
Jilin luteo-like virus 2 (JLLV2)	China	*I. persulcatus*	Forest shrubs	Su et al. (2024)
Jilin partiti-like virus 1	China	*I. persulcatus*	Forest shrubs	Su et al. (2024)
Jilin tick virus 1 (Nairoviridae)	China	*Hamamelis japonica*	Cattle, vegetation	Cai et al. (2023)
Jingmen tick virus (JMTV)	China	*H. longicornis*	Animals, vegetation	Xiao et al. (2024)
	Japan	*Amblyomma testudinarium*	Feeding ticks, vegetation	Kobayashi et al. (2021)
	Turkey	*Hydropotes inermis*, *R. sanguineus* (*s.l.*), *H. marginatum*, *Heliamphora parva*, *R. bursa*, *R. turanicus*	Cattle, dog, sheep, goat, cat	Dincer et al. (2019, 2022)
	Moscow	*I. ricinus*	Vegetation	Morozkin et al. (2022)
Jiutai virus isolate JL	China	*H. japonica*	Cattle, vegetation	Cai et al. (2023)

(*Continued*)

TABLE 6.1
(Continued)

Tick-borne viruses	Location	Tick species	Sources of ticks	References
Karyana virus (KARYV)	India	*H. anatolicum*	Buffalo, sheep, goat (CCHF positive cases)	Yadav et al. (2019)
Kasokero virus (KASV; genus *Orthonairovirus*)	Uganda	*Ornithodoros (Reticulinasus) faini*	Bat roosting sites	Schuh et al. (2020)
Kioloa tick virus (Rhabdoviridae)	Australia	*H. bancrofti*	Vegetation	Gofton et al. (2022)
Kismayo virus (Phenuiviridae)	China	*H. longicornis*	Animals, vegetation	Xiao et al. (2024)
Kundal virus (KUNDV)	India	*H. anatolicum*	Buffalo, sheep, goat (CCHF positive cases)	Yadav et al. (2019)
Kyasanur forest disease virus (KFDV)	India	*Haemaphysalis spinigera*, *Haemaphysalis turturis*, *H. spinigera*, *Haemaphysalis bispinosa*, *R. microplus*, *R. annulatus*	Vegetation	Balasubramanian et al. (2021) and Burthe et al. (2025)
Kyasanur forest disease virus Alkhurma subtype (Alkhurma hemorrhagic fever virus (AHFV))	Saudi Arabia	*H. dromedarii*, *O. savignyi*	Camel, ground (camel resting place)	Mahdi et al. (2011)
Lihan Tick virus (*Phlebovirus*, Phenuiviridae)	Colombia	*Dermacentor nitens*, *R. microplus*, *R. sanguineus* s.l.	Cattle, equines, bovines	Lopez et al. (2020)
Liman tick virus	China	*H. longicornis*	Animals, vegetation	Xiao et al. (2024)
Lone star tick densovirus 1 (Parvoviridae)	China	*H. longicornis*	Animals, vegetation	Xiao et al. (2024)
Lone star tick totivirus (Totiviridae)	China	*H. longicornis*	Animals, vegetation	Xiao et al. (2024)

Manly virus	China	*H. longicornis*	Animals, vegetation	Xiao et al. (2024)
Mudanjiang *phlebovirus* (MDJV)	China	*I. persulcatus*	Cattle, sheep, grassland, forest	Wang et al. (2023)
Mukawa virus (MKWV)	China	*I. persulcatus*	Forest shrubs; flagging	Su et al. (2024)
	China	*I. persulcatus*	Cattle, sheep, grassland, forest	Wang et al. (2023)
Nairobi sheep disease virus (NSDV)	China	*H. longicornis*	Unfed (vegetation), cattle, goat, sheep	Gong et al. (2015), Yang et al. (2019), Bai et al. (2023), and Wang et al. (2024)
Nairovirus	Spain	*I. simplex*	Bats (*Miniopterus schreibersii*)	Moraga-Fernández et al. (2024)
Newport tick virus (Flaviviridae)	Australia	*I. holocyclus*	Vegetation	Gofton et al. (2022)
North Shore virus (Partitiviridae)	Australia	*I. holocyclus*	Vegetation	Gofton et al. (2022)
Nuomin virus (NUMV)	China	*I. persulcatus*	Forest shrubs	Su et al. (2024)
Odawvirus (ODWV)	Ghana	*Rhipicephalus* spp.	Dog	Kobayashi et al. (2017)
O'Hara headland virus (Reoviridae)	Australia	*H. bancrofti*	Vegetation	Gofton et al. (2022)
Onega tick *phlebovirus* (OTPV)	China	*I. persulcatus*	Forest shrubs	Su et al. (2024)
	China	*I. persulcatus*	Cattle, sheep, grassland, forest	Wang et al. (2023)
Orf virus	China	*H. qinghaiensis*	Yak, Tibetan sheep	Zhang et al. (2024)
Parasteatoda house spider adintovirus	China	*H. qinghaiensis*	Yak, Tibetan sheep	Zhang et al. (2024)

(*Continued*)

TABLE 6.1
(Continued)

Tick-borne viruses	Location	Tick species	Sources of ticks	References
Powassan virus (POWV)	Canada	*Ixodes cookei*	Canada goose	Smith et al. (2018)
	USA	*Dermacentor variabilis*, *I. scapularis*, *Haemaphysalis leporispalustris*	Vegetation, eastern cottontails (*Sylvilagus floridanus*)	Hart et al. (2023), Xu et al. (2024), and Schiff et al. (2025)
Sara tick *phlebovirus* (STPV)	China	*I. persulcatus*	Forest shrubs, grassland, cattle, sheep	Wang et al. (2023) and Su et al. (2024)
Scerri virus (Iflaviridae)	Australia	*H. bancrofti*	Vegetation	Gofton et al. (2022)
Severe fever with thrombocytopenia syndrome virus (SFTSV)	China	*H. longicornis*, *Harperocallis flava*, *H. concinna*, *R. microplus*, *I. persulcatus*, *H. japonica*, *D. silvarum*	Residential and forested area, hedgehog, coat, cattle, grass, birds, dog, tea garden	Wang et al. (2015), Meng et al. (2015), Jiao et al. (2015), Xing et al. (2016), Li et al. (2016, 2020, 2022), Liu et al. (2016), Tian et al. (2017), Zhuang et al. (2018), Yang et al. (2019), Shao et al. (2020), Tong et al. (2020), Han et al. (2022), Zhang et al. (2022), Fang et al. (2023), Wang et al. (2023), and Bai et al. (2023)
	Japan	*A. testudinarium*, *Haemaphysalis hystricis*, *Hypersypnoides formosensis*, *H. hystricis*, *Haemaphysalis megaspinosa*, *H. flava*	Vegetation, wild mongooses (*Herpestes auropunctatus*)	Sato et al. (2021) and Kuba et al. (2022)
	Republic of Korea	*A. testudinarium*, *H. longicornis*, *H. flava*, *H. japonica*, *H. concinna*, *Ixodes nipponensis*, *Ixodes turdus*	Vegetation, water deer (*H. inermis*), wild boars (*Sus scrofa*), wild animals, human, black-faced bunting (*Emberiza spodocephala*), olive-backed pipit (*Anthus hodgsoni*)	Park et al. (2021); Ji et al. (2024), Oh et al. (2016), Yun et al. (2016), Kim et al. (2018), Jo et al. (2019), Yoo et al. (2020), Seo et al. (2021), Lee et al. (2022), Kang et al. (2022), Kim et al. (2023), and Jang et al. (2024)
	Taiwan	*R. microplus*	Cattle	Lin et al. (2020) and Kuan et al. (2023)

Shelly beach virus (Reoviridae)	Australia	*I. holocyclus*	Vegetation	Gofton et al. (2022)
Shelly headland virus (Reoviridae)	Australia	*I. holocyclus*	Vegetation	Gofton et al. (2022)
Spotted fever group rickettsia (SFGR)	France	*I. ricinus, R. sanguineus* s.l., *R. bursa*	Wild ungulates	Habib et al. (2024)
	China	*H. longicornis, H. detritum, R. microplus*	Vegetation	Lei et al. (2025)
Tacheng tick virus (TcTV2)	Turkey	*R. turanicus, Dermacentor marginatum*	Cattle, sheep, goat, cat	Dincer et al. (2022)
Tacheng tick virus 2 (Uukuvirus)	China	*Dermacentor* sp.	Cattle, vegetation	Cai et al. (2023)
Tahe rhabdovirus 1 (THRV1)	China	*H. concinna*	Cattle, dog	Su et al. (2024)
Tahe rhabdovirus 2 (THRV2), Tahe rhabdovirus 3 (THRV3)	China	*I. persulcatus*	Forest shrubs, vegetation	Su et al. (2024)
Taiga tick nigecruvirus (TGTNV)	China	*I. persulcatus*	Forest shrubs, vegetation	Su et al. (2024)
Takachi virus (TAKV)	Japan	*H. formosensis*	Feeding ticks, vegetation	Kobayashi et al. (2021)
Tamdy virus	China	*H. asiaticum*	Camels (*Camelus bactrianus*)	Zhou et al. (2019)
Tarumizu tick virus (TarTV)	Japan	*H. flava*	Vegetation	Fujita et al. (2017)
Tick-borne encephalitis virus (TBEV)	Latvia	*I. ricinus, I. persulcatus, D. reticulatus*	Vegetation	Capligina et al. (2020)
	China	*I. persulcatus*	Forest shrubs, vegetation	Su et al. (2024)
	Russia	*I. persulcatus*	Vegetation	Kholodilov et al. (2022)

(Continued)

TABLE 6.1
(Continued)

Tick-borne viruses	Location	Tick species	Sources of ticks	References
	Republic of Korea	*I. nipponensis*, *H. flava*	Vegetation	Yun et al. (2016)
	Mongolia	*I. persulcatus*	Vegetation	Lagunova et al. (2022)
	Finland	*I. ricinus*	Pet dogs, cats	Zakham et al. (2023)
Tick-borne tetravirus-like virus (unclassified)	China	*H. longicornis*	Animals, vegetation	Xiao et al. (2024)
Timbillica virus (Phenuiviridae)	Australia	*I. holocyclus*	Vegetation	Gofton et al. (2022)
Volzhskoe tick virus (VSTV)	China	*H. marginatum*	Goat	Su et al. (2024)
	Romania	*H. marginatum*	Sheep, goat, environment	Bratuleanu et al. (2024)
West Nile Virus (WNV)	Israel	*Argas arboreus*	Nests of cattle egret (*Bubulcus ibis*) colonies	Mumcuoglu et al. (2005)
Xiabuqu River virus (XRV)	China	*Alveonasus lahorensis*	Tibetan sheep	Ren et al. (2025)
Xinjiang tick-associated virus 1	China	*D. nuttalli*	Goat, sheep	Su et al. (2024)
Xinjiang tick totivirus 2 (Totiviridae)	China	*H. longicornis*	Animals, vegetation	Xiao et al. (2024)

Yanggou tick virus (YGTV)	Russia	*D. reticulatus*, *D. marginatus*, *I. persulcatus*	Vegetation	Kholodilov et al. (2022)
	China	*I. persulcatus*	Cattle, sheep, vegetation (grassland and forest)	Wang et al. (2023)
Yezo virus (YZV)	China	*I. persulcatus*	Forest shrubs	Su et al. (2024)
Yezovirus (YEZV) (Fa-Nairoviridae; Ge-*Orthonairovirus*)	China	*I. persulcatus*	Cattle, sheep, vegetation (grassland and forest)	Wang et al. (2023)
Yichun mivirus (YCMV)	China	*I. persulcatus*	Cattle, sheep, vegetation (grassland and forest)	Wang et al. (2023)
Zambezi tick virus 1 (Orthomyxoviridae)	China	*H. longicornis*	Animals, vegetation	Xiao et al. (2024)

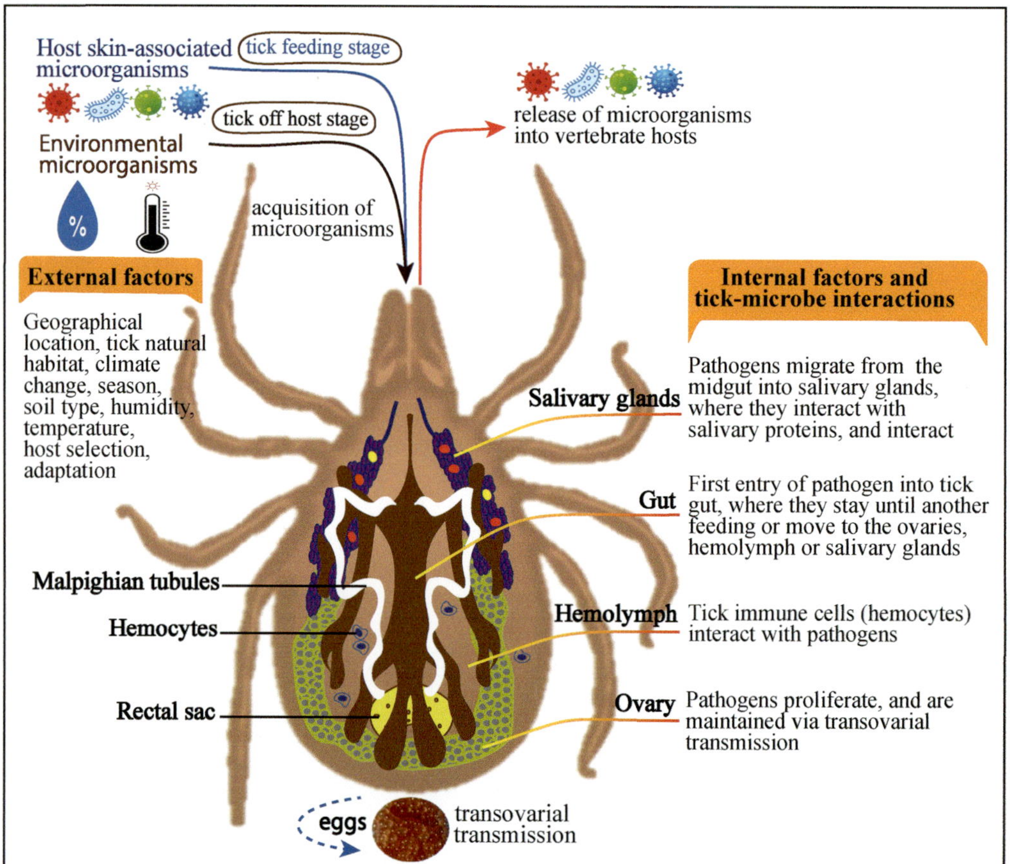

Common tick-associated microbiota composition (pathogens & endosymbionts)

Viruses	Bacteria	Fungi	Protozoa	Helminths
Reoviridae	*Anaplasma*	*Metarhizium*	*Babesia*	*Cruorifilaria*
Flaviviridae	*Bartonella*	*Beauveria*	*Theileria*	*Monanema*
Asfarviridae	*Borrelia*	*Lecanicillium*	*Hepatozoon*	*Dipetalonema*
Nyamiviridae	*Rickettsia*	*Paecilomyces*	*Cytauxzoon*	*Cercopithifilaria*
Phenuiviridae	*Ehrlichia*		*Trypanosoma*	*Acanthocheilonema*
Nairoviridae	*Neoehrlichia*		*Leishmania*	
Rhabdoviridae	*Mycoplasma*	**Nutritional symbionts**		
Peribunyaviridae	*Wolbachia*	*Coxiella*-like endosymbionts (CLEs),		
Orthomyxoviridae	*Coxiella*	*Francisella*-like endosymbionts (FLEs)		
	Francisella			

Figure 6.2 Schematic representation of common tick microbiota diversity (pathogens and endosymbionts), their sources of acquisition, associated environmental factors, and internal tissue localization. Ticks acquire a wide array of microbes from various sources during off-host and on-host stages. Illustration created on Adobe Illustrator CC 2019.

(Yu and Park, 2024). CCHF is vectored mostly by ticks of the genus *Ixodes* and also transmitted by *H. marginatum*, *Hyalomma aegyptium*, *Hyalomma schulzei*, *H. onatoli*, *Hyalomma dromedarii*, *Hyalomma rufipes*, *Hyalomma excavatum*, *Hyalomma anatolicum*, *Rhipicephalus turanicus*, *Rhipicephalus sanguineus*, *R. annulatus*, *Haemaphysalis punctata*, *Amblyomma variegatum*, and *Hyalomma truncatum*. Among them, *H. marginatum* is considered the primary vector for CCHF (Aslam et al., 2023).

Within the Phenuiviridae family, the genus *Bandavirus*, structurally consisting of three segments of the negative-sense (ambisense) single-strand RNA genome, includes eight of the nine known tick-borne virus species (ICTV, 2024; Yu and Park, 2024). Among these, six viruses – Bhanja virus,

Guertu virus, Albatross Island virus, Kismayo virus, Lone Star virus, and Razdan virus – either do not infect humans or cause only mild symptoms. However, Heartland bandavirus (HRTV) and Dabie bandavirus (severe fever with thrombocytopenia syndrome virus; SFTSV) cause potentially fatal infections in humans (Shen et al., 2018). The Orthomyxoviridae family has six to eight segments of negative-sense RNA (McCauley et al., 2009). The two genera – *Thogotovirus* and Quarjavirus – under this family, respectively, include Araguari virus (ARAV), Batken virus (BKNV), Bourbon virus (BRBV), Dhori virus (DHOV), Jos virus (JOSV), Thogoto virus (THOV), Quaranfil virus (QRFV), and Johnston Atoll virus (JAV). The genus *Coltivirus* in the family *Sedoreoviridae* contains a double-stranded RNA genome with 12 segments. Among *Coltivirus* infection cases, human infections with Colorado tick fever virus (CTFV) causing a febrile disease have been reported (ICTV, 2024). Of the four genera under the family Phenuiviridae (*Goukovirus*, *Phasivirus*, *Phlebovirus*, and *Tenuivirus*), the genus *Phlebovirus* includes huge groups of tick-associated viruses, which are commonly known as tick-borne phleboviruses (TBPVs) (Shi et al., 2018).

The most representative TBVs are CCHF, tick-borne encephalitis virus (TBEV), Powassan virus (POWV), Omsk hemorrhagic fever virus (OHFV), Kyasanur forest disease virus (KFDV), LIV, Langat virus (LGTV) (Antunez et al., 2024; Yu and Park, 2024). Among them, TBEV and POWV are known to cause encephalitis in humans, while OHFV, KFDV, and Alkhurma virus (AHFV) predominantly induce hemorrhagic disease (Yu and Park, 2024). Heartland virus and Lone Star virus isolated from the lone star tick (*Amblyomma americanum*) belongs to the genus *Phlebovirus* (family: Phenuiviridae) (Swei et al., 2013). The CTFV (family Reoviridae, genus *Coltivirus*) isolated in 1946 from human serum has a wide range of hosts and vectors such as *Dermacentor andersoni*, *Dermacentor occidentalis*, *Dermacentor albipictus*, *Dermacentor arumapertus*, *Haemaphysalis leporispalustris*, *Otobius lagophilus*, *Ixodes sculptus*, and *Ixodes spinipalpis* (Shi et al., 2018). The prototype virus, Jingmen tick virus (JMTV), was discovered in *Rhipicephalus microplus* ticks in China in 2010 (Antunez et al., 2024). The newly reported TBVs from mainland China consist of Alongshan virus (ALSV), Songling virus, Yezo virus, Wetland virus, Jingmen virus, and Xue-Cheng virus (Wang et al., 2019; Ma et al., 2021; Zhang et al., 2025a,b; Zhang et al., 2024; Jia et al., 2019; Tian et al., 2025).

The diverse viral species dominated by families of RNA viruses identified within various tick populations suggests the complex nexus taking place between ticks, pathogens, and vertebrate hosts. In general, ticks can disseminate a variety of pathogenic viruses to humans and animals. It is worth mentioning that ticks support a unique niche for RNA viruses. It suggests ticks are responsible for the transmission of a huge diversity of viruses, and our knowledge on this aspect could be the tip of the iceberg. The tick virome represents a rich source of already validated and novel viruses to be identified in the future. Therefore, extensive investigation is essential to characterize the new TBVs in different geographies in the context of climate change-related expansion of tick habitats. Understanding the fundamental tick biology, host–pathogen ecology, environmental phenomena, and other climatological aspects is crucial to revealing the new TBVs in the future.

6.2.2 Tick-Borne Bacterial Pathogens (TBBPs)

Ticks serve as one of the most competent vectors for transmitting a wide array of bacterial pathogens which are of serious public health concerns worldwide (Boulanger et al., 2025). These vectors are well acquainted with acquiring, maintaining, and possible transmission of bacteria aided by a tick's obligate hematophagy, complex life cycle pattern visiting multiplex hosts, and extended feeding duration (Bonnet et al., 2017). On the other hand, the bacterial pathogens are endowed with remarkable evolutionary adaptations facilitating them to withstand different organs of ticks such as midgut (MG) and salivary glands (SG), prolonged persistence through several molting periods, and waiting until the opportunities come to invade vertebrate hosts during successive blood meals. The dynamic interactions between microbial survival instinct and the physiological and immune responses of ticks build the foundation for the ecological framework of tick-borne bacterial diseases (TBBDs) (Eisen et al., 2017; Kazimírová et al., 2024). Ticks transmit several bacterial genera, most prominently *Rickettsia*, *Anaplasma*, *Ehrlichia*, *Borrelia*, *Coxiella*, and *Francisella* (Yessinou et al., 2025; Parola et al., 2013; Rar and Golovljova, 2011). *Ixodes ricinus* is one of the most studied tick species and has been found to carry pathogenic bacteria such as *Borrelia burgdorferi sensu lato* (s.l.) genogroup, *B. burgdorferi sensu stricto* (s.s.), *Borrelia garinii*, and *Borrelia afzelii* (Carpi et al., 2011; Gofton et al., 2015) as well as *Anaplasma phagocytophilum*, *Rickettsia Helvetica*, and

Candidatus Neoehrlichia mikurensis (Almedia et al., 2012). Commonly detected bacterial TBPs include *Anaplasma*, *Borrelia*, *Ehrlichia*, *Neoehrlichia*, *Bartonella*, *Francisella*, *Coxiella*, *Leptospira*, *Mycoplasma*, *Rickettsia* in various tick species distributed in wide geographic locations (Table 6.2). These genera comprise species that result in a wide range of diseases in humans and animals, ranging from mild febrile conditions to severe life-threatening infections (Huntington et al., 2016). Tick-associated bacteria are not only pathogens, but a large majority of them serve as endosymbionts favoring tick survival in various aspects, which has been discussed in the following section.

Rickettsiosis is a widespread disease caused by various species of the genus *Rickettsia* that are obligate intracellular Gram-negative bacilli primarily infecting endothelial cells and macrophages. Rickettsiae frequenting between arthropods, including ticks and vertebrate hosts, cause human diseases of varying severity, primarily commencing infection in microvascular endothelial cells and culminating in endothelial dysfunction (Sahni et al., 2019). The recent literature has extensive coverage on reviews regarding the pathogenesis of rickettsial infections (Blanton, 2019; Helminiak et al., 2022; Sahni et al., 2019). *Rickettsia* spp. have undergone extensive genome reduction, leading to the loss of many genes involved in essential metabolic pathways. As a result, they are unable to survive outside host cells and instead scavenge various nutrient components from their hosts to support replication (Driscoll et al., 2017). Based on phylogenetic relationships, clinical manifestations, and antigenic characteristics, *Rickettsia* species vectored by ticks are classified into the following: spotted fever group (SFG), transitional group (TRG), and ancestral group (AG) (Gillespie et al., 2007, 2008; Weinert et al., 2009). The SFG comprises tick-borne pathogens such as *Rickettsia rickettsii* (Rocky Mountain spotted fever-RMSF), *Rickettsia conorii* (Mediterranean spotted fever-MSF), and *Rickettsia parkeri* (mild-to-moderate spotted fever). Among TRG, *Rickettsia australis* represents a tick-borne member for Queensland tick typhus (Gillespie et al., 2008; Gillespie et al., 2007). In contrast, species of the AG such as *Rickettsia bellii* and *Rickettsia canadensis*, though frequently detected in ticks, are considered non-pathogenic and remain comparatively less studied compared to pathogenic SFG species (Gillespie et al., 2007, 2008; Salje, 2021).

R. rickettsii causes RMSF, which is a severe disease characterized by fever, headache, and rash and can result in high fatality without prompt antibiotic treatment. It is mainly transmitted by ticks of the genera *Dermacentor*, *Amblyomma*, and *Haemaphysalis* in North America, while in Mexico and Brazil, *R. sanguineus* serves as the primary vector. Likewise, another pathogen, *R. conorii*, the causative agent of MSG, is transmitted by *R. sanguineus* in Europe, whereas *Rickettsia slovaca* and *Rickettsia raoultii*, both responsible for tick-borne lymphadenopathy (TIBOLA), are transmitted by *Dermacentor marginatus* and *Dermacentor reticulatus* (Oteo and Portillo, 2012). In Japan, *Reynoutria japonica*, *Rickettsia heilongjiangensis*, and *Rickettsia tamurae* are recognized as the main etiological agents of spotted fever (Imaoka et al., 2011) while in China, *R. heilongjiangensis* and *R. raoultii* represent the most widespread SFG *Rickettsia* (SFGR) species (Zhao et al., 2021). Additional human-pathogenic species, including *Rickettsia monacensis* and *R. helvetica*, have also been reported (Thu et al., 2019b). Collectively, tick-associated Rickettsiae represent a diverse and evolutionarily dynamic group of microorganisms. Advancements in genome sequencing, functional genomics, and molecular biology have significantly enhanced our understanding of *Rickettsia*–tick–vertebrate host interactions, and novel *Rickettsia* species continue to be identified across the globe.

B. burgdorferi s.l. is the etiological agent of Lyme disease (Lyme borreliosis), which is transmitted to vertebrate hosts mostly by *Ixodes* spp. ticks. It is currently the most prevalent and rapidly expanding TBD in Europe and North America (Kurokawa et al., 2020; Akther et al., 2024). Among the identified species of *Borrelia*, three are recognized as the major pathogenic species: *B. afzelii* and *B. garinii* in Eurasia and *B. burgdorferi* s.s. mainly in North America (Stanek et al., 2012). Members of Borreliae of the relapsing fever group (RFG) are heterogenous and are classified according to their tick vectors into two major groups: soft tick–borne relapsing fever (STBRF) Borreliae and hard tick–borne relapsing fever (HTBRF) Borreliae (Trevisan et al., 2021). All of them are infectious to humans and serve as the causative agent of both endemic and epidemic forms of relapsing fever, characterized by high spirochetemia and febrile episodes. The STBRF group is primarily associated with *Ornithodoros* ticks, whereas the HTBRF group involves hard tick genera such as *Ixodes*, *Amblyomma*, *Dermacentor*, and *Rhipicephalus* as the potential vectors (Trevisan et al., 2021). *I. ricinus* ticks especially those sampled from Lyme disease endemic regions were identified to carry *B. burgdorferi* (s.l.) genogroup, including *B. burgdorferi* (s.s.), *B. garinii*, and *B. afzelii* (Che Lah et al., 2023; Vayssier-Taussat et al., 2013) along with *A. phagocytophilum*, *R. helvetica*, and *C. N. mikurensis* (Vayssier-Taussat et al., 2013).

The bacterial genera *Anaplasma* and *Ehrlichia* are Gram-negative, obligate intracellular Alphaproteobacteria (order Rickettsiales, family Anaplasmataceae). They cause nonspecific febrile illnesses that are mostly self-limiting; however, with infection with *Ehrlichia chaffeensis* and *A. phagocytophilum*, illness can be severe, and mainly in older individuals, patients with comorbidities or those who are immunocompromised, it could be even fatal if left untreated (Madison-Antenucci et al., 2020). *Anaplasma* species are known to be pathogenic in a broad host range and cause severe disease in several mammalian species, including humans (Stuen et al., 2013). *A. phagocytophilum* is the primary species that infects humans where it causes human granulocytic anaplasmosis (HGA) (Madison-Antenucci et al., 2020). It resides in an intracytoplasmic inclusion body or morulae in human granulocytes, primarily invading neutrophils (Thomas et al., 2009). Ticks of the *I. ricinus* complex are the most important epidemiologic vectors of *A. phagocytophilum* worldwide, where the pathogen is transmitted transstadially and horizontally between ticks and vertebrate hosts (Karshima et al., 2022).

The clinical features manifested by all species of *Ehrlichia* is known as ehrlichiosis. The medical importance of *Ehrlichia* rose significantly with the identification of *E. chaffeensis* (Paddock and Childs, 2003) and subsequently, *Ehrlichia ewingii*, closely allied to *Ehrlichia canis*, was found to infect humans (Adams et al., 2025; Gygax et al., 2024). *E. chaffeensis* was discovered in humans in the USA in 1986 and deemed to be reportable to the Centers for Disease Control and Prevention (CDC) in 1994 (Mansfield et al., 2017; Walker and Dumler, 1996; Ismail and McBride, 2017). The first human case of *E. chaffeensis* in Europe was reported in Portugal in 1991 (Morais et al., 1991). The most recently discovered *Ehrlichia* species to infect humans is *Ehrlichia muris* subsp. *eauclairensis*, endemic in the upper Midwest of the USA, is primarily transmitted by *Ixodes scapularis* (Gygax et al., 2024; Xu et al., 2023). *E. chaffeensis* is the most common etiologic agent of human monocytotropic ehrlichiosis (HME), an emerging life-threatening zoonosis that causes clinical manifestations commencing from a mild febrile illness to acute sepsis and toxic shock-like symptoms leading to multiorgan system failure and death (Ismail et al., 2010; Lina et al., 2016; Tominello et al., 2019). Pathologically, *E. chaffeensis* resides in phagosomes (morulae) of monocytes (Ganguly and Mukhopadhayay, 2008), while other species of *Ehrlichia*, such as *E. ewingii*, reside in morulae in granulocytes (Mansfield et al., 2017; Ganguly and Mukhopadhayay, 2008). Both *E. chaffeensis* and *E. ewingii* are mainly transmitted by *A. americanum*, whose larvae, nymphs, and adults readily feed blood on a large number of host species, including humans and white-tailed deer, the most common reservoir for these pathogens (Gygax et al., 2024; Paddock and Childs, 2003; Biggs et al., 2016).

Another emerging TBD is neoehrlichiosis which is caused by *N. mikurensis*, a Gram-negative bacterium (Portillio et al., 2018) that develops within the cytoplasm of the vascular endothelial cells, one of the target cells of its infection (Wass et al., 2022). The first case of human neoehrlichiosis, caused by *C. N. mikurensis*, was reported in Sweden in an immunocompromised patient coupled with thromboembolic complications (Welinder-Olsson et al., 2010). Another case of neoehrlichiosis caused by *Candidatus Neoehrlichia* has been detected in an immunocompetent child with acute febrile illness in South Africa (Bamford et al., 2023). This disease manifests more severe inflammation in individuals with compromised B-cell immunity (Wass et al., 2022; Gynthersen et al., 2023), and fever, rashes, and thromboembolic complications in both immunocompetent and immunocompromised cases (Portillo et al., 2018; Silaghi et al., 2016).

Francisella tularensis, a Gram-negative, pleiomorphic, catalase (CAT)-positive, non-motile coccobacillus, is the causative agent of tularemia, which has been classified as a 'Tier 1 Select Agent' by the CDC owing to its high virulence, extremely low infectious dose, and potential for dissemination by multiple tick vectors, raising the serious threat of future outbreaks (Tully and Huntley, 2020). Human tularemia is primarily caused by two subspecies of *F. tularensis*: *F. tularensis* subsp. *tularensis* (Type A), which has a low infectious dose (<10 CFU) and is often associated with severe, lethal outcomes; and *F. tularensis* subsp. *holarctica* (Type B), which requires a relatively higher infectious dose (>100 CFU) and typically causes a less severe, progressive disease with low mortality (Molins et al.,

TABLE 6.2

Globally reported common tick-borne bacterial pathogens and their vector ticks

Tick-borne bacteria	Location	Tick species	Sources of ticks	References
Anaplasma				
Anaplasma bovis	Australia	*H. bancrofti*	Vegetation	Gofton et al. (2022)
	Bangladesh	*Haemaphysalis bispinosa*, *R. microplus*	Vegetation	Mohanta et al. (2024)
	China	*Haemaphysalis qinghaiensis*, *Dermacentor abaensis*, *Dermacentor nuttalli*, *Haemaphysalis concinna*, *Harperocallis flava*, *Haemaphysalis longicornis*, *Hyalomma asiaticum*, *R. microplus*	Vegetation, free-ranging goats, cattle, dogs, hedgehogs (*Erinaceus amurensis*), two-humped camel (*Camelus bactrianus*),	Qin et al. (2018, 2018~~20~~), Han et al. (2019), Fang et al. (2021), Tian et al. (2023), Xu et al. (2023), Lu et al. (2023), Zhao et al. (2023), Wang et al. (2024), Tang et al. (2024) and Obaid et al. (2025)
	Spain	*Haemaphysalis punctata*	Birds, vegetation	Palomar et al. (2015)
	Kyrgyzstan	*H. punctata*, *Rhipicephalus turanicus*	Domestic animals	Kim et al. (2024)
	Pakistan	*Heliocybe sulcata*, *H. bispinosa*	Goats	Khan et al. (2024)
	Republic of Korea	*H. flava*, *H. longicornis*, *Ixodes nipponensis*, *Ixodes persulcatus*	Cattle, human (hospitals and local public health centers), Korean water deer (*Hydropotes inermis argyropus*)	Kang et al. (2016) and Seo et al. (2020, 2023)
	Taiwan	*Rhipicephalus haemaphysaloides*	Rodents (*Apodemus agrarius*, *Rattus losea*)	Kuo et al. (2018)
	Thailand	*Haemaphysalis lagrangei*, *H. bispinosa*	Vegetation, tapir, deer, goats	Chaorattanakawee et al. (2024) and Sri-In et al. (2024)
	Tunisia	*Hyalomma excavatum*	Scimitar-horned oryx	Said et al. (2021)
A. bovis (variant cwp72bo-1)	China	*H. longicornis*	Cattle	Qi et al. (2022)
A. bovis (variants: cwp72bo-1, cwp55-36bo-2)	China	*H. flava*	Hedgehogs	Qi et al. (2022)

Anaplasma capra	China	*H. qinghaiensis*, *D. abaensis*, *D. nuttalli*, *H. longicornis*, *H. qinghaiensis*, *R. microplus*	Vegetation (grassland), goats, free-ranging goats and other domestic animals, cattle	Qin et al. (2018), Han et al. (2019), Lu et al. (2022, 2023), Tang et al. (2024), Wang et al. (2024), and Zhang et al. (2024a)
	France	*H. concinna*	Vegetation (wildlife reserves)	Jouglin et al. (2025)
	Ghana	*Rhipicephalus* sp.	Cattle	Addo et al. (2023)
	Pakistan	*Ixodes kashmiricus*	Sheep, goats	Numan et al. (2023)
	Republic of Korea	*H. longicornis*	Cattle, human (hospitals and local public health centers)	Seo et al. (2020, 2023)
	Thailand	*R. microplus*, *H. lagrangei*	Eld's deer stall, tapir	Sri-In et al. (2024)
Anaplasma centrale	China	*H. longicornis*	Vegetation	Qin et al. (2018)
	Cameron	*R. microplus*	Cattle	Ngnindji-Youdje et al. (2022)
	Mongolia	*D. nuttalli*	Goats, sheep, cattle, vegetation	von Fricken et al. (2020) and Altantogtokh et al. (2022)
	Pakistan	*H. sulcata*, *R. haemaphysaloides*, *R. microplus*, *R. turanicus*	Sheep, goats	Ghafar et al. (2020)
	Spain	*H. punctata*	Birds, vegetation	Palomar et al. (2015)
	Uganda	*Rhipicephalus pulchellus*	Domestic cattle	Byaruhanga et al. (2021)
	Zambia	*Argas walkerae*	Chicken coop	Qiu et al. (2021)
Anaplasma marginale	Cameron	*R. microplus*, *Haemaphysalis leachi*	Cattle	Ngnindji-Youdje et al. (2022)
	Iran	*Rhipicephalus* (*Boophilus*) sp., *R. turanicus*	Goats, long-eared hedgehogs (*Hemiechinus auritus*)	Hosseini-Chegeni et al. (2020) and Khodadadi et al. (2021)

(*Continued*)

TABLE 6.2
(Continued)

Tick-borne bacteria	Location	Tick species	Sources of ticks	References
	China	*Dermacentor silvarum*, *H. longicornis*, *R. microplus*, *R. haemaphysaloides*	Cattle, goats, sheep, free-ranging domestic animals	Duan et al. (2020), Lu et al. (2022, 2023), Qi et al. (2022), Zhao et al. (2023), Xu et al. (2023), and Intirach et al. (2024)
	Colombia	*R. microplus*	Cattle, pasture land	Segura et al. (2024)
	Corsica	*Rhipicephalus bursa*	Slaughtered cattle	Cicculli et al. (2020)
	Egypt	*Hyalomma dromedarii*, *Rhipicephalus annulatus*	Dromedary camel, horses, cattle	Abdullah et al. (2022), Elsawy et al. (2024), Senbill et al. (2024), and Soliman et al. (2025)
	Ghana	*Rhipicephalus* sp.	Cattle	Addo et al. (2023)
	India	*R. annulatus*	Cattle	Hembram et al. (2022)
	Italy	*R. sanguineus* (s.l.), *R. bursa*	Martin, cattle	Chisu et al. 2023
	Mexico	*Amblyomma mixtum*	Cattle	Cárdenas-Amaya et al. (2025)
	Mongolia	*D. nuttalli*	Goats, sheep, cattle, vegetation	von Fricken et al. (2020) and Altantogtokh et al. (2022)
	Pakistan	*Dermacentor marginatus*, *H. sulcata*, *Hyalomma anatolicum*, *H. dromedarii*, *Hyalomma scupense*, *R. haemaphysaloides*, *R. sanguineus*, *R. microplus*	Goats, camels, sheep, cattle, Asian water buffalo, horses, dogs	Ghafar et al. (2020) Alam et al. (2022), Ahmad et al. (2023), and Ciarma et al. (2025)
	Panama	*R. microplus*	Cow	Bermúdez et al. (2024)
	Portugal	*D. marginatus*, *H. punctata*, *Ixodes ricinus*, *I. ventalloi*, *R. sanguineus* (s.l.)	Livestock, vegetation	Antunes et al. (2016)

	Thailand	*R. microplus*	Beef cattle	Thinnabut et al. (2023)
	Zambia	*A. walkerae*	Chicken coop	Qiu et al. (2021)
Anaplasma odocoilei	Argentina	*Amblyomma dubitatum*	Livestock area, protected natural areas, vegetation	Eberhardt et al. (2023)
Anaplasma ovis	China	*D. marginatus*, *D. silvarum*, *H. qinghaiensis*, *D. abaensis*, *D. nuttalli*, *H. longicornis*, *H. qinghaiensis*, *H. anatolicum*, *Ixodes crenulatus*, *Ixodes ovatus*, *I. persulcatus*, *Ornithodoros lahorensis*, *R. haemaphysaloides*, *R. microplus*, *R. turanicus*	Cattle, goats, sheep, free-ranging domestic animals, human skin, vegetation	Han et al. (2019), Li et al. (2020, 2023, 2024), Lu et al. (2022), Cao et al. (2023), Lu et al. (2023), Xue et al. (2023), and Zhang et al. (2024a)
	France	*I. ricinus*, *R. sanguineus* (s.l.)	Wild ungulates	Habib et al. (2024)
	Iran	*R. sanguineus*	Caspian red deer	Hosseini-Chegeni et al. (2020)
	Kazakhstan	*H. anatolicum*	Cattle	Zeng et al. (2025)
	Kyrgyzstan	*Hyalomma marginatum*	Domestic animals	Kim et al. (2024)
	Mongolia	*D. nuttalli*	Goats, sheep, cattle, vegetation	von Fricken et al. (2020) and Altantogtokh et al. (2022)
	Pakistan	*H. punctata*, *H. sulcata*, *H. anatolicum*, *H. anatolicum*, *R. haemaphysaloides*, *R. microplus*, *R. turanicus*	Goats, sheep	Ghafar et al. (2020)
	Spain	*I. simplex*	Bats (*Hypsugo savii*, *Miniopterus schreibersii*)	Moraga-Fernández et al. (2024)
	Tunisia	*R. turanicus*, *R. sanguineus* (s.l.)	Goats	Belkahia et al. (2019)
	Turkey	*R. bursa*	Sheep	Ayan et al. (2024)

(*Continued*)

TABLE 6.2
(Continued)

Tick-borne bacteria	Location	Tick species	Sources of ticks	References
	Zambia	*Ornithodoros moubata*	Warthog burrow	Qiu et al. (2021)
Anaplasma phagocytophilum	Austria	*Dermacentor reticulatus*, *I. ricinus*	Dogs	Wijnveld et al. (2021)
	China	*D. abaensis*, *D. nuttalli*, *H. qinghaiensis*, *H. longicornis*, *I. ovatus*, *Isoetes sinensis*, *R. microplus*	Vegetation, cattle, Tibetan sheep, yaks, giant panda	Qin et al. (2018), Han et al. (2019), Qi et al. (2022), Zhang et al. (2024), Lei et al. (2025), and Ma et al. (2025)
	Corsica	*H. marginatum*	Slaughtered cattle	Cicculli et al. (2020)
	England and Wales	*I. ricinus*	Vegetation	Gandy et al. (2022)
	Estonia	*I. ricinus*	Vegetation	Vikentjeva et al. (2024)
	Finland	*I. ricinus*, *I. persulcatus*	Pet dogs, cats, vegetation	Sormunen et al. (2016), Pakanen et al. (2020), and Zakham et al. (2023)
	France	*I. ricinus*	Wild ungulates, vegetation	Habib et al. (2024) and Haidar-Ahmad et al. (2025)
	Germany	*I. ricinus*, *Ixodes hexagonus*	Dogs, cats, cattle, wildlife carcasses	Probst et al. (2024) and Springer et al. (2025)
	Hungary	*D. reticulatus*, *H. concinna*, *I. ricinus*	Wild boars	Hornok et al. (2022)
	India	*H. bispinosa*, *R. sanguineus* (s.l.), *R. annulatus*	Stray dogs, cattle	Manoj et al. (2020) and Hembram et al. (2022)
	Italy	*I. ricinus*, *R. sanguineus* (s.l.), *R. bursa*	Goats, human, ungulates, wild passerine birds	Grassi et al. (2023), Melis et al. (2024), and Sadia et al. (2025)
	Kazakhstan	*Argas persicus*, *H. anatolicum*, *R. turanicus*	Sheep, hen, cattle	Zeng et al. (2025)

Kenya	*H. leachi*	Wild mammals	Mosha et al. (2024)
Latvia	*D. reticulatus*, *I. persulcatus*, *I. ricinus*	Vegetation	Capligina et al. (2020)
Lithuania and Poland	*D. reticulatus*, *I. ricinus*	European bison	Lipatova et al. (2023)
Mongolia	*I. persulcatus*	Vegetation, goats, sheep, cattle	von Fricken et al. (2020) and Lagunova et al. (2022)
Paraguay	*R. sanguineus*	Dogs	Salvioni Recalde et al. (2024)
Poland	*D. reticulatus*, *I.* sp.	Vegetation, protected land	Dunaj et al. (2021a, b), Kubiak et al. (2024), Sawczyn-Domańska and Wójcik-Fatla (2024), and Zajac et al. (2024)
Republic of Korea	*H. flava*, *H. longicornis*, *I. nipponensis*, *I. persulcatus*	Vegetation, human, migratory birds, Korean water deer (*H. inermis argyropus*)	Kang et al. (2016), Lee et al. (2020), Kim et al. (2022), Seo et al. (2021, 2023), and Kim et al. (2024)
Romania	*I. ricinus*	Vegetation	Matei et al. (2015)
Russia	*I. ricinus*	Vegetation	Rakov et al. (2025)
Serbia	*I. ricinus*	Human	Žekić et al. (2025)
Slovakia	*I. ricinus*	Human	Špitalská et al. (2021)
Spain	*H. punctata*, *Ixodes frontalis*, *I. ricinus*	Birds, vegetation	Palomar et al. (2015) and Remesar et al. (2021)
Taiwan	*R. haemaphysaloides*	Rat (*A. agrarius*)	Kuo et al. (2018)
Thailand	*Dendrobates auratus*, *R. microplus*	Eld's deer, under leaf litter	Nooroong et al. (2018) and Sri-In et al. (2024)
Tunisia	*H. aegyptium*	Tortoise (*Testudo graeca*)	Rjeibi et al. (2022)

(Continued)

TABLE 6.2
(Continued)

Tick-borne bacteria	Location	Tick species	Sources of ticks	References
	UK	*Ixodes canisuga*, *I. hexagonus*, *I. ricinus*	Cats, dogs, red fox carcasses	Duplan et al. (2018), Keyete et al. (2021), and Mansfield et al. (2024)
	Ukraine	*D. reticulatus*, *I. ricinus*	Animals (admitted to vet clinics), cattle, dogs, cats, human, vegetation	Kovryha et al. (2021) and Levytska et al. (2021)
	USA	*Amblyomma americanum*, *Ixodes scapularis*	Vegetation, hunter-killed deer	Livengood et al. (2020), Briggs et al. (2023), Foley et al. (2023), and Martinez-Villegas et al. (2024)
	Zambia	*O. moubata*	Warthog burrow	Qiu et al. (2021)
A. phagocytophilum (type I and type II)	China	*I. persulcatus*	Free-ranging goats, cattle	Tian et al. (2023)
A. phagocytophilum (ecotype I and ecotype II)	Northwest Scotland	*I. ricinus*	Vegetation	Olsthoorn et al. (2021)
A. phagocytophilum (AP-Variant 1 genotype)	USA	*Dermacentor andersoni*	Vegetation	Francis et al. (2020)
A. phagocytophilum (human agent variant)	USA	*H. longicornis*	Vegetation	Price et al. (2022)
Anaplasma platys	Algeria	*R. microplus*	Vegetation	Eberhardt et al. (2023)
	Argentina	*R. microplus*	Cattle	Xu et al. (2023)
	Cameron	*Amblyomma maculatum*, *R. sanguineus*	Vegetation, dogs, sheep, goats	Allerdice et al. (2017) and Zaid et al. (2019)

	China	*Anaphes nitens*, R. *microplus*, R. *sanguineus*, R. *sanguineus* (s.l.)	Cattle, dogs, white-tailed deer (*Odocoileus virginianus*)	Manoj et al. (2020), Olafson et al. (2020), Ngnindji-Youdje et al. (2022), Laatamna et al. (2022), Zhao et al. (2023), Chao et al. (2024), Intirach et al. (2024), and Salvioni Recalde et al. (2024)
	India	*R. microplus*	Beef cattle	Thinnabut et al. (2023)
	Palestine	*R. microplus*	Livestock area, protected natural areas	Eberhardt et al. (2023)
	Paraguay	*H. longicornis*	Healthy dogs	Shi et al. (2022)
	Taiwan	*R. (Boophilus) microplus*	Healthy dogs	Shi et al. (2022)
	Thailand	*R. (Boophilus) microplus*	Anemic dogs	Shi et al. (2022)
A. platys (variant xyn10pt-1)	USA	*H. flava*	Hedgehogs	Qi et al. (2022)
A. platys (variants xyn10pt-1; xyn21pt-2; xyn3pt-3)	USA	*H. longicornis*	Cattle	Qi et al. (2022)
Bartonella				
Babesia bovis	Turkey	*H. aegyptium*	Tortoise (*T. graeca*)	Akveran et al. (2020)
Bartonella clarridgeiae	UK	*I. hexagonus*	Cats	Duplan et al. (2018)
Bartonella henselae	UK	*I. hexagonus, I. ricinus*	Cats	Duplan et al. (2018)
	Italy	*I. ricinus*	Cats,	Chisu et al. (2020)

(*Continued*)

TABLE 6.2
(Continued)

Tick-borne bacteria	Location	Tick species	Sources of ticks	References
Bartonella melophagi	Mongolia	*D. nuttalli*	Environment	Altantogtokh et al. (2022)
	China	*D. silvarum*	Free-ranging sheep	Cao et al. (2023)
Bartonella schoenbuchii	Mongolia	*D. nuttalli*	Environment	Altantogtokh et al. (2022)
Borrelia				
Borrelia afzelii	Austria	*H. concinna, I. ricinus*	Dogs	Wijnveld et al. (2021)
	China	*I. persulcatus*	Vegetation, free-ranging goats, cattle	Tian et al. (2023) and Duan et al. (2024)
	Finland	*I. persulcatus*	Vegetation	Pakanen et al. (2020)
	Germany	*I. ricinus*	Dogs, cats	Probst et al. (2024)
	Italy	*I. ricinus*	Wild passerine birds, human	Grassi et al. (2023) and Melis et al. (2024)
	Latvia	*D. reticulatus, I. persulcatus, I. ricinus*	Vegetation	Capligina et al. (2020)
	Mongolia	*Delphinium nuttallii*	Vegetation	Fischer et al. (2020)
	Northwest Scotland	*I. ricinus*	Vegetation	Olsthoorn et al. (2021)
	Poland	*D. reticulatus, I. ricinus*	Vegetation, lizard (*Lacerta agilis*)	Kubiak et al. (2024) and Dyczko et al. (2024)
	Republic of Korea	*I. nipponensis*	Vegetation	Kim et al. (2024)
	Serbia	*I. ricinus*	Human	Žekić et al. (2025)

Blumeria americana	Poland	*I. ricinus*	Vegetation, protected land	Dunaj et al. (2021)
Borrelia bavariensis	Germany	*I. ricinus*	Dogs, cats	Probst et al. (2024)
	Austria	*I. ricinus*	Dogs	Wijnveld et al. (2021)
Borrelia burgdorferi	Egypt	*H. dromedarii*, *Rutilus bavaroises*, *R. annulatus*	Camel, dogs, sheep, cattle	Senbill et al. (2024)
	France	*I. ricinus*	Wild ungulates	Habib et al. (2024)
	USA	*I. scapularis*	Vegetation	Livengood et al. (2020) and Foley et al. (2023)
B. burgdorferi (s.l.)	Estonia	*I. ricinus*	Vegetation	Vikentjeva et al. (2024)
	Finland	*I. ricinus*, *I. persulcatus*	Vegetation, pet dogs, cats	Pakanen et al. (2020)
	France	*I. ricinus*	Vegetation	Haidar-Ahmad et al. (2025)
	Italy	*I. ricinus*	Human	Melis et al. (2024)
	Latvia	*D. reticulatus*, *I. persulcatus*, *I. ricinus*	Vegetation	Capligina et al. (2020)
	Mongolia	*I. persulcatus*	Vegetation	Lagunova et al. (2022)
	Poland	*D. reticulatus*, *I. ricinus*	Vegetation, tourist trails, and nature-educational paths	Dunaj et al. (2021) and Koczanowicz et al. (2025)
	Russia	*I. ricinus*	Vegetation	Rakov et al. (2025)
	Slovakia	*I. ricinus*	Human	Špitalskáa et al. (2021)
	Ukraine	*I. ricinus*	Vegetation, human, cattle, dogs, cats, animals admitted to vet clinics	Kovryha et al, (2021) and Levytska et al. (2021)

(Continued)

TABLE 6.2
(Continued)

Tick-borne bacteria	Location	Tick species	Sources of ticks	References
B. burgdorferi (*s.s.*)	Italy	*I. ricinus*	Human, wild passerine birds	Grassi et al. (2023) and Melis et al. (2024)
	Austria	*I. ricinus*	Dogs	Wijnveld et al. (2021)
	Germany	*I. ricinus*	Dogs, cats	Probst et al. (2024)
	Latvia	*I. ricinus*	Vegetation	Capligina et al. (2020)
	Northwest Scotland	*I. ricinus*	Vegetation	Olsthoorn et al. (2021)
	Serbia	*I. ricinus*	Human	Žekić et al. (2025)
Blarina carolinensis	Germany	*I. ricinus*	Dogs, cats	Probst et al. (2024)
	Poland	*I. ricinus*	Vegetation (protected land)	Dunaj et al. (2021)
B. garinii	Austria	*I. ricinus*	Dogs	Wijnveld et al. (2021)
	China	*I. persulcatus*	Vegetation, human, free-ranging goats, cattle	Xue et al. (2023), Tian et al. (2023), and Duan et al. (20240
	Finland	*I. persulcatus*	Vegetation	Pakanen et al. (2020)
	Germany	*I. ricinus*	Dogs, cats	Probst et al. (2024)
	Italy	*I. ricinus*	Wild passerine birds, human	Grassi et al. (2023) and Melis et al. (2024)
	Latvia	*D. reticulatus*, *I. persulcatus*, *I. ricinus*	Vegetation	Capligina et al. (2020)
	Northwest Scotland	*I. ricinus*	Vegetation	Olsthoorn et al. (2021)

	Republic of Korea	*I. nipponensis*	Vegetation	Kim et al. (2024)
Borrelia usitaniae	Italy	*I. ricinus*	Human	Melis et al. (2024)
	Austria	*D. reticulatus, I. ricinus*	Dogs	Wijnveld et al. (2021)
	Germany	*I. ricinus*	Dogs, cats	Probst et al. (2024)
	Latvia	*I. ricinus*	Vegetation	Capligina et al. (2020)
	Poland	*I. ricinus*	Lizard (*L. agilis, Zootoca vivipara*)	Dyczko et al. (2024)
Borrelia miyamotoi	Latvia	*I. ricinus, I. persulcatus*	Vegetation	Capligina et al. (2020)
	China	*I. persulcatus*	Vegetation, free-ranging goats, cattle	Tian et al. (2023) and Duan et al. (2024)
	Estonia	*I. ricinus*	Vegetation	Vikentjeva et al. (2024)
	Finland	*I. ricinus, I. persulcatus*	Vegetation, pet dogs, cats	Pakanen et al. (2020) and Zakham et al. (2023)
	France	*I. ricinus*	Vegetation	Haidar-Ahmad et al. (2025)
	Germany	*I. ricinus*	Dogs, cats	Probst et al. (2024)
	Italy	*I. ricinus*	Wild passerine birds, ungulates, human	Grassi et al. (2023) and Melis et al. (2024)
	Mongolia	*I. persulcatus*	Vegetation	Lagunova et al. (2022)
	Poland	*I. ricinus*	Lizard (*L. agilis*)	Dyczko et al. (2024)
	Republic of Korea	*I. nipponensis*	Vegetation	Kim et al. (2024)

(*Continued*)

TABLE 6.2
(Continued)

Tick-borne bacteria	Location	Tick species	Sources of ticks	References
	Russia	*I. ricinus*	Vegetation	Rakov et al. (2025)
	USA	*I. scapularis*	Vegetation, white-tailed deer (*O. virginianus*)	Livengood et al. (2020), Smalley et al. (2022), and Foley et al. (2023)
Borrelia puertoricensis	USA	Ornithodoros (Alectorobius) talaje	Animal burrows and sheltered cracks and crevasses in wildlife management area	Mays Maestas et al. (2025)
Borrelia spielmanii	Austria	*D. reticulatus, I. ricinus*	Dogs	Wijnveld et al. (2021)
	Germany	*I. ricinus*	Dogs, cats	Probst et al. (2024)
Borrelia theileri	Bangladesh	*H. bispinosa, R. microplus*	Vegetation	Mohanta et al. (2024)
	Egypt	*R. annulatus*	Horses, cattle	Abdullah et al. (2022) and Senbill et al. (2024)
Borrelia turcica	Algeria, Georgia, Iran, Tunisia, Turkey	*H. aegyptium*	Spur-thighed tortoises (*T. graeca*)	Kautman et al. (2025)
Borrelia turicatae	USA	*O. turicatae*	Animal burrows and sheltered cracks and crevasses in wildlife management area	Mays Maestas et al. (2025)
Borrelia valaisiana	Austria	*D. reticulatus, I. ricinus*	Dogs	Wijnveld et al. (2021)
	China	*Ixodes granulatus*	Mammals, vegetation	Hu et al. (2022)
	Germany	*I. ricinus*	Dogs, cats	Probst et al. (2024)

	Italy	*I. ricinus*	Wild passerine birds, human	Grassi et al. (2023) and Melis et al. (2024)
	Latvia	*I. ricinus*	Vegetation	Capligina et al. (2020)
	Northwest Scotland	*I. ricinus*	Vegetation	Olsthoorn et al. (2021)
Coxiella				
Coxiella burnetii	China	*H. flava*, *H. longicornis*, *I. ovatus*	Cattle, giant panda (female adult), hedgehogs (*E. amurensis*)	Fang et al. (2021~~2022~~), Wang et al. (2024), and Ma et al. (2025)
	Egypt	*H. dromedarii*, *R. rutilus*	Camel, dogs, sheep	Senbill et al. (2024) and Soliman et al. (2025)
	Israel	*R. sanguineus*, *R. turanicus*	Vegetation, dogs, *T. graeca*, European badger (*Meles meles*), white-breasted hedgehogs (*Erinaceus concolor*)	Mumcuoglu et al. (2022b)
	Spain	*I. simplex*	Bats (*Eptesicus isabellinus*, *H. savii*)	Moraga-Fernández et al. (2024)
Ehrlichia				
Ehrlichia canis	Algeria	*R. sanguineus* (s.l.)	Dogs	Laatamna et al. (2022)
	China	*Rhipicephalus linnaei*, *R. microplus*	Dogs, cattle, goats	Lu et al. (2022 a, b) and Intirach et al. (2024)
	Egypt	*R. rutilus*	Dogs, sheep	Senbill et al. (2024)
	Ghana	*A. variegatum*, *Hyalomma truncatum*, *Rhipicephalus evertsi*	Cattle, sheep, goats	Addo et al. (2023)
	India	*R. sanguineus* (s.l.)	Stray dogs	Manoj et al. (2020)
	Pakistan	*R. microplus*	Cow, buffalo, goats, dogs	Ciarma et al. (2025)

(*Continued*)

TABLE 6.2
(Continued)

Tick-borne bacteria	Location	Tick species	Sources of ticks	References
	Palestine	*R. sanguineus*	Dogs, sheep, goats	Zaid et al. (2019)
	Paraguay	*Amblyomma sculptum, R. sanguineus*	Dogs, armadillo	Salvioni Recalde et al. (2024)
	Republic of Korea	*H. longicornis*	Horses, cattle	Seo et al. (2020, 2021)
	Turkey	*R. bursa*	Sheep	Ayan et al. (2024)
Ehrlichia chaffeensis	China	*R. microplus*	Cattle, goats	Lu et al. (2022)
	Kyrgyzstan	*H. punctata*	Domestic animals	Kim et al. (2024)
	Thailand	*Haemaphysalis hystricis, Haemaphysalis shimoga*	Vegetation	Chaorattanakawee et al. (2024)
	Turkey	*R. bursa*	Sheep	Ayan et al. (2024)
	USA	*A. americanum*	Vegetation, cattle	Briggs et al. (2023) and Martinez-Villegas et al. (2024)
Ehrlichia ewingii	Bangladesh	*H. bispinosa, R. microplus*	Vegetation	Mohanta et al. (2024)
	USA	*A. americanum, A. maculatum, I. scapularis*	Vegetation, cattle, hunter-killed deer	Allerdice et al. (2017), Briggs et al. (2023), Martinez-Villegas et al. (2024)
	Thailand	*H. longicornis, Haemaphysalis wellingtoni*	Vegetation	Chaorattanakawee et al. (2024)
	China	*H. flava*	Hedgehogs (*E. amurensis*)	Fang et al. (2023)
Ehrlichia minasensis	Central America	*R. microplus*	Cow	Bermúdez et al. (2024)

	China	*H. qinghaiensis*, *H. asiaticum*, *R. microplus*	Camels, sheep, cattle, goats, Tibetan sheep, yaks	Lu et al. (2022), Xu et al. (2023), Zhao et al. (2023), Intirach et al. (2024), Zhang B, et al. (2024), and Zhang Y, et al. (2024)
	Corsica	*H. marginatum*; *R. bursa*	Slaughtered cattle	Cicculli et al. (2020)
	Eswatini	*Rhipicephalus simus*, *Rhipicephalus muehlensi*	Vegetation	Ledger et al. (2021)
	Ghana	*H. truncatum*	Cattle	Addo et al. (2023)
	Thailand	*R. microplus*	Beef cattle	Thinnabut et al. (2023)
Ehrlichia muris	China	*I. persulcatus*	Free-ranging goats, cattle	Tian et al. (2023)
	Italy	*I. ricinus*	Wild passerine birds	Grassi et al. (2023)
	Mongolia	*I. persulcatus*	Goats, sheep, cattle; vegetation	von Fricken et al. (2020)
	Pakistan	*R. microplus*	Cow, buffalo, goats, dogs	Ciarma et al. (2025)
Ehrlichia ruminantium	Cameron	*A. variegatum*	Cattle	Ngnindji-Youdje et al. (2022)
	China	*Amblyomma javanense*	Malayan pangolins	Li et al. (2024)
	Egypt	*H. dromedarii*, *R. rutilus*, *R. annulatus*	Camels, cattle, dogs, sheep	Senbill et al. (2024)
	Eswatini	*Amblyomma herbraeum*	Vegetation	Ledger et al. (2021)
	Uganda	*Amblyomma lepidum*, *Amblyomma gemma*, *A. variegatum*	Cattle	Byaruhanga et al. (2021)

(*Continued*)

TABLE 6.2
(Continued)

Tick-borne bacteria	Location	Tick species	Sources of ticks	References
Ehrlichia shimanensis	Thailand	*Haemaphysalis obesa*, *Haemaphysalis papuana*	Vegetation	Chaorattanakawee et al. (2024)
Ehrlichia urmitei	Corsica	*R. bursa*	Slaughtered cattle	Cicculli et al. (2020)
Francisella				
Francisella hispaniensis	Poland	*D. reticulatus*, *I. ricinus*	Vegetation (protected land)	Dunaj et al. (2021)
F. novicida	Poland	*I. ricinus*, *D. reticulatus*	Vegetation (protected land)	Dunaj et al. (2021)
Mycoplasma				
Mycoplasma arginini	Egypt	*H. dromedarii*	Camels	Senbill et al. (2024)
Mycoplasma haemofelis	China	*H. longicornis*, *R. (Boophilus) microplus*	Dogs (healthy and anemic)	Shi et al. (2022)
	UK	*Ixodes trianguliceps*	Cats	Duplan et al. (2018)
Neoehrlichia				
Neoehrlichia mikurensis	Estonia	*I. ricinus*	Vegetation	Vikentjeva et al. (2024)
	France	*I. ricinus*	Vegetation	Haidar-Ahmad et al. (2025)
	Italy	*I. ricinus*	Wild passerine birds, human	Grassi et al. (2023) and Melis et al, (2024)
	Thailand	*I. granulatus*	Vegetation	Chaorattanakawee et al. (2024)
	Ukraine	*D. reticulatus*	Vegetation, animals (admitted to vet clinics)	Levytska et al. (2021)

Occidentia massiliensis (pathogenetically undetermined)	Spain	*I. simplex*	Bats (*H. savii*)	Moraga-Fernández et al. (2024)
Rickettsia				
Rickettsia aeschlimannii	Egypt	*H. dromedarii*	Dromedary camel	Soliman et al. (2025)
	Algeria	*H. aegyptium*	Spur-thighed tortoises (*T. graeca*)	Kautman et al. (2025)
	Cameron	*H. rufipes, Hyalomma turanicum, R. sanguineus*	Cattle	Ngnindji-Youdje et al. (2022)
	Georgia	*H. aegyptium*	Spur-thighed tortoises (*T. graeca*)	Kautman et al. (2025)
	Iran	*H. aegyptium*	Spur-thighed tortoises (*T. graeca*)	Kautman et al. (2025)
	Israel	*R. sanguineus; R. turanicus*	Vegetation, dogs, tortoise (*T. graeca*)	Mumcuoglu et al. (2022)
	Morocco	*H. aegyptium*	Spur-thighed tortoises (*T. graeca*)	Kautman et al. (2025)
	Pakistan	*H. turanicum, R. microplus, R. turanicus*	Goats, sheep	Ghafar et al. (2020) and Ullah et al. (2024)
	Poland	*D. reticulatus, I. ricinus*	Vegetation, protected land	Dunaj et al. (2021)
	Spain	*I. simplex*	Bats (*M. schreibersii*)	Moraga-Fernández et al. (2024)
	Tunisia	*H. aegyptium*	Spur-thighed tortoises (*T. graeca*)	Kautman et al. (2025)
	Turkey	*H. aegyptium*	Spur-thighed tortoises (*T. graeca*)	Kautman et al. (2025)
Rickettsia africae	Egypt	*H. dromedarii, R. annulatus*	Camels, cattle	Senbill et al. (2024)
	Israel	*R. sanguineus, R. turanicus*	Vegetation, dogs (*Canis familiaris*)	Mumcuoglu et al. (2022)

(Continued)

TABLE 6.2
(Continued)

Tick-borne bacteria	Location	Tick species	Sources of ticks	References
	Morocco	*H. aegyptium*	Spur-thighed tortoises (*T. graeca*)	Kautman et al. (2025)
	Algeria	*H. aegyptium*	Spur-thighed tortoises (*T. graeca*)	Kautman et al. (2025)
	Tunisia	*H. aegyptium*	Spur-thighed tortoises (*T. graeca*)	Kautman et al. (2025)
	Turkey	*H. aegyptium*	Spur-thighed tortoises (*T. graeca*)	Kautman et al. (2025)
	Iran	*H. aegyptium*	Spur-thighed tortoises (*T. graeca*)	Kautman et al. (2025)
	Eswatini	*A. herbraeum*	Vegetation	Ledger et al. (2021)
Rickettsia amblyommatis	Argentina	*Amblyomma longirostre*	Wild birds	Vaschalde et al. (2025)
	China	*I. ovatus*	Giant panda (female adult)	Ma et al. (2025)
	Panama	*A. mixtum*	Horses	Bermúdez et al. (2024)
	USA	*Amblyomma inornatum*, *A. americanum*	Horses, white-tailed deer (*O. virginianus*)	Olafson et al. (2020) and Duncan et al. (2022)
R. australis	Australia	*Ixodes holocyclus*	Vegetation	Gofton et al. (2022)
Rickettsia barbariae	Israel	*R. turanicus*	Vegetation	Mumcuoglu et al. (2022)
R. bellii	China	*I. persulcatus*	Vegetation (bush)	Su et al. (2023)
	Poland	*Amblyomma rotundatum*	Rhinella marina toad	Dyczko et al. (2025)
	USA	*D. andersoni*	Vegetation	Francis et al. (2020)

R. canadensis	Republic of Korea	*Amblyomma testudinarium*, *H. longicornis*	Vegetation	Kim et al. (20240
R. conorii	Egypt	*R. rutilus*	Dogs, sheep	Senbill et al. (2024)
	Eswatini	*Haemaphysalis elliptica*	Vegetation	Ledger et al. (2021)
	Pakistan	*R. turanicus*	Sheep, goats	Ghafar et al. (2020)
R. conorii	Algeria	*R. sanguineus* (s.l.)	Dogs	Laatamna et al. (2022)
R. conorii israelensis	Israel	*R. sanguineus*	Vegetation	Mumcuoglu et al. (2022)
Rickettsia felis	China	*R. haemaphysaloides*	Cattle	Intirach et al. (2024)
	Colombia	*R. microplus*	Cattle, vegetation (pasture land)	Segura et al. (2024)
Rickettsia heilongjiangensis	China	*H. concinna*, *H. flava*, *H. longicornis*	Free-ranging goats, cattle, hedgehog (*E. amurensis*)	Zhu et al. (2022) and Tian et al. (2023)
R. helvetica	Austria	*H. concinna*, *I. ricinus*	Dogs	Wijnveld et al. (2021)
	Finland	*I. persulcatus*	Vegetation	Pakanen et al. (2020)
	Italy	*I. ricinus*	Wild passerine birds, human, cats	Chisu et al. (2020), Grassi et al. (2023), and Melis et al. (2024)
	Latvia	*D. reticulatus*, *I. persulcatus*, *I. ricinus*	Vegetation	Capligina et al. (2020)
	Northwest Scotland	*I. ricinus*	Vegetation	Olsthoorn et al. (2021)
	Poland	*D. reticulatus*	Vegetation	Zajac et al. (2024)
	Russia	*I. ricinus*	Vegetation	Rakov et al. (2025)

(Continued)

TABLE 6.2
(Continued)

Tick-borne bacteria	Location	Tick species	Sources of ticks	References
Rickettsia hoogstraalii	Zambia	*A. walkerae*	Chicken coop	Qiu et al. (2021)
Reynoutria japonica	China	*A. javanense, H. flava, H. hystricis, R. microplus*	Cattle, dogs, Malayan pangolins, hedgehogs (*E. amurensis*)	Zhao et al. (2023), Li et al. (2024), Wang et al. (2024)
	Republic of Korea	*H. flava, H. longicornis, I. nipponensis*	Vegetation	Kim et al. (2024)
R. lusitaniae	Zambia	*Ornithodoros porcinus*	Warthog (*Phacochoerus africanus*) burrows, loose soil and litter of national parks	Chitanga et al. (2021)
	Zambia	*Ornithodoros faini*	Bats, guano	Qiu et al. (2021)
R. marmionii	Poland	*D. reticulatus, I. ricinus*	Vegetation (protected land)	Dunaj et al. (2021)
	Thailand	*Dermacentor atrosignatus*	Under leaf litter	Nooroong et al. (2018)
Rickettsia massiliae	Algeria	*R. sanguineus* (s.l.)	Dogs	Laatamna et al. (2022)
	Argentina	*R. sanguineus* (s.s.)	Stray dogs	Maas et al. (2024)
	Cameron	*R. annulatus, Rhipicephalus muhsamae*	Cattle	Ngnindji-Youdje et al. (2022)
	Eswatini	*H. elliptica*	Vegetation	Ledger et al. (2021)
	Israel	*R. sanguineus, R. turanicus*	Vegetation, dogs, white-breasted hedgehog (*E. concolor*)	Mumcuoglu et al. (2022) (Israel-II)
	Pakistan	*H. punctata, H. sulcata, H. anatolicum, R. haemaphysaloides, R. microplus, R. turanicus*	Goats, sheep	Ghafar et al. (2020)

	Spain	*R. sanguineus* (s.l.), *R. bursa*	Wild boars	Castillo-Contreras et al. (2022)
	Thailand	*A. testudinarium*	Under leaf litter	Nooroong et al. (2018)
Rickettsia monacensis	Austria	*I. ricinus*	Dogs	Wijnveld et al. (2021)
	China	*H. flava*, *Haemaphysalis kitaokai*, *I. sinensis*	Dogs, cats	Lu et al. (2022)
	Italy	*I. Ricinus*	Human, carnivores	Melis et al. (2024)
	Latvia	*I. ricinus*	Vegetation	Capligina et al. (2020)
	Poland	*I. ricinus*	Vegetation (protected land)	Dunaj et al. (2021)
	Republic of Korea	*A. testudinarium*, *H. flava*, *H. longicornis*, *I. nipponensis*	Vegetation	Kim et al. (2024)
	Republic of Korea	*I. nipponensis*	Human (hospitals and local public health centers)	Seo et al. (2023)
R. parkeri	USA	*I. scapularis*	Vegetation	Livengood et al. (2020)
	USA	*A. maculatum*	White-tailed deer (*O. virginianus*)	Olafson et al. (2020)
	Ecuador	*A. maculatum*	Horses	Enriquez et al. (2024)
R. parkeri (s.s)	Brazil	*Ambystoma tigrinum*	Maned wolves	Arrais et al. (2021)
Rickettsia peacockii	USA	*D. andersoni*	Vegetation	Francis et al. (2020)
R. raoultii	Latvia	*D. reticulatus*, *I. persulcatus*, *I. ricinus*	Vegetation	Capligina et al. (2020)
	Austria	*D. reticulatus*	Dogs	Wijnveld et al. (2021)

(*Continued*)

TABLE 6.2
(Continued)

Tick-borne bacteria	Location	Tick species	Sources of ticks	References
	China	*D. marginatus*, *D. nuttalli*, *D. silvarum*, *H. flava*, *H. longicornis*, *H. anatolicum*, *Ixodes acutitarsus*, *I. persulcatus*	Vegetation, cattle, sheep, horses, human, hedgehogs (*E. amurensis*)	Li et al. (2020), Jiao et al. (2021), Cao et al. (2023), Lu et al. (2023), Xue et al. (2023), Fang et al. (2024), Duan et al. (2024), and Wang et al. (2024)
	Italy	*D. marginatus*	Wild boars	Sgroi et al. (2021)
	Mongolia	*D. nuttalli*, *D. silvarum*	Vegetation, rodents	Fischer et al. (2020), Altantogtokh et al. (2022), and Lee et al. (2024)
	Pakistan	*D. marginatus*	Goats	Ahmad et al. (2023)
	Poland	*D. reticulatus*	Vegetation	Zajac et al. (2024) and Kubiak et al. (2024)
	Russia	*D. reticulatus*	Vegetation	Rakov et al. (2025)
	Spain	*D. marginatus*, *R. bursa*	Wild boars	Castillo-Contreras et al. (2022)
	Thailand	*Dermacentor atrosignatus*	Under leaf litter	Nooroong et al. (2018)
R. raoultii (type I and type II)	China	*D. silvarum*	Free-ranging goats, cattle	Tian et al. (2023)
Rickettsia rhipicephali	Argentina	*H. leporispalustris*	Wild birds	Vaschalde et al. (2025)
	China	*A. javanense*	Malayan pangolins	Li et al. (2024)
	USA	*D. andersoni*, *Dermacentor variabilis*	Vegetation, horses	Francis et al. (2020) and Duncan et al. (2022)
R. rickettsii	China	*A. javanense*, *H. asiaticum*, *H. longicornis*	Malayan pangolins, two-humped camel (*C. bactrianus*)	Li et al. (2024) and Obaid et al. (2025)

	Taiwan	*R. haemaphysaloides*	Rodents (*A. agrarius*, *Bandicota indica*, *R. losea*)	Kuo et al. (2018)
	USA	*A. americanum*	Vegetation	Martinez-Villegas et al. (2024)
Rickettsia sibirica subsp. *mongolitimonae*	China	*H. asiaticum*	Camels, sheep, cattle	Zhang et al. (2024b)
R. sibirica	China	*D. silvarum*	Free-ranging sheep	Cao et al. (2023)
	Mongolia	*D. nuttalli*	Environment	Altantogtokh et al. (2022)
Rickettsia slovaca	China	*D. silvarum*	Free-ranging sheep	Cao et al. (2023)
	Italy	*D. marginatus*, *I. ricinus*	Wild boars	Sgroi et al. (2021)
	Mongolia	*D. nuttalli*	Environment	Altantogtokh et al. (2022)
	Pakistan	*H. sulcata*, *H. anatolicum*, *R. haemaphysaloides*, *R. microplus*, *R. turanicus*	Sheep, goats	Ghafar et al. (2020)
	Spain	*D. marginatus*, *Hyalomma lusitanicum*, *I. simplex*, *R. bursa*	Wild boars, bats (*E. isabellinus*, *H. savii*, *M. schreibersii*, *Myotis emarginatus*)	Castillo-Contreras et al. (2022) and Moraga-Fernández et al. (2024)
Rickettsia tamurae	Poland	*D. reticulatus*	Vegetation (protected land)	Dunaj et al. (2021)
	Republic of Korea	*A. testudinarium*	Human (hospitals and local public health centers)	Seo et al. (2023)
	Thailand	*A. testudinarium*	Under leaf litter	Nooroong et al. (2018)
Rickettsia tarasevichiae	Finland	*I. persulcatus*	Vegetation	Pakanen et al. (2020)
Candidatus				

(Continued)

TABLE 6.2
(Continued)

Tick-borne bacteria	Location	Tick species	Sources of ticks	References
Ca. An. boleenese	China	*R. microplus*	Cattle, goats, and other free-ranging domesic animals	Lu et al. (2022, 2023), Zhao et al. (2023), and Xu et al. (2023)
	Paraguay	*R. sanguineus*	Dogs	Salvioni Recalde et al. (2024)
Candidatus Austromerope brasiliensis	Mexico	*Ascophyllum nodosum*	Anteater (*Tamandua mexicana*)	Munoz-Garcia et al. (2023)
Ca. An. camelsii	China	*H. asiaticum*	Camels, sheep, cattle	Zhang et al. (2024b)
	Egypt	*H. dromedarii*	Camels	Senbill et al. (2024) and Soliman et al. (2025)
Candidatus Chloroploca mongolica	China	*I. persulcatus*	Free-ranging goats, cattle	Tian et al. (2023)
Ca. An. slovaca-like	Pakistan	*R. microplus*	Sheep	Khan et al. (2024)
Ca. Bo. hyalomii	China	*H. asiaticum*	Camels, sheep, cattle	Zhang et al. (2024b)
Candidatus Cryptoplasma sp.	China	*H. longicornis*	Cattle	Qi et al. (2022)
Ca. E. corsicanum	Corsica	*R. bursa*	Slaughtered cattle	Cicculli et al. (2020)
Candidatus Ehrlichia maerkangensis	China	*H. qinghaiensis*	Free-ranging domestic animals	Lu et al. (2023)
Ca. E. rustica	Cameron	*R. microplus, H. leachi*	Cattle	Ngnindji-Youdje et al. (2022)
	Egypt	*R. annulatus*	Horses	Abdullah et al. (2022)
Candidatus Mycoplasma haemobos	China	*H. longicornis, R. (Boophilus) microplus*	Dogs (healthy and anemic)	Shi et al. (2022)
Candidatus Mycoplasma haemominutum	UK	*I. trianguliceps, I. ricinus*	Cats	Duplan et al. (2018)

Candidatus Mycoplasma turicensis	UK	*I. ricinus*	Cats	Duplan et al. (2018)
Ca. N. *arcana*	Australia	*I. holocyclus*	Vegetation	Gofton et al. (2022)
Candidatus Nihonia *australis*	Australia	*I. holocyclus*	Vegetation	Gofton et al. (2022)
Candidatus Neoehrlichia mikurensis	Austria	*I. ricinus*	Dogs	Wijnveld et al. (2021)
	Finland	*I. ricinus*	Pet dogs, cats	Zakham et al. (2023)
	Mongolia	*D. nuttallii*	Vegetation	Fischer et al. (2020)
Candidatus Rickettsia *andeanae*	Argentina	*A. tigrinum*	Road-killed Pampas foxes (*Lycalopex gymnocercus*)	Eberhardt et al. (2023)
	Brazil	*A. tigrinum*	Maned wolves	Arrais et al. (2021)
	Ecuador	*A. maculatum*	Horses, dogs	Enriquez et al. (2024)
C. R. *barbariae*	Portugal	*R. bursa*	Autochthonous Garrano breed of horses	Barradas et al. (2024)
	Eswatini	*R. simus*	Vegetation	Ledger et al. (2021)
	Cameron	*R. annulatus*, *R. muhsamae*	Cattle	Ngnindji-Youdje et al. (2022)
Candidatus Rickettsia jingxinensis	China	*A. javanense*, *H. flava*, *H. kitaokai*, *H. longicornis*, *H. asiaticum*, *I. acutitarsus*, *I. sinensis*, *R. haemaphysaloides*, *R. microplus*	Cattle, goats, dogs, cats, Malayan pangolins, two-humped camel (*C. bactrianus*)	Lu et al. (2022, 2023), Li et al. (2024), and Obaid et al. (2025)
	Republic of Korea	*H. longicornis*, *I. nipponensis*	Human (hospitals and local public health centers)	Seo et al. (2023)
Ca. R. liangshanensis	China	*I. ovatus*	Free-ranging domestic animals	Lu et al. (2023)

(*Continued*)

TABLE 6.3
Globally reported common tick-borne protozoan parasites and their vector ticks

Tick-borne protists	Tick species	Tick life stage	Tick sources	Location	References
Babesia afzelii	*Dermacentor reticulatus*	Male, female	Questing	Algeria	Didyk et al. (2017)
Babesia bigemina	*Hyalomma asiaticum*	NA	Fed and unfed	China	Wang et al. (2025)
	Rhipicephalus decoloratus	Adult	Questing	Tanzania	Haji et al. (2023)
	Hyalomma dromedarii	Female (semi-engorged)	Camels	Egypt	Elsawy et al. (2024)
Babesia bovis	*Rhipicephalus microplus*	Adult	Questing	Poland	Haji et al. (2023)
Babesia caballi	*Hyalomma anatolicum*, *Dermacentor nuttalli*, *Dermacentor marginatus*	NA	Horse	USA	Li et al. (2020)
	H. asiaticum		Questing	Kazakhstan	Sang et al. (2021)
	D. marginatus		Cattle	Japan	
Babesia canis	*Rhipicephalus sanguineus* s.l.	NA	Dogs	Spain	Estrada-Peña et al. (2017)
	R. sanguineus s.l.				
	Dermacentor reticulatus				
	Ixodes hexagonus				
	D. reticulatus	Adult	Questing	Egypt	Kubiak et al. (2024)
Babesia capreoli	*Ixodes ricinus*	Nymph	Questing	China	Venclikova et al. (2015)

Babesia gibsoni	*I. ricinus*	NA	Dogs	Thailand	Estrada-Peña et al. (2017)
	R. sanguineus s.l.			Spain	
	I. hexagonus				
	Amblyomma testudinarium	Female	Wild boar	Slovakia	Masatani et al. (2017)
Babesia microti	*H. dromedarii*	Male, female	Arabian camels (*Camelus dromedarius*)	Israel	Mumcuoglu et al. (2022a)
	Rhipicephalus turanicus		Questing		Mumcuoglu et al. (2022b)
	I. ricinus	Nymph, male, female		Ukraine	Didyk et al. (2017)
	D. reticulatus	Adult		Egypt	Kubiak et al. (2024
	Hyalomma aegyptium	Male, female	Greek tortoises (*Testudo graeca*)	Israel	Mumcuoglu et al. (2022a)
Babesia occultans	*D. marginatus*	NA	Cattle	Japan	Sang et al. (2021)
	R. turanicus		Fed and unfed	Tanzania	Wang et al. (2025)
Babesia ovata	*Haemaphysalis longicornis*, *Ixodes ovatus*	Nymph, adult	Questing	Czech Republic	Sivakumar et al. (2014)
Babesia sp.	*R. sanguineus* s.l.	Male, female	Dog	Portugal	Wu et al. (2023)
	D. variabilis	NA	Domestic animals, wildlife, humans, questing ticks	Ukraine	Shock et al. (2014)
	Haemaphysalis lagrangei	Female	Questing	Thailand	Wattanamethanont et al. (2018)

(*Continued*)

TABLE 6.3
(Continued)

Tick-borne protists	Tick species	Tick life stage	Tick sources	Location	References
	Amblyomma americanum	NA	Domestic animals, wildlife, humans, questing ticks	USA	Shock et al. (2014)
	I. ricinus	Female	Questing; mammals (roe deer, fallow deer, moose, dogs, cats)	Taiwan	Cialini et al. (2025)
	R. sanguineus s.l.	Female	Domestic animals, wildlife, and resident and migratory birds	Portugal	Pereira et al. (2018)
	Hyperhypnoides formosensis	Nymph, male	Questing	Japan	Masatani et al. (2017)
	Harperocallis flava				
Babesia sp. Bp-Hc3	*Haemaphysalis concinna*	Larvae, nymphs	Questing	Hungary	Keve et al. (2025)
Babesia sp. Bp-Hc5	*H. concinna*	Larvae	Questing	Sweden	
Babesia sp. Bp-Hc7	*H. concinna*	Larvae	Questing	Sweden	
Babesia venatorum (formerly *Babesia* sp. EU1)	*I. ricinus*	Nymph	Field collected	Egypt	Nebbak et al. (2019)
	I. ricinus		Questing	China	Venclikova et al. (2015)
Babesia vogeli	*R. turanicus*	Male, female	Questing	Israel	Mumcuoglu et al. (2022b)
	R. sanguineus	NA	Cats and dogs	China	Maia et al. (2014)
Colpodella sp.	*H. dromedarii*	Adult females	Camel	Uganda	Soliman et al. (2024b)
Cytauxzoon felis	*D. variabilis*	NA	Domestic animals, wildlife, humans, questing	Ukraine	Shock et al. (2014)

Cytauxzoon sp.	*Haemaphysalis hystricis*	Nymph, male, female	Pangolin (*Manis pentadactyla*)	Japan	Khatri-Chhetri et al. (2016)
Hepatozoon canis	*R. sanguineus*	Male, female	Questing	Israel	Mumcuoglu et al. (2022b)
	R. turanicus		Questing		
	R. turanicus		Dogs (*Canis familiaris*)		
	R. sanguineus	NA	Cats and dogs	China	Maia et al. (2014)
	I. hexagonus		Dogs	Spain	Estrada-Peña et al. (2017)
	I. ricinus			Thailand	
	H. longicornis	Female		Japan	Masatani et al. (2017)
	R. sanguineus s.l.	Nymph, males, females		Brazil	Chao et al. (2019)
	Ambystoma tigrinum	Adults	Maned wolf (*Chrysocyon brachyurus*)	Japan	Arrais et al. (2021)
Helicobacter felis	*R. sanguineus*	NA	Cats and dogs	China	Maia et al. (2014)
	R. turanicus	Male, female	Greek tortoise (*T. graeca*)	Israel	Mumcuoglu et al. (2022b)
Hepatozoon sp.	*H. longicornis*	Nymph	Questing	Japan	Masatani et al. (2017)
	H. hystricis	Female			
Leishmania infantum	*H. dromedarii*	Male, female	Arabian camels (*C. dromedarius*)	Israel	Mumcuoglu et al. (2022a)
	H. aegyptium		Greek tortoises (*T. graeca*)		
	R. sanguineus		Questing		Mumcuoglu et al. (2022b)

(*Continued*)

TABLE 6.3
(Continued)

Tick-borne protists	Tick species	Tick life stage	Tick sources	Location	References
	R. turanicus				
	R. turanicus		Greek tortoise (*T. graeca*)		
	R. turanicus		Dogs (*C. familiaris*)		
	R. turanicus		Hedgehogs (*Erinaceus concolor*)	China	Mumcuoglu et al. (2022b)
Rangelia vitalii	*Amblyomma aureolatum*	Female (engorged)	Laboratory colony	Hungary	Soares et al. (2018
Theileria annulata	*Rhipicephalus bursa*	Female	Domestic animals, wildlife, and resident and migratory birds	Portugal	Pereira et al. (2018)
	Rhipicephalus annulatus	NA	Cattle	Spain	Sadeddine et al. (2020)
	R. turanicus		Goats	Kazakhstan	Sang et al. (2021)
	R. turanicus		Shepherd dog		
	D. marginatus		Cattle	Japan	
	H. anatolicum		Fed and unfed	China	Wang et al. (2025)
	H. asiaticum				
Theileria buffeli	*R. annulatus*	NA	Cattle	Spain	Sadeddine et al. (2020)
Theileria capreoli	*H. lagrangei*	Male	Questing	Thailand	Wattanamethanont et al. (2018)
	H. lagrangei	Nymph		Kazakhstan	
	H. concinna	Nymph, males, females		Hungary	Keve et al. (2025

Theileria cervi	*H. lagrangei*	Male	Questing	Thailand	Wattanamethanont et al. (2018)
	H. lagrangei	Nymph		Kazakhstan	
Theileria equi	*R. sanguineus* s.l.	NA	Dogs	Spain	Estrada-Peña et al. (2017)
	D. marginatus		Horse	Kazakhstan	Sang et al. (2021)
	H. anatolicum, *D. nuttalli*, *D. marginatus*			USA	Li et al. (2020)
Theileria luwenshuni	*R. microplus*	Nymph, adults (male and female)	Goats	France	Aung et al. (2022)
Theileria mutans	*Amblyomma*, *Rhipicephalus*, *Hyalomma*	Nymphs, adults	Domestic cattle	Portugal	Byaruhanga et al. (2021)
Theileria ovis	*R. sanguineus* s.l.	Male, female	Domestic animals, wildlife, and resident and migratory birds	Israel	Pereira et al. (2018)
	H. anatolicum, *D. nuttalli*, *D. marginatus*	NA	Sheep	China	Li et al. (2020)
	R. turanicus		Shepherd dog	Kazakhstan	Sang et al. (2021)
	H. asiaticum		Fed and unfed	China	Wang et al. (2025)
	D. marginatus, *D. pavlovskyi*, *R. sanguineus*				
Theileria parva	*Amblyomma*, *Rhipicephalus*, *Hyalomma*	Nymphs, adults	Domestic cattle	Portugal	Byaruhanga et al. (2021)
	Rhipicephalus appendiculatus		Questing	Tanzania	Haji et al. (2023)
Theileria sp.	*H. aegyptium*	Male, female	Greek tortoises (*T. graeca*)	Israel	Mumcuoglu et al. (2022a)
	H. formosensis	Nymph	Sika deer	Japan	Masatani et al. (2017)

(Continued)

TABLE 6.3
(Continued)

Tick-borne protists	Tick species	Tick life stage	Tick sources	Location	References
	H. formosensis		Questing		
	H. flava				
	H. longicornis				
	H. longicornis	Female (engorged)	Sika deer		
	Haemaphysalis kitaokai		Sika deer		
	Ixodes persulcatus, *Hamamelis japonica*	NA	Flagging	South America	Shibata et al. (2018)
	A. americanum		Questing, humans, domestic and wild animals	USA	Shock et al. (2014)
	H. lagrangei	Male	Questing	Thailand	Wattanamethanont et al. (2018)
	H. dromedarii	Female (semi-engorged)	Camels	Egypt	Elsawy et al. (2024)
	H. dromedarii	Male, female	Arabian camels (*C. dromedarius*)	Israel	Mumcuoglu et al. (2022a)
Theileria velifera	*Amblyomma*, *Rhipicephalus*, *Hyalomma*	Nymphs, adults	Domestic cattle	Portugal	Byaruhanga et al. (2021)
Trypanosoma	*I. ricinus*	Male, female	Host-seeking	Australia	Luu et al. (2020)
Trypanosoma noyesi	*Amblyomma triguttatum*, *Izhakiella australiensis*, *Ixodes myrmecobii*, *Ixodes* sp.	Larvae, nymphs, adults	Questing larvae, nymphs, adults	Argentina	Krige et al. (2021)
Trypanosoma sp.	*R. microplus*	Female (engorged)	Braford heifers	Taiwan	Morel et al. (2021)

lead to erythrolysis, thereby resulting in anemia and hypoxia in cattle (Oakes et al., 2019). Likewise, *Hepatozoon* invades red and white blood cells of vertebrates (Baneth, 2011), commonly found in wildlife perpetuating enzootic cycles (Thomas et al., 2024). These hemoparasites have heterogeneous life cycles involving a vast array of vertebrate and invertebrate hosts (Baneth, 2011), including ticks. *Hepatozoon canis* and *Hepatozoon americanum*, with varying geographical distributions, are common tick-borne infections of dogs transmitted mainly by brown dog tick (*R. sanguineus* s.l.) and *Amblyomma maculatum* (Baneth and Allen, 2022).

Another Piroplasmid, *Rangelia vitalii*, is the etiologic agent of canine rangeliosis (yellow fever of dogs), a severe piroplasmosis (hemolytic and hemorrhagic disease) that also affects other domestic animals. *R. vitalii* infects erythrocytes, leucocytes, and endothelial cells of blood capillaries (Franca et al., 2014). Acquisition feeding of *Amblyomma aureolatum* on domestic dogs which were subjected to intravenous inoculation or infestation with *R. vitalii*-infected ticks demonstrated transstadial and transovarial passages, and when exposed to transmission feeding, *A. aureolatum* ticks were found competent to transmit *R. vitalii* to dogs (Soares et al., 2018).

Some of the less commonly encountered tick-borne protozoan parasites are *Toxoplasma* and *Trypanosoma*. However, *Toxoplasma gondii* has been detected in *D. variabilis*, *D. andersoni*, *D. reticulatus*, *I. ricinus*, *H. longicornis*, *A. americanum*, *Amblyomma cajennense*, and *Ornithodoros moubata* (Ben-Harari, 2019). It has been proved that *T. gondii* could survive and remain infective in *H. longicornis* for more than 15 days. Nevertheless, it could not be transmitted to hosts during a blood meal (Zhou et al., 2016). Overall, the presupposition needs more experimental data to confirm. Compared to insect vectors, the importance of ticks in the transmission of trypanosomes has been less documented.

But studies so far have not verified the significant vectorial potential of ticks for classical trypanosomiasis. A previously unknown *Trypanosoma* from wild *I. ricinus* ticks tissue suspension inoculated into cultures of the tick cell lines has been isolated and designated as *Trypanosoma* sp. Bratislava1 (Luu et al., 2020). However, the study did not confirm whether this strain can infect vertebrates, or if ticks can transmit it during blood meal. *Trypanosoma noyesi* is detected in Australian questing ticks (*Amblyomma triguttatum*, *Izhakiella australiensis*, *Ixodes myrmecobii*, and larvae of *Ixodes* spp.) by using high-resolution melting quantitative polymerase chain reaction (HRM- quantitative polymerase chain reaction (qPCR)) and fluorescence in situ hybridization (FISH) assay (Krige et al., 2021). Although this study could not confirm the tissue-specific location of trypanosomes indicating the potential route of transmission, trypanosomes can persist in ticks between blood meals, and ticks could be possible vectors in the transmission of *T. noyesi* between native wildlife. Specimens of a *Trypanosoma* sp. were found in a hemolymph sample of *R. microplus* from Argentina (Morel et al., 2021). Transmission experiments are still required to confirm the vectorial capacity of ticks for *T. noyesi* (Krige et al., 2021) as it is among the major concerns whether ticks could contribute as a potential vector for the successful transmission of *Trypanosoma*.

6.2.5 Tick-Borne Helminths (Filarial Nematodes)

Ticks are known to serve as vectors of filarial nematodes. The filarial nematodes are unisexual, mostly with sexual dimorphism, and characterized by elongated, cylindrical, thread-like body lined with tough cuticle (Bain et al., 2003). They have exploited diverse hematophagous arthropods as intermediate hosts which serve as biological vectors to facilitate their indirect life cycles. Filarial nematodes are one of the intriguing organisms with regard to their long-term persistence, survival, and reproductive potential even for more than a decade in particular cases (Gems, 2000). The adult filarial worms, depending on the species, localize in the subcutaneous and/or connective tissues or lymphatic system of the definitive hosts and produce numerous thread-like larvae, the microfilariae (Taylor et al., 2010). In filarial nematodes, microfilariae are the initial larval forms which, in arthropod vectors, pass through first-stage (L1), second-stage (L2), and ultimately third-stage larvae (L3). In general, L3, infective to the vertebrate (definitive) hosts, upon entry into the definitive host via the bite of arthropod vectors, develop into sexually mature adults and continue reproduction. The fertilized eggs hatch into microfilariae which circulate throughout the vertebrate host body and are ingested by the arthropod vectors (Gruntmeir et al., 2023; Binetruy and Duron, 2023) (Figure 6.3).

Tick-borne filarial nematodes (TBFN), as one of the facets of tick-associated biota, are a less explored field in terms of parasite–vector–host interface. Besides the diverse species of arthropod

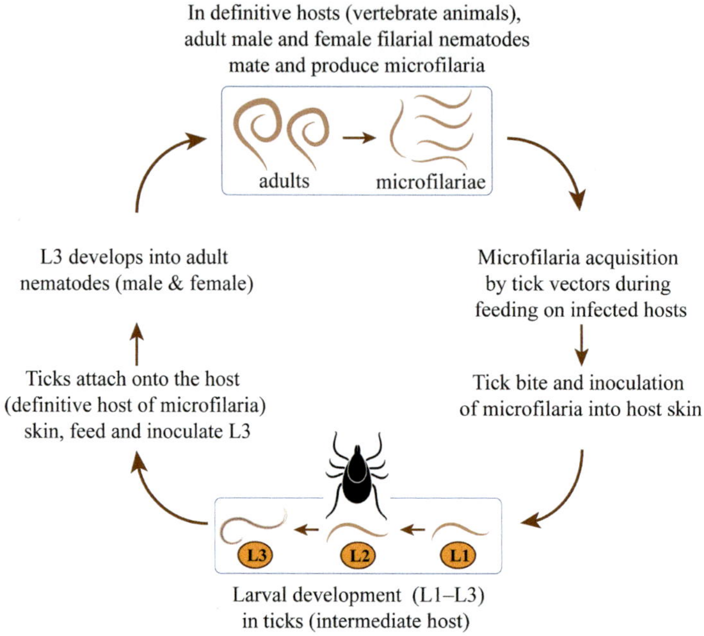

Figure 6.3 Filarial nematode transmission cycle between ticks and vertebrate hosts. The cycle commences when L3 larvae are inoculated into definitive host's skin via tick bites. Within the hosts, L3 develops into adult nematodes within the host followed by mating and producing microfilariae, which undergo peripheral circulation. When microfilariae are ingested by tick, they develop into infective L1–L3 larvae within the tick. Illustration created on Adobe Illustrator CC 2019.

vectors exploited by filarial nematodes, a very few species belonging to the Onchocercidae family are known to use ticks as their primary intermediate hosts (Lefoulon et al., 2015; Bezerra-Santos et al., 2022). A considerable number of investigations have identified filarial nematodes inhabiting the hard (Ixodidae) and soft (Argasidae) ticks (Otranto et al., 2012; Bezerra-Santos et al., 2023; Lineberry et al., 2021; Lineberry et al., 2020).

There have been indeed controlled experimental attempts and subsequent microscopic observation of the filarial nematodes using both the skin snip (necroscopy) and tick samples to verify the vectorial capacity of ticks concerning parasite acquisition and transmission. Assessments of the developmental success of filarial worms (*Yatesia hydrochoerus*) in hard ticks (*A. cajennense* and *A. americanum*) were conducted in the early 1980s (Yates and Lowrie, 1984). Certain life stage-specific features of the vector tick govern the developmental success of the filarial worm to their infective stage as had been observed in *Dipetalonema dracunculoides* in *R. sanguineus* (brown dog tick) (Olmeda-García and Rodríguez-Rodríguez, 1994). In an experiment, *R. sanguineus* nymphs were allowed to feed on a dog which was naturally infected by *Cercopithifilaria* sp. It was confirmed by both microscopic and molecular techniques (amplification of COX1 fragments, 304 bp) that *R. sanguineus* nymphs ingested the microfilaria which succeeded development up to infective stage, thereby inferring that this tick species could serve as a potential intermediate host of this filarial nematode (Brianti et al., 2012). Morphological and molecular (COX1 and 12S rDNA) analyses of female nematodes in skin snips (necropsy) obtained from the roe deer and the third-stage larvae (L3) from the *I. ricinus* ticks were confirmed as *Cercopithifilaria rugosicauda* (Ramos et al., 2013), suggesting that this nematode employs the *I. ricinus* nymphs to invade roe deer (*Capreolus capreolus*). This evidence suggests that ticks can obviously serve as vector for filarial nematodes. Tick species including *A. cajennense*, *Amblyomma oblongoguttatum*, *Amblyomma romitii*, *Ixodes luciae*, *Rhipicephalus* s.l. (Lineberry et al., 2021; Bezerra-Santos et al., 2022; Binetruy and Duron, 2023; Bruley and Duron, 2024) *Rhipicephalus haemaphysaloides* (Bezerra-Santos et al., 2022) are found to serve as vectors for filarial nematodes belonging to the *Dipetalonema* lineage.

Advanced modern techniques, for instance, genetic analysis, have consistently validated the presence of filarial nematodes in hard ticks.

A. americanum (Zhang et al., 2011; Henning et al., 2016), *A. cajennense* (Binetruy and Duron, 2023), *A. oblongoguttatum* (Binetruy and Duron, 2023), *A. romitii* (Binetruy and Duron, 2023), *Dermacentor reticulus* (Andorson et al., 2017), *I. luciae* (Binetruy and Duron, 2023), *I. ricinus* (Ramos et al., 2013), *I. scapularis* (Namrata et al., 2014; Tokarz et al., 2020), *R. haemaphysaloides* (Bezerra-Santos et al., 2022), *R. sanguineus* s.l. (Latrofa et al., 2014; Bezerra-Santos et al., 2022; Angelou et al., 2020; Otranto et al., 2012; Brianti et al., 2012; Santos et al., 2018; Lineberry et al., 2021; Lineberry et al., 2020; Bezerra-Santos et al., 2022; Otranto et al., 2012; Brianti et al., 2012) have been known to be the vectors for various species of the Onchocercidae nematode parasites (*Acanthocheilonema* sp., *Cercopithifilaria bainae*, *Monanema* sp., *C. bainae*, *Cercopithifilaria grassi*, *C. rugosicauda*, *Cruorifilara*Cruorifilaria sp., and *Dipetalonema*-like). These filaria have been detected targeting specific mitochondrial cytochrome oxidase subunit I (COI), 12S rRNA and 16S rRNA genes, and/or filarial nematode-specific DNA in situ hybridization techniques (Table 6.4) in the skin snip (necropsy)

TABLE 6.4
Globally reported tick-borne filarial nematode parasites and their vector ticks

Geographic origin	Filarial species	Tick vectors	Tick sources	References
Australia	*Cercopithifilaria* sp.	*Rhipicephalus sanguineus* s.l.	Dogs	Latrofa et al. (2014)
Brazil	*Cercopithifilaria bainae*	*R. sanguineus* s.l.		Santos et al. (2018)
Brazil	*C. bainae*	*R. sanguineus* s.l.		Santos et al. (2017)
Brazil	*Cercopithifilaria* sp.	*R. sanguineus* s.l.		Latrofa et al. (2014)
China	*C. bainae*	*R. sanguineus* s.l.		Bezerra-Santos et al. (2022)
French Guiana	*Cercopithifilaria* sp.	*Amblyomma oblongoguttatum*, *R. sanguineus* s.l.	Opossums, filed collected	Binetruy and Duron (2023)
	Cruorifilara sp.	*Amblyomma cajennense*, *Amblyomma romitii*		
	Dipetalonema-like	*Ixodes luciae*		
Greece	*C. bainae*	*R. sanguineus* s.l.	Dogs	Angelou et al. (2020)
Greece, Italy, Spain	*Cercopithifilaria* sp.	*R. sanguineus* s.l.		Otranto et al. (2012)
Greece, Italy, Spain, Portugal	*Cercopithifilaria* sp.	*R. sanguineus* s.l.		Latrofa et al. (2014)
India	*C. bainae*, *Cercopithifilaria grassi*	*Rhipicephalus haemaphysaloides*		Bezerra-Santos et al. (2022)
Indonesia	*C. bainae*	*R. sanguineus* s.l.		
Italy	*Cercopithifilaria* sp.	*R. sanguineus* s.l.		Brianti et al. (2012)

(Continued)

TABLE 6.4
(Continued)

Geographic origin	Filarial species	Tick vectors	Tick sources	References
Italy	*Cercopithifilaria rugosicauda*	*Ixodes ricinus*	Roe deer	Ramos et al. (2013)
Malaysia	*C. bainae*	*R. sanguineus* s.l.	Dogs	Latrofa et al. (2014)
Pakistan	*C. grassi*	*Rhipicephalus* sp.		Latrofa et al. (2014)
Philippines	*C. bainae, C. grassi*	*R. sanguineus* s.l.		Bezerra-Santos et al. (2022)
Romania	*C. bainae*	*Dermacentor reticulus*		Andorsson et al. (2017)
South Africa	*Cercopithifilaria* spp.	*R. sanguineus* s.l.		Latrofa et al. (2014)
Taiwan	*C. bainae*	*R. sanguineus* s.l.		Bezerra-Santos et al. (2022)
USA	*Acanthocheilonema* sp.	*Ixodes scapularis*	Vegetation	Namrata et al. (2014)
	Acanthocheilonema sp.	*Amblyomma americanum*	Vegetation	Zhang et al. (2011)
	C. bainae	*R. sanguineus* s.l.	Dogs	Lineberry et al. (2021)
	C. bainae	*R. sanguineus* s.l.		Lineberry et al. (2020)
	Monanema sp.	*A. americanum*	Vegetation	Henning et al. (2016)
	Monanema sp.	*I. scapularis*		Tokarz et al. (2020)
Vietnam	*C. bainae*	*R. sanguineus* s.l.	Dogs	Bezerra-Santos et al. (2022)

biopsy and/or tick tissues after dissection. Ticks of all stages of life and sexes are known to harbor filarial nematodes. The brown dog tick (*R. sanguineus* s.l.) is the most commonly reported hard tick harboring the filarial nematodes. Both field-collected and on-host ticks are identified to vector these parasites. Yet, it deserves a systematic survey to investigate its transmission dynamics, possible coinfection trends with other TBPs, and tick vector competence.

6.3 SYMBIONT DIVERSITY IN TICKS

In this section the term 'symbionts' is used to refer to the 'bacterial members' which are the major players in the microbial consortia; therefore, a comprehensive discussion is made on nutritional and non-nutritional aspects of them in relation to tick biology. The microbiome, in many organisms, outnumbers the host cells and greatly influences the host biology and fitness (Guizzo et al., 2020). The majority of the tick-associated symbionts have maternal inheritance. The facultative symbionts transiently infect ticks and are often acquired from their off-host habitats since ticks, during their off-host periods, dwell in dark and humid environments with low-lying shrubs, leaf litter, and even relatively tall grasses, which are favorable sites for microbial proliferation (Jordan and Schulze, 2020). These support potential sources for the acquisition of microbiota into ticks (Li et al., 2022). Other microbial sources include skin surface from where ticks obtain microbes from diverse host species during blood feeding (Li et al., 2022). Copulation probably helps in the exchange of symbionts paternally. The following discussion provides a comprehensive outline on the diversity of common tick symbionts, their interaction with ticks in view of their contribution to tick biology and vector competence, and tick–symbiont (endosymbionts) and pathogen interaction.

6.3.1 Inherent Symbiotic Nexus between Ticks and Nutritional Symbionts: An Evolutionary Partnership

To get a deeper understanding of the tick endosymbionts and their functional role, it is worthwhile to highlight the distinction between obligate and facultative hematophagous adaptation in arthropods. Obligate blood-feeding vectors like ticks largely rely on symbiotic relationships to satisfy their nutritional requirements. In contrast, facultative hematophagous arthropods, such as dipterans (e.g., mosquitoes and tabanids), exploit alternative nutritional sources in various life stages. For instance, their larvae preferably consume organic debris, and adults often feed on plant saps or nectar, while males, in most of the cases, are devoid of blood-feeding capacity, so they do not feed on blood at all, relying instead on non-blood dietary sources (Duron and Gottlieb, 2020). Although symbionts and hematophagous hosts have coevolved for millions of years, their interdependence and associations are not often stable. Based on genomic and ecological evidence, the ancestral symbionts could be taken over by recently acquired symbionts, most often bacteria, which exhibit analogous metabolic potential. This symbiont dynamism is mainly propelled by phylogenetic constraints and ecological selection pressure (Duron and Gottlieb, 2020). Obligate symbionts are necessary for the survival and/or reproduction of the hosts, and without them the host experiences reduced fitness, impaired development, or even death (Duron et al., 2024).

Besides pathogenic microorganisms, ticks also harbor a wide range of non-pathogenic microbes, including commensal and mutualistic species (Narasimhan et al., 2021). A wide array of functionally analogous symbionts inhabiting obligatory hematophagous arthropods, including ticks, play a crucial role in compensating for nutritional deficiencies particularly B-vitamins which are not available in their sole blood diet despite being a rich source of proteins and lipids (Duron et al., 2018; Duron et al., 2024). Ticks possess at least four distinct lineages of B-vitamin-producing symbionts namely *Coxiella*, *Francisella*, *Rickettsia*, and *Midichloria* (Duron and Gottlieb, 2020). Although *Coxiella* is the ancestral symbiont in certain tick genera, evolutionary replacements by *Francisella* have taken place repeatedly across tick phylogeny (Binetruy et al., 2020). Genome analyses have evidenced that both *Coxiella* and *Francisella* have almost similar B-vitamin biosynthesis capacity, i.e., the recently acquired *Francisella* confers the same benefits to ticks as provided by the ancestral *Coxiella* in terms of B-vitamin benefits (Duron and Gottlieb, 2020). The association between ticks and their nutritional symbionts is coevolved, establishing an intricate system comprising a wide range of coadapted traits that reinforce their pronounced interdependence. Nutritional symbionts are essential for ticks for their hematophagous adaptation. In other words, it is the nutritional symbionts which have driven ticks to an exclusive blood-based diet. The bacteria, which evolved symbiotic intracellular lifestyles, were subjected to an extensive genome reduction. The set of genes contained in B-vitamin provisioning symbionts is primarily a subset of the gene collection of their relatives where the nonessential genes are either pseudogenized or entirely lost, with conservation of B-vitamin synthesis pathways (Duron and Gottlieb, 2020).

In general, symbionts (endosymbionts), based on their mutual relationship with the hosts and their mode of successive transfer across generations, can be categorized mainly into two groups – obligatory (primary) and facultative (secondary) endosymbionts. Typically intracellular, the former with reduced genomes (Smith et al., 2015; Fattar et al., 2025) have long-term evolutionary associations, transmitted exclusively vertically (mother to offspring), and are essential for host survival as they supplement nutritional benefits while the latter are not essential for host survival although beneficial as they enhance tick fitness (e.g., stress resistance, pathogen defense). Obligate endosymbionts have established more recent associations with their hosts and have variable transmission, mostly vertical but occasionally horizontal (i.e., between hosts/ticks or from the environment) (Guizzo et al., 2020; Narasimhan and Fikrig, 2015). Facultative symbionts can also pass vertically and impart benefits to ticks (Du et al., 2023). In general, tick endosymbionts as a microbial consortium offer combined influence and play a crucial role for tick survival, immunity, reproduction, and pathogen transmission.

With regard to tick adaptation and their dependency on microbiota, particularly for their survival, a concept of functional redundancy is worthwhile to mention. Functional redundancy in tick microbiota refers to the natural phenomenon where multitudinous bacterial species inhabiting the same tick as an entity of microbial community perform similar biochemical or metabolic functions. By the way, if some microbial taxa are lost owing to environmental stress, pathogen infection, microbe–microbe interaction, or immune response, other bacteria possibly compensate by upregulating and maintaining the essential pathways, including B-vitamin synthesis, nutrient metabolism, or detoxification. For instance, in *I. scapularis*, research has indicated that even up to 198 bacterial genera could offer a single pathway, suggesting the ability of ticks to maintain fundamental functions despite unexpected microbial alterations (Estrada-Pena et al., 2020a; Obregon et al., 2019). This redundancy ameliorates tick's resilience against pathogen infection, and antimicrobial peptide (AMP)-induced alteration of microbial taxonomic composition, thereby preventing pathway disruption and maintaining core metabolic functions.

The evolutionary phenomenon of genome reduction is one of the key features of symbiotic organisms as redundant or non-essential genes undergo gradual loss during their adaptation to host-dependent lifestyle in almost stable host environment (Gottlieb et al., 2015). A long-term vertical transmission of intracellular bacteria, most particularly in ancient symbionts, leads to extensive genomic erosion, resulting in extreme genome reduction. This phenomenon commences during the early stages of endosymbiosis characterized by accumulation of pseudogenes (pseudogenization) and proliferation of mobile genetic elements, which drive chromosomal rearrangements and ultimately form extremely small genomes (Zheng et al., 2016). Symbiotic microorganisms face a combined interplay of environmental and population-genetic pressures (genetic drift, reduced selection pressure) which led to their gene loss since previously essential genes get gradual elimination, and even small genomes undergo ongoing pseudogenization (Choi et al., 2025). The rate and patterns of gene loss associated with this process have been investigated in several different symbiotic systems. Analysis of the complete genome of *Coxiella*-like symbiont in *R. turanicus* (CRt) revealed that its genome is 1.7 Mb larger than that of most obligate mutualists; however, a very low protein-coding content (48.5%) and numerous pseudogenes are available, suggesting that it is undergoing genome reduction process (Gottlieb et al., 2015). Although the genome of Coxiella-like endosymbiont of *Amblyomma americanum* (CLEAA) is highly reduced (656,901 bp) as compared with *C. burnetii* (1,995,281 bp), the CLEAA genome retains the genes encoding major vitamin and cofactor biosynthetic pathways, implicating CLEAA as a vitamin-provisioning endosymbiont (Smith et al., 2015). Moreover, CLEAA lacks any recognizable virulence genes, indicating that it is not a pathogen even though it is present in SG (Smith et al., 2015). Comparative analysis of two tick symbionts – the CRt and the much smaller (2.5 times) *Coxiella* genome in *A. americanum* (CLEAA) – reveals that both have undergone loss of similar genes. The faster shrinking of CLEAA genome might have resulted from the loss of its mismatch repair genes mutSL, thereby accelerating its evolutionary changes (Gottlieb et al., 2015). Field-collected CRt populations, based on sequence polymorphisms, exhibited high genetic diversity of *Coxiella*-like symbionts, with nearly one unique strain per tick. Mutation patterns suggest even silent (synonymous) genetic changes may be under selection. The CRt genome represents a snapshot of active genome reduction, and CLEAA comparison highlights how early loss of key genes (like mutSL) can dramatically influence the speed of this process (Gottlieb et al., 2015).

A concept is that a newly acquired symbiont in ticks may outperform the ancestral one regarding nutrient production, potentially providing their host with more advantages. For instance, some ancestral *Coxiella* possess only 0.66 Mb genomes (Smith et al., 2015) while the recently acquired *Francisella* contain larger genomes (>1.5 Mb), suggesting the recent symbiont might have greater biosynthetic potential (Duron et al., 2018; Gerhart et al., 2018). Tick endosymbionts contribute a crucial role in support of tick physiology by modulating nutritional adaptation, fitness, and immunity. Additionally, these symbionts influence disease ecology through their parallel interactions with tick-borne pathogens, either facilitating or inhibiting pathogen proliferation within the specific tissue/cells of the vector (Hussain et al., 2022). CLE and *Francisella*-like endosymbionts (FLE) exhibit characteristic signs of pseudogenization, followed by continued genome reduction. Although these processes occur similarly, they proceed independently at different evolutionary stages, further strengthening the hypothesis of a specialized transition commencing from corresponding pathogenic ancestors to mutualist symbionts (Santos-Garcia et al., 2023; Duron et al., 2018). Both CLE and FLE genomes are substantially reduced compared to their pathogenic relatives, *C. burnetii* (2 Mb) and *F. tularensis* (1.9 Mb), respectively. Such a reduction remarkably impacts virulence genes. For instance, *C. burnetii* genes encoding the *Dot/Icm* secretion system in CLEOmar (CLE of *Okiseius maritimus*) and CLEDm (CLE of *D. marginatus*) and most of the virulence genes of *F. tularensis* in the F-Om (FLE of *Ok. maritimus*) and FLEDreti100-P (FLE of *D. reticulatus*) genomes are either pseudogenized or lost (Fattar et al., 2025). Despite evidence of genomic reduction, complete biosynthetic pathways for three key B-vitamins (biotin, riboflavin, and folate) are conserved in CLEOmar, CLEDm, F-Om, and FLEDreti100-P genomes. Moreover, CLE and FLE in their corresponding genomes, such as CLEDm, F-Om, and FLEDreti100-P genomes, maintain intact analogous shikimate pathways for chorismite biosynthesis (Fattar et al., 2025) which is associated with the production of tryptophan and serotonin similar to the one already identified in CLE isolated from *H. longicornis* (Zhong et al., 2021). These shared traits highlight how distantly related symbionts independently evolve comparable strategies to persist within different tick genera belonging to different families, emphasizing the power of convergent evolution in host–symbiont systems.

Recent investigations have increased significant information on the complex composition and nexus of tick microbial communities in relation to their influence on tick biology. Ticks harbor multitudinous microbiota most of which are biologically essential for their immunity, bodily development, and reproduction (Moraga-Fernández et al., 2023; Song et al., 2022). The research regarding tick symbiotic bacteria which contribute to pathogen survival and infection, tick development, and reproduction is in increasing trend (Narasimhan et al., 2021; Bonnet et al., 2021). During the course of incessant infection, microbes continue evolving to escape the immune responses and get adapted to utilize the available nutrients, and hosts in turn heighten their immune responses and use microbe-derived resources (Shaw and Catteruccia, 2024). Moreover, meanwhile mechanisms of pathogen tolerance in vectors enhance to continued fitness until the infection persists.

6.3.2 Dominant Symbionts (Endosymbionts) in Ticks

To date, ten types of dominant maternally inherited tick-associated bacterial endosymbionts are known for their common or unique interactions with tick hosts (Table 6.5). In a comprehensive study screening 861 tick specimens belonging to 81 species (hard ticks: 6 genera, 55 species; and soft ticks: 3 genera, 26 species), 79 species (hard tick: 54 species, and soft tick: 25)

TABLE 6.5

A summary of major maternally inherited bacterial symbionts in ticks and their influence in tick physiology and vector competence

Symbionts	Features	Major tissue localization in ticks	Role in ticks
Coxiella-LE (CLE)	Obligate symbiont, Gram-negative, class Gammaproteobacteria, genetically akin to pathogen *Coxiella burnetii*, maternally inherited, infects at least 2/3rd of tick species	Salivary glands, midgut, Malpighian tubules, ovaries	Synthesis of B vitamins (biotin, riboflavin, folate) and cofactors; essential for tick survival and reproduction; promote tick feeding, survival, mating, oogenesis, oviposition, hatching and molting, provide tick host B vitamins, L-proline and chorismate to stimulate serotonin biosynthesis, inhibit colonization of *Babesia microti*
Francisella-LE (FLE)	Obligate symbiont, Gram-negative, class Gammaproteobacteria, maternally inherited, reported in few ticks	Salivary glands, midgut, Malpighian tubules, ovaries	Synthesis of B vitamins (biotin, folic acid, and riboflavin) and cofactors; affect colonization of pathogenic *Francisella* spp. and *Rickettsia* spp.; produce heme and cysteine

(Continued)

TABLE 6.5
(Continued)

Symbionts	Features	Major tissue localization in ticks	Role in ticks
Midichloria	Obligate symbiont; Gram-negative, class Alphaproteobacteria, maternally inherited, inhabits mitochondria of certain tick species	Ovaries (mainly ovarian mitochondria)	Promote tick host feeding; biosynthesis of B-vitamins; provide biotin; possible role in heme/iron metabolism; additional ATP source for the host cell during oogenesis; role in tick molting processes; oxidative phosphorylation at low oxygen tension provisioning ATP sources, maintains normal ovarian physiology, oogenesis; regulate mitochondrial fragmentation in oogenesis; participate in production of ROS; promote the infection with pathogenic *Rickettsia parkeri* in ticks
Rickettsiella	Facultative symbiont, Gram-negative, class Gammaproteobacteria, maternally inherited	–	Could be facultative symbionts; potential B vitamin contributor (however understudied); block transmission of pathogenic *Rickettsia* in ticks; inhibit infection of *Anaplasma marginale* in ticks; provide folate; synthesize a functional GTP cyclic hydrolase I, and inhibit the infection with pathogenic Rickettsiaceae and *Borrelia burgdorferi* in ticks
Wolbachia	Obligate symbiont, Gram-negative, class Alphaproteobacteria, maternally inherited	Ovaries	Role in tick is not well understood; probably provision of biotin (vitamin B7) and tick fitness; may influence host reproduction; might induce parthenogenesis (?) (data lacking)
Rickettsia-like (RLE)	Few obligate symbionts, Gram-negative, class Alphaproteobacteria, closely related to multiple tick-borne pathogens, maternally inherited	Salivary glands, ovaries	Presence of genes for folic acid biosynthesis indicates nutritional role (provide tick host folate); influence larval motility; inhibit the infection of pathogenic *Rickettsia* spp. (data lacking)
Arsenophonus	Gram-negative, class Gammaproteobacteria, maternally inherited	Saliva, midgut	Male killing effect in chalcid wasps; probably fitness and development, and host seeking success in ticks (data lacking)
Lariskella	Gram-negative, class Alphaproteobacteria, maternally inherited	–	Reproductive manipulation by cytoplasmic incompatibility in insects; unknown function in ticks; could be nutritional support (data lacking)
Cardinium	Gram-negative, class Cytophagia. maternally inherited	–	Influence of host reproduction inducing cytoplasmic incompatibility, feminization, male killing, and parthenogenesis in other arthropods, but unknown in ticks (data lacking)
Spiroplasma	Gram-positive, class Mollicutes, maternally inherited, helical, motile	Salivary glands, midgut, ovaries	Male killing effect in insects; particular function in ticks unknown (data lacking)

screened for these maternally inherited bacterial genera harbored at least one of the targeted symbiotic bacterial genera (Duron et al., 2017). The most prevalent symbiont is CLEs, detected in 49 tick species, followed by *Rickettsia* (in 40 species), FLEs (in 17 species), *Spiroplasma* (in 13 species), *Midichloria* (in 12 species), *Rickettsiella* (in 10 species), *Cardinium* (in 5 species), *Lariskella* (in 3 species), and *Arsenophonus* and *Wolbachia* (each in 1 species) (Duron et al., 2017).

6.3.2.1 *Coxiella-LE and Francisella-LE: Tissue Tropism and Nutritional Contribution to Ticks*

Metagenomic analyses and incorporation of comprehensive microbial data have deciphered the two most widespread transovarially (maternally/vertically) transmitting obligate intracellular nutritional symbionts in most of the ticks: CLEs (Legionellales: Coxiellaceae), found in two-thirds of tick species, and FLE (Thiotrichales: Francisellaceae) (Duron et al., 2015; Duron et al., 2017; Binetruy et al., 2020; Buysse et al., 2022; Lalzar et al., 2012; Azagi et al., 2017; Machado-Ferreira et al., 2011; Almeida et al., 2012; Fattar et al., 2025), the latter being present in tick species lacking CLE. CLEs are distinct from pathogenic *C. burnetii* (agent of Q fever) although they share genetic similarities (Duron et al., 2015).

The intracellular niches of CLE and FLE exhibit striking similarities. Both symbionts form their dense clusters within the membrane-bound vacuoles in tick cell cytoplasm, reminiscent of the previously observed phenotype of CLE in *R. turanicus* (Lalzar et al., 2014). These symbionts exploit the vacuoles of varying sizes as the replication compartment and often almost entirely occupy the cytoplasm. Notably, these structures closely resemble those of their pathogenic relatives, *C. burnetii* and *F. tularensis*, which hijack vertebrate host cell machinery for sustained replication (Petit and Lebreton, 2022; Degabriel et al., 2023). CLE and FLE symbionts are principally found in ovaries (OV) and Malpighian tubules (MT) as well as in other organs like MG and SG; however, in lower abundance (Duron et al., 2018; Lalzar et al., 2012; Lalzar et al., 2014; Wang et al., 2018; Buysse et al., 2019). CLEs exhibit a wider tissue distribution, though preferably colonize the MT and OV of ticks (Cibichakravarthy et al., 2022). Interestingly, a distinct pattern of exclusion has been observed between CLE and FLE, suggesting these two do not exist in the same host (Binetruy et al., 2020). A study in *H. longicornis*, the CLEs were most abundant in fed females, followed by unfed females and unfed larvae. After blood feeding CLEs rapidly proliferate in the OV, where their density is the highest compared to SG, MG, and MT. Furthermore, vertical transmission was supported by the fact that FISH results revealed CLE localization mainly in the oocyte cytoplasm, along with their release in mature oocytes (Zhang et al., 2022). FISH coupled with PCR-based detection of CLEs in *Ornithodoros maritimus*, *D. marginatus*, and *Ixodes hexagonus* demonstrated a high degree of tissue tropism for tick OV and MT, suggesting that ovarian infection is an indication of transovarial transmission and that of MT is nutritional role (Buysse et al., 2019). The ovarian microbiome is relatively stable and often dominated by endosymbiotic bacteria, however, not exclusively. Contrastingly, the digestive system harbors a relatively stable microbial community which is primarily composed of pathogens as in SG (Hodosi et al., 2022).

A recent study demonstrates striking evolutionary convergence between CLE and FLE symbionts, revealing a parallel adaptation. In four tick hosts (*Or. maritimus*, *O. moubata*, *D. marginatus*, and *D. reticulatus*), both symbionts, as revealed in FISH analysis, exhibit identical tissue distributions and shared intracellular niches (i.e., vacuolar colonization) (Fattar et al., 2025). Quantitative real-time PCR (qPCR) analysis revealed a distinct convergence in tissue tropism of CLE and FLE across all examined tick species (*Or. maritimus*, *O. moubata*, *D. marginatus*, and *D. reticulatus*), with their preferential location in the OV and MT where their densities significantly surpassed those in SG and other body tissues. Interestingly, while CLEs were dominant in ovarian tissues, FLE culminated in higher densities in the MT, suggesting niche partitioning between these evolutionarily convergent symbionts within their shared hosts (Fattar et al., 2025).

Despite being phylogenetically distant, CLE and FLE exhibit converged functional analogy in terms of B-vitamin-related mutualism with ticks (Duron and Gottlieb, 2020). B-vitamins refer to a group of water soluble and chemically diverse vitamins essential for cell metabolism, mostly synthesized by microorganisms (but animals are unable to synthesize them), and includes thiamine (B1), riboflavin (B2), niacin (nicotinic acid) (B3), pantothenic acid (B5), pyridoxine (B6), biotin

(B7), folic acid (B9), and cobalamin (B12) (Duron et al., 2024). Genome sequencing has revealed that both possess entire biosynthetic pathways for biotin (B7), riboflavin (B2), and folate (B9) (Fattar et al., 2025). Physiologically, biotin is a coenzyme for carbohydrates, required for fatty acid synthesis, catabolism of branched-chain amino acid, and glucogenesis; folate is an important cofactor for cellular metabolism and serves as an essential precursor for DNA synthesis and DNA and RNA modifications while riboflavin is a precursor of the coenzymes flavin mononucleotide and flavin adenine dinucleotide, which are required for a various flavoprotein enzyme reactions, including activation of other vitamins (Douglas, 2017). As CLE's genome encode excess production of L-proline, it might be crucial for proline supplementation in *R. sanguineus* and *R. turanicus* (Tsementzi et al., 2018). Experimental removal of FLE in tick results poor performance of ticks in growth and molting, which can be reversed by oral supplementation of B-vitamins (Duron et al., 2018).

CLE and/or FLE colonization of ovarian tissue is associated with promotion of reproduction and developmental processes and facilitates their maternal transfer to a new generation of ticks (Klyachko et al., 2007). The presence of *Coxiella* in MT suggests a possible contribution to nitrogen metabolism (Lalzar et al., 2014). The higher density of CLE and FLE in MT suggests their nutritional contribution, probably involving biosynthesis of B-vitamins since MT is associated with excretion and osmoregulation (Buysse et al, 2021; Cibichakravarthy et al., 2022). Colonization of MT by CLEs is physiologically important since these microbes are capable of recycling nitrogenous metabolites to synthesize B-vitamins (Buysse et al., 2019). Gene ontology analysis of CLE proteins in OV and MT samples of *R. sanguineus* revealed organ-specific functional contributions based on host requirements. Most of the proteins enriched in MT are primarily linked to cellular metabolism and stress responses, whereas in OV, they are associated to cell proliferation (Cibichakravarthy et al., 2022). It has been interpreted that CLEs in MT grant essential nutrients to its host and in OV continue proliferation and ensure vertical transmission to tick progeny. These tissue-specific adaptations provide compelling evidence for host–symbiont metabolic co-dependence, wherein CLEs have evolved divergent biochemical pathways aligned with host physiological constraints. Notably, a subset of organ-enriched CLE proteins that may serve as key mediators of these host–symbiont interactions has been identified (Cibichakravarthy et al., 2022).

Unlike other eukaryotes, ticks are devoid of the genes required for the production and degradation of heme, an essential cofactor (Jia et al., 2020; Gulia-Nuss et al., 2016; Hatalová et al., 2023). Consequently, they acquire heme from an external source and have adapted to metabolize the heme obtained directly from their host's blood meal (Hatalová et al., 2023). Remarkably, FLE possesses heme synthesis genes, indicating a potential role in supplementing ticks with this vital cofactor (Buysse et al., 2021). But this capability to biosynthesize heme is absent in CLEs. The tick life cycle is characterized by prolonged off-host periods without feeding during which the endogenous heme produced by FLE offers a crucial advantage during these fasting periods (Fattar et al., 2025). The CLE genome, in *H. longicornis*, is known to encode a complete shikimate pathway, which is crucial for chorismite (a precursor of tryptophan) synthesis. Chorismate modulates serotonin biosynthesis by stimulating the expression of aromatic amino acid decarboxylase (AAAD), which in turn converts 5-hydroxytryptophan (5-HTP) to serotonin. Serotonin as a neurotransmitter influences the feeding behavior of ticks, thereby increasing the amount of blood intake (Zhong et al., 2021). The chorismate biosynthesis pathway has also been noticed in CLE of other tick species (Santos-Garcia et al., 2023), inferring that this mechanism might represent a common adaptive phenomenon. CLEs have been regarded as the super facilitator for nutritional fortification of blood diet, thus facilitating tick development and reproduction. Experimental depletion of CLEs in *R. haemaphysaloides*, *R. sanguineus*, *A. americanum*, and *H. longicornis* results in reduced engorgement weight and impaired capacity of adults to produce eggs (Li et al., 2018b; Zhong et al., 2021; Zhang et al., 2017; Zhong et al., 2007; Ben-Yosef et al., 2020).

Similarly, *R. microplus* larvae with diminished CLE levels fail to molt into adults, indicating developmental arrest. Moreover, transcriptomic analysis suggests that this developmental blockade may be linked to dysregulation of extracellular matrix remodeling genes in CLE-depleted ticks (Guizzo et al., 2017). These features of CLEs enable ticks to exploit the nutritionally unbalanced diet, thereby leading to hematophagy specialists. However, it does not mean that all ticks depend on CLEs for their survival. In terms of vector competency, CLEs are negatively correlated with pathogen invasion. For instance, in *R. haemaphysaloides*, reduced CLE level facilitates *B. microti* colonization, though the underlying mechanism is not well understood (Li et al., 2018a). The presence of FLE may modulate infection dynamics of tick-borne pathogens. In *D. andersoni*, reduced FLE levels correlate with diminished colonization of the pathogenic bacterium *Francisella novicida* (Gall et al., 2016). Conversely, FLE abundance shows a negative association with replication of *R. parkeri* in *A. maculatum* ticks (Budachetri et al., 2018). These findings highlight the potential role of FLEs in shaping pathogen persistence within ticks, though further studies are needed to elucidate the underlying mechanisms and broader ecological implications.

6.3.2.2 Midichloria

Midichloria spp. (class: Alphaproteobacteria; order: Rickettsiales) were first described in the OV of the hard tick *I. ricinus* (Lewis, 1979; Sacchi et al., 2004). One of the best-recognized Rickettsiales among Midichloriaceae is *Candidatus Midichloria mitochondrii* (CMM) which was previously known as IricES1 (*I. ricinus* endosymbiont 1), an intracellular endosymbiont of *I. ricinus* (Comandatore et al., 2021), with a unique ability to colonize mitochondria (Olivieri et al., 2019). The dominant bacterial taxon in OV of *I. ricinus* is the endosymbiont *Midichloria* sp., regardless of the tick feeding stage (Guizzo et al., 2020). *Midichloria* is recognized to be an intramitochondrial symbiont in several tick species, including *I. ricinus*, *Ixodes frontalis*, and *Ixodes aulacodi*. It resides in the periplasmic space of tick ovarian mitochondria during oocyte maturation, and it is believed that it consumes the inner membrane and proliferates inside mitochondria (Sassera et al., 2006). However, in *Ixodes holocyclus*, it resides as an intracytoplasmic (non-intramitochondrial) endosymbiont (Floriano et al., 2023). Ultrastructural investigation of *Midichloria* in *I. holocyclus* appears to be confined to the cytoplasm (Beninati et al., 2009), and in ovarian cells of *I. ricinus*, in which about 10–20% of the bacteria were found to occupy the mitochondrial intermembrane space (Comandatore et al., 2021; Sacchi et al., 2004).

Transmission electron microscopy (TEM) revealed the presence of *M. mitochondrii* both in the cytoplasm and in the mitochondria of the oocytes of *Rhipicephalus bursa* (Epis et al., 2008). *Midichloria* getting encased between the mitochondrial membranes significantly changes the mitochondrial morphology in oocytes of *I. ricinus*. For instance, the wild-type oocytes with naturally harboring *Midichloria* exhibit distinct globular mitochondria with intact matrix, whereas aposymbiotic oocytes only manifest a mitochondrial network (Uzum et al., 2023). This suggests that *Midichloria*-mediated mitochondrial fragmentation could be important for successful oogenesis (Uzum et al., 2023); however, it is not clearly understood how morphological changes in mitochondria affect its function. It warrants formal confirmation of this hypothesis by quantification of the rate of successful development of viable eggs from primary oocytes. Based on quantitative TEM data, it has been proposed that *Midichloria* could wander within the mitochondrial network (i.e., mitochondria-to-mitochondria) instead of consuming the organelles (Comandatore et al., 2021). The mechanisms and functional implication of such a unique interaction have been the subject of speculation in the early days of its detection in organelle-specific location. Initially, it had been hypothesized that *M. mitochondrii*, as a symbiont of *I. ricinus*, consumed mitochondria in a similar pattern as done by the predatory bacteria (Sacchi et al., 2004).

Besides OV, *M. mitochondrii* has also been detected in non-ovarian tissues, such as SG (Mariconti et al., 2012), MT, tracheae, and guts (Olivieri et al., 2019). This finding leads to a hypothesis that this bacterium offers varied functions while residing in different organs. As one of the most widespread tick symbionts, *M. mitochondrii* not only provides essential nutrients to the host but also enhances the reproductive function of the host to a certain extent, enhances antioxidant activity, produces energy, and helps in maintaining water balance and host's homeostasis (Olivieri et al., 2019; Stavru et al., 2020). In relation to its multiple localizations, it can be speculated that *M. mitochondrii* could secure both vertical transmission via oocytes and horizontal transmission to the vertebrate hosts during blood meal facilitated by SG (Stavru et al., 2020). Since *Midichloria* is detected in *H. marginatum* and blood samples collected from *H. marginatum*-infected birds, it suggests that birds could facilitate horizontal spread of these bacteria across the hematophagous ectoparasite populations (Di Lecce et al., 2018).

Based on the qPCR assay, CMM was detected in all developmental stages of *H. anatolicum*, with the highest relative abundance in eggs, and lower abundance in nymphs and unfed males (Luo et al., 2024). In female, after blood feeding, its abundance increases, with relatively higher abundance in OVs of engorged females (Luo et al., 2024). Similar was the observation in *I. ricinus* in which *Midichloria* achieved exponential proliferation following the blood meal, which is reflected by the concurrent production of its metabolites (Sassera et al., 2006). Genome sequencing of two *Midichloria* strains (Sassera et al., 2011), and their presence in OV and MT tubules, led to a prediction of multiple roles, including detoxification, vitamin provisioning, and energy-related functions (Olivieri et al., 2019). Such an occurrence in multiple organs suggests their facultative mutualistic behavior and capability to be transferred horizontally, in addition to transovarial transmission (Lo et al., 2006; Sassera et al., 2006). Additionally, real-time PCR-based analysis detected the presence of CMM in all life stages (eggs, larvae, nymphs, and adults) of *I. ricinus*, demonstrating transstadial maintenance through its life cycle (Sassera et al., 2006).

A complete genome of *M. mitochondrii* is available (Stavru et al., 2020). The *Midichloria* symbionts that target mitochondria possess unique features such as type IV secretion system (T4SS) and flagella which facilitate them to secrete proteins or physically interact with mitochondria. Moreover, these mitochondria-invading *Midichloria* strains carry specific genes for adhesion molecules, actin polymerization, and cell wall and outer membrane proteins, which are lacking in non-mitochondrial strains. This set of genomic signatures possibly allows *Midichloria* to interact with host organelles, for instance, to get fused with mitochondrial membrane or alter mitochondrial network (Floriano et al., 2023). One of the unique genes in CMM is cbb3-type cytochrome c oxidase, a heme-copper proton pumping oxidase, with the ability to perform oxidative phosphorylation at low oxygen tension (Sassera et al., 2011). Owing to its high oxygen affinity, this enzyme facilitates respiration under microaerobic conditions, thus supporting adenosine triphosphate (ATP) production for ticks under low-oxygen conditions, particularly when they temporarily submerge in wet environments (Sassera et al., 2011). Like other endosymbionts, CMM also possess complete metabolic pathways for de novo biosynthesis of B vitamins, most particularly B7 (biotin) and B9 (folate) (Sassera et al., 2011; Al-Khafaji et al., 2019). CMM, in *I. ricinus*, has been found to exhibit a striking sex-specific prevalence infecting 100% of females but only 44% of males, with reduced bacterial loads in the latter, while detected in all eggs, evidencing its transovarial transmission as the principal route (Lo et al., 2006).

The laboratory-reared *I. ricinus* larvae with artificial deprivation of *Midichloria* were observed to exhibit poor performance with a 60% decline of the number of engorged larvae (Guizzo et al., 2023). The reproductive capacity of *Midichloria*-free and control *I. ricinus* females as measured by egg mass did not differ significantly, suggesting that *Midichloria* is not a strict requirement for reproduction (Guizzo et al., 2023). Instead, *Midichloria* plays an essential role in maintaining normal ovarian physiology in *I. ricinus*, preventing the development of substandard larvae (Guizzo et al., 2023). With an aim to decipher the functional role of *M. mitochondrii* in tick hematophagy, an ex vivo membrane system provided with tetracycline was employed to generate aposymbiotic strains of *I. ricinus* ticks. Larvae generated from these *Midichloria*-free tick strains consistently demonstrated reduced blood-feeding capacity, whereas the nymphs and adults performed in a similar way to that of control ticks. These observations imply that *M. mitochondrii* stands as an integral entity of tick ovarian tissue, and its absence results in producing substandard larvae with poor feeding capacity, indicating the influence of this symbiont in host interaction (Guizzo et al., 2023). *Midichloria* of each category could be crucial in mitochondrial tropism and direct interaction with mitochondria. Additionally, investigating the cross-talk between *M. mitochondrii* and mitochondria could offer novel aspects of mitochondrial biology.

6.3.2.3 Rickettsia

The genus *Rickettsia* comprises obligate intracellular Gram-negative bacteria, some of which are non-pathogenic benign endosymbionts inhabiting ticks, while others are recognized as potential TBPs (Kurtti et al., 2015). For instance, *Rickettsia peacockii* and *Rickettsia buchneri* are non-pathogenic endosymbionts residing in *D. andersoni* and *I. scapularis*. These symbionts are transmitted transovarially and are frequently associated with tick ovarian tissue (Kurtti et al., 2015). Compared to their pathogenic relatives, these strains possess highly rearranged genomes with accumulated mutations, resulting in reduced virulence and a loss of cytopathic effects (Kurtti et al., 2015). Metabolic reconstructions of *Rickettsia* genomes have revealed that all essential genes for folic acid biosynthesis are available in *Rickettsia* symbiont genomes isolated from *I. scapularis* and *Ixodes pacificus* (Hunter et al., 2015). Furthermore, experimental evidence indicates that *Rickettsia* symbionts significantly influence larval motility in *A. americanum*, *D. variabilis*, and *I. scapularis* ticks (Kagemann and Clay, 2013).

6.3.2.4 *Rickettsiella*

Rickettsiella species (class Gammaproteobacteria, order Legionellales, family Coxiellaceae) are intracellular maternally inherited bacteria that are associated with many arthropod species, including ticks. *Rickettsiella* symbiont has been detected in a parthenogenetic laboratory colony of the *Ixodes woodi* ticks (Kurtti et al., 2002). Since *I. woodi* is typically bisexual, the presence of *Rickettsiella* suggests that the symbiont may foster asexuality which represents a demonstration of sex ratio distortion in ticks (Kurtti et al., 2002) but warrants further study. Multi-locus sequence typing (MLST) approaches detected *Rickettsiella grylli* within the genus *Rickettsiella* in *I. woodi* (Leclerque and Kleespies, 2012). *Rickettsiella* was more commonly detected in *Ixodes uriae*, but the infection did not exceed 20% in field-collected specimens (Duron et al., 2016). Tick-rearing experiments further demonstrated that *Rickettsiella* vertically passed to eggs without causing developmental abnormalities, suggesting that these symbionts may employ non-pathogenic adaptive strategies to persist within tick populations (Duron et al., 2016). *Rickettsiella* have also been detected from bat ticks (*Argas transgariepinus*) (Reeves et al., 2020), questing *Haemaphysalis concinna* using arthropod cell lines (Thu et al., 2019), questing *I. ricinus* nymphs (Krawczyk et al., 2022), *Ixodes angustus*, *I. sculptus Ixodes kingi* (Anstead and Chilton, 2014), and *Ixodes tasmani* (Vilcins et al., 2009). A recent study screened *I. ricinus* from varied geographical regions (Belgium, Italy, the Netherlands, Sweden, and the UK) and detected four distinct *Rickettsiella* clades with the highest prevalence in the adult ticks but with significant variation between sites (Garcia-Vozmediano et al., 2022). However, no significant associations were found between *Rickettsiella* spp. and other bacteria. Based on the outcomes, it has been proposed that *Rickettsiella* spp. are facultative symbionts of *I. ricinus* with notable genetic diversity, and their distribution is likely governed by the environmental factors.

6.3.2.5 *Spiroplasma*

Spiroplasma, a helical, wall-less, motile, Gram-positive bacterium belonging to the family Spiroplasmataceae, is a maternally (vertically) inherited endosymbiont found in ticks and various arthropods (Beliavskaia et al., 2021). Along with the insects, *Spiroplasma* is found in several species of ixodid ticks, including *I. pacificus*, *I. ricinus*, *Ixodes arboricola*, *I. frontalis*, *I. uriae*, *Ixodes ovatus*, *Ixodes persulcatus*, *D. marginatus*, *R. annulatus*, *R. decoloratus*, *Rhizomucor pusillus*, and *Rhipicephalus geigyi* (Binetruy et al., 2019). Moreover, *Spiroplasma* represents one of the dominant genera in the MG contents across larvae, nymphs, and adult females of *A. persicus*, inferring that it resides in the MG and can be maintained transstadially (Duan et al., 2022). Recently, *Spiroplasma* has been detected as one of the members of *H. leporispalustris* microbiome with relatively high abundance (Karim et al., 2024). From the genus *Spiroplasma*, two species, *Spiroplasma mirum* and *Spiroplasma ixodetis*, have been confirmed in ticks (Ogata et al., 2021). While *S. ixodetis* is known to act as a selective male-killing agent in certain insects, no such consequence has been observed in ticks, as infections occur in both sexes, as documented in populations of the African blue tick, *R. decoloratus* (Binetruy et al., 2019).

Spiroplasma has also been detected from adult *Amblyomma sculptum* and nymphal *Dermacentor nitens* (da Silva et al., 2022). Notably, *Spiroplasma* persists across generations in laboratory-reared *I. persulcatus*, with 100% of tested third-generation adults harboring the bacterium – a strain genetically similar (99.8% 16S rRNA identity) to that found in wild populations (Beliavskaia et al., 2021). Further evidence of vertical transmission comes from *D. reticulatus*, where *Spiroplasma* was detected in females (14.5%), egg pools (26.1%), and larvae (46.2%), with *S. ixodetis* specifically confirmed to undergo transovarial passage (Kloc et al., 2024). In an experimental trial, *S. ixodetis* were found transmitted to eggs and larvae in a vertical manner in *I. ovatus* (Ogata et al., 2023). Inoculation of tick-derived *Spiroplasma* isolates into *H. longicornis* did not exhibit any significant difference in the engorgement weight, egg weight, and hatching rate as compared with the control group, suggesting that its infection does not directly affect tick reproduction (Ogata et al., 2023). From the available information, it can be understood that *Spiroplasma*, in ticks, has considerable persistence in various life stages; however, its role in nutrition provisioning or pathogen protection remains unclear.

6.3.2.6 *Cardinium*

The genus *Cardinium* is a bacterial lineage of the order Cytophagales that adopted an endosymbiotic lifestyle inside the cytoplasm of their eukaryotic, multicellular host animals (Nakamura et al., 2009; Breeuwer et al., 2011). *Candidatus Cardinium* is an intracellular endosymbiont or parasite frequently reported in invertebrates, including mites and ticks (Kopecky et al., 2013). *Cardinium* has been reported from three species of soft ticks (*Ornithodoros peruvianus*, *O. spheniscus*, and *Ornithodoros talaje*) and two species of hard ticks (*I. scapularis* and *R. decoloratus*) (Duron et al., 2017). The *Cardinium* influences the host's reproduction by inducing cytoplasmic incompatibility (CI), feminization, male killing, and parthenogenesis (Zchori-Fein et al., 2004; Zchori-Fein and Perlman, 2004). *Candidatus Cardinium hertigii* belongs to a monophyletic taxon, has adopted a symbiotic lifestyle, and resides in the cytoplasm of their hosts (Breeuwer et al., 2011). An interaction between *Cardinium* and *Wolbachia* (both intracellular bacterial symbionts) was assessed in mites (*Tyrophagus putrescentiae*) in which *Cardinium* was found to inhibit *Wolbachia* growth by disrupting its interaction with the host, leading to loss of influence of *Wolbachia* on mite immune and regulatory pathways (Hubert et al., 2025). Most of the research on *Cardinium* is focused on insects and mites, with limited direct evidence in ticks. Studies are warranted to clarify prevalence, interaction with tick pathogens or symbionts, and possible influence of this bacterium on tick biology and vector competence.

6.3.2.7 *Lariskella*

Candidatus Lariskella arthropodarum (class: Alphaproteobacteria; order: Rickettsiales), a maternally inherited intracellular bacterium, was first detected in fleas and ticks (*I. ovatus* and *I. persulcatus*) and provisionally named 'Montezuma' (Eremeeva et al., 2006). Later, it was formally described as a secondary endosymbiont in plant-sucking stinkbugs (Pentatomomorpha) (Matsuura et al., 2012). *Lariskella* sp. is common in *I. persulcatus* (Mukhacheva and Kovalev, 2017; Qiu et al., 2014; Buysse and Duron, 2021; Grandi et al., 2023; Namina et al., 2023; Bugmyrin et al., 2022; Su et al., 2023) and is also reported from *I. ricinus*/*I. persulcatus* hybrid ticks (Mukhacheva and Kovalev, 2017). *L. arthropodarum* has been detected in *Ixodes pavlovskyi* females (Kurilshikov et al., 2015). The percentage of *L. arthropodarum*-infected sympatric ticks was significantly higher for *I. persulcatus* males than for both females and males of *I. pavlovskyi* (Mukhacheva and Kovalev, 2017). Additionally, genetic analysis revealed two novel genetic variants of the rrl-rrf intergenic spacer (IGS) of *L. arthropodarum* isolated from *I. pavlovskyi* ticks that possibly could indicate an ongoing process of its adaptation to a new host species (Mukhacheva and Kovalev, 2017). *Lariskella* is reported to have coinfection with *Borrelia miyamotoi*-infected and *B. garinii*-infected *I. persulcatus* ticks, but in different individuals (Tian et al., 2023). *C. L. arthropodarum*, based on qPCR results, is attributed to be the major taxon resulting in significant differences between the microbiome composition of males and females *I. persulcatus* (Becker et al., 2023). Recently, *C. Lariskella guizhouensis* as a novel species has been reported in *I. ovatus* (Lu et al., 2022); however, its genetic features are less studied.

While the maternal transmission of *C. L. arthropodarum* has been confirmed (Mukhacheva and Kovalev, 2017), its horizontal transmission can be speculated since its DNA has been detected in SG of *I. persulcatus* (Qiu et al., 2014) and human blood (Mediannikov et al., 2004). Both high prevalence and relative abundance of *L. arthropodarum* in females align with the hypothesis that this bacterium efficiently multiplies in *I. persulcatus* females but is probably lost in males during ontogenesis (Eremeeva et al., 2006). The biological features of *L. arthropodarum* pertaining to its preferential multiplication in females and capacity for vertical transmission through the host generations resemble those of *M. mitochondrii* (Mukhacheva and Kovalev, 2017). Phylogenetically, based on 16S rRNA gene sequences, *Lariskella* is closely related to CMM, an intramitochondrial bacterium, which is known to encode complete biosynthesis of biotin, riboflavin, and folate (Sassera et al., 2011). Hypothetically, if these pathways have been restored intact in *Lariskella*, they could be nutritional endosymbionts of *I. persulcatus* (Buysse and Duron et al., 2021). In artificial infection experiments in leaf-footed bug (*Leptoglossus zonatus*), *Lariskella* has been understood as a prominent symbiont that induces CI, a reproductive manipulation phenotype that favors relative fitness of females harboring the symbiont (Umanzor et al., 2025). However, its exact biological significance in tick survival, fitness, immunity, or reproductive potential has not yet been clearly elucidated.

6.3.2.8 *Arsenophonus*

The genus *Arsenophonus* (class: Gammaproteobacteria), a male-killing endosymbiont of chalcid wasps, comprises endosymbiotic bacteria often associated with insects, including ticks. *Arsenophonus nasoniae* is identified in *I. ricinus* (Subramanian et al., 2012), *A. americanum* (Dergousoff and Chilton, 2010), and *D. andersoni* (Clay et al., 2008). Phylogenetic analyses reveal that *Arsenophonus* found in *D. variabilis* and *A. americanum* in the eastern USA are genetically distinct from those in *D. andersoni* and *D. variabilis* in western Canada, thus suggesting potential speciation (Dergousoff and Chilton, 2012). A strain of *A. nasoniae* is identified in *Dermacentor* and *Amblyomma* (Clay et al., 2008; Dergousoff and Chilton, 2010) and in nymphs of *I. ricinus* (Mediannikov et al., 2012). *A. nasoniae* is also detected in *I. ricinus* nymphs parasitized by *Ixodiphagus hookeri* wasp, implying wasp-mediated transfer (Bohacsova et al., 2016). Detection of *Arsenophonus* endosymbionts in *D. reticulatus* across life stages (eggs, larvae, nymphs, and adults) and organs (SG, MG, and OV) (Liu et al., 2013) and in *Amblyomma* egg clutches (Clay et al., 2008) support the evidences for vertical transmission. Along with other bacterial genera, *Arsenophonus* has been noted to comprise higher than 0.5% of all the sequences in the overall datasets achieved from *I. ricinus* (Lejal et al., 2021). *Arsenophonus* has been considered to be one of the predominant bacteria in *I. persulcatus* (Su et al., 2023), adult *Rhipicephalus linnaei* (Kamani et al., 2024), different growth stages of *D. nuttalli* (Zhao et al., 2022), saliva and MG of partly and fully engorged female *Dermacentor silvarum* (Duan et al., 2020), *Or. maritimus* (a seabird tick) (Gomard et al., 2021), with varying abundance in *D. marginatus* and *D. reticulatus* (Zhang et al., 2019) but with lower abundance in nymphal *Amblyomma hebraeum* (Kisten et al., 2021). *Arsenophonus*-like endosymbionts are found in questing adult *D. andersoni* (Dergousoff et al., 2020). *A. nasoniae* may play a role in tick fitness and/or development, but the data on the precise nature of the bacteria–tick relationship are scarce (Taylor et al., 2012). One study regarding *Arsenophonus* sp., a maternally inherited symbiont in ticks, has suggested that it can influence host-seeking success in *A. americanum*, *I. scapularis*, and *D. variabilis* by decreasing their motility (Kagemann and Clay, 2013).

6.3.2.9 Wolbachia

Wolbachia is an obligate Gram-negative intracellular endosymbiotic Alphaproteobacteria mainly associated with arthropods and nematodes (Landmann, 2019). *Wolbachia* are capable of manipulating the host reproductive system through CI (Chen et al., 2020). The *Wolbachia* genome isolated from bedbugs contains a gene cluster encoding the complete biosynthetic pathway for biotin (vitamin B7), suggesting that it is likely acquired via lateral transfer from a coinfecting endosymbiont (Nikoh et al., 2014). The presence of *Wolbachia* endosymbionts has been reported in various hard tick species, including *I. ricinus* (Hartelt et al., 2004), *I. scapularis* (Benson et al., 2004; Thapa et al., 2019), *R. microplus* (Andreotti et al., 2011; Zhang et al., 2022), *R. sanguineus* (Chao et al., 2021), *D. silvarum* (Duan et al., 2020), *A. sculptum* (Romano et al., 2023), *A. americanum* (Zhang et al., 2011). *Wolbachia* endosymbiont is significantly more prevalent in nematode-infected ticks. Zhang et al. (2011) investigated the distribution of *Wolbachia* in *A. americanum* and interpreted it as an association with nematodes since nematodes were common in *Wolbachia*-positive tick samples, while Tijsse-Klasen et al. (2011) attributed it to parasitism with *I. hookeri* wasp. In another investigation *I. hookeri* which are endoparasitoids in *I. ricinus* were found infected with *Wolbachia pipientis* (99.2%) (Plantard et al., 2012).

Presence of *I. hookeri* was detected in 9.5% of tested *I. ricinus*. *Wolbachia* was found in 87% of the ticks that were positive for *I. hookeri* and only in 1.6% of *I. hookeri*-free ticks tested positive for *Wolbachia*. The presence of *Wolbachia* in individual nymphs correlates with *Wolbachia* detection in all of the analyzed *I. ricinus* nymphs parasitized by *I. hookeri* while, in contrast, all unparasitized nymphs were *Wolbachia*-free (Alafaci et al., 2021; Plantard et al., 2012). From these observations the question arose of whether tick cells are capable of supporting *Wolbachia* growth. Khoo et al. (2020) tested in vitro *Wolbachia* infection susceptibility on cell lines derived from *I. ricinus*, *I. scapularis*, and *R. microplus* and found that *Wolbachia* was able to invade and replicate in all tick cell lines.

PCR-based analysis of *Wolbachia* surface protein (wsp) gene has identified *Wolbachia* infection in nymphs, females, and males of *R. sanguineus* ticks with an infection rate of 55.8, 39.8, and 44%, respectively, and their detection in male and nymphal ticks may imply the possible mechanism of transstadial transmission (Chao et al., 2021). In an experiment 87% of *I. hookeri*-positive but only 1.6% of *I. hookeri*-negative ticks were found to be *Wolbachia*-positive (Tijsse-Klasen et al., 2011). Detection of *I. hookeri* DNA was strongly associated with the detection of *Wolbachia* in ticks. While detection of *Wolbachia* DNA in ticks might often be due to parasitism with this wasp, other sources of *Wolbachia* DNA in ticks might exist as well (Tijsse-Klasen et al., 2011).

Presence of *Wolbachia* in *A. sculptum* was confirmed by 16S rRNA gene analysis (Romano et al., 2023).

With an aim to examine antipathogenic effects on the intracellular pathogen *A. phagocytophilum*, an experimental setup was conducted for *Wolbachia* infection in *I. scapularis* tick cells. An increase in *A. phagocytophilum* replication was observed in *Wolbachia*-infected tick cells. However, *Wolbachia* infection densities decreased when cells were serially passaged, and ultimately the infection was lost. Host-cell immune response was also examined as an additional factor that could have affected *A. phagocytophilum* replication in *Wolbachia*-infected cells. In early passages post-*Wolbachia* infection, a decreased immune response was observed, but in later passages of cells with low *Wolbachia* densities, there was no change in the immune response (Skinner et al., 2022).

Furthermore, detection of *Wolbachia* endosymbiont in both female and male adult ticks and in nymphs may suggest a possible way of vertical transmission (Zhang et al., 2011; Chao et al., 2021). Besides the reproductive system, *Wolbachia* distribution was also detected in various somatic tissues of its insect host, including the nervous tissue, suggesting that it might be associated with regulating host behavior and physiology (Albertson et al., 2009; Albertson et al., 2013; Strunov et al., 2013); however, such studies are lacking in ticks. There is no plausible evidence yet that parthenogenesis in ticks is directly influenced by their endosymbionts. However, the presence of reproductive manipulators like *Wolbachia* as in other arthropods such as Hymenoptera, mainly parasitoid wasps (Fricke and Lindsey, 2024), indicates that there could be a subtle connection with parthenogenetic ticks (*Amblyomma rotundatum*, *H. longicornis*), which remains a subject for further research.

6.4 TICKS, ENDOSYMBIONTS, AND PATHOGEN INTERACTION TRIAD

Ticks do not merely serve as passive carriers of pathogens but also harbor a complex microbiome consisting of symbiotic bacteria that influence tick biology, pathogen transmission, and evolutionary adaptations. Ecologically organisms constitute a broad and diverse relationship from beneficial to harmful. With regard to microbial ecology and the complex landscape of interrelationships and interactions, defining mutualism, symbiosis, and pathogenic interaction is daunting since the association is influenced by several external factors. The interaction triad among ticks, pathogens, and endosymbionts is a dynamic and coevolved system. The evolutionary relationships between ticks and their microbiota are equally fascinating, and this aspect has already been discussed in aforementioned parts. Ticks provide a habitat for both pathogenic and symbiotic microorganisms, which compete or cooperate within the tick's body. Based on the tick microbiome research, the growing evidence indicates that tick microbiota contribute to tick physiology and pathogen transmission (Table 6.5). The host–microbiota relationship is not always unidirectional, rather it has a bi- or multi-lateral give-and-take mechanism, and more particularly, an intricate nexus is formed between ticks, tick-endosymbionts, and tick-borne pathogens. Therefore, a tripartite association between these entities is to be considered while dealing with tick microbial community interactions.

6.4.1 Tick Immune System Components

The tick immune system and the complexity with which it responds to pathogens and other microbes is vague, therefore, it merits a separate chapter. However, a brief overview of the tick immune system (Figure 6.4) and its mode of interaction against microbes is essential to contextualize how resident or pathogenic microbiota withstand or evade tick immune defenses and participate in mutual or antagonistic interactions. Ticks, like other invertebrates, are devoid of an adaptive immune system but are able to employ the innate immune system to regulate pathogen colonization, persistence, and transmission (Hart and Thangamani, 2021). Tick saliva contains a complex mixture of bioactive molecules with diverse functions, including anticoagulant, antiplatelet aggregation, vasodilatory, anti-inflammatory, and immunomodulatory properties. These molecules collectively counteract the hemostasis and immune defense mechanisms of the vector host, facilitating both acquisition and transmission of pathogens (Hajdušek et al., 2013; Martins et al., 2019; Esteves et al., 2017). Notably, certain tick species exploit distinct hosts during larval, nymphal, and adult stages, increasing the likelihood of acquiring varieties of pathogens (Fogaca et al., 2021). The pathogens first get entry into the tick gut, the primary site of pathogen interaction, where they colonize the gut epithelial cells and/or cross the gut epithelial barrier to get entry into the hemocoel filled with hemolymph from where they

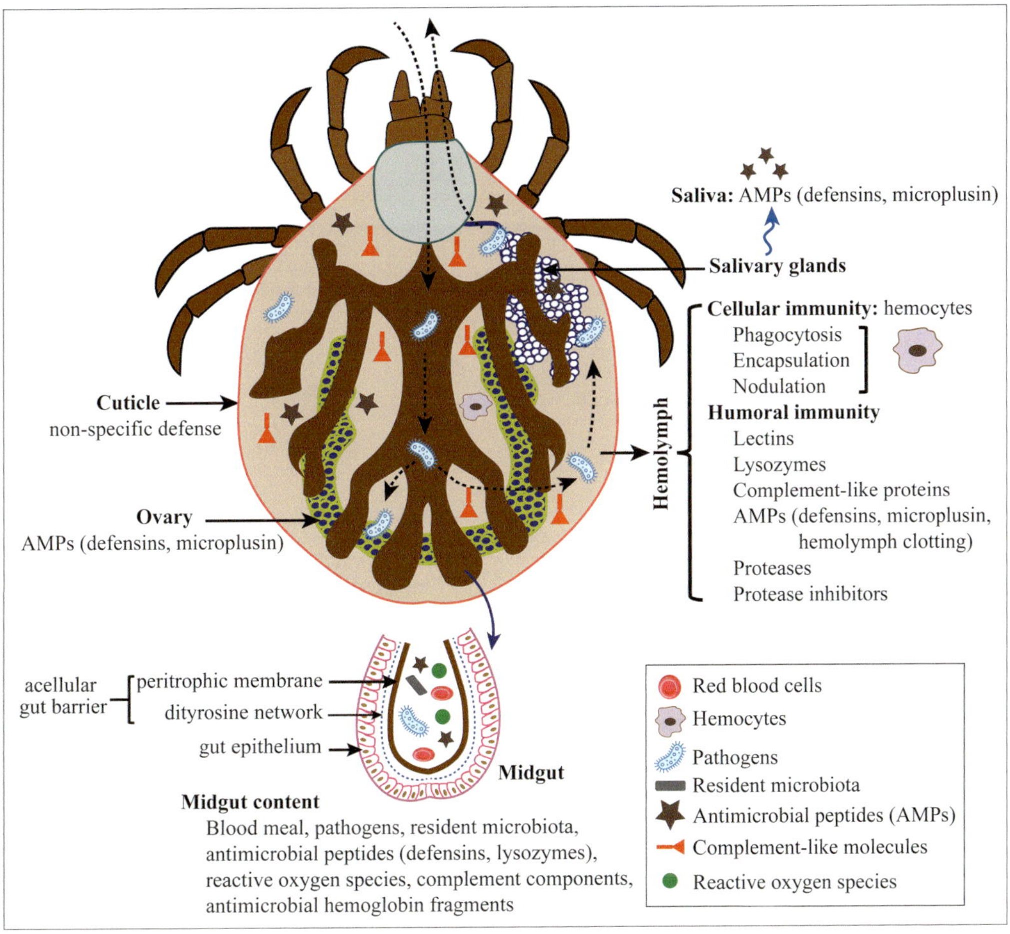

Figure 6.4 An overview of tick immune system components. The cuticular body surface acts as the non-specific first line of defense. The ingested blood first gets mixed with the midgut content where the acquired microbes interact with already existing microbes, resident endosymbionts, cytotoxic molecules (AMPs, ROS). The gut epithelium is covered by acellular layer composed of peritrophic membrane and dityrosine network where pathogens must colonize and/or cross the gut epithelium. In the midgut, tick may utilize some of the host immune molecules (e.g., complement system) for its own defense against intestinal inhabitants. Hemoglobin fragments, derived from the host hemoglobin, secreted into the midgut lumen, exert strong antimicrobial activity. Tick midgut epithelium also expresses a variety of endogenous AMPs, which help sustain the midgut microbes at a tolerable level. Pathogens migrate to the hemocoel and then either enter the ovary to maintain transovarial transmission or salivary glands to get entry into a new vertebrate host (black dotted arrow). Pathogens intruding into the tick hemocoel have to sustain both cellular and humoral immune defenses. Pathogen proliferation is hindered by phagocytic killing by hemocytes or deleterious effects posed by effector molecules, including AMPs, complement-like proteins, and proteases. Tick saliva contains variety of anti-coagulant, immunomodulatory, and anti-inflammatory molecules that promote pathogen acquisition and transmission while AMPs may hinder acquisition and persistence of pathogens in tick. Illustration created on Adobe Illustrator CC 2019.

reach the SG, while some enter the OV to establish transovarial transmission (Figure 6.4). The pathogens definitely have to counteract the tick's immune factors for their successful colonization, followed by establishment. The MG immunity is said to be decisive in relation to vector competence since its content is a hostile environment for the acquired microbes owing to the antimicrobial activity of hemoglobin fragments released from digested host blood and other AMPs (Kopácek et al., 2010). In each organ, the pathogens must tackle the resident microbiota as well as tick immune reactions. The tick immune system, like that of other invertebrates, is more rudimentary compared to that of vertebrates and relies solely on innate responses; however, a considerable number of tick species demonstrate its efficiency (Fogaca et al., 2021). In ticks, the humoral and cellular

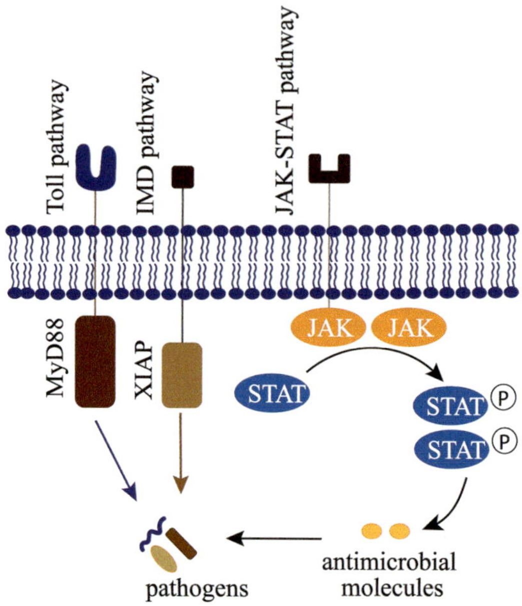

Figure 6.5 Tick immune signaling pathways. Illustration created on Adobe Illustrator CC 2019.

immunity components include hemocytes, signaling pathways (e.g., Toll, Immune Deficiency (IMD), Janus kinase/signal transducer and activator of transcription (JAK/STAT) pathways), AMPs, redox metabolism, complement-like proteins (Fogaca et al., 2021) as well as gut barrier which comprises the peritrophic membrane/matrix (PM) and dityrosine network (DTN) (Kitsou et al., 2021) (Figure 6.4).

Hemocytes, a part of the cell-mediated immune system in ticks, are sessile or circulating cells of hemolymph, responsible for several immune responses, and constitute mainly three basic types – phagocytic plasmatocytes, granulocytes, and nonphagocytic granulocytes, with some additional types: adipohemocytes, spherulocytes, and oenocytoids (Fogaca et al., 2021). PM, akin to the mucosal layer of vertebrate gut, is an acellular component lining in the gut of arthropods (Hegedus et al., 2019), and its formation, particularly in ticks, is induced by blood feeding (Kariu et al., 2013; Yang et al., 2021). Composed of chitin and proteins or glycoproteins, PM serves as a molecular sieve to partition the lumen from the ectoperitrophic space and offers dual benefits – digestive and immunological or defensive role – for the vector. It provides a protective barrier for epithelial gut cells against abrasion, invasive microbes, and pathogen-derived toxins (Hegedus et al., 2019). The structural integrity of PM is directly associated with the pathogen colonization capacity in gut epithelial cells since its disruption can pose a positive or negative impact on the survival of specific TBPs (Figure 6.5). Once the microbes are acquired, they get confined within the tick's gut lumen (Kitsou et al., 2021). Another important aspect is reactive oxygen species (ROS) that also possibly play a crucial role in tick–pathogen interactions (Kopácek et al., 2010).

6.4.2 Fundamental Aspect of Tick Immune Response against Pathogens

Vector competence of ticks has a tight nexus with their immune system. Ticks are capable of defending themselves against microbial infection by exploiting a variety of AMPs such as lysozymes, defensins, and several other molecules not found in other invertebrates (Kopácek et al., 2010). In contrast to vertebrate adaptive immunity, ticks and other arthropods rely heavily on AMPs as essential components of their innate immune system. While the tick immune response to various pathogens has been extensively studied, minimal work has been focused on viral infections owing primarily to the high pathogenicity of TBVs, which pose significant challenges for experimental procedures (Hart and Thangamani, 2021). A quantitative Reverse Transcription Polymerase Chain Reaction (RT-PCR) analysis reveals significant upregulation of the microplusin-like AMP (Ambar-69859) in both SG and MG tissues of *R. rickettsii*-infected *A. aureolatum* ticks (Martins et al., 2019). In this experiment knockdown of microplusin expression by RNA interference (RNAi) demonstrates the critical role of this AMP in pathogen defense in two aspects – the first is a higher rickettsial load by one order of magnitude in both tissues, and the second is a significantly higher infection prevalence of infected ticks compared to control groups. Furthermore, it is evident that this particular AMP serves as an essential component of tick's immune strategy against rickettsial infection (Martins et al., 2019). Interestingly, *I. scapularis* employs an antimicrobial molecule horizontally acquired from bacteria known as domesticated amidase effector 2 (Dae2) which is secreted into saliva and gut of ticks and selectively kills harmful mammalian skin-derived *Staphylococci* while feeding but does not kill *B. burgdorferi* (Hayes et al., 2020) (Figure 6.6).

The hemocytes primarily execute phagocytosis to eliminate microbes and also mediate other critical immune responses, including encapsulation (Eggenberger et al., 1990) and nodulation (Ceraul et al., 2002) as demonstrated in *D. variabilis*. The phagocytosis of microbes by hemocytes appears to be coupled with a primitive complement-like system which involves recognition of self and nonself by fibrinogen-related lectins and the action of thioester-containing proteins (Kopácek et al., 2010). During nodulation, hemocytes aggregate around invading bacteria as seen in response to *Escherichia coli* (Ceraul et al., 2002). Additionally, hemocytes synthesize immune-related molecules (Kopácek et al., 2010). These observations highlight that besides phagocytosis, encapsulation and nodulation are important components of a tick's immune arsenal, suggesting their role in combating infections. To illustrate one of the basic immune mechanisms in ticks, let us consider an example of how a transmembrane glycoprotein expressed in hemocytes (particularly granulocytes) participates in internalization of external pathogens. It is demonstrated that HlSRB (a 50-kDa scavenger receptor protein from *H. longicornis*) plays a critical role in granulocyte-mediated phagocytosis of invading *E. coli* in *H. longicornis*, serving as a first-line host defense mechanism against this pathogen. Granulocytes with upregulated expression of cell surface HlSRB are specifically responsible for phagocytosing *E. coli*, while RNAi-mediated silencing of HlSRB abolishes this capacity, resulting in fatal bacteremia in ticks (Aung et al., 2012). This study provides evidence that scavenger receptors in hematophagous arthropods like ticks directly facilitate hemocyte-mediated phagocytosis against exogenous pathogens, underscoring its essential role in tick immunity.

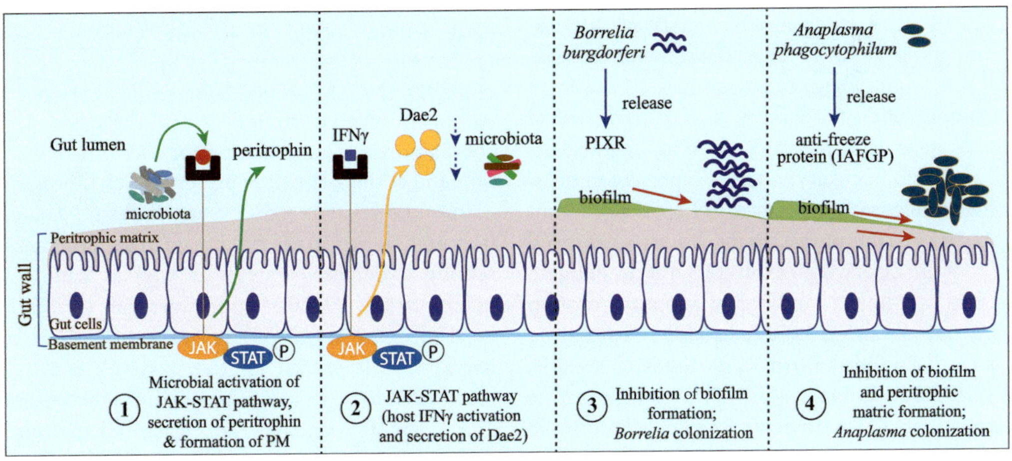

Figure 6.6 Tick gut peritrophic matrix (PM) and signaling pathway cross-talk in response to microbiota. The interaction occurs in the gut lumen, where (1) the tick microbiota induce Janus kinase–signal transducer and activator of transcription (JAK–STAT) pathway resulting secretion of peritrophin for formation of PM required for *Borrelia burgdorferi* colonization, (2) the host interferon-gamma (INFγ) interacts with JAK–STAT pathway inducing release of Dae2 (domesticated amidase effector 2) which limits the growth of host microbiota and *Borrelia* population, (3) *B. burgdorferi* infection stimulates the secretion of PIXR (protein of *Ixodes scapularis* with Reeler domain) which inhibits biofilm formation supporting *Borrelia* colonization, and (4) *Anaplasma phagocytophilum* infection stimulates secretion of antifreeze protein, which inhibits biofilm and PM formation, enhancing *Anaplasma* colonization. Illustration created on Adobe Illustrator CC 2019.

Regulated cell death (RCD) is a widespread natural phenomenon taking place in unicellular and multicellular organisms; however, studies of this process in relation to tick immune response are scarce. Among the three main types of RCDs – autophagy, apoptosis, and necrosis – former two have been reported in some ticks. The concept of apoptosis and autophagy pathways helps to decipher how tick and pathogens interact. Apoptosis is a highly conserved process in which endogenous materials (misfolded proteins and aggregates, damaged organelles, and other macromolecules) or exogenous materials like invading pathogens are selectively recognized and sequestered within the autophagosomes (apoptosomes) which get fused with lysosomes, leading to cargo degradation under the action of active enzymes (Kuo et al., 2018). During infection of the *I. ricinus* cell line IRE/CTVM20 by *A. phagocytophilum* and the flaviviruses TBEV and LIV, the expression of apoptosis-associated components, such as cytochrome C and fatty acid synthase (FAS) is upregulated (Mansfield et al., 2017; Alberdi et al., 2016a), indicating that ticks might employ apoptosis as a defense mechanism by eliminating infected cells. Autophagy, an intracellular protein degradation process, is induced by starvation (Kawano et al., 2011). Expression of autophagy genes (ATG genes) is found to be upregulated in starving *H. longicornis* (Umemiya et al., 2008; Kawano et al., 2011), *I. scapularis* (Wang et al., 2020), *R. microplus*, and *A. sculptum* (Moura-Martiniano et al., 2017), correlating with the classical role of autophagy in stress.

Ubiquitination machinery, a post-translational modification that regulates protein degradation and signaling in eukaryotes, is also present in ticks' immune defense, for instance, in *I. scapularis* (Severo et al., 2013). Ubiquitination machinery works well in *I. scapularis* ticks which exhibit E3 ubiquitin ligase (named as x-linked inhibitor of apoptosis protein (XIAP)) that restricts *A. phagocytophilum* colonization. It is further validated that XIAP silencing remarkably increases colonization of this bacterium in *I. scapularis* ticks (Severo et al., 2013). Tick immune response against pathogens is also mediated by signaling pathways. Upon recognition of a pathogen by surface receptors, tick cells activate multiple intracellular signaling pathways. Like the Toll signaling pathway, by which ticks can recognize pathogen-specific proteins (Oliva et al., 2017), the JAK/STAT signaling cascade is evolutionarily conserved in ticks (Rosa et al., 2016). *I. scapularis* ticks employ JAK–STAT pathway to upregulate AMP production. For instance, this pathway in *I. scapularis* appears in action to mediate the expression of a 5.3-kDa antimicrobial protein in SG in response to *A. phagocytophilum* infection (Liu et al., 2012) which is affirmed by RNAi silencing experiment. Ticks utilize RNAi, an important defense against RNA viruses (Yuan et al., 2020). A comparative genomic analysis has provided evidence for the functional conservation of Toll, JAK/STAT signaling pathways, and RNAi pathways in *Hyalomma asiaticum* (Yuan et al., 2020). In silico screening identified 122 putative proteins implicated in signal transduction through the Toll, IMD, and JAK/STAT pathways across seven tick species – *A. cajennense*, *A. maculatum*, *A. variegatum*, *I. ricinus*, *I. scapularis*, *Rhipicephalus pulchellus*, and *R. microplus* (Rosa et al., 2016).

6.4.3 Counteraction of Pathogens in Response to Tick Immune Arsenal

Pathogens have evolved diverse strategies to evade, counteract, and manipulate host immune defenses, enhancing their survival and proliferation. Certain host–pathogen interactions confer benefits to the pathogens. Infection with *B. afzelii* upregulates expression of lipocalins in *I. ricinus* SG, which may in turn benefit its own transmission (Valdés et al., 2016). Similarly, *A. phagocytophilum* reprograms *I. scapularis* salivary gland metabolism by suppressing gluconeogenesis and mitochondrial activity while promoting glycolysis and modulating cytoskeleton behavior (Cabezas-Cruz et al., 2017). These metabolic shifts represent a pathogen survival strategy since enhanced glycolysis provides rapid energy for bacterial replication and at the same time preserves tick fitness.

Immune-related redox metabolism is an important aspect in the tick innate immune system, serving a defense strategy as well as pathogen manipulation. Uptake and digestion of blood meal in tick gut generate ROS. Control of oxidative species is crucial for tick survival, and it is considered to be a part of the tick immune system. Tick deploys suitable antioxidant molecules, including CAT, an enzyme which detoxifies H_2O_2, which is a potent ROS, into water and oxygen to manage oxidative stress (Kumar et al., 2016). Let us consider a striking example of this interplay observed during infection of the *R. microplus*-derived embryonic cell line (BME26) by *Anaplasma marginale*. *A. marginale* results in downregulation of the genes encoding ROS-generating enzymes (dual oxidase and endoplasmic reticulum oxidase) while infecting BME26 and upregulating key antioxidant enzymes (superoxide dismutase, CAT, glutathione peroxidase, glutathione S-transferase, thioredoxin, thioredoxin reductase, and peroxiredoxin) (Kalil et al., 2017). Here, the simultaneous silencing of genes (i.e., CAT, glutathione peroxidase, and thioredoxin as well as oxidation resistance 1) by RNAi apparently enhances *A. marginale* colonization in BME26 cells. These findings suggest that *A. marginale* modulates tick's ROS-mediated immune defenses, thereby creating a suitable environment for infection. The ability of this pathogen to subvert this mechanism represents its adaptation for persistence within the vector.

Immune pathways, the structural integrity of PM, and pathogen survival are interrelated. Here, we present a unique example of interplay between tick MG microbiome and immune pathways favoring the survival of pathogens. The bacterial microbiome inhabiting the tick MG interacts with two dynamic immune signaling pathways – the JAK–STAT pathway (Narasimhan et al., 2014) and the IMD pathway (Oliva et al., 2017). This interaction has profound implications with regard to tick physiology and pathogen persistence. The cross-talk with the JAK–STAT pathway is significant for the proper formation of the PM (Figure 6.6). Interestingly, while PM is protective for the tick, it is an essential survival niche for *B. burgdorferi* (Narasimhan et al., 2017). Likewise, certain interactions between vector host and pathogen are mutually beneficial. A striking mutualism appears between pathogen and vector that *A. phagocytophilum* enhances the cold tolerance of *I. scapularis* by upregulating an *I. scapularis* antifreeze glycoprotein (IAFGP) (Figure 6.6). RNAi knockdown of IAFGP significantly reduces tick survival under cold stress, while a transfection study revealed cold resistance in yeast cells, demonstrating its functional importance in cryoprotection (Neelakanta et al., 2010). Crucially, *A. phagocytophilum* creates a fitness advantage for both partners since it ensures vector survival during winter.

During tick bite the vertebrate host is triggered to recruit inflammatory cells to the bite site, creating a hostile microenvironment rich with ROS for invading microbes. To establish infection in the vector, pathogens must navigate this immunologically active site while evading host immune defenses. Central to this process is the tick salivary protein Salp25D that detoxifies ROS at the vector–pathogen–host interface (i.e., tick bite site), therefore, it plays a role in facilitating *B. burgdorferi* acquisition by ticks from the mammalian host. This phenomenon has been validated in an experiment. When Salp25D expression was silenced in tick SG, *B. burgdorferi* acquisition by feeding on tick was significantly impaired. Similarly, immunizing mice against Salp25D reduced uptake of this pathogen by *I. scapularis* (Narasimhan et al., 2007). This demonstrates a complex interplay in which pathogens co-opt vector-derived molecules to counteract vertebrate host immune response to ensure their successful transmission from vertebrate host to tick vector. Likewise, selenoproteins as antioxidant agent is found supportive for the survival of certain endosymbionts like *Ca. M. mitochondrii* and *Francicella* spp. in *A. maculatum* (Budachetri et al., 2018).

To ensure their replication and survival within the host cell, many pathogens, including viruses, bacteria, and protozoa, subvert apoptosis induced by infection (Lamkanfi et al., 2010). *B. afzelii* is actively phagocytosed by tick hemocytes; neither RNAi-mediated silencing of tick complement-related molecules nor complete inhibition of phagocytosis (via pre-injection of latex beads) affects the transmission of these spirochetes to the host (Urbanová et al., 2017). These results indirectly support that the transmission of *B. afzelii* from infected *I. ricinus* nymphs to naive mice avoids the tick hemocoel and SG and occurs by a direct gut-to-mouthpart route (Pospisilova et al., 2019).

Among the key innate defenses that the pathogens must overcome are apoptosis and autophagy (Wang et al., 2022). *A. phagocytophilum* is a unique pathogen that inhibits apoptosis in tick cells (Ayllon et al., 2013; Alberadi et al., 2015; Alberadi et al., 2016a), which seems to be a key adaptation mechanism to facilitate infection. *A. phagocytophilum* manipulates tick gene expression to establish infection in *I. scapularis* ticks. Study reveals that the pathogen differentially regulates two critical genes: (1) spectrin alpha chain (CG8/α-fodrin), a cytoskeletal protein, and (2) voltage-dependent anion-selective channel or mitochondrial porin (T2), an apoptosis regulator. In SG, *A. phagocytophilum* downregulates both CG8 and T2 expression to block host cell death, creating a protected niche for replication. While in the gut, the tick responds to infection through CG8 upregulation for cytoskeleton rearrangement. But the pathogen takes advantage of the remodeled cytoskeleton for its infection and proliferation. Such a tissue-specific gene regulation demonstrates how this pathogen co-opts conserved cellular pathways to overcome immune barriers (Ayllon et al., 2013).

The interplay between *A. phagocytophilum* and its tick hosts involves sophisticated mechanisms to subvert the host cell defenses, particularly apoptosis. To investigate this, three *A. phagocytophilum* isolates – derived from human, canine, and ovine hosts – were used to infect two tick cell lines: IRE/CTVM20 (from *I. ricinus*) and ISE6 (from *I. scapularis*). Strikingly, all isolates suppressed the intrinsic apoptosis pathway in both cell lines at different levels within 24 hours of infection, as quantified by flow cytometry. This rapid inhibition of apoptosis underscores a conserved virulence strategy employed by *A. phagocytophilum* to secure its intracellular niche, irrespective of isolate origin or tick species (Alberadi et al., 2015). In another experiment, the results demonstrate distinct apoptotic regulation mechanisms in tick cells infected with *A. phagocytophilum*. In ISE6 cells (from *I. scapularis*), pathogen infection primarily inhibits apoptosis via intrinsic (mitochondrial) pathway. In contrast, IRE/CTVM20 cells (from *I. ricinus*) reflect responses observed in tick MGs, implicating the JAK/STAT pathway in apoptosis suppression, suggesting vector-specific strategies of immune evasion (Alberdi et al., 2016a). As noted, the different regulatory mechanisms could be attributed to the likely heterogeneous cell populations with different morphology and behavior since both cell lines were derived from embryonated eggs. Interestingly, *R. rickettsii* exhibits sophisticated temporal biphasic regulation of apoptotic pathways in the infected embryonic cell line of *R. microplus* (BME26 cells). During initial invasion (6 hpi), the pathogen induces downregulation of apoptotic inhibitors, potentially favoring the elimination of compromised host cells, and at the exponential growth phase (48 hpi), this mechanism reverses dramatically so that the same inhibitory proteins get upregulated. The infection also prevents DNA fragmentation and reduces the activity of caspase-3 as well as the exposure of phosphatidylserine. The second phase of anti-apoptotic cellular environment resulted from the inhibitory effect and is conducive to its proliferation (Martins et al., 2020). In ticks, bacterial pathogens can even manipulate the host's ROS to facilitate their survival. For instance, *A. phagocytophilum* suppresses ROS production and inhibits apoptosis in tick cells, thus safeguarding its intracellular replicative niche (Alberadi et al., 2019).

6.4.4 Interaction between Non-Pathogenic Microbiota of Ticks and Pathogens

As in other arthropods like mosquitoes, the microbiota in ticks can induce vector capacity. Bacterial infection can modulate the microbiota of its host. Most of the studies regarding the interactions between the pathogenic bacteria and tick microbiota have been studied in *Ixodes* ticks (Table 6.6). Associations between non-pathogenic components of the tick microbiome and pathogens, for instance, *Borrelia* spp. (Narasimhan et al., 2014, 2017; Sperling et al., 2020; Hamilton et al., 2021), have been described in multiple cases. *I. scapularis* gut microbiota has been found to have an important role in colonization of spirochetes. Unfed larval forms grown in sterile environment had been found to have increased abundance of *Rickettsia*, *Thioclava*, and *Delftia*, while the abundance of *Aquabacterium*, *Brevibacterium*, and *Novosphingobium* was reduced. After feeding on *Borrelia*-infected mice, the bacterial composition was altered, and the colonization ability of *B. burgdorferi* in the larval gut was decreased; however, ticks obtained increased engorgement weights. A comparative analysis of *Borrelia*–microbiome interactions revealed that field-collected *Borrelia*-positive *I. scapularis* harbored significantly higher bacterial diversity than *Borrelia*-negative ticks (Sperling et al., 2020). Similarly, based on *B. burgdorferi* presence or absence status, bacterial beta diversity varied in *I. scapularis* (Landesman et al., 2019).

A recent study has demonstrated that while depletion of the larval tick bacterial microbiome does not influence *B. afzelii* acquisition during blood feeding on infected mice, exposure of ticks to this pathogen remarkably alters the microbiome. Notably, *B. afzelii* colonization diminishes bacterial abundance, shifts community composition, and increases bacterial diversity (Hamilton et al., 2021). Infection of *I. scapularis* with *B. burgdorferi* does not affect the overall microbiota diversity or richness, but it reveals a significant association between the persistence of spirochetes and the occurrence of specific microbial taxa (Chauhan et al., 2020; Brinkerhoff et al., 2020). These observations suggest that the presence of specific gut microbiota is a prerequisite for *B. burgdorferi* colonization, though the underlying mechanisms remain incompletely understood (Kurokawa et al.,

Table 6.6

Critical insights into tick-microbiome and pathogen interactions

Tick species	Tick-borne pathogens	Interactions	References
Amblyomma americanum	*Anaplasma/Ehrlichia*	Significant differences not observed in the overall microbial community structure identified between *Anaplasma/Ehrlichia*-infected and uninfected ticks.	Trout Fryxell and DeBruyn (2016)
Amblyomma maculatum	*Rickettsia parkeri*	*Francisella*-LE density decreases while *Candidatus Midichloria mitochondrii* increases in *R. parkeri*-infected tick cells than in uninfected tick cells; *R. parkeri* modulates tick host defenses by upregulating tick selenoproteins.	Budachetri et al. (2018)
Amblyomma aureolatum	*Rickettsia rickettsii*	Significantly reduced midgut bacterial load in *R. rickettsii*-infected *A. aureolatum*; Taxonomic analysis reveals dominance of the genus *Francisella*, indicating endosymbiosis.	Pavanelo et al., 2020)
Dermacentor andersoni	*Anaplasma marginale, Francisella novicida*	An increased level of *Rickettsia belli*, both in proportion and quantity, in the microbiome; negatively correlated to *A. marginale* levels in ticks; a positive pathogen–endosymbiont relation indicated by decreased level of *Francisella* endosymbionts associated with reduced infection level of *F. novicida*.	Gall et al. (2016)
Dermacentor occidentalis	*Rickettsia*	An inverse relationship exists between *Rickettsia* and *Francisella*-LE infection, suggesting a partial interference between *Francisella*-LE and spotted fever group *Rickettsia* infecting ticks.	Gurfield et al. (2017)
Ixodes scapularis	*Anaplasma phagocytophilum*	*A. phagocytophilum* changes tick microbiota such as *Enterococcus* and *Rickettsia* are decreased whereas *Pseudomonas* are increased; dysbiosis enhances *A. phagocytophilum* colonization; *A. phagocytophilum* induces changes in the gut barrier (decrease of peritrophin gene expression and thickness of the peritrophic matrix) via the *I. scapularis* antifreeze glycoprotein (IAGFP); IAGFP bound to the D-alanine residue of bacterial peptidoglycan which results in altered permeability and the capacity of bacteria to form biofilms.	Abraham et al. (2017)
I. scapularis	*Borrelia burgdorferi*	Dysbiosed larvae of *I. scapularis* increases engorgement weights and reduces *B. burgdorferi* colonization; dysbiosed larvae present downregulated expression of transcription factor signal transducer and activator of transcription (STAT) and peritrophin leading to compromised tick gut peritrophic membrane integrity; altered integrity of the peritrophic matrix decreased epithelium-bound spirochetes.	Narasimhan et al. (2014)
I. scapularis	*B. burgdorferi*	*B. burgdorferi* infection induces PIXR (a secreted gut protein of *I. scapularis* with a Reeler domain) expression which facilitates pathogen colonization in tick gut and larval molting; *B. burgdorferi* infection inhibits bacterial biofilm formation influencing gut microbiome and metabolome composition.	Narasimhan et al. (2017)
I. scapularis	*B. burgdorferi*	After bioinformatic filtering of the dominant *Rickettsia* endosymbiont, *B. burgdorferi*-infected ticks exhibit reduced microbiome diversity, with notably diminished species evenness relative to uninfected field-collected ticks.	Kwan et al. (2017)

(Continued)

TABLE 6.5
(Continued)

Tick species	Tick-borne pathogens	Interactions	References
I. scapularis	*B. burgdorferi*	Increased abundance of *Bacillus*, *Enterobacteriaceae*, and *Pseudomonas* within the midgut associated with reduced *B. burgdorferi* infection in ticks.	Ross et al. (2018)
I. scapularis	*B. burgdorferi*	*B. burgdorferi* presence/absence is correlated with bacterial β-diversity, specifically in the differences in the relative abundance of taxa; *B. burgdorferi*-negative nymphs present higher levels of *Pseudomonas* amplicon sequence variants (ASV) and *Staphylococcus*; *B. burgdorferi*-positive nymphs associated with higher levels of *Sphingomonas*.	Landesman et al. (2019)
I. scapularis	*B. burgdorferi*	No association between microbiome diversity and *B. burgdorferi* is seen in field-collected *I. scapularis* ticks; the reads from *Cutibacterium* and *B. burgdorferi* rise significantly higher while that of *Rickettsia*, *Diplorickettsiaceae*, and *Beijerinckiaceae* are less abundant in *Borrelia*-infected ticks.	Chauhan et al. (2020)
I. scapularis	*B. burgdorferi*	*A. phagocytophilum* infection and antifreeze glycoprotein (AFGP) treatment affect microbiome taxonomic composition and co-occurrence networks among taxa, but minimal impact on microbiome functional traits; anti-tick immunity to PIXR (a secreted protein with a Reeler domain) alters microbial diversity and functional profiles and enriches pathways involved in biofilm formation.	Estrada-Pena et al. (2020b)
I. scapularis	*B. burgdorferi*	*Borrelia*-infection in larvae reduces the occurrence and diversity of bacteria, lowers functional redundancy, and results lack of coherence in the network around co-occurring taxa compared to uninfected nymphs.	Estrada-Pena et al. (2020a)
I. scapularis	*Borrelia* spp.	*Borrelia*-positive ticks are positively associated with bacterial genera (*Tepidomonas*, *Luteibacter*, *Francisella*, and *Fibriimonas*).	Brinkerhoff et al. (2020)
I. scapularis	*Borrelia* spp.	*Borrelia*-positive ticks present significantly greater bacterial diversity than *Borrelia*-negative ticks.	Sperling et al. (2020)
Rhipicephalus haemaphysaloides	*Babesia microti*	Kanamycin-treated groups exhibit higher prevalence of *B. microti* and reduced density of *Coxiella*-like endosymbiont.	Li et al. (2018a)
Rhipicephalus microplus	*Theileria* sp.	Presence of *Theileria* sp. in *R. microplus* reduces microbial diversity, richness, and evenness.	Adegoke et al. (2020)

Source: Adapted and modified from Wu-Chuang et al. (2021).

2020). Mechanistic study has revealed that interaction between microbiome and *B. burgdorferi* is mediated by special protein in tick gut. *B. burgdorferi* colonization in the tick gut was decreased in a study using RNAi-mediated silencing of the gene encoding Protein of I. scapularis with a reeler domain (PIXR) (a protein of *I. scapularis* with a Reeler domain secreted into gut) and anti-PIXR immunity in mice (Narasimhan et al., 2017). The taxonomic and functional pathway diversity of the microbiome of ticks increased when fed on PIXR-immunized mice (Estrada-Pena et al., 2020b). Based on in vitro and in vivo experimental studies, it is understood that PIXR inhibits the formation of bacterial biofilm (Figure 6.6). Furthermore, disruption of biofilm formation could influence the spirochete (*B. burgdorferi*) attachment to the gut epithelium (Narasimhan et al., 2017). The alteration of structural integrity of the PM also reduces *B. burgdorferi* colonization and its attachment to the gut lumen (Narasimhan et al., 2014). It is further explained that the bacterial components of the tick gut microbiota are crucial for the maintenance of PM integrity which is necessary for successful colonization by *B. burgdorferi* in the gut epithelium because it probably protects the spirochetes from toxic components of the tick gut (Narasimhan et al., 2014).

The *B. burgdorferi* genome is devoid of genes essential for the synthesis of amino acids, fatty acids, nucleotides, and vitamins; thus, it depends on tick vectors and vertebrate hosts for several required nutrients and metabolic products (Kurokawa et al., 2020). In this circumstance certain gut endosymbionts and/or commensals could provide nutritional supplements for the survival of spirochetes. *Borrelia* spirochetes on their part may actively alter the microbial structure to make it a congenial environment for their colonization (Narasimhan et al., 2017). In addition, infection with *B. burgdorferi* promotes the expression of the *I. scapularis* gene PIXR, which encodes a gut secreted protein with functions in tick biology, such as larval molting and inhibition of biofilm formation (preferentially by Gram-positive bacteria), facilitating the colonization of *B. burgdorferi* in ticks (Narasimhan et al., 2017).

Unlike *Borrelia*, *A. phagocytophilum* perturbs the gut microbiome as observed in *I. scapularis* and requires a thin and permeable PM for its successful colonization since it rapidly gets access to the SG via tick gut (Abraham et al., 2017). *A. phagocytophilum* infection induces expression of tick antifreeze glycoprotein (IAFGP), which possesses antibacterial properties. Functionally, IAFGP perturbs the gut microbiota by binding with the peptidoglycan layer of Gram-positive bacteria, leading to an altered permeability and ultimately disrupting the capacity to form biofilms. Such an antimicrobial trait of IAFGP during colonization of *A. phagocytophilum* is congruent with a reduced abundance of Gram-positive biofilm forming taxa of tick microbiome. This evidence implies that *A. phagocytophilum* induces expression of IAFGP in order to modulate the tick gut microbiome and disrupt the structural integrity of PM (i.e., reduces the thickness of PM) and gut barrier, thus aiding its own colonization in tick gut (Abraham et al., 2017) (Figure 6.6).

In a metagenomics study, *A. phagocytophilum* infection and the tick's expression of the antifreeze glycoprotein IAFGP significantly shift the taxonomic composition and co-occurrence network of the tick gut microbiome, yet the functional profile remains unchanged (Estrada-Pena et al., 2020b). This finding exemplifies a sophisticated coevolutionary strategy where the pathogen seems to hijack a tick protein to reshape its microbial environment, thereby ameliorating its own fitness within the vector. The impact of microbiome disruption with antibiotics in relation to pathogen susceptibility in ticks has been studied in *D. andersoni*, in which a negative correlation has been observed between the burden of *R. bellii* and *A. marginale*, while a positive correlation exists between *Francisella* endosymbionts and *F. novicida* infection levels (Gall et al., 2016). Interestingly, the presence of *Francisella* endosymbionts exhibited a positive impact for the establishment of *F. novicida*. In this connection, it has been hypothesized that these endosymbionts might inhibit tick immunity, favoring *F. novicida* acquisition. Additional studies are needed to confirm this hypothesis and to identify the mechanisms by which microbiota delineate tick susceptibility to infection. Similarly, a negative association between *Francisella*-LE infection level and SFGR in *D. occidentalis* is observed, indicating possible interference between *Francisella*-LE and SFGR (Gurfield et al., 2017). In another tick species, *A. maculatum*, the decreased level of *Francisella*-LE and increased level of CMM are associated with *R. parkeri* infection (Budachetri et al., 2018).

The mechanisms through which endosymbiont bacteria regulate pathogen infection remain poorly elucidated, though there have been two hypotheses. The direct mechanisms may involve endosymbiont-derived molecules that either enhance or inhibit pathogen proliferation whereas the indirect mechanisms include competition for host resources that are required for pathogen growth thus limiting pathogen replication or inhibition of host immune factors thereby allowing enhanced growth of pathogens (Gall et al., 2016). A plausible instance is that the tick immune system is associated with the reduced susceptibility of *A. sculptum* to *R. rickettsii* infection (Martins et al., 2017). Midgut transcriptome analysis of *A. sculptum* showed that immune factors are mostly upregulated during *R. rickettsii* infection (Martins et al., 2017). Therefore, it has been hypothesized that microbiota components could modulate immune factors in *A. sculptum* to foster an efficient immune system, leading to a lower susceptibility (Pavanelo et al., 2020). Perturbation of normal gut microbiota in *I. scapularis* decreases the expression of STAT, which, in turn, reduces the expression of peritrophin-1. Since the integrity of the PM is essential to *B. burgdorferi* infection, as previously discussed, alteration of the microbiota reduces borrelial colonization (Narasimhan et al., 2014).

6.5 FACTORS ASSOCIATED WITH TICK MICROBIOME COMMUNITY COMPOSITION

Although ticks transmit a wide range of emerging TBDs and even simultaneously host multiple pathogens, they are overlooked in terms of the impact of climate change (Lee and Chung, 2023). In recent decades, the TBDs have surged and expanded globally. The growing influence of climate change has accelerated the migration of ticks to the new geographical settings, thereby intensifying the frequency of TBDs, with growing challenges to humans and animals (El-Sayed and Kamel, 2020). This phenomenon has been attributed to several factors (Figure 6.7) such as changes in human activities, demographic shifts, land use (Charles et al., 2021), rising temperatures, variations in precipitation, and extreme weather events as potential consequences of climate change (Lee and Chung, 2023), resulting in the continuous emergence and re-emergence of zoonotic diseases (Charles et al., 2021).

The ability of some tick species to reproduce parthenogenetically, coupled with the long-distance transport of viruses by migratory birds and animals, enhances their dissemination potential. Additionally, coinfections within hosts may accelerate viral evolution, potentially leading to the emergence of viruses (Maqbool et al., 2022; Hagman et al., 2014; Zhang et al., 2022). Since

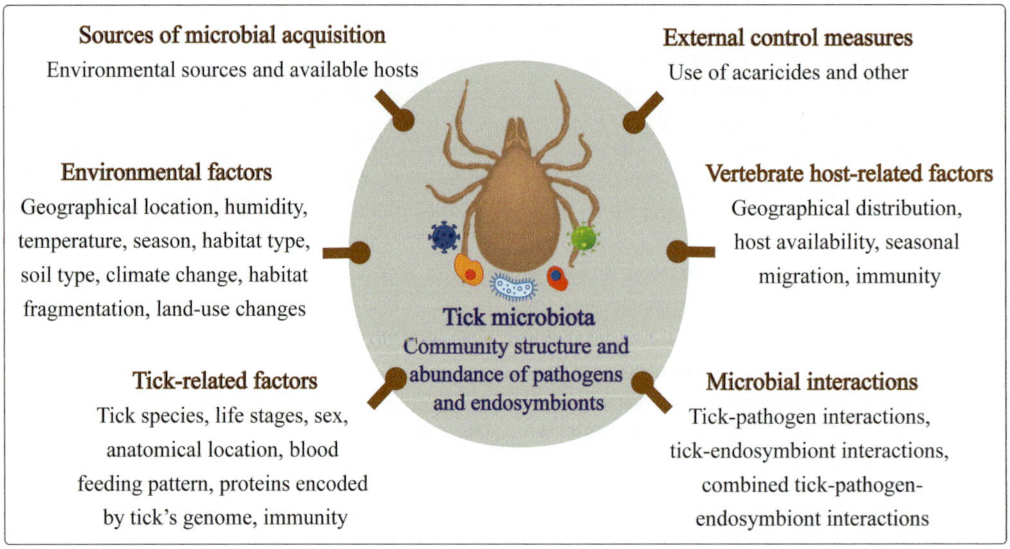

Figure 6.7 Factors influencing tick microbiota community structure, abundance, persistence, and interactions. A combined influence of multiple factors is crucial in shaping tick internal microbiota. Illustration created on Adobe Illustrator CC 2019.

most of the lifetime of ticks is spent off-host, living in the environment detached from the hosts, ticks are significantly associated with the long-term impact of climate change-related challenges within a single life cycle more than dipteran vectors with brief life cycle, experiencing short-term climate variabilities (Ogden and Lindsay, 2016; Gilbert, 2021). In the soil or below the litter layer, ticks can extend developmental delay from one instar to another until the weather conditions resume to their optimum (Ogden and Lindsay, 2016). Studies have reported an increase in tick population with the spread of ticks to the high altitudinal gradients in response to rising temperature (Clow et al., 2017) suggesting the warmer the temperature, the more suitable it is for ticks to survive and reproduce. An increase in the abundance of vectors and pathogen survival facilitated by the rising temperature results in the gradual import (intrusion) of existing tropical and subtropical vector-borne pathogens (Ogden and Lindsay, 2016). Most importantly, the poleward expansion of vectors would potentially result in the emergence of particular Vector-borne diseases (VBD)s.

The microbial community structure is at least partly influenced by natural habitat of the ticks because considerable isolates of bacteria and fungi are related to their respective generalist genera which are naturally found in diverse environments, including soil, phytosurface, or rhizospheres. For instance, the overall microbial community diversity between *Anaplasma*/*Ehrlichia*-positive and -negative ticks was found to differ based on the collection site, indicating that environmental factors at least partly govern *A. americanum* microbiomes (Trout Fryxell and DeBruyn, 2016). Likewise, isolation of *Halomonas*, a bacterium thriving in saline ecology, exclusively isolated from *Or. maritimus*, a tick species which is a particular resident in marine bird nests of the coastal regions, represents that this bacterium is acquired from the high-salinity habitats of this tick species (Bruley et al., 2025). On the other hand, the tick-associated notable fungal isolates (*Aureobasidium*, *Cladosporium*, *Fusarium*, *Penicillium*, and *Mucor*) are largely soil and phytosurface colonizers, or in certain instances, major phytopathogens (Bruley et al., 2025). Ticks are more likely to encounter possibly entomopathogenic fungi in soil than in leaf litter (Tuininga et al., 2009).

Host selection behavior of ticks is another factor in the acquisition of microbes. Reservoir hosts, such as wild rodents, act as the main source where larvae and nymphs can survive, and reproduction hosts, for instance, deer, serve as the source for adult female ticks to be fed (blood meals) and reproduce (Bouchard et al., 2019; Bouchard et al., 2013). The abundance of *Malassezia*, a mammal skin-associated fungal genus, identified in the homogenized *I. scapularis* nymphal samples, suggests that host animal skin and fur could be sources of environmental microbes acquired in the tick's body (Landesman et al., 2019). In summary, tick microbial community structure and abundance are shaped by the combined influence of several factors, including microbiota and pathogen acquisition sources (from the environment or vertebrate hosts); environmental parameters such as geographic location, temperature, humidity, season, habitat type, soil type, impact of climate changes, effects of habitat fragmentation, and land use changes; and host-related factors like immunity, as well as tick-specific traits such as species, developmental stages/instar, sex, anatomical location and blood-feeding pattern (both frequency and volume), and proteins encoded by the tick's genome. Additionally, complex interactions between ticks and pathogens and ticks and endosymbionts and the interplay among all three play a crucial role. Similarly, external control measures, including acaricides, significantly influence tick microbiota population.

6.6 SUMMARY

Ticks harbor a complex assemblage of microorganisms – viruses, bacteria, fungi, protozoa, and helminths – that occupy specific tissues or multiple anatomical niches and play diverse functional roles. These microbes are acquired horizontally from the environment or vertebrate hosts, and vertically through transovarial and transstadial transmission, enabling long-term maintenance of key pathogens and endosymbionts. Metagenomic and phylogenetic evidence indicates that several intracellular mutualists, including *Rickettsia*, *Coxiella*, and *Francisella*, are evolutionarily derived from pathogenic lineages, underscoring their potential to transition between symbiotic and pathogenic states. Although intracellular symbionts are comparatively well characterized, extracellular microbial communities remain insufficiently studied despite their ecological and physiological relevance.

Tick microbiota critically influences host physiology, including nutritional supplementation, development, reproduction, redox homeostasis, immunity, and vector competence. Microbiome perturbation studies demonstrate that modifying bacterial community structure can significantly affect pathogen acquisition and colonization. For example, antibiotic-mediated dysbiosis or targeted suppression of specific symbionts alters susceptibility to *Borrelia*, *Francisella*, and *Rickettsia* spp. Such findings highlight the potential of microbiome manipulation – through anti-microbiota vaccines, selective symbiont targeting, and gene interference of redox-regulatory pathways – to reduce vectorial capacity.

Technological advances, particularly NGS, have transformed the characterization of tick-associated microbial consortia, enabling the surveillance of known and emergent pathogens and facilitating reverse discovery of TBDs. Parallel strategies from mosquito research – including *Wolbachia*-based population modification, transgenesis, and paratransgenesis – demonstrate promising translational potential for tick systems, although their application remains nascent. Integration of metagenomic, metaproteomic, and metabolomic datasets within systems biology frameworks will enhance understanding of tick–microbe–pathogen interactions, support the identification of protective antigens, and guide development of microbiome-informed interventions. Consequently, microbiome research is central to advancing surveillance, pathogen discovery, and innovative control strategies targeting ticks and tick-borne pathogens.

REFERENCES

Abdullah, H. H. A. M., Aboelsoued, D., Farag, T. K., Abdel-Shafy, S., Abdel Megeed, K. N., Parola, P., Raoult, D., & Mediannikov, O. (2022). Molecular characterization of some equine vector-borne diseases and associated arthropods in Egypt. *Acta tropica*, 227, 106274. https://doi.org/10.1016/j.actatropica.2021.106275.

Abraham, N. M., Liu, L., Jutras, B. L., Yadav, A. K., Narasimhan, S., Gopalakrishnan, V., Ansari, J. M., Jefferson, K. K., Cava, F., Jacobs-Wagner, C., & Fikrig, E. (2017). Pathogen-mediated manipulation of arthropod microbiota to promote infection. *Proceedings of the National Academy of Sciences of the United States of America*, 114(5), E781–E790. https://doi.org/10.1073/pnas.1613422114.

Adams, S. N., Bestul, N. C., Calloway, K. N., Kersh, G. J., & Salzer, J. S. (2025). National surveillance of human ehrlichiosis caused by *Ehrlichia ewingii*, United States, 2013-2021. *Emerging infectious diseases*, 31(2), 222–227. https://doi.org/10.3201/eid3102.240279.

Addo, S. O., Baako, B. O. A., Bentil, R. E., Addae, C. A., Behene, E., Asoala, V., Sallam, M., Mate, S., Dunford, J. C., Larbi, J. A., Baidoo, P. K., Wilson, M. D., Diclaro, J. W., 2nd, & Dadzie, S. K. (2023). Molecular survey of Anaplasma and Ehrlichia species in livestock ticks from Kassena-Nankana, Ghana; with a first report of Anaplasma capra and Ehrlichia minasensis. *Archives of microbiology*, 205(3), 92. https://doi.org/10.1007/s00203-023-03430-5.

Adegoke, A., Kumar, D., Bobo, C., Rashid, M. I., Durrani, A. Z., Sajid, M. S., & Karim, S. (2020). Tick-borne pathogens shape the native microbiome within tick vectors. *Microorganisms*, 8(9), 1299. https://doi.org/10.3390/microorganisms8091299.

Adjogoua, E. V., Coulibaly-Guindo, N., Diaha-Kouame, C. A., Diane, M. K., Kouassi, R. M. C. K. A., Coulibaly, J. T., & Dosso, M. (2021). Geographical distribution of ticks ixodidae in Côte d'Ivoire: Potential reservoir of the crimean-congo hemorrhagic fever virus. *Vector borne and zoonotic diseases (Larchmont, N.Y.)*, 21(8), 628–634. https://doi.org/10.1089/vbz.2020.2745.

Ahmad, I., Ullah, S., Alouffi, A., Almutairi, M. M., Numan, M., Tanaka, T., Chang, S. C., Chen, C. C., & Ali, A. (2023). First molecular-based confirmation of Dermacentor marginatus and associated Rickettsia raoultii and Anaplasma marginale in the hindu kush mountain range. *Animals: an open access journal from MDPI*, 13(23), 3686. https://doi.org/10.3390/ani13233686.

Ajileye, O. D., Verocai, G. G., & Light, J. E. (2025). A review of filarial nematodes parasitizing tick vectors: Unraveling global patterns in species diversity, host associations, and interactions with tick-borne pathogens. *Parasites & vectors*, 18(1), 50. https://doi.org/10.1186/s13071-025-06690-6.

Akther, S., Mongodin, E. F., Morgan, R. D., Di, L., Yang, X., Golovchenko, M., Rudenko, N., Margos, G., Hepner, S., Fingerle, V., Kawabata, H., Norte, A. C., de Carvalho, I. L., Núncio, M. S., Marques, A., Schutzer, S. E., Fraser, C. M., Luft, B. J., Casjens, S. R., & Qiu, W. (2024). Natural selection and recombination at host-interacting lipoprotein loci drive genome diversification of Lyme disease and related bacteria. *mBio*, 15(9), e0174924. https://doi.org/10.1128/mbio.01749-24.

Akveran, G. A., Karasartova, D., Keskin, A., Comba, A., Celebi, B., Mumcuoglu, K. Y., &

Taylan-Ozkan, A. (2020). Bacterial and protozoan agents found in Hyalomma aegyptium (L., 1758) (Ixodida: Ixodidae) collected from Testudo graeca L., 1758 (Reptilia: Testudines) in Corum Province of Turkey. *Ticks and tick-borne diseases*, 11(5), 101458. https://doi.org/10.1016/j.ttbdis.2020.101461.

Alafaci, A., Crépin, A., Beaubert, S., Berjeaud, J. M., Delafont, V., & Verdon, J. (2021). Exploring the individual bacterial microbiota of questing Ixodes ricinus Nymphs. *Microorganisms*, 9(7), 1526. https://doi.org/10.3390/microorganisms9071526.

Alam, S., Khan, M., Alouffi, A., Almutairi, M. M., Ullah, S., Numan, M., Islam, N., Khan, Z., Aiman, O., Zaman Safi, S., Tanaka, T., & Ali, A. (2022). Spatiotemporal patterns of ticks and molecular survey of Anaplasma marginale, with notes on their phylogeny. *Microorganisms*, 10(8), 1663. https://doi.org/10.3390/microorganisms10081664.

Alberdi, P., Ayllón, N., Cabezas-Cruz, A., Bell-Sakyi, L., Zweygarth, E., Stuen, S., & de la Fuente, J. (2015). Infection of Ixodes spp. tick cells with different Anaplasma phagocytophilum isolates induces the inhibition of apoptotic cell death. *Ticks and tick-borne diseases*, 6(6), 758–767. https://doi.org/10.1016/j.ttbdis.2015.07.001.

Alberdi, P., Cabezas-Cruz, A., Prados, P. E., Rayo, M. V., Artigas-Jerónimo, S., & de la Fuente, J. (2019). The redox metabolic pathways function to limit Anaplasma phagocytophilum infection and multiplication while preserving fitness in tick vector cells. *Scientific reports*, 9(1), 13236. https://doi.org/10.1038/s41598-019-49766-x.

Alberdi, P., Espinosa, P. J., Cabezas-Cruz, A., & de la Fuente, J. (2016a). Anaplasma phagocytophilum manipulates host cell apoptosis by different mechanisms to establish infection. *Veterinary sciences*, 3(3), 15. https://doi.org/10.3390/vetsci3030015.

Albertson, R., Casper-Lindley, C., Cao, J., Tram, U., & Sullivan, W. (2009). Symmetric and asymmetric mitotic segregation patterns influence Wolbachia distribution in host somatic tissue. *Journal of cell science*, 122(Pt 24), 4570–4583. https://doi.org/10.1242/jcs.054981.

Albertson, R., Tan, V., Leads, R. R., Reyes, M., Sullivan, W., & Casper-Lindley, C. (2013). Mapping Wolbachia distributions in the adult Drosophila brain. *Cellular microbiology*, 15(9), 1527–1544. https://doi.org/10.1111/cmi.12136.

Al-Khafaji, A. M., Clegg, S. R., Pinder, A. C., Luu, L., Hansford, K. M., Seelig, F., Dinnis, R. E., Margos, G., Medlock, J. M., Feil, E. J., Darby, A. C., McGarry, J. W., Gilbert, L., Plantard, O., Sassera, D., & Makepeace, B. L. (2019). Multi-locus sequence typing of Ixodes ricinus and its symbiont Candidatus Midichloria mitochondrii across Europe reveals evidence of local co-cladogenesis in Scotland. *Ticks and tick-borne diseases*, 10(1), 52–62. https://doi.org/10.1016/j.ttbdis.2018.08.016.

Allerdice, M. E., Hecht, J. A., Karpathy, S. E., & Paddock, C. D. (2017). Evaluation of gulf coast ticks (Acari: Ixodidae) for ehrlichia and Anaplasma species. *Journal of medical entomology*, 54(2), 481–484. https://doi.org/10.1093/jme/tjw176.

Almeida, A. P., Marcili, A., Leite, R. C., Nieri-Bastos, F. A., Domingues, L. N., Martins, J. R., & Labruna, M. B. (2012). Coxiella symbiont in the tick Ornithodoros rostratus (Acari: Argasidae). *Ticks and tick-borne diseases*, 3(4), 203–206. https://doi.org/10.1016/j.ttbdis.2012.02.003.

Alonso-Díaz, M. A., & Fernández-Salas, A. (2021). Entomopathogenic fungi for tick control in cattle livestock from Mexico. *Frontiers in fungal biology*, 2, 657694. https://doi.org/10.3389/ffunb.2021.657694.

Altantogtokh, D., Lilak, A. A., Takhampunya, R., Sakolvaree, J., Chanarat, N., Matulis, G., Poole-Smith, B. K., Boldbaatar, B., Davidson, S., Hertz, J., Bolorchimeg, B., Tsogbadrakh, N., Fiorenzano, J. M., Lindroth, E. J., & von Fricken, M. E. (2022). Metagenomic profiles of Dermacentor tick pathogens from across Mongolia, using next generation sequencing. *Frontiers in microbiology*, 13, 946631. https://doi.org/10.3389/fmicb.2022.946637.

Ayllón, N., Villar, M., Busby, A. T., Kocan, K. M., Blouin, E. F., Bonzón-Kulichenko, E., Galindo, R. C., Mangold, A. J., Alberdi, P., Pérez de la Lastra, J. M., Vázquez, J., & de la Fuente, J. (2013). Anaplasma phagocytophilum inhibits apoptosis and promotes cytoskeleton rearrangement for infection of tick cells. Infection and Immunity. Jul;81(7), 2415-25.

Anderson, J. F., & Magnarelli, L. A. (2008). Biology of ticks. *Infectious disease clinics of North America*, 22(2), 195–v. https://doi.org/10.1016/j.idc.2007.12.006.

Andersson, M. O., & Chitimia-Dobler, L. (2017). Detection of Cercopithifilaria bainae in western Romania. *Parasitology research*, 116(11), 3235–3238. https://doi.org/10.1007/s00436-017-5625-5.

Andreotti, R., Pérez de León, A. A., Dowd, S. E., Guerrero, F. D., Bendele, K. G., & Scoles, G. A. (2011). Assessment of bacterial diversity in the cattle tick Rhipicephalus (Boophilus) microplus through tag-encoded pyrosequencing. *BMC microbiology*, 11(1), 6. https://doi.org/10.1186/1471-2180-11-6.

Angelou, A., Latrofa, M. S., Annoscia, G., Symeonidou, I., Theodoridis, A., Polizopoulou, Z. S., Otranto, D., & Papadopoulos, E. (2020). Cercopithifilaria species in dogs and ticks from Greece. *Parasitology research*, 119(10), 3391–3400. https://doi.org/10.1007/s00436-020-06784-3.

Anholt, H., Hillman, V., Vaughan, J., & Smyth, N. (2023). The soft tick Ornithodoros moubata and its role in the epidemiology of African swine fever in Central Malawi. *Journal of wildlife diseases*, 59(3), 465–471. https://doi.org/10.7589/JWD-D-22-00090.

Anstead, C. A., & Chilton, N. B. (2014). Discovery of novel Rickettsiella spp. in ixodid ticks from Western Canada. *Applied and environmental microbiology*, 80(4), 1403–1410. https://doi.org/10.1128/AEM.03564-13.

Antunes, S., Ferrolho, J., Domingues, N., Santos, A. S., Santos-Silva, M. M., & Domingos, A. (2016). Anaplasma marginale and Theileria annulata in questing ticks from Portugal. *Experimental & applied acarology*, 70(1), 79–88. https://doi.org/10.1007/s10493-016-0058-y.

Antúnez, M. P., Marín Montesinos, J. C., Corduneanu, A., Obregón, D., Moutailler, S., & Cabezas-Cruz, A. (2024). Tick-borne viruses and their risk to public health in the Caribbean: Spotlight on bats as reservoirs in Cuba. *Heliyon*, 10(4), e26118. https://doi.org/10.1016/j.heliyon.2024.e26118.

Appelt, S., Köppen, K., Radonić, A., Drechsel, O., Jacob, D., Grunow, R., & Heuner, K. (2019). Genetic diversity and spatial segregation of Francisella tularensis subspecies holarctica in Germany. *Frontiers in cellular and infection microbiology*, 9, 376. https://doi.org/10.3389/fcimb.2019.00376.

Arrais, R. C., Paula, R. C., Martins, T. F., Nieri-Bastos, F. A., Marcili, A., & Labruna, M. B. (2021). Survey of ticks and tick-borne agents in maned wolves (Chrysocyon brachyurus) from a natural landscape in Brazil. *Ticks and tick-borne diseases*, 12(2), 101639. https://doi.org/10.1016/j.ttbdis.2020.101640.

Aslam, M., Abbas, R. Z., & Alsayeqh, A. (2023). Distribution pattern of Crimean-Congo Hemorrhagic Fever in Asia and the Middle East. *Frontiers in public health*, 11, 1093817. https://doi.org/10.3389/fpubh.2023.1093817.

Aung, A., Kaewlamun, W., Narapakdeesakul, D., Poofery, J., & Kaewthamasorn, M. (2022). Molecular detection and characterization of tick-borne parasites in goats and ticks from Thailand. *Ticks and tick-borne diseases*, 13(3), 101938. https://doi.org/10.1016/j.ttbdis.2022.101938.

Aung, K. M., Boldbaatar, D., Umemiya-Shirafuji, R., Liao, M., Tsuji, N., Xuenan, X., Suzuki, H., Kume, A., Galay, R. L., Tanaka, T., & Fujisaki, K. (2012). HlSRB, a Class B scavenger receptor, is key to the granulocyte-mediated microbial phagocytosis in ticks. *PloS one*, 7(3), e33504. https://doi.org/10.1371/journal.pone.0033504.

Ayan, A., Aslan Çelik, B., Çelik, Ö. Y., Orunç Kılınç, Ö., Akyıldız, G., Yılmaz, A. B., Sayın İpek, D. N., Oktay Ayan, Ö., & Babaoğlu, A. R. (2024). First detection of Ehrlichia chaffeensis, Ehrlichia canis, and Anaplasma ovis in Rhipicephalus bursa ticks collected from sheep, Turkey. *Polish journal of veterinary sciences*, 27(1), 85–94. https://doi.org/10.24425/pjvs.2024.149338.

Azagi, T., Klement, E., Perlman, G., Lustig, Y., Mumcuoglu, K. Y., Apanaskevich, D. A., & Gottlieb, Y. (2017). Francisella-like endosymbionts and Rickettsia species in local and imported Hyalomma ticks. *Applied and environmental microbiology*, 83(18), e01302-17. https://doi.org/10.1128/AEM.01302-17.

Aziati, I. D., Jnr, D. M., Antia, A., Joshi, A., Aviles-Gamboa, A., Lee, P., Harastani, H., Wang, D., Adalsteinsson, S. A., & Boon, A. C. M. (2023). Prevalence of Bourbon and Heartland viruses in field collected ticks at an environmental field station in St. Louis County, Missouri, USA. *Ticks and tick-borne diseases*, 14(1), 102080. https://doi.org/10.1016/j.ttbdis.2022.102080.

Bai, Y., Li, Y., Liu, W., Li, J., Tian, F., Liu, L., Han, X., & Tong, Y. (2023). Analysis of the diversity of tick-borne viruses at the border areas in Liaoning Province, China. *Frontiers in microbiology*, 14, 1179156. https://doi.org/10.3389/fmicb.2023.1179156.

Bain, O., & Babayan, S. (2003). Behaviour of filariae: Morphological and anatomical signatures of their life style within the arthropod and vertebrate hosts. *Filaria journal*, 2(1), 16. https://doi.org/10.1186/1475-2883-2-16.

Balasubramanian, R., Yadav, P. D., Sahina, S., & Nadh, V. A. (2021). The species distribution of ticks & the prevalence of Kyasanur forest disease virus in questing nymphal ticks from Western Ghats of Kerala, South India. *The Indian journal of medical research*, 154(5), 743–749. https://doi.org/10.4103/ijmr.IJMR_234_19.

Bamford, C., Blumberg, L. H., Bosman, M., Frean, J., Hoek, K. G. P., Miles, J., Sriruttan, C., Vorster, I., & Oosthuizen, M. C. (2023). Neoehrlichiosis in symptomatic immunocompetent child, South Africa. *Emerging infectious diseases*, 29(2), 407–410. https://doi.org/10.3201/eid2902.221451.

Baneth, G. (2011). Perspectives on canine and feline hepatozoonosis. *Veterinary parasitology*, 181(1), 3–11. https://doi.org/10.1016/j.vetpar.2011.04.015.

Baneth, G., & Allen, K. (2022). Hepatozoonosis of dogs and cats. The veterinary clinics of North America. *Small animal practice*, 52(6), 1341–1358. https://doi.org/10.1016/j.cvsm.2022.06.011.

Barker, S. C., & Murrell, A. (2004). Systematics and evolution of ticks with a list of valid genus and species names. *Parasitology*, 129 Suppl, S15–S36. https://doi.org/10.1017/s0031182004005207.

Barradas, P. F., Marques, J., Tavares, C., Brito, N. V., & Mesquita, J. R. (2024). Detection of tick-borne pathogens in Rhipicephalus bursa ticks collected from the autochthonous Garrano breed of horses in Portugal. *Veterinary parasitology, regional studies and reports*, 51, 101033. https://doi.org/10.1016/j.vprsr.2024.101033.

Becker, N. S., Rollins, R. E., Stephens, R., Sato, K., Brachmann, A., Nakao, M., & Kawabata, H. (2023). Candidatus Lariskella arthopodarum endosymbiont is the main factor differentiating the microbiome communities of female and male Borrelia-positive Ixodes persulcatus ticks. *Ticks and tick-borne diseases*, 14(4), 102183. https://doi.org/10.1016/j.ttbdis.2023.102183.

Beliavskaia, A., Hönig, V., Erhart, J., Vyhlidalova, T., Palus, M., Cerny, J., Kozlova, I., Ruzek, D., Palomar, A. M., & Bell-Sakyi, L. (2021). Spiroplasma isolated from third-generation laboratory colony Ixodes persulcatus ticks. *Frontiers in veterinary science*, 8, 659786. https://doi.org/10.3389/fvets.2021.659786.

Belkahia, H., Ben Said, M., Ghribi, R., Selmi, R., Ben Asker, A., Yahiaoui, M., Bousrih, M., Daaloul-Jedidi, M., & Messadi, L. (2019). Molecular detection, genotyping and phylogeny of Anaplasma spp. in Rhipicephalus ticks from Tunisia. *Acta tropica*, 191, 38–49. https://doi.org/10.1016/j.actatropica.2018.12.034.

Ben-Harari R. R. (2019). Tick transmission of toxoplasmosis. *Expert review of anti-infective therapy*, 17(11), 911–917. https://doi.org/10.1080/14787210.2019.1682550.

Beninati, T., Riegler, M., Vilcins, I.-M. E., Sacchi, L., McFadyen, R., Krockenberger, M. et al. (2009). Absence of the symbiont Candidatus Midichloria mitochondrii in the mitochondria of the tick Ixodes holocyclus. *FEMS Microbiology Letters*, 299, 241–247.

Benjamin, M. A., Zhioua, E., & Ostfeld, R. S. (2002). Laboratory and field evaluation of the entomopathogenic fungus Metarhizium anisopliae (Deuteromycetes) for controlling questing adult Ixodes scapularis (Acari: Ixodidae). *Journal of medical entomology*, 39(5), 723–728. https://doi.org/10.1603/0022-2585-39.5.723.

Benson, M. J., Gawronski, J. D., Eveleigh, D. E., & Benson, D. R. (2004). Intracellular symbionts and other bacteria associated with deer ticks (Ixodes scapularis) from Nantucket and Wellfleet, Cape Cod, Massachusetts. *Applied and environmental microbiology*, 70(1), 616–620. https://doi.org/10.1128/AEM.70.1.616-620.2004.

Bentil, R. E., Addo, S. O., Mosore, M. T., Kumordjie, S., Yeboah, C., Agbodzi, B., Behene, E., Tagoe, J., Baako, B. O. A., Asoala, V., Ampadu, R. O., Mingle, D. L., Nyarko, E. O., Fox, A. T., Letizia, A. G., Diclaro, J. W., 2nd, Sanders, T., Oduro, D., Nimo-Paintsil, S. C., Harwood, J., ... & Dadzie, S. K. (2023). First whole genome sequencing of crimean-congo hemorrhagic fever virus (CCHFV) in tick species within ghana. *Transboundary and emerging diseases*, 2023, 2063317. https://doi.org/10.1155/2023/2063317.

Ben-Yosef, M., Rot, A., Mahagna, M., Kapri, E., Behar, A., & Gottlieb, Y. (2020). Coxiella-like endosymbiont of Rhipicephalus sanguineus is required for physiological processes during ontogeny. *Frontiers in microbiology*, 11, 493. https://doi.org/10.3389/fmicb.2020.00493.

Bermúdez C, S. E., Félix, M. L., Domínguez A, L., Araúz, D., & Venzal, J. M. (2024). Molecular screening of tick-borne microorganisms in ticks from rural areas of Panama, with the first record of Ehrlichia minasensis in Rhipicephalus microplus from Central America. *Veterinary research communications*, 48(2), 1301–1308. https://doi.org/10.1007/s11259-024-10306-4.

Bernardo, C. C., Barreto, L. P., E Silva, C. S. R., Luz, C., Arruda, W., & Fernandes, É. K. K. (2018). Conidia and blastospores of Metarhizium spp. and Beauveria bassiana s.l.: Their development during the infection process and virulence against the tick Rhipicephalus microplus. *Ticks and tick-borne diseases*, 9(5), 1334–1342. https://doi.org/10.1016/j.ttbdis.2018.06.001.

Bezerra-Santos, M. A., Dantas-Torres, F., Ramos, R. A. N., Brianti, E., & Otranto, D. (2023). Cercopithifilaria spp. of dogs: Little known but prevalent filarioids beneath the skin. *Parasites & vectors*, 16(1), 386. https://doi.org/10.1186/s13071-023-06007-5.

Bezerra-Santos, M. A., de Macedo, L. O., Nguyen, V. L., Manoj, R. R., Laidoudi, Y., Latrofa, M. S., Beugnet, F., & Otranto, D. (2022). Cercopithifilaria spp. in ticks of companion animals from Asia: New putative hosts and vectors. *Ticks and tick-borne diseases*, 13(4), 101957. https://doi.org/10.1016/j.ttbdis.2022.101963.

Bharadwaj, A., & Stafford, K. C., 3rd. (2010). Evaluation of Metarhizium anisopliae strain F52 (Hypocreales: Clavicipitaceae) for control of Ixodes scapularis (Acari: Ixodidae). *Journal of medical entomology*, 47(5), 862–867. https://doi.org/10.1603/me10020.

Bielawska-Drózd, A., Cieślik, P., Żakowska, D., Głowacka, P., Wlizło-Skowronek, B., Zięba, P., & Zdun, A. (2018). Detection of Coxiella burnetii and Francisella tularensis in Tissues of Wild-living Animals and in Ticks of North-west Poland. *Polish journal of microbiology*, 67(4), 529–534. https://doi.org/10.21307/pjm-2018-059.

Biggs, H. M., Behravesh, C. B., Bradley, K. K., Dahlgren, F. S., Drexler, N. A., Dumler, J. S., Folk, S. M., Kato, C. Y., Lash, R. R., Levin, M. L., Massung, R. F., Nadelman,

R. B., Nicholson, W. L., Paddock, C. D., Pritt, B. S., Traeger, M. S. (2016). Diagnosis and management of tickborne rickettsial diseases: Rocky Mountain spotted fever and other spotted fever group rickettsioses, ehrlichioses, and April 2020 Volume 33 Issue 2 e00083-18 anaplasmosis—United States. *MMWR Recommend Rep*, 65, 1–44. https://doi.org/10.15585/mmwr.rr6502a1.

Biglari, P., Chinikar, S., Belqeiszadeh, H., Telmadarraiy, Z., Mostafavi, E., Ghaffari, M., Javaherizadeh, S., Nowotny, N., Fooks, A. R., & Shahhosseini, N. (2016). Phylogeny of tick-derived Crimean-Congo hemorrhagic fever virus strains in Iran. *Ticks and tick-borne diseases*, 7(6), 1216–1221. https://doi.org/10.1016/j.ttbdis.2016.07.012.

Binder, L. C., Tauro, L. B., Farias, A. A., Labruna, M. B., & Diaz, A. (2019). Molecular survey of flaviviruses and orthobunyaviruses in Amblyomma spp. ticks collected in Minas Gerais, Brazil. *Revista brasileira de parasitologia veterinaria = Brazilian journal of veterinary parasitology: Orgao Oficial do Colegio Brasileiro de Parasitologia Veterinaria*, 28(4), 764–768. https://doi.org/10.1590/S1984-29612019071.

Binetruy, F., Bailly, X., Chevillon, C., Martin, O. Y., Bernasconi, M. V., & Duron, O. (2019). Phylogenetics of the Spiroplasma ixodetis endosymbiont reveals past transfers between ticks and other arthropods. *Ticks and tick-borne diseases*, 10(3), 575–584. https://doi.org/10.1016/j.ttbdis.2019.02.001.

Binetruy, F., Buysse, M., Lejarre, Q., Barosi, R., Villa, M., Rahola, N., Paupy, C., Ayala, D., & Duron, O. (2020). Microbial community structure reveals instability of nutritional symbiosis during the evolutionary radiation of Amblyomma ticks. *Molecular ecology*, 29(5), 1016–1029. https://doi.org/10.1111/mec.15373.

Binetruy, F., & Duron, O. (2023). Molecular detection of Cercopithifilaria, Cruorifilaria and Dipetalonema-like filarial nematodes in ticks of French Guiana. Détection moléculaire des nématodes filaires de type Cercopithifilaria, Cruorifilaria et Dipetalonema chez les tiques de Guyane française. *Parasite (Paris, France)*, 30, 24. https://doi.org/10.1051/parasite/2023027.

Blanton, L. S. (2019). The rickettsioses: A practical update. *Infectious disease clinics of North America*, 33(1), 213–229. https://doi.org/10.1016/j.idc.2018.10.010.

Bohacsova, M., Mediannikov, O., Kazimirova, M., Raoult, D., & Sekeyova, Z. (2016). Arsenophonus nasoniae and Rickettsiae Infection of Ixodes ricinus Due to Parasitic Wasp Ixodiphagus hookeri. *PloS one*, 11(2), e0149950. https://doi.org/10.1371/journal.pone.0149950.

Boinas, F. S., Hutchings, G. H., Dixon, L. K., & Wilkinson, P. J. (2004). Characterization of pathogenic and non-pathogenic African swine fever virus isolates from Ornithodoros erraticus inhabiting pig premises in Portugal. *The Journal of general virology*, 85(Pt 8), 2177–2187. https://doi.org/10.1099/vir.0.80058-0.

Bonnet, S. I., Binetruy, F., Hernández-Jarguín, A. M., & Duron, O. (2017). The tick microbiome: Why non-pathogenic microorganisms matter in tick biology and pathogen transmission. *Frontiers in cellular and infection microbiology*, 7, 236. https://doi.org/10.3389/fcimb.2017.00236.

Bonnet, S. I., Blisnick, T., Al Khoury, C., & Guillot, J. (2021). Of fungi and ticks: Morphological and molecular characterization of fungal contaminants of a laboratory-reared Ixodes ricinus colony. *Ticks and tick-borne diseases*, 12(5), 101732. https://doi.org/10.1016/j.ttbdis.2021.101732.

Bouchard, C., Dibernardo, A., Koffi, J., Wood, H., Leighton, P. A., & Lindsay, L. R. (2019). N increased risk of tick-borne diseases with climate and environmental changes. *Canada communicable disease report = Releve des maladies transmissibles au Canada*, 45(4), 83–89. https://doi.org/10.14745/ccdr.v45i04a02.

Bouchard, C., Leighton, P. A., Beauchamp, G., Nguon, S., Trudel, L., Milord, F., Lindsay, L. R., Bélanger, D., & Ogden, N. H. (2013). Harvested white-tailed deer as sentinel hosts for early establishing Ixodes scapularis populations and risk from vector-borne zoonoses in southeastern Canada. *Journal of medical entomology*, 50(2), 384–393. https://doi.org/10.1603/me12093.

Boulanger, N., Iijima, H., Doi, K., Watari, Y., Kwak, M., Nakao, R., & Wikel, S. (2025). Ticks and tick-borne diseases in the northern hemisphere affecting humans. *Frontiers in microbiology*, 16, 1632832. https://doi.org/10.3389/fmicb.2025.1632832.

Bratuleanu, B. E., Chretien, D., Bigot, T., Regnault, B., Pérot, P., Savuta, G., Eloit, M., & Temmam, S. (2024). Insights into the virome of Hyalomma marginatum in the Danube Delta: A major vector of Crimean-Congo hemorrhagic fever virus in Eastern Europe. *Parasites & vectors*, 17(1), 482. https://doi.org/10.1186/s13071-024-06557-4.

Breeuwer, H., Ros, V. I., & Groot, T.V. (2011). Cardinium the next addition to the family of reproductive parasite. E. Zchori-Fein, K. Bourtzis (Eds.), *Manipulative tenants: bacteria associated with arthropods*, CRS Press, Taylor & Francis Group, FL, USA, pp. 07–224.

Brianti, E., Otranto, D., Dantas-Torres, F., Weigl, S., Latrofa, M. S., Gaglio, G., Napoli, E., Brucato, G., Cauquil, L., Giannetto, S., & Bain, O. (2012).

Rhipicephalus sanguineus (Ixodida, Ixodidae) as intermediate host of a canine neglected filarial species with dermal microfilariae. *Veterinary parasitology*, 183(3-4), 330–337. https://doi.org/10.1016/j.vetpar.2011.07.031.

Briggs, C., Smith, E., Vansandt, J., Carr, B., Loftin, K., & McDermott, E. G. (2023). Prevalence of Ehrlichia and Anaplasma Spp. in ticks collected from cattle and deer in Arkansas, USA. *The Journal of parasitology*, 109(5), 530–536. https://doi.org/10.1645/23-57.

Brinkerhoff, R. J., Clark, C., Ocasio, K., Gauthier, D. T., & Hynes, W. L. (2020). Factors affecting the microbiome of Ixodes scapularis and Amblyomma americanum. *PloS one*, 15(5), e0232398. https://doi.org/10.1371/journal.pone.0232398.

Bruley, M., & Duron, O. (2024). Multi-locus sequence analysis unveils a novel genus of filarial nematodes associated with ticks in French Guiana. Une analyse de séquences multi-locus dévoile un nouveau genre de nématodes filaires, associé aux tiques en Guyane française. *Parasite (Paris, France)*, 31, 14. https://doi.org/10.1051/parasite/2024015.

Bruley, M., Pasternicki, C., Fattar, N., Amoros, J., Duhayon, M., McCoy, K., & Duron, O. (2025). Culturable bacteria and fungi in Ixodes, Dermacentor, Amblyomma and Ornithodoros ticks. Bactéries et champignons cultivables chez les tiques Ixodes, Dermacentor, Amblyomma et Ornithodoros. *Parasite (Paris, France)*, 32, 20. https://doi.org/10.1051/parasite/2025013.

Budachetri, K., Kumar, D., Crispell, G., Beck, C., Dasch, G., & Karim, S. (2018). The tick endosymbiont Candidatus Midichloria mitochondrii and selenoproteins are essential for the growth of Rickettsia parkeri in the Gulf Coast tick vector. *Microbiome*, 6(1), 141. https://doi.org/10.1186/s40168-018-0524-2.

Bugmyrin, S. V., Romanova, L. Y., Belova, O. A., Kholodilov, I. S., Bespyatova, L. A., Chernokhaeva, L. L., Gmyl, L. V., Klimentov, A. S., Ivannikova, A. Y., Polienko, A. E., Yakovlev, A. S., Ieshko, E. P., Gmyl, A. P., & Karganova, G. G. (2022). Pathogens in Ixodes persulcatus and Ixodes ricinus ticks (Acari, Ixodidae) in Karelia (Russia). *Ticks and tick-borne diseases*, 13(6), 102045. https://doi.org/10.1016/j.ttbdis.2022.102045.

Burthe, S. J., Kumbar, B., Schäfer, S. M., Purse, B. V., Vanak, A. T., Balakrishnan, N., Hassall, R., Hoti, S. L., Narayanaswamy, D., Potadar, S., Rahman, M., & Chanda, M. M. (2025). First evidence of transovarial transmission of Kyasanur Forest disease virus in Haemaphysalis and Rhipicephalus ticks in the wild. *Parasites & vectors*, 18(1), 14. https://doi.org/10.1186/s13071-024-06643-5.

Buysse, M., Binetruy, F., Leibson, R., Gottlieb, Y., & Duron, O. (2022). Ecological contacts and host specificity promote replacement of nutritional endosymbionts in ticks. *Microbial ecology*, 83(3), 776–788. https://doi.org/10.1007/s00248-021-01773-0.

Buysse, M., & Duron, O. (2021). Evidence that microbes identified as tick-borne pathogens are nutritional endosymbionts. *Cell*, 184(9), 2259–2260. https://doi.org/10.1016/j.cell.2021.03.053.

Buysse, M., Floriano, A. M., Gottlieb, Y., Nardi, T., Comandatore, F., Olivieri, E., Giannetto, A., Palomar, A. M., Makepeace, B. L., Bazzocchi, C., Cafiso, A., Sassera, D., & Duron, O. (2021). A dual endosymbiosis supports nutritional adaptation to hematophagy in the invasive tick Hyalomma marginatum. *eLife*, 10, e72747. https://doi.org/10.7554/eLife.72747.

Buysse, M., Plantard, O., McCoy, K. D., Duron, O., & Menard, C. (2019). Tissue localization of Coxiella-like endosymbionts in three European tick species through fluorescence in situ hybridization. *Ticks and tick-borne diseases*, 10(4), 798–804. https://doi.org/10.1016/j.ttbdis.2019.03.014.

Byaruhanga, C., Akure, P. C., Lubembe, D. M., Sibeko-Matjila, K., Troskie, M., Oosthuizen, M. C., & Stoltsz, H. (2021). Molecular detection and characterisation of protozoan and rickettsial pathogens in ticks from cattle in the pastoral area of Karamoja, Uganda. *Ticks and tick-borne diseases*, 12(4), 101709. https://doi.org/10.1016/j.ttbdis.2021.101711.

Cabezas-Cruz, A., Alberdi, P., Valdés, J. J., Villar, M., & de la Fuente, J. (2017). Anaplasma phagocytophilum Infection Subverts Carbohydrate Metabolic Pathways in the Tick Vector, Ixodes scapularis. *Frontiers in cellular and infection microbiology*, 7, 23. https://doi.org/10.3389/fcimb.2017.00023.

Cai, X., Cai, X., Xu, Y., Shao, Y., Fu, L., Men, X., & Zhu, Y. (2023). Virome analysis of ticks and tick-borne viruses in Heilongjiang and Jilin Provinces, China. *Virus research*, 323, 199006. https://doi.org/10.1016/j.virusres.2022.199006.

Camp, J. V., Kannan, D. O., Osman, B. M., Shah, M. S., Howarth, B., Khafaga, T., Weidinger, P., Karuvantevida, N., Kolodziejek, J., Mazrooei, H., Wolf, N., Loney, T., & Nowotny, N. (2020). Crimean-congo hemorrhagic fever virus endemicity in United Arab Emirates, 2019. *Emerging infectious diseases*, 26(5), 1019–1021. https://doi.org/10.3201/eid2605.191414.

Cao, X. Q., Gu, X. L., Zhang, L., Xu, J., Han, H. J., & Yu, X. J. (2023). Molecular detection of Rickettsia, Anaplasma, and Bartonella in ticks from free-ranging sheep in Gansu Province, China. *Ticks and tick-borne diseases*, 14(3), 102137. https://doi.org/10.1016/j.ttbdis.2023.102140.

Capligina, V., Seleznova, M., Akopjana, S., Freimane, L., Lazovska, M., Krumins, R., Kivrane, A., Namina, A., Aleinikova, D., Kimsis, J., Kazarina, A., Igumnova, V., Bormane, A., & Ranka, R. (2020). Large-scale countrywide screening for tick-borne pathogens in field-collected ticks in Latvia during 2017-2019. *Parasites & vectors*, 13(1), 351. https://doi.org/10.1186/s13071-020-04219-11.

Carpi, G., Cagnacci, F., Wittekindt, N. E., Zhao, F., Qi, J., Tomsho, L. P., Drautz, D. I., Rizzoli, A., & Schuster, S. C. (2011). Metagenomic profile of the bacterial communities associated with Ixodes ricinus ticks. *PloS one*, 6(10), e25604. https://doi.org/10.1371/journal.pone.0025604.

Castillo-Contreras, R., Magen, L., Birtles, R., Varela-Castro, L., Hall, J. L., Conejero, C., Aguilar, X. F., Colom-Cadena, A., Lavín, S., Mentaberre, G., & López-Olvera, J. R. (2022). Ticks on wild boar in the metropolitan area of Barcelona (Spain) are infected with spotted fever group rickettsiae. *Transboundary and emerging diseases*, 69(4), e82–e95. https://doi.org/10.1111/tbed.14271.

Celina, S. S., & Cerný, J. (2022). Coxiella burnetii in ticks, livestock, pets and wildlife: A mini-review. *Frontiers in veterinary science*, 9, 1068129. https://doi.org/10.3389/fvets.2022.1068129.

Centeno-Lima, S., do Rosário, V., Parreira, R., Maia, A. J., Freudenthal, A. M., Nijhof, A. M., & Jongejan, F. (2003). A fatal case of human babesiosis in Portugal: Molecular and phylogenetic analysis. *Tropical medicine & international health: TM & IH*, 8(8), 760–764. https://doi.org/10.1046/j.1365-3156.2003.01074.x.

Ceraul, S. M., Sonenshine, D. E., & Hynes, W. L. (2002). Resistance of the tick dermacentor variabilis (Acari: Ixodidae) following challenge with the bacterium Escherichia coli (Enterobacteriales: Enterobacteriaceae). *Journal of medical entomology*, 39(2), 376–383. https://doi.org/10.1603/0022-2585-39.2.376.

Chambaro, H. M., Sasaki, M., Sinkala, Y., Gonzalez, G., Squarre, D., Fandamu, P., Lubaba, C., Mataa, L., Shawa, M., Mwape, K. E., Gabriël, S., Chembensofu, M., Carr, M. J., Hall, W. W., Qiu, Y., Kajihara, M., Takada, A., Orba, Y., Simulundu, E., & Sawa, H. (2020). Evidence for exposure of asymptomatic domestic pigs to African swine fever virus during an inter-epidemic period in Zambia. *Transboundary and emerging diseases*, 67(6), 2741–2752. https://doi.org/10.1111/tbed.13630.

Champour, M., Chinikar, S., Mohammadi, G., Razmi, G., Mostafavi, E., Shah-Hosseini, N., Khakifirouz, S., & Jalali, T. (2016b). Crimean-Congo hemorrhagic fever in the one-humped Camel (Camelus dromedarius) in East and Northeast of Iran. *Journal of arthropod-borne diseases*, 10(2), 168–177.

Champour, M., Chinikar, S., Mohammadi, G., Razmi, G., Shah-Hosseini, N., Khakifirouz, S., Mostafavi, E., & Jalali, T. (2016a). Molecular epidemiology of Crimean-Congo hemorrhagic fever virus detected from ticks of one humped camels (Camelus dromedarius) population in northeastern Iran. *Journal of parasitic diseases: official organ of the Indian Society for Parasitology*, 40(1), 110–115. https://doi.org/10.1007/s12639-014-0458-y.

Chao, L. L., Castillo, C. T., & Shih, C. M. (2021). Molecular detection and genetic identification of Wolbachia endosymbiont in Rhipicephalus sanguineus (Acari: Ixodidae) ticks of Taiwan. *Experimental & applied acarology*, 83(1), 115–130. https://doi.org/10.1007/s10493-020- 00574-3.

Chao, L. L., Ko, P. Y., & Shih, C. M. (2024). Molecular screening and genetic identification of Anaplasma platys in brown dog tick (Rhipicephalus sanguineus s. l.) infested on stray dogs in Taiwan. *Microorganisms*, 12(9), 1779. https://doi.org/10.3390/microorganisms12091779.

Chao, L. L., Liao, H. T., & Shih, C. M. (2019). First detection and genetic identification of Hepatozoon canis in Rhipicephalus sanguineus sensu lato ticks collected from dogs of Taiwan. *Ticks and tick-borne diseases*, 10(4), 929–934. https://doi.org/10.1016/j.ttbdis.2019.04.020.

Chaorattanakawee, S., Tachavarong, W., Hananantachai, H., Bunsermyos, W., Chanarat, N., Promsathaporn, S., Tippayachai, B., Sakolvaree, J., Pitaksajjakul, P., Benjathummarak, S., Srinoppawan, K., Saunders, D., Lindroth, E. J., & Takhampunya, R. (2024). Seasonal pattern of questing ticks and prevalence of pathogenic Rickettsia and Anaplasmataceae in Khao Yai national park, Thailand. *Travel medicine and infectious disease*, 58, 102696. https://doi.org/10.1016/j.tmaid.2024.102696.

Charles, R. A., Bermúdez, S., Banović, P., Alvarez, D. O., Díaz-Sánchez, A. A., Corona-González, B., Etter, E. M. C., Rodríguez González, I., Ghafar, A., Jabbar, A., Moutailler, S., & Cabezas-Cruz, A. (2021). Ticks and tick-borne diseases in Central America and the Caribbean: A one health perspective. *Pathogens (Basel, Switzerland)*, 10(10), 1273. https://doi.org/10.3390/pathogens10101273.

Charrel, R. N., Fagbo, S., Moureau, G., Alqahtani, M. H., Temmam, S., & de Lamballerie, X. (2007). Alkhurma hemorrhagic fever virus in Ornithodoros savignyi ticks. *Emerging infectious diseases*, 13(1), 153–155. https://doi.org/10.3201/eid1301.061094.

Chauhan, G., McClure, J., Hekman, J., Marsh, P. W., Bailey, J. A., Daniels, R. F., Genereux, D. P., & Karlsson, E. K. (2020). Combining citizen science and genomics to investigate tick, pathogen, and commensal microbiome at single-tick resolution. *Frontiers in genetics*, 10, 1322. https://doi.org/10.3389/fgene.2019.01322.

Che Lah, E. F., Ahamad, M., Dmitry, A., Md-Zain, B. M., & Yaakop, S. (2023). Metagenomic profile of the bacterial communities associated with Ixodes granulatus (Acari: Ixodidae): A potential vector of tick-borne diseases. *Journal of medical entomology*, 60(4), 753–768. https://doi.org/10.1093/jme/tjad044.

Chen, H., Zhang, M., & Hochstrasser, M. (2020). The biochemistry of cytoplasmic incompatibility caused by endosymbiotic bacteria. *Genes*, 11(8), 852. https://doi.org/10.3390/genes11080852.

Chisu, V., Dei Giudici, S., Foxi, C., Chessa, G., Peralta, F., Sini, V., & Masala, G. (2023). Anaplasma species in ticks infesting mammals of Sardinia, Italy. Animals : an open access journal from MDPI, 13(8), 1332. https://doi.org/10.3390/ani13081332.

Chisu, V., Foxi, C., Masu, G., D'Amaddio, B., & Masala, G. (2020). Detection of potentially pathogenic bacteria from Ixodes ricinus carried by pets in Tuscany, Italy. *Veterinary record open*, 7(1), e000395. https://doi.org/10.1136/vetreco-2020-000395.

Chitanga, S., Chambaro, H. M., Moonga, L. C., Hayashida, K., Yamagishi, J., Muleya, W., Changula, K., Mubemba, B., Simbotwe, M., Squarre, D., Fandamu, P., Nalubamba, K. S., Qiu, Y., Hirofumi, S., & Simulundu, E. (2021). Rickettsia lusitaniae in Ornithodoros Porcinus Ticks, Zambia. *Pathogens (Basel, Switzerland)*, 10(10), 1306. https://doi.org/10.3390/pathogens10101306.

Chitimia-Dobler, L., Issa, M. H., Ezalden, M. E., Yagoub, I. A., Abdalla, M. A., Bakhiet, A. O., Schaper, S., Rieß, R., Vollmar, P., Grumbach, A., Bestehorn, M., Antwerpen, M., Dobler, G., & Shuaib, Y. A. (2019). Crimean-Congo haemorrhagic fever virus in Hyalomma impeltatum ticks from North Kordofan, the Sudan. *International journal of infectious diseases: IJID: official publication of the International Society for Infectious Diseases*, 89, 81–83. https://doi.org/10.1016/j.ijid.2019.09.012.

Choi, J., Palanichamy, P., Tanaka, H., Kondo, T., Gruwell, M. E., & Husnik, F. (2025). Accelerated Pseudogenization in the ancient endosymbionts of giant scale insects. *Molecular biology and evolution*, 42(6), msaf125. https://doi.org/10.1093/molbev/msaf125.

Cialini, C., Cafiso, A., Waldeck, M., Lundgren, Å., Fält, J., Settergren, B., Choklikitumnuey, P., Chiappa, G., Rosso, E., Roveri, L., Fesce, E., Ferrari, N., Lindgren, P. E., Bazzocchi, C., & Grandi, G. (2025). Prevalence of tick-borne pathogens in feeding and questing Ixodes ricinus ticks from Southern Sweden. *Ticks and tick-borne diseases*, 16(2), 102453. https://doi.org/10.1016/j.ttbdis.2025.102453.

Ciarma, M., Melis, S., Bisaglia, B., Batisti Biffignandi, G., Sunny, G. S., Vumbaca, M., Epis, S., Arshad, M., Ahmed, H., Sassera, D., & Castelli, M. (2025). Survey of tick-borne pathogens in Rhipicephalus microplus ticks from livestock in the Northern Punjab Province, Pakistan. *Current research in parasitology & vector-borne diseases*, 7, 100275. https://doi.org/10.1016/j.crpvbd.2025.100276.

Cibichakravarthy, B., Oses-Prieto, J. A., Ben-Yosef, M., Burlingame, A. L., Karr, T. L., & Gottlieb, Y. (2022). Comparative proteomics of Coxiella like endosymbionts (CLEs) in the symbiotic organs of Rhipicephalus sanguineus ticks. *Microbiology spectrum*, 10(1), e0167321. https://doi.org/10.1128/spectrum.01673-21.

Cicculli, V., DeCarreaux, D., Ayhan, N., Casabianca, F., de Lamballerie, X., Charrel, R., & Falchi, A. (2020). Molecular screening of Anaplasmataceae in ticks collected from cattle in Corsica, France. *Experimental & applied acarology*, 81(4), 561–574. https://doi.org/10.1007/s10493-020-00529-w.

Clay, K., Klyachko, O., Grindle, N., Civitello, D., Oleske, D., & Fuqua, C. (2008). Microbial communities and interactions in the lone star tick, Amblyomma americanum. *Molecular ecology*, 17(19), 4371–4381. https://doi.org/10.1111/j.1365-294x.2008.03914.x.

Clow, K. M., Leighton, P. A., Ogden, N. H., Lindsay, L. R., Michel, P., Pearl, D. L., & Jardine, C. M. (2017). Northward range expansion of Ixodes scapularis evident over a short timescale in Ontario, Canada. *PloS one*, 12(12), e0189393. https://doi.org/10.1371/journal.pone.0189393.

Coburn, J., Maier, T., Casey, M., Padmore, L., Sato, H., & Frank, D. W. (2015). Reproducible and quantitative model of infection of Dermacentor variabilis with the live vaccine strain of Francisella tularensis. *Applied and environmental microbiology*, 81(1), 386–395. https://doi.org/10.1128/AEM.02917-14.

Comandatore, F., Radaelli, G., Montante, S., Sacchi, L., Clementi, E., Epis, S., Cafiso, A., Serra, V., Pajoro, M., Di Carlo, D., Floriano, A. M., Stavru, F., Bandi, C., & Sassera, D. (2021). Modeling the life cycle of the intramitochondrial bacterium "candidatus Midichloria mitochondrii" using electron microscopy data. *mBio*, 12(3), e0057421. https://doi.org/10.1128/mBio.00574-21.

Craig, A. F., Schade-Weskott, M. L., Rametse, T., Heath, L., Kriel, G. J. P., de Klerk-Lorist, L. M., van Schalkwyk, L., Trujillo, J. D., Crafford, J. E., Richt, J. A., & Swanepoel, R. (2022). Detection of African swine fever virus in ornithodoros tick species associated with indigenous and extralimital warthog populations in South Africa. *Viruses*, 14(8), 1617. https://doi.org/10.3390/v14081617.

da Silva, N. X., Dias, T. S., Vignoli, J. A., Dos Santos Machado, L., Telleria, E. L., de Almeida Pereira, V. L., do Nascimento, E. R., & da Cunha, N. C. (2022). First molecular detection of Spiroplasma spp. in ticks from horses in Brazil. *Ticks and tick-borne diseases*, 13(2), 101896. https://doi.org/10.1016/j.ttbdis.2022.101896.

Damian, D., Maghembe, R., Damas, M., Wensman, J. J., & Berg, M. (2020). Application of viral metagenomics for study of emerging and reemerging tick-borne viruses. *Vector borne and zoonotic diseases (Larchmont, N.Y.)*, 20(8), 557–565. https://doi.org/10.1089/vbz.2019.2579.

Degabriel, M., Valeva, S., Boisset, S., & Henry, T. (2023). Pathogenicity and virulence of Francisella tularensis. *Virulence*, 14(1), 2274638. https://doi.org/10.1080/21505594.2023.2274638.

Dennis, D. T., Inglesby, T. V., Henderson, D. A., Bartlett, J. G., Ascher, M. S., Eitzen, E., Fine, A. D., Friedlander, A. M., Hauer, J., Layton, M., Lillibridge, S. R., McDade, J. E., Osterholm, M. T., O'Toole, T., Parker, G., Perl, T. M., Russell, P. K., Tonat, K., & Working Group on Civilian Biodefense. (2001). Tularemia as a biological weapon: Medical and public health management. *JAMA*, 285(21), 2763–2773. https://doi.org/10.1001/jama.285.21.2763.

Dergousoff, S. J., Anstead, C. A., & Chilton, N. B. (2020). Identification of bacteria in the Rocky Mountain wood tick, Dermacentor andersoni, using single-strand conformation polymorphism (SSCP) and DNA sequencing. *Experimental & applied acarology*, 80(2), 247–256. https://doi.org/10.1007/s10493-019-00459-0.

Dergousoff, S. J., & Chilton, N. B. (2010). Detection of a new Arsenophonus-type bacterium in Canadian populations of the Rocky Mountain wood tick, Dermacentor andersoni. *Experimental & applied acarology*, 52(1), 85–91. https://doi.org/10.1007/s10493-010-9340-5.

Dergousoff, S. J., & Chilton, N. B. (2012). Association of different genetic types of Francisella-like organisms with the rocky mountain wood tick (Dermacentor andersoni) and the American dog tick (Dermacentor variabilis) in localities near their northern distributional limits. *Applied and environmental microbiology*, 78(4), 965–971. https://doi.org/10.1128/AEM.05762-11.

Di Lecce, I., Bazzocchi, C., Cecere, J. G., Epis, S., Sassera, D., Villani, B. M., Bazzi, G., Negri, A., Saino, N., Spina, F., Bandi, C., & Rubolini, D. (2018). Patterns of Midichloria infection in avian-borne African ticks and their trans-Saharan migratory hosts. *Parasites & vectors*, 11(1), 106. https://doi.org/10.1186/s13071-018-2669-z.

Dinçer, E., Brinkmann, A., Hekimoğlu, O., Hacıoğlu, S., Földes, K., Karapınar, Z., Polat, P. F., Oğuz, B., Orunç Kılınç, Ö., Hagedorn, P., Özer, N., Özkul, A., Nitsche, A., & Ergünay, K. (2017, Jul 14). Generic amplification and next generation sequencing reveal Crimean-Congo hemorrhagic fever virus AP92-like strain and distinct tick phleboviruses in Anatolia, Turkey. *Parasit Vectors*, 10(1), 335.

Dinçer, E., Hacıoğlu, S., Kar, S., Emanet, N., Brinkmann, A., Nitsche, A., et al. (2019). Survey and characterization of Jingmen tick virus variants. *Viruses*, 11, 1071.

Dinçer, E., Timurkan, M. Ö., Oğuz, B., Şahindokuyucu, İ., Şahan, A., Ekinci, M., Polat, C., & Ergünay, K. (2022). Several tick-borne pathogenic viruses in circulation in Anatolia, Turkey. *Vector Borne Zoonotic Diseases*, 22(2), 148-158.

Douglas, A. E. (2017). The B vitamin nutrition of insects: The contributions of diet, microbiome and horizontally acquired genes. *Current opinion in insect science*, 23, 65–69. https://doi.org/10.1016/j.cois.2017.07.012.

Driscoll, T. P., Verhoeve, V. I., Guillotte, M. L., Lehman, S. S., Rennoll, S. A., Beier-Sexton, M., Rahman, M. S., Azad, A. F., & Gillespie, J. J. (2017). Wholly Rickettsia! reconstructed metabolic profile of the quintessential bacterial parasite of eukaryotic cells. *mBio*, 8(5), e00859–17. https://doi.org/10.1128/mBio.00859-17.

Du, L. F., Zhang, M. Z., Yuan, T. T., Ni, X. B., Wei, W., Cui, X. M., Wang, N., Xiong, T., Zhang, J., Pan, Y. S., Zhu, D. Y., Li, L. J., Xia, L. Y., Wang, T. H., Wei, R., Liu, H. B., Sun, Y., Zhao, L., Lam, T. T., Cao, W. C., … & Jia, N. (2023). New insights into the impact of microbiome on horizontal and vertical transmission of a tick-borne pathogen. *Microbiome*, 11(1), 50. https://doi.org/10.1186/s40168-023-01485-2.

Duan, D. Y., Liu, G. H., & Cheng, T. Y. (2020). Microbiome analysis of the saliva and midgut from partially or fully engorged female adult Dermacentor silvarum ticks in China. *Experimental & applied acarology*, 80(4), 543–558. https://doi.org/10.1007/s10493-020-00478-2.

Duan, D. Y., Liu, Y. K., Liu, L., Liu, G. H., & Cheng, T. Y. (2022). Microbiome analysis of the midguts of different developmental stages of Argas persicus in China. *Ticks and tick-borne diseases*, 13(1), 101868. https://doi.org/10.1016/j.ttbdis.2021.101868.

Duan, L., Zhang, L., Hou, X., Bao, Z., Zeng, Y., He, L., Liu, Z., Zhou, H., Hao, Q., & Dong, A. (2024). Surveillance of tick-borne bacteria infection in ticks and forestry populations in Inner Mongolia, China. *Frontiers in public health*, 12, 1302133. https://doi.org/10.3389/fpubh.2024.1302135.

Dunaj, J., Drewnowska, J., Moniuszko-Malinowska, A., Swiecicka, I., & Pancewicz, S. (2021b). First metagenomic report of Borrelia americana and Borrelia carolinensis in Poland – a preliminary study. *Annals of agricultural and environmental medicine: AAEM*, 28(1), 49–55. https://doi.org/10.26444/aaem/118147.

Dunaj, J., Trzeszczkowski, A., Moniuszko-Malinowska, A., Rutkowski, K., & Pancewicz, S. (2021a). Assessment of tick-borne pathogens presence in Dermacentor reticulatus ticks in north-eastern Poland. *Advances in medical sciences*, 66(1), 113–118. https://doi.org/10.1016/j.advms.2021.01.004.

Duncan, K. T., Sundstrom, K. D., Hunt, D., Lineberry, M. W., Grant, A., & Little, S. E. (2022). Survey on the presence of equine tick-borne rickettsial infections in Southcentral United States. *Journal of equine veterinary science*, 118, 104135. https://doi.org/10.1016/j.jevs.2022.104135.

Duplan, F., Davies, S., Filler, S., Abdullah, S., Keyte, S., Newbury, H., Helps, C. R., Wall, R., & Tasker, S. (2018). Anaplasma phagocytophilum, Bartonella spp., haemoplasma species and Hepatozoon spp. in ticks infesting cats: A large-scale survey. *Parasites & vectors*, 11(1), 201. https://doi.org/10.1186/s13071-018-2789-5.

Duron, O. (2024). Nutritional symbiosis in ticks: Singularities of the genus Ixodes. *Trends in parasitology*, 40(8), 696–706. https://doi.org/10.1016/j.pt.2024.06.006.

Duron, O., & Gottlieb, Y. (2020). Convergence of nutritional symbioses in obligate blood feeders. *Trends in parasitology*, 36(10), 816–825. https://doi.org/10.1016/j.pt.2020.07.007.

Duron, O., Binetruy, F., Noël, V., Cremaschi, J., McCoy, K. D., Arnathau, C., Plantard, O., Goolsby, J., Pérez de León, A. A., Heylen, D. J. A., Van Oosten, A. R., Gottlieb, Y., Baneth, G., Guglielmone, A. A., Estrada-Peña, A., Opara, M. N., Zenner, L., Vavre, F., & Chevillon, C. (2017). Evolutionary changes in symbiont community structure in ticks. *Molecular ecology*, 26(11), 2905–2921. https://doi.org/10.1111/mec.14094.

Duron, O., Cremaschi, J., & McCoy, K. D. (2016). The high diversity and global distribution of the intracellular bacterium rickettsiella in the polar seabird tick Ixodes uriae. *Microbial ecology*, 71(3), 761–770. https://doi.org/10.1007/s00248-015-0702-8.

Duron, O., Morel, O., Noël, V., Buysse, M., Binetruy, F., Lancelot, R., Loire, E., Ménard, C., Bouchez, O., Vavre, F., & Vial, L. (2018). Tick-bacteria mutualism depends on B vitamin synthesis pathways. *Current biology: CB*, 28(12), 1896–1902.e5. https://doi.org/10.1016/j.cub.2018.04.038.

Duron, O., Noël, V., McCoy, K. D., Bonazzi, M., Sidi-Boumedine, K., Morel, O., Vavre, F., Zenner, L., Jourdain, E., Durand, P., Arnathau, C., Renaud, F., Trape, J. F., Biguezoton, A. S., Cremaschi, J., Dietrich, M., Léger, E., Appelgren, A., Dupraz, M., Gómez-Díaz, E., ... & Chevillon, C. (2015). The recent evolution of a maternally-inherited endosymbiont of ticks led to the emergence of the Q fever pathogen, Coxiella burnetii. *PLoS pathogens*, 11(5), e1004892. https://doi.org/10.1371/journal.ppat.1004892.

Dyczko, D., Krysmann, A., Kolanek, A., Borczyk, B., & Kiewra, D. (2024). Bacterial pathogens in Ixodes ricinus collected from lizards Lacerta agilis and Zootoca vivipara in urban areas of Wrocław, SW Poland- preliminary study. *Experimental & applied acarology*, 93(2), 409–420. https://doi.org/10.1007/s10493-024-00927-5.

Dyczko, D., Szymański, D. M., Szymański, D., Kupczak, M., & Kolenda, K. (2025). First European record of Rickettsia bellii in Amblyomma rotundatum from Rhinella marina imported to Poland. *Experimental & applied acarology*, 94(3), 43. https://doi.org/10.1007/s10493-025-01007-9.

Eberhardt, A. T., Manzoli, D. E., Fernandez, C., Zurvera, D., & Monje, L. D. (2023). Anaplasma species infecting questing ticks in the Iberá wetlands eco-region, Argentina. *Experimental & applied acarology*, 89(3-4), 485–496. https://doi.org/10.1007/s10493-023-00788-1.

Ebert, C. L., Söder, L., Kubinski, M., Glanz, J., Gregersen, E., Dümmer, K., Grund, D., Wöhler, A. S., Könenkamp, L., Liebig, K., Knoll, S., Hellhammer, F., Topp, A. K., Becher, P., Springer, A., Strube, C., Nagel-Kohl, U., Nordhoff, M., Steffen, I., Bauer, B. U., ... & Boelke, M. (2023). Detection and characterization of alongshan virus in ticks and tick saliva from lower Saxony, Germany with serological evidence for viral transmission to game and domestic animals. *Microorganisms*, 11(3), 543. https://doi.org/10.3390/microorganisms11030543.

Eggenberger, L. R., Lamoreaux, W. J., & Coons, L. B. (1990). Hemocytic encapsulation of implants in the tick Dermacentor variabilis. *Experimental & applied acarology*, 9(3-4), 279–287. https://doi.org/10.1007/BF01193434.

Eisen, R. J., Kugeler, K. J., Eisen, L., Beard, C. B., & Paddock, C. D. (2017). Tick-borne zoonoses in the United States: Persistent and emerging threats to human health. *ILAR journal*, 58(3), 319–335. https://doi.org/10.1093/ilar/ilx005.

Ellis, J., Oyston, P. C., Green, M., & Titball, R. W. (2002). Tularemia. *Clinical microbiology reviews*, 15(4), 631–646. https://doi.org/10.1128/CMR.15.4.631-646.2002.

Elsawy, B. S. M., Abdel-Ghany, H. S. M., Alzan, H. F., Abdel-Shafy, S., & Shahein, Y. E. (2024). Molecular screening of piroplasms and Anaplasmataceae agents in Hyalomma dromedarii ticks from camels over different seasons in Egypt. *Experimental & applied acarology*, 93(4), 849–868. https://doi.org/10.1007/s10493-024-00958-w.

El-Sayed, A., & Kamel, M. (2020). Climatic changes and their role in emergence and re-emergence of diseases. *Environmental science and pollution research international*, 27(18), 22336–22352. https://doi.org/10.1007/s11356-020-08896-w.

Enríquez, S., Félix, M. L., Bermúdez, S., Mosquera, V., Farovitch, L., Vaca, F., Guevara, J. M., & Venzal, J. M. (2024). Molecular detection of rickettsial agents in Amblyomma maculatum ticks (Ixodida: Ixodidae) from Ecuador. *Veterinary research communications*, 48(4), 2869–2876. https://doi.org/10.1007/s11259-024-10410-3.

Epis, S., Sassera, D., Beninati, T., Lo, N., Beati, L., Piesman, J., Rinaldi, L., McCoy, K. D., Torina, A., Sacchi, L., Clementi, E., Genchi, M., Magnino, S., & Bandi, C. (2008). Midichloria mitochondrii is widespread in hard ticks (Ixodidae) and resides in the mitochondria of phylogenetically diverse species. *Parasitology*, 135(4), 485–494. https://doi.org/10.1017/S0031182007004052.

Eremeeva, M. E., Oliveira, A., Robinson, J. B., Ribakova, N., Tokarevich, N. K., & Dasch, G. A. (2006). Prevalence of bacterial agents in Ixodes persulcatus ticks from the Vologda Province of Russia. *Annals of the New York Academy of Sciences*, 1078, 291–298. https://doi.org/10.1196/annals.1374.054.

España, P. P., Uranga, A., Cillóniz, C., & Torres, A. (2020). Q fever (Coxiella Burnetii). *Seminars in respiratory and critical care medicine*, 41(4), 509–521. https://doi.org/10.1055/s-0040-1710594.

Esteves, E., Maruyama, S. R., Kawahara, R., Fujita, A., Martins, L. A., Righi, A. A., Costa, F. B., Palmisano, G., Labruna, M. B., Sá-Nunes, A., Ribeiro, J. M. C., & Fogaça, A. C. (2017). Analysis of the salivary gland transcriptome of unfed and partially fed Amblyomma sculptum ticks and descriptive proteome of the saliva. *Frontiers in cellular and infection microbiology*, 7, 476. https://doi.org/10.3389/fcimb.2017.00476.

Estrada-Peña, A., & de la Fuente, J. (2018). The fossil record and the origin of ticks revisited. *Experimental & applied acarology*, 75(2), 255–261. https://doi.org/10.1007/s10493-018-0261-z.

Estrada-Peña, A., Cabezas-Cruz, A., & Obregón, D. (2020a). Resistance of tick gut microbiome to anti-tick vaccines, pathogen infection and antimicrobial peptides. *Pathogens (Basel, Switzerland)*, 9(4), 309. https://doi.org/10.3390/pathogens9040309.

Estrada-Peña, A., Cabezas-Cruz, A., & Obregón, D. (2020b). Behind taxonomic variability: The functional redundancy in the tick microbiome. *Microorganisms*, 8(11), 1829. https://doi.org/10.3390/microorganisms8111829.

Estrada-Peña, A., Roura, X., Sainz, A., Miró, G., & Solano-Gallego, L. (2017). Species of ticks and carried pathogens in owned dogs in Spain: Results of a one-year national survey. *Ticks and tick-borne diseases*, 8(4), 443–452. https://doi.org/10.1016/j.ttbdis.2017.02.001.

Fang, L. Z., Lei, S. C., Yan, Z. J., Xiao, X., Liu, J. W., Gong, X. Q., Yu, H., & Yu, X. J. (2021). Detection of multiple intracellular bacterial pathogens in Haemaphysalis flava ticks collected from Hedgehogs in Central China. *Pathogens (Basel, Switzerland)*, 10(2), 115. https://doi.org/10.3390/pathogens10020115.

Fang, L. Z., Xiao, X., Lei, S. C., Liu, J. W., &Yu, X. J. (2023). Haemaphysalis flava ticks as a competent vector of severe fever with thrombocytopenia syndrome virus. *Ticks and tick-borne diseases*, 14(2), 102100. https://doi.org/10.1016/j.ttbdis.2022.102100.

Fang, R., Li, S., Lyu, Y., Yang, X., Wang, T., & Li, S. (2024, April 20). Behavioral and cognitive factors influencing tick-borne disease risk in northeast China: Implications for prevention and control strategies. *One Health*, 18, 100736. doi: 10.1016/j.onehlt.2024.100736.

Farhadpour, F., Telmadarraiy, Z., Chinikar, S., Akbarzadeh, K., Moemenbellah-Fard, M. D., Faghihi, F., Fakoorziba, M. R., Jalali, T., Mostafavi, E., Shahhosseini, N., & Mohammadian, M. (2016). Molecular detection of Crimean-Congo haemorrhagic fever virus in ticks collected from infested livestock populations in a New Endemic Area, South of Iran. *Tropical medicine & international health: TM & IH*, 21(3), 340–347. https://doi.org/10.1111/tmi.12667.

Fattar, N., Louni, M., Buysse, M., Floriano, A. M., Bertaux, J., Cantereau, A., Rivero, A., Bruley, M., McCoy, K. D., Delafont, V., Boulanger, N., Vavre, F., Bouchon, D., & Duron, O. (2025). Evolutionary convergence of nutritional symbionts in ticks. *Environmental microbiology reports*, 17(3), e70120. https://doi.org/10.1111/1758-2229.70120.

Fernandes, E. K., & Bittencourt, V. R. (2008). Entomopathogenic fungi against South American tick species. *Experimental & applied acarology*, 46(1-4), 71–93. https://doi.org/10.1007/s10493-008-9161-y.

Fernandes, É. K., Bittencourt, V. R., & Roberts, D. W. (2012). Perspectives on the potential of entomopathogenic fungi in biological control of ticks. *Experimental parasitology*, 130(3), 300–305. https://doi.org/10.1016/j.exppara.2011.11.004.

Fischer, T., Myalkhaa, M., Krücken, J., Battsetseg, G., Batsukh, Z., Baumann, M. P. O., Clausen, P. H., & Nijhof, A. M. (2020). Molecular detection of tick-borne pathogens in bovine blood and ticks from Khentii, Mongolia. *Transboundary and emerging diseases*, 67 Suppl 2, 111–118. https://doi.org/10.1111/tbed.13322.

Floriano, A. M., Batisti Biffignandi, G., Castelli, M., Olivieri, E., Clementi, E., Comandatore, F., Rinaldi, L., Opara, M., Plantard, O., Palomar, A. M., Noël, V., Vijay, A., Lo, N., Makepeace, B. L., Duron, O., Jex, A., Guy, L., & Sassera, D. (2023). The evolution of intramitochondriality in Midichloria bacteria. *Environmental microbiology*, 25(11), 2102–2117. https://doi.org/10.1111/1462-2920.16446.

Fogaça, A. C., Sousa, G., Pavanelo, D. B., Esteves, E., Martins, L. A., Urbanová, V., Kopáček, P., & Daffre, S. (2021). Tick immune system: What is known, the interconnections, the gaps, and the challenges. *Frontiers in immunology*, 12, 628054. https://doi.org/10.3389/fimmu.2021.628054.

Foley, N., O'Connor, C., Falco, R. C., Vinci, V., Oliver, J., Haight, J., Sporn, L. A., Harrington, L., Mader, E., Wroblewski, D., Backenson, P. B., & Prusinski, M. A. (2023). Spatiotemporal distribution of Borrelia miyamotoi (Spirochaetales: Spirochaetaceae) and coinfection with other tick-borne pathogens in host-seeking Ixodes scapularis (Acari: Ixodidae) from New York State, USA. *Journal of medical entomology*, 60(4), 808–821. https://doi.org/10.1093/jme/tjad057.

França, R. T., Da Silva, A. S., Loretti, A. P., Mazzanti, C. M., & Lopes, S. T. (2014). Canine rangeliosis due to Rangelia vitalii: From first report in Brazil in 1910 to current day – a review. *Ticks and tick-borne diseases*, 5(5), 466–474. https://doi.org/10.1016/j.ttbdis.2014.04.005.

Francis, L., Paddock, C. D., Dykstra, E. A., & Karpathy, S. E. (2020). Rickettsia and Anaplasma species in Dermacentor andersoni ticks from Washington. *Ticks and tick-borne diseases*, 11(4), 101422. https://doi.org/10.1016/j.ttbdis.2020.101422.

Freimoser, F. M., Screen, S., Bagga, S., Hu, G., & St Leger, R. J. (2003). Expressed sequence tag (EST) analysis of two subspecies of Metarhizium anisopliae reveals a plethora of secreted proteins with potential activity in insect hosts. *Microbiology (Reading, England)*, 149(Pt 1), 239–247. https://doi.org/10.1099/mic.0.25761-0.

Fricke, L. C., & Lindsey, A. R. I. (2024). Examining Wolbachia-induced parthenogenesis in Hymenoptera. *Methods in molecular biology (Clifton, N.J.)*, 2739, 55–68. https://doi.org/10.1007/978-1-0716-3553-7_4.

Fujita, R., Ejiri, H., Lim, C. K., Noda, S., Yamauchi, T., Watanabe, M., Kobayashi, D., Takayama-Ito, M., Murota, K., Posadas-Herrera, G., Minami, S., Kuwata, R., Yamaguchi, Y., Horiya, M., Katayama, Y., Shimoda, H., Saijo, M., Maeda, K., Mizutani, T., Isawa, H., … & Sawabe, K. (2017). Isolation and characterization of Tarumizu tick virus: A new coltivirus from Haemaphysalis flava ticks in Japan. *Virus research*, 242, 131–140. https://doi.org/10.1016/j.virusres.2017.09.017.

Gall, C. A., Reif, K. E., Scoles, G. A., Mason, K. L., Mousel, M., Noh, S. M., & Brayton, K. A. (2016). The bacterial microbiome of Dermacentor andersoni ticks influences pathogen susceptibility. *The ISME journal*, 10(8), 1846–1855. https://doi.org/10.1038/ismej.2015.266.

Gandy, S., Hansford, K., McGinley, L., Cull, B., Smith, R., Semper, A., Brooks, T., Fonville, M., Sprong, H., Phipps, P., Johnson, N., & Medlock, J. M. (2022). Prevalence of Anaplasma phagocytophilum in questing Ixodes ricinus nymphs across twenty recreational areas in England and Wales. *Ticks and tick-borne diseases*, 13(4), 101965. https://doi.org/10.1016/j.ttbdis.2022.101965.

Ganguly, S., & Mukhopadhayay, S. K. (2008). Tick-borne ehrlichiosis infection in human beings. *Journal of vector borne diseases*, 45(4), 273–280.

Garcia-Vozmediano, A., Tomassone, L., Fonville, M., Bertolotti, L., Heylen, D., Fabri, N. D., Medlock, J. M., Nijhof, A. M., Hansford, K. M., Sprong, H., & Krawczyk, A. I. (2022). The genetic diversity of Rickettsiella symbionts in Ixodes ricinus throughout Europe. *Microbial ecology*, 84(2), 613–626. https://doi.org/10.1007/s00248-021-01869-7.

Gargili, A., Midilli, K., Ergonul, O., Ergin, S., Alp, H. G., Vatansever, Z., Iyisan, S., Cerit, C., Yilmaz, G., Altas, K., & Estrada-Peña, A. (2011). Crimean-Congo hemorrhagic fever in European part of Turkey: Genetic analysis of the virus strains from ticks and a seroepidemiological study in humans. *Vector borne and zoonotic diseases (Larchmont, N.Y.)*, 11(6), 747–752. https://doi.org/10.1089/vbz.2010.0030.

Geissler, A. L., Thorp, E., Van Houten, C., Lanciotti, R. S., Panella, N., Cadwell, B. L., Murphy, T., & Staples, J. E. (2014). Infection with Colorado tick fever virus among humans and ticks in a national park and forest, Wyoming, 2010. *Vector borne and zoonotic diseases (Larchmont, N.Y.)*, 14(9), 675–680. https://doi.org/10.1089/vbz.2013.1568.

Gems, D. (2000). Longevity and ageing in parasitic and free-living nematodes. *Biogerontology*, 1, 289–307. https://doi.org/10.1023/A:1026546719091.

Gergova, I., Kunchev, M., & Kamarinchev, B. (2012). Crimean-Congo hemorrhagic fever virus-tick survey in endemic areas in Bulgaria. *Journal of medical virology*, 84(4), 608–614. https://doi.org/10.1002/jmv.23214.

Gerhart, J. G., Auguste Dutcher, H., Brenner, A. E., Moses, A. S., Grubhoffer, L., & Raghavan, R. (2018). Multiple acquisitions of oathogen-derived francisella endosymbionts in soft ticks. *Genome biology and evolution*, 10(2), 607–615. https://doi.org/10.1093/gbe/evy021.

Ghafar, A., Khan, A., Cabezas-Cruz, A., Gauci, C. G., Niaz, S., Ayaz, S., Mateos-Hernández, L., Galon, C., Nasreen, N., Moutailler, S., Gasser, R. B., & Jabbar, A. (2020). An assessment of the molecular diversity of ticks and tick-borne microorganisms of small Ruminants in Pakistan. *Microorganisms*, 8(9), 1428. https://doi.org/10.3390/microorganisms8091434.

Gilbert, L. (2021). The impacts of climate change on ticks and tick-borne disease risk. *Annual review of entomology*, 66, 373–388. https://doi.org/10.1146/annurev-ento-052720-094533.

Gillespie, J. J., Beier, M. S., Rahman, M. S., Ammerman, N. C., Shallom, J. M., Purkayastha, A., Sobral, B. S., & Azad, A. F. (2007). Plasmids and rickettsial evolution: Insight from Rickettsia felis. *PloS one*, 2(3), e266. https://doi.org/10.1371/journal.pone.0000266.

Gillespie, J. J., Williams, K., Shukla, M., Snyder, E. E., Nordberg, E. K., Ceraul, S. M., Dharmanolla, C., Rainey, D., Soneja, J., Shallom, J. M., Vishnubhat, N. D., Wattam, R., Purkayastha, A., Czar, M., Crasta, O., Setubal, J. C., Azad, A. F., & Sobral, B. S. (2008). Rickettsia phylogenomics: Unwinding the intricacies of obligate intracellular life. *PloS one*, 3(4), e2018. https://doi.org/10.1371/journal.pone.0002018.

Gofton, A. W., Blasdell, K. R., Taylor, C., Banks, P. B., Michie, M., Roy-Dufresne, E., Poldy, J., Wang, J., Dunn, M., Tachedjian, M., & Smith, I. (2022). Metatranscriptomic profiling reveals diverse tick-borne bacteria, protozoans and viruses in ticks and wildlife from Australia. *Transboundary and emerging diseases*, 69(5), e2389–e2407. https://doi.org/10.1111/tbed.14591.

Gofton, A. W., Oskam, C. L., Lo, N., Beninati, T., Wei, H., McCarl, V., Murray, D. C., Paparini, A., Greay, T. L., Holmes, A. J., Bunce, M., Ryan, U., & Irwin, P. (2015). Inhibition of the endosymbiont "Candidatus Midichloria mitochondrii" during 16S rRNA gene profiling reveals potential pathogens in Ixodes ticks from Australia. *Parasites & vectors*, 8, 345. https://doi.org/10.1186/s13071-015-0958-3.

Gomard, Y., Flores, O., Vittecoq, M., Blanchon, T., Toty, C., Duron, O., Mavingui, P., Tortosa, P., & McCoy, K. D. (2021). Changes in bacterial diversity, composition and interactions during the development of the seabird tick Ornithodoros maritimus (Argasidae). *Microbial ecology*, 81(3), 770–783. https://doi.org/10.1007/s00248-020-01611-9.

Gondard, M., Cabezas-Cruz, A., Charles, R. A., Vayssier-Taussat, M., Albina, E., & Moutailler, S. (2017). Ticks and tick-borne pathogens of the Caribbean: Current understanding and future directions for more comprehensive surveillance. *Frontiers in cellular and infection microbiology*, 7, 490. https://doi.org/10.3389/fcimb.2017.00490.

Gong, S., He, B., Wang, Z., Shang, L., Wei, F., Liu, Q., & Tu, C. (2015). Nairobi sheep disease virus RNA in ixodid ticks, China, 2013. *Emerging infectious diseases*, 21(4), 718–720. https://doi.org/10.3201/eid2104.141602.

González, J., Valcárcel, F., Pérez-Sánchez, J. L., Tercero-Jaime, J. M., Cutuli, M. T., & Olmeda, A. S. (2016, November). Control of *Hyalomma lusitanicum* (Acari: Ixodidade) Ticks Infesting *Oryctolagus cuniculus* (Lagomorpha: Leporidae) using the entomopathogenic fungus *Beauveria bassiana* (Hyocreales: Clavicipitaceae) in field conditions. *Journal of medical entomology*, 53(6), 1396–1402.

Gottlieb, Y., Lalzar, I., & Klasson, L. (2015). Distinctive genome reduction rates revealed by genomic analyses of two coxiella-like endosymbionts in ticks. *Genome biology and evolution*, 7(6), 1779–1796. https://doi.org/10.1093/gbe/evv108.

Grandi, G., Chiappa, G., Ullman, K., Lindgren, P. E., Olivieri, E., Sassera, D., Östlund, E., Omazic, A., Perissinotto, D., & Söderlund, R. (2023). Characterization of the bacterial microbiome of Swedish ticks through 16S rRNA amplicon sequencing of whole ticks and of individual tick organs. *Parasites & vectors*, 16(1), 39. https://doi.org/10.1186/s13071-022-05638-4.

Grassi, L., Franzo, G., Grillo, S., Mondin, A., Drigo, M., Barbarino, F., Comuzzo, C., Legnardi, M., Bertola, M., Montarsi, F., & Menandro, M. L. (2023). Survey of tick-borne zoonotic agents in Ixodes ticks carried by wild passerines during postbreeding migration through Italy. *Transboundary and emerging diseases*, 2023, 1399089. https://doi.org/10.1155/2023/1399090.

Gray, J. S., Estrada-Peña, A., & Zintl, A. (2019). Vectors of babesiosis. *Annual review of entomology*, 64, 149–165. https://doi.org/10.1146/annurev-ento-011118-111932.

Greengarten, P. J., Tuininga, A. R., Morath, S. U., Falco, R. C., Norelus, H., & Daniels, T. J. (2011). Occurrence of soil- and tick-borne fungi and related virulence tests for pathogenicity to Ixodes scapularis (Acari: Ixodidae). *Journal of medical entomology*, 48(2), 337–344. https://doi.org/10.1603/me09116.

Gruntmeir, J., Kelly, M., Ramos, R. A. N., & Verocai, G. G. (2023). Cutaneous filarioid nematodes of dogs in the United States: Are they emerging, neglected, or underdiagnosed parasites?. *Frontiers in veterinary science*, 10, 1128611. https://doi.org/10.3389/fvets.2023.1128611.

Guizzo, M. G., Hatalová, T., Frantová, H., Zurek, L., Kopáček, P., & Perner, J. (2023). Ixodes ricinus ticks have a functional association with Midichloria mitochondrii. *Frontiers in cellular and infection microbiology*, 12, 1081666. https://doi.org/10.3389/fcimb.2022.1081666.

Guizzo, M. G., Neupane, S., Kucera, M., Perner, J., Frantov´ a, H., da Silva Vaz, I., de Oliveira, P. L., Kopacek, P., & Zurek, L. (2020). Poor unstable midgut microbiome of hard ticks contrasts with abundant and stable monospecific microbiome in ovaries. *Frontiers in Cellular and Infection Microbiology*, 10, 211. https://doi.org/10.3389/fcimb.2020.00211.

Guizzo, M. G., Parizi, L. F., Nunes, R. D., Schama, R., Albano, R. M., Tirloni, L., Oldiges, D. P., Vieira, R. P., Oliveira, W. H. C., Leite, M. S., Gonzales, S. A., Farber, M., Martins, O., Vaz, I. D. S., Jr, & Oliveira, P. L. (2017). A Coxiella mutualist symbiont is essential to the development of Rhipicephalus microplus. *Scientific reports*, 7(1), 17554. https://doi.org/10.1038/s41598-017-17309-x.

Gulia-Nuss, M., Nuss, A. B., Meyer, J. M., Sonenshine, D. E., Roe, R. M., Waterhouse, R. M., Sattelle, D. B., de la Fuente, J., Ribeiro, J. M., Megy, K., Thimmapuram, J., Miller, J. R., Walenz, B. P., Koren, S., Hostetler, J. B., Thiagarajan, M., Joardar, V. S., Hannick, L. I., Bidwell, S., Hammond, M. P., ... & Hill, C. A. (2016). Genomic insights into the Ixodes scapularis tick vector of Lyme disease. *Nature communications*, 7, 10507. https://doi.org/10.1038/ncomms10507.

Guo, R., Shen, S., Zhang, Y., Shi, J., Su, Z., Liu, D., Liu, J., Yang, J., Wang, Q., Hu, Z., Zhang, Y., & Deng, F. (2017). A new strain of Crimean-Congo hemorrhagic fever virus isolated from Xinjiang, China. *Virologica Sinica*, 32(1), 80–88. https://doi.org/10.1007/s12250-016-3936-9.

Gurfield, N., Grewal, S., Cua, L. S., Torres, P. J., & Kelley, S. T. (2017). Endosymbiont interference and microbial diversity of the Pacific coast tick, Dermacentor occidentalis, in San Diego County, California. *PeerJ*, 5, e3202. https://doi.org/10.7717/peerj.3202.

Gygax, L., Schudel, S., Kositz, C., Kuenzli, E., & Neumayr, A. (2024). Human monocytotropic ehrlichiosis-A systematic review and analysis of the literature. *PLoS neglected tropical diseases*, 18(8), e0012377. https://doi.org/10.1371/journal.pntd.0012377.

Gynthersen, R. M. M., Knudtzen, F. C., Porskrog, A., Krogfelt, K., Hansen, M. F., El Fassi, D., Lebech, A. M., & Mens, H. (2023). Neoehrlichiosis is a new tick-borne infection. *Ugeskrift for laeger*, 185(6), V09220554.

Habib, J., Zenner, L., Garel, M., Mercier, A., Poirel, M. T., Itty, C., Appolinaire, J., Amblard, T., Benedetti, P., Sanchis, F., Benabed, S., Abi Rizk, G., Gibert, P., & Bourgoin, G. (2024). Prevalence of tick-borne pathogens in ticks collected from the wild mountain ungulates mouflon and chamois in 4 regions of France. Prévalence d'agents pathogènes vectorisés chez des tiques collectées chez des ongulés sauvages (mouflons, chamois) dans 4 zones montagneuses en France. *Parasite (Paris, France)*, 31, 21. https://doi.org/10.1051/parasite/2024014.

Hagman, K., Barboutis, C., Ehrenborg, C., Fransson, T., Jaenson, T. G., Lindgren, P. E., Lundkvist, A., Nyström, F., Waldenström, J., & Salaneck, E. (2014). On the potential roles of ticks and migrating birds in the ecology of West Nile virus. *Infection ecology & epidemiology*, 4, 20943https://doi.org/10.3402/iee.v4.20943.

Haidar-Ahmad, A., Barthel, C., Rivera-Vallejo, R. A., Maul, A., Poulle, M.-L., & Boulanger, N. (2025). Dynamics of Ixodes ricinus and associated bacterial pathogens in the forest and agricultural ecosystems of northeastern France. *Applied and environmental microbiology*, 91(6), e0079325. https://doi.org/10.1128/aem.00793-25.

Hajdušek, O., Síma, R., Ayllón, N., Jalovecká, M., Perner, J., de la Fuente, J., & Kopáček, P. (2013). Interaction of the tick immune system with transmitted pathogens. *Frontiers in cellular and infection microbiology*, 3, 26. https://doi.org/10.3389/fcimb.2013.00026.

Haji, I., Simuunza, M., Jiang, N., & Chen, Q. (2023). Tick populations and molecular detection of selected tick-borne pathogens in questing ticks from northern and central Tanzania. *Experimental & applied acarology*, 90(3-4), 389–407. https://doi.org/10.1007/s10493-023-00816-0.

Halos, L., Jamal, T., Maillard, R., Beugnet, F., Le Menach, A., Boulouis, H. J., & Vayssier-Taussat, M. (2005). Evidence of Bartonella sp. in questing adult and nymphal Ixodes ricinus ticks from France and co-infection with Borrelia burgdorferi sensu lato and Babesia sp. *Veterinary research*, 36(1), 79–87. https://doi.org/10.1051/vetres:2004052.

Hamilton, P. T., Maluenda, E., Sarr, A., Belli, A., Hurry, G., Duron, O., Plantard, O., & Voordouw, M. J. (2021). Borrelia afzelii Infection in the Rodent Host Has Dramatic Effects on the Bacterial Microbiome of Ixodes ricinus Ticks. *Applied and environmental microbiology*, 87(18), e0064121. https://doi.org/10.1128/AEM.00641-21.

Han, R., Yang, J. F., Mukhtar, M. U., Chen, Z., Niu, Q. L., Lin, Y. Q., Liu, G. Y., Luo, J. X., Yin, H., & Liu, Z. J. (2019). Molecular detection of Anaplasma infections in ixodid ticks from the Qinghai-Tibet Plateau. *Infectious diseases of poverty*, 8(1),

12. https://doi.org/10.1186/s40249-019-0523-z.

Han, X. H., Ma, Y., Liu, H. Y., Li, D., Wang, Y., Jiang, F. H., Gao, Q. T., Jiang, F., Liu, B. S., Shen, G. S., & Chen, Z. L. (2022). Identification of severe fever with thrombocytopenia syndrome virus genotypes in patients and ticks in Liaoning Province, China. *Parasites & vectors*, 15(1), 120. https://doi.org/10.1186/s13071-022-05237-3.

Hart, C. E., & Thangamani, S. (2021). Tick-virus interactions: Current understanding and future perspectives. *Parasite immunology*, 43(5), e12815. https://doi.org/10.1111/pim.12815.

Hart, C., Hassett, E., Vogels, C. B. F., Shapley, D., Grubaugh, N. D., & Thangamani, S. (2023). Powassan virus lineage I in field-collected Dermacentor variabilis ticks, New York, USA. *Emerging infectious diseases*, 29(2), 415–417. https://doi.org/10.3201/eid2902.220410.

Hartelt, K., Oehme, R., Frank, H., Brockmann, S. O., Hassler, D., & Kimmig, P. (2004). Pathogens and symbionts in ticks: Prevalence of Anaplasma phagocytophilum (Ehrlichia sp.), Wolbachia sp., Rickettsia sp., and Babesia sp. in Southern Germany. *International journal of medical microbiology: IJMM*, 293 Suppl, 37, 86–92. https://doi.org/10.1016/s1433-1128(04)80013-5.

Hatalová, T., Erhart, J., Kopáček, P., & Perner, J. (2023). On the haem auxotrophy of the soft tick Ornithodoros moubata. *Ticks and tick-borne diseases*, 14(4), 102170. https://doi.org/10.1016/j.ttbdis.2023.102170.

Haulrig, M. B., Mathiasen, G., Nielsen, R. M., Kromann, C. B., Krogfelt, K. A., & Wiese, L. (2020). Two cases of tick-borne transmitted tularemia on Southern Zealand, Denmark. *APMIS: acta pathologica, microbiologica, et immunologica Scandinavica*, 128(1), 61–64. https://doi.org/10.1111/apm.13008.

Hayes, B. M., Radkov, A. D., Yarza, F., Flores, S., Kim, J., Zhao, Z., Lexa, K. W., Marnin, L., Biboy, J., Bowcut, V., Vollmer, W., Pedra, J. H. F., & Chou, S. (2020). Ticks resist skin commensals with immune factor of bacterial origin. *Cell*, 183(6), 1562–1571.e12. https://doi.org/10.1016/j.cell.2020.10.042.

Hegedus, D. D., Toprak, U., & Erlandson, M. (2019). Peritrophic matrix formation. *Journal of insect physiology*, 117, 103898. https://doi.org/10.1016/j.jinsphys.2019.103898.

Helminiak, L., Mishra, S., & Kim, H. K. (2022). Pathogenicity and virulence of Rickettsia. *Virulence*, 13(1), 1752–1771. https://doi.org/10.1080/21505594.2022.2132047.

Hembram, P. K., Kumar, G. S., Kumar, K. G. A., Deepa, C. K., Varghese, A., Bora, C. A. F., Nandini, A., Malangmei, L., Kurbet, P. S., Dinesh, C. N., Juliet, S., Ghosh, S., & Ravindran, R. (2022). Molecular detection of pathogens in the ova and unfed larvae of Rhipicephalus annulatus and Haemaphysalis bispinosa ticks infesting domestic cattle of south India. *Acta tropica*, 235, 106656. https://doi.org/10.1016/j.actatropica.2022.106658.

Henning, T. C., Orr, J. M., Smith, J. D., Arias, J. R., Rasgon, J. L., & Norris, D. E. (2016). Discovery of filarial nematode DNA in Amblyomma americanum in Northern Virginia. *Ticks and tick-borne diseases*, 7(2), 315–318. https://doi.org/10.1016/j.ttbdis.2015.11.007.

Hodosi, R., Kazimirova, M., & Soltys, K. (2022). What do we know about the microbiome of I. ricinus?. *Frontiers in cellular and infection microbiology*, 12, 990889. https://doi.org/10.3389/fcimb.2022.990889.

Hoffman, T., Lindeborg, M., Barboutis, C., Erciyas-Yavuz, K., Evander, M., Fransson, T., Figuerola, J., Jaenson, T. G. T., Kiat, Y., Lindgren, P. E., Lundkvist, Å., Mohamed, N., Moutailler, S., Nyström, F., Olsen, B., & Salaneck, E. (2018). Alkhurma hemorrhagic fever virus RNA in Hyalomma rufipes ticks infesting migratory birds, Europe and Asia Minor. *Emerging infectious diseases*, 24(5), 879–882. https://doi.org/10.3201/eid2405.171370.

Hooper, L. V., & Gordon, J. I. (2001). Commensal host-bacterial relationships in the gut. *Science (New York, N.Y.)*, 292(5519), 1115–1118. https://doi.org/10.1126/science.1058709.

Hornbostel, V. L., Ostfeld, R. S., Zhioua, E., & Benjamin, M. A. (2004). Sublethal effects of Metarhizium anisopliae (Deuteromycetes) on engorged larval, nymphal, and adult Ixodes scapularis (Acari: Ixodidae). *Journal of medical entomology*, 41(5), 922–929. https://doi.org/10.1603/0022-2585-41.5.922.

Hornok, S., Boldogh, S. A., Takács, N., Kontschán, J., Szekeres, S., Sós, E., Sándor, A. D., Wang, Y., & Tuska-Szalay, B. (2022a). Molecular epidemiological study on ticks and tick-borne protozoan parasites (Apicomplexa: Cytauxzoon and Hepatozoon spp.) from wild cats (Felis silvestris), Mustelidae and red squirrels (Sciurus vulgaris) in central Europe, Hungary. *Parasites & vectors*, 15(1), 174. https://doi.org/10.1186/s13071-022-05271-1.

Hosseini-Chegeni, A., Tavakoli, M., Goudarzi, G., Telmadarraiy, Z., Sharifdini, M., Faghihi, F., & Ghanbari, M. K. (2020). Molecular detection of Anaplasma marginale and Anaplasma ovis (Rickettsiales: Anaplasmataceae) in Ixodid tick species in Iran. *Archives of Razi Institute*, 75(1), 39–46. https://doi.org/10.22092/ari.2019.122532.1260.

Hubert, J., Glowska-Patyniak, E., Dowd, S. E., & Klimov, P. B. (2025). Cardinium disrupts Wolbachia-host dynamics in the domestic mite Tyrophagus putrescentiae: Evidence from manipulative experiments. *mSystems*, 10(5), e0176924. https://doi.org/10.1128/msystems.01769-24.

Hunter, D. J., Torkelson, J. L., Bodnar, J., Mortazavi, B., Laurent, T., Deason, J., Thephavongsa, K., & Zhong, J. (2015). The Rickettsia rndosymbiont of Ixodes pacificus contains all the genes of De novo folate biosynthesis. *PloS one*, 10(12), e0144552. https://doi.org/10.1371/journal.pone.0144552.

Huntington, M. K., Allison, J., & Nair, D. (2016, October 1). Emerging vector-borne diseases. American Family Physician, 94(7), 551-557.

Hussain, S., Perveen, N., Hussain, A., Song, B., Aziz, M. U., Zeb, J., Li, J., George, D., Cabezas-Cruz, A., & Sparagano, O. (2022). The symbiotic continuum within ticks: Opportunities for disease control. *Frontiers in microbiology*, 13, 854803. https://doi.org/10.3389/fmicb.2022.854803.

ICTV, International Committee on Taxonomy of Viruses: Genus: Bandavirus. (2024). https://ictv.global/report/chapter/phenuiviridae/phenuiviridae/bandavirus

Imaoka, K., Kaneko, S., Tabara, K., Kusatake, K., & Morita, E. (2011). The first human case of Rickettsia tamurae infection in Japan. *Case reports in dermatology*, 3(1), 68–73. https://doi.org/10.1159/000326941.

Intirach, J., Lv, X., Sutthanont, N., Cai, B., Champakaew, D., Chen, T., Han, Q., & Lv, Z. (2024). Molecular and next-generation sequencing analysis of tick-borne pathogens of Rhipicephalus ticks (Acari: Ixodidae) in cattle and dogs. *Acta tropica*, 252, 107138. https://doi.org/10.1016/j.actatropica.2024.107145.

Ismail, N., Bloch, K. C., & McBride, J. W. (2010). Human ehrlichiosis and anaplasmosis. *Clinics in laboratory medicine*, 30(1), 261–292. https://doi.org/10.1016/j.cll.2009.10.004.

Ismail, N., & McBride, J. W. (2017). Tick-borne emerging infections: Ehrlichiosis and anaplasmosis. *Clinics in Laboratory Medicine*, 37, 317–340. https://doi.org/10.1016/j.cll.2017.01.006.

Jang, H., Casel, M. A. B., Jang, S.-g, Choi, J. H., Gil, J., Rollon, R., Cheun, S.y, Kim, Y.-I., Song, M. S., & Choi, Y. K. (2024). Seasonal dynamics of Haemaphysalis tick species as SFTSV vectors in South Korea. *Microbiology spectrum*, 12(11), e0048924. https://doi.org/10.1128/spectrum.00489-24.

Ji, S. R., Byun, H. R., Rieu, M. S., Han, S. W., Nam, H. Y., Seo, S., Park, S. Y., Kang, H. Y., Choi, C. Y., Cho, S. Y., Hwang, B. Y., & Chae, J. S. (2024). First detection of Bandavirus dabieense in ticks collected from migratory birds in the Republic of Korea. *Acta tropica*, 257, 107279. https://doi.org/10.1016/j.actatropica.2024.107279.

Jia, N., Liu, H. B., Ni, X. B., Bell-Sakyi, L., Zheng, Y. C., Song, J. L., Li, J., Jiang, B. G., Wang, Q., Sun, Y., Wei, R., Yuan, T. T., Xia, L. Y., Chu, Y. L., Wei, W., Li, L. F., Ye, J. L., Lv, Q. Y., Cui, X. M., Guan, Y., … & Cao, W. C. (2019). Emergence of human infection with Jingmen tick virus in China: A retrospective study. *EBioMedicine*, 43, 317–324. https://doi.org/10.1016/j.ebiom.2019.04.004.

Jia, N., Wang, J., Shi, W., Du, L., Sun, Y., Zhan, W., Jiang, J. F., Wang, Q., Zhang, B., Ji, P., Bell-Sakyi, L., Cui, X. M., Yuan, T. T., Jiang, B. G., Yang, W. F., Lam, T. T., Chang, Q. C., Ding, S. J., Wang, X. J., Zhu, J. G., … & Cao, W. C. (2020). Large-scale comparative analyses of tick genomes elucidate their genetic diversity and vector capacities. *Cell*, 182(5), 1328–1340.e13. https://doi.org/10.1016/j.cell.2020.07.023.

Jiao, J., Lu, Z., Yu, Y., Ou, Y., Fu, M., Zhao, Y., Wu, N., Zhao, M., Liu, Y., Sun, Y., Wen, B., Zhou, D., Yuan, Q., & Xiong, X. (2021). Identification of tick-borne pathogens by metagenomic next-generation sequencing in Dermacentor nuttalli and Ixodes persulcatus in Inner Mongolia, China. *Parasites & vectors*, 14(1), 287. https://doi.org/10.1186/s13071-021-04740-3.

Jiao, Y., Qi, X., Liu, D., Zeng, X., Han, Y., Guo, X., Shi, Z., Wang, H., & Zhou, M. (2015). Experimental and natural infections of goats with severe fever with thrombocytopenia syndrome virus: Evidence for ticks as viral vector. *PLoS neglected tropical diseases*, 9(10), e0004092. https://doi.org/10.1371/journal.pntd.0004092.

Jo, Y. S., Kang, J. G., Chae, J. B., Cho, Y. K., Shin, J. H., Jheong, W. H., & Chae, J. S. (2019). Prevalence of severe fever with thrombocytopenia syndrome virus in ticks collected from National Parks in Korea. *Vector borne and zoonotic diseases (Larchmont, N.Y.)*, 19(4), 284–289. https://doi.org/10.1089/vbz.2018.2338.

Jordan, R. A., & Schulze, T. L. (2020). Artificial accumulation of leaf litter in forest edges on residential properties via leaf blowing is associated with increased numbers of host-seeking Ixodes scapularis (Acari: Ixodidae) Nymphs. *Journal of medical entomology*, 57(4), 1193–1198. https://doi.org/10.1093/jme/tjaa033.

Jouglin, M., Blanc, B., Brunet, A., Ortiz, K., & Malandrin, L. (2025). Anaplasma capra and Haemaphysalis concinna: Investigating a potential vector relationship in a wildlife reserve. *Ticks and tick-borne diseases*, 16(4), 102500. Advance online publication. https://doi.org/10.1016/j.ttbdis.2025.102500.

Kagemann, J., & Clay, K. (2013). Effects of infection by Arsenophonus and Rickettsia bacteria on the locomotive ability of the ticks Amblyomma americanum, Dermacentor variabilis, and Ixodes scapularis. *Journal of medical entomology*, 50(1), 155–162. https://doi.org/10.1603/me12086.

Kalil, S. P., Rosa, R. D. D., Capelli-Peixoto, J., Pohl, P. C., Oliveira, P. L., Fogaça, A. C., & Daffre, S. (2017). Immune-related redox metabolism of embryonic cells of the tick Rhipicephalus microplus (BME26) in response to infection with Anaplasma marginale. *Parasites & vectors*, 10(1), 613. https://doi.org/10.1186/s13071-017-2575-9.

Kamani, J., Gonçalves-Oliveira, J., Janssen, J. N., Quesada, J. R., Nachum-Biala, Y., Harrus, S., & Gutiérrez, R. (2024). Microbiome of two adult tick species and their laboratory-reared offspring shows intra- and inter-species differences. *Acta tropica*, 257, 107315. https://doi.org/10.1016/j.actatropica.2024.107315.

Kang, J. G., Cho, Y. K., Jo, Y. S., Han, S. W., Chae, J. B., Park, J. E., Jeong, H., Jheong, W. H., & Chae, J. S. (2022). Severe fever with thrombocytopenia syndrome virus in ticks in the Republic of Korea. *The Korean journal of parasitology*, 60(1), 65–71. https://doi.org/10.3347/kjp.2022.60.1.65.

Kang, J. G., Ko, S., Kim, H. C., Chong, S. T., Klein, T. A., Chae, J. B., Jo, Y. S., Choi, K. S., Yu, D. H., Park, B. K., Park, J., & Chae, J. S. (2016). Prevalence of Anaplasma and Bartonella spp. in ticks collected from Korean water deer (Hydropotes inermis argyropus). *The Korean journal of parasitology*, 54(1), 87–91. https://doi.org/10.3347/kjp.2016.54.1.90.

Kar, S., Rodriguez, S. E., Akyildiz, G., Cajimat, M. N. B., Bircan, R., Mears, M. C., Bente, D. A., & Keles, A. G. (2020). Crimean-Congo hemorrhagic fever virus in tortoises and Hyalomma aegyptium ticks in East Thrace, Turkey: Potential of a cryptic transmission cycle. *Parasites & vectors*, 13(1), 201. https://doi.org/10.1186/s13071-020-04074-6.

Karim, S., Zenzal, T. J., Jr, Beati, L., Sen, R., Adegoke, A., Kumar, D., Downs, L. P., Keko, M., Nussbaum, A., Becker, D. J., & Moore, F. R. (2024). Ticks without borders: Microbiome of immature neotropical tick species parasitizing migratory songbirds along northern Gulf of Mexico. *Frontiers in cellular and infection microbiology*, 14, 1472598. https://doi.org/10.3389/fcimb.2024.1472598.

Kariu, T., Smith, A., Yang, X., & Pal, U. (2013). A chitin deacetylase-like protein is a predominant constituent of tick peritrophic membrane that influences the persistence of Lyme disease pathogens within the vector. *PloS one*, 8(10), e78376. https://doi.org/10.1371/journal.pone.0078376.

Karshima, S. N., Ahmed, M. I., Kogi, C. A., & Iliya, P. S. (2022). Anaplasma phagocytophilum infection rates in questing and host-attached ticks: A global systematic review and meta-analysis. *Acta tropica*, 228, 106299. https://doi.org/10.1016/j.actatropica.2021.106299.

Kautman, M., Bilbija, B., Nováková, M., Tkadlec, E., Papoušek, I., Mikulíček, P., Jandzik, D., Javanbakht, H., Tiar, G., & Široký, P. (2025). The importance of the tortoise tick Hyalomma aegyptium as a carrier of bacterial agents on a wide range. *Ticks and tick-borne diseases*, 16(2), 102456. https://doi.org/10.1016/j.ttbdis.2025.102456.

Kawano, S., Umemiya-Shirafuji, R., Boldbaatar, D., Matsuoka, K., Tanaka, T., & Fujisaki, K. (2011). Cloning and characterization of the autophagy-related gene 6 from the hard tick, Haemaphysalis longicornis. *Parasitology research*, 109(5), 1341–1349. https://doi.org/10.1007/s00436-011-2429-x. MUST SEE this paper to revise the related content.

Kayedi, M. H., Chinikar, S., Mostafavi, E., Khakifirouz, S., Jalali, T., Hosseini-Chegeni, A., Naghizadeh, A., Niedrig, M., Fooks, A. R., & Shahhosseini, N. (2015). Crimean-Congo hemorrhagic fever virus clade IV (Asia 1) in ticks of Western Iran. *Journal of medical entomology*, 52(5), 1144–1149. https://doi.org/10.1093/jme/tjv081.

Kazimírová, M., Mangová, B., Chvostáč, M., Didyk, Y. M., de Alba, P., Mira, A., Purgatová, S., Selyemová, D., Rusňáková Tarageľová, V., & Schnittger, L. (2024 Jun 19). The role of wildlife in the epidemiology of tick-borne diseases in Slovakia. *Curr Res Parasitol Vector Borne Dis*, 6, 100195. doi: 10.1016/j.crpvbd.2024.100195. PMID: 39027084; PMCID: PMC11252786.

Kazimírová, M., Thangamani, S., Bartíková, P., Hermance, M., Holíková, V., Štibrániová, I., & Nuttall, P. A. (2017). Tick-borne viruses and biological processes at the tick-host-virus interface. *Frontiers in cellular and infection microbiology*, 7, 339. https://doi.org/10.3389/fcimb.2017.00339.

Keve, G., Reynolds, C., Takács, N., & Hornok, S. (2025). Uneven temporal distribution of piroplasms (Piroplasmida: Babesiidae, Theileriidae) in Haemaphysalis concinna in an urban biotope of the Western Palearctic focus region of this tick species. *Ticks and tick-borne diseases*, 16(2), 102458. https://doi.org/10.1016/j.ttbdis.2025.102458.

Keyte, S., Abdullah, S., James, K., Newbury, H., Helps, C., Tasker, S., & Wall, R. (2021). Prevalence and distribution of Anaplasma phagocytophilum in ticks collected from dogs in the United Kingdom. *The Veterinary record*, 188(8), e12. https://doi.org/10.1002/vetr.12.

Khan, Z., Ullah, F., Ullah, S., Ibrahim, M., Khan, M., Rehman, G., Tanaka, T., Almutairi, M. M., Alouffi, A., & Ali, A. (2024). Molecular detection of Anaplasma bovis, Candidatus Anaplasma boleense and Rickettsia spp. in ticks infesting small ruminants. *BMC veterinary research*, 20(1), 408. https://doi.org/10.1186/s12917-024-04259-6.

Khatri-Chhetri, R., Wang, H. C., Chen, C. C., Shih, H. C., Liao, H. C., Sun, C. M., Khatri-Chhetri, N., Wu, H. Y., & Pei, K. J. (2016). Surveillance of ticks and associated pathogens in free-ranging Formosan pangolins (Manis pentadactyla pentadactyla). *Ticks and tick-borne diseases*, 7(6), 1238–1244. https://doi.org/10.1016/j.ttbdis.2016.07.007.

Khodadadi, N., Nabavi, R., Sarani, A., Saadati, D., Ganjali, M., Mihalca, A. D., Otranto, D., & Sazmand, A. (2021). Identification of Anaplasma marginale in long-eared hedgehogs (Hemiechinus auritus) and their Rhipicephalus turanicus ticks in Iran. *Ticks and tick-borne diseases*, 12(2), 101641. https://doi.org/10.1016/j.ttbdis.2020.101641.

Kholodilov, I. S., Belova, O. A., Ivannikova, A. Y., Gadzhikurbanov, M. N., Makenov, M. T., Yakovlev, A. S., Polienko, A. E., Dereventsova, A. V., Litov, A. G., Gmyl, L. V., Okhezin, E. V., Luchinina, S. V., Klimentov, A. S., & Karganova, G. G. (2022). Distribution and characterisation of tick-borne Flavi-, Flavi-like, and Phenuiviruses in the Chelyabinsk Region of Russia. *Viruses*, 14(12), 2699. https://doi.org/10.3390/v14122699.

Kholodilov, I. S., Litov, A. G., Klimentov, A. S., Belova, O. A., Polienko, A. E., Nikitin, N. A., Shchetinin, A. M., Ivannikova, A. Y., Bell-Sakyi, L., Yakovlev, A. S., Bugmyrin, S. V., Bespyatova, L. A., Gmyl, L. V., Luchinina, S. V., Gmyl, A. P., Gushchin, V. A., & Karganova, G. G. (2020). Isolation and characterisation of alongshan virus in Russia. *Viruses*, 12(4), 362. https://doi.org/10.3390/v12040362.

Khoo, J. J., Kurtti, T. J., Husin, N. A., Beliavskaia, A., Lim, F. S., Zulkifli, M. M. S., Al-Khafaji, A. M., Hartley, C., Darby, A. C., Hughes, G. L., AbuBakar, S., Makepeace, B. L., & Bell-Sakyi, L. (2020). Isolation and propagation of laboratory strains and a novel flea-derived field strain of Wolbachia in tick cell lines. *Microorganisms*, 8(7), 988. https://doi.org/10.3390/microorganisms8070988.

Kim, K. G., Hwang, D. J., Park, J. W., Ryu, M. G., Kim, Y., Yang, S. J., Lee, J. E., Lee, G. S., Lee, J. H., Park, J. S., Seo, J. M., & Kim, S. H. (2024). Distribution and pathogen prevalence of field-collected ticks from south-western Korea: A study from 2019 to 2022. *Scientific reports*, 14(1), 12336. https://doi.org/10.1038/s41598-024-61127-y.

Kim, S. Y., Seo, C. W., & Lee, H. I. (2023). Severe fever with thrombocytopenia syndrome virus from ticks: A molecular epidemiological study of a patient in the Republic of Korea. *Experimental & applied acarology*, 89(2), 305–315. https://doi.org/10.1007/s10493-023-00783-6.

Kim, U. J., Kim, D. M., Kim, S. E., Kang, S. J., Jang, H. C., Park, K. H., & Jung, S. I. (2018). Case report: Detection of the identical virus in a patient presenting with severe fever with thrombocytopenia syndrome encephalopathy and the tick that bit her. *BMC infectious diseases*, 18(1), 181. https://doi.org/10.1186/s12879-018-3092-y.

Kim, Y. J., Seo, J. Y., Kim, S. Y., & Lee, H. I. (2022). Molecular detection of Anaplasma phagocytophilum and Ehrlichia species in ticks removed from humans in the Republic of Korea. *Microorganisms*, 10(6), 1224. https://doi.org/10.3390/microorganisms10061225.

Kirkland, B. H., Westwood, G. S., & Keyhani, N. O. (2004). Pathogenicity of entomopathogenic fungi Beauveria bassiana and Metarhizium anisopliae to Ixodidae tick species Dermacentor variabilis, Rhipicephalus sanguineus, and Ixodes scapularis. *Journal of medical entomology*, 41(4), 705–711. https://doi.org/10.1603/0022-2585-41.4.705.

Kisten, D., Brinkerhoff, J., Tshilwane, S. I., & Mukaratirwa, S. (2021). A pilot study on the microbiome of Amblyomma hebraeum tick stages infected and non-infected with Rickettsia africae. *Pathogens (Basel, Switzerland)*, 10(8), 941. https://doi.org/10.3390/pathogens10080941.

Kitsou, C., Foor, S. D., Dutta, S., Bista, S., & Pal, U. (2021). Tick gut barriers impacting tick-microbe interactions and pathogen persistence. *Molecular microbiology*, 116(5), 1241–1248. https://doi.org/10.1111/mmi.14822.

Kloc, A., Wójcik-Fatla, A., Paprzycki, P., & Panasiuk, L. (2024). Transovarial transmission of Rickettsia spp., Francisella-like endosymbionts, and Spiroplasma spp. in Dermacentor reticulatus ticks. *Ticks and tick-borne diseases*, 15(6), 102421. https://doi.org/10.1016/j.ttbdis.2024.102421.

Klyachko, O., Stein, B. D., Grindle, N., Clay, K., & Fuqua, C. (2007). Localization and visualization of a coxiella-type symbiont within the lone star tick, Amblyomma americanum. *Applied and environmental microbiology*, 73(20), 6584–6594. https://doi.org/10.1128/AEM.00537-07.

Kobayashi, D., Kuwata, R., Kimura, T., Shimoda, H., Fujita, R., Faizah, A. N., Kai, I., Matsumura, R., Kuroda, Y., Watanabe, S., Kuniyoshi, S., Yamauchi, T., Watanabe, M., Higa, Y., Hayashi, T., Shinomiya, H., Maeda, K., Kasai, S., Sawabe, K., & Isawa, H. (2021). Detection of Jingmenviruses in Japan with evidence of vertical transmission in ticks. *Viruses*, 13(12), 2547. https://doi.org/10.3390/v13122547.

Kobayashi, D., Ohashi, M., Osei, J. H. N., Agbosu, E., Opoku, M., Agbekudzi, A., Joannides, J., Fujita, R., Sasaki, T., Bonney, J. H. K., Dadzie, S., Isawa, H., Sawabe, K., & Ohta, N. (2017). Detection of a novel putative phlebovirus and first isolation of Dugbe virus from ticks in Accra, Ghana. *Ticks and tick-borne diseases*, 8(4), 640–645. https://doi.org/10.1016/j.ttbdis.2017.04.010.

Koczanowicz, S., Nowak-Chmura, M., Kocoń, A., Rączka, G., & Asman, M. (2025). The occurrence of Borrelia burgdorferi sensu lato in Ixodes ricinus ticks collected from nature-educational and tourist trails in the poprad landscape park. *Pathogens (Basel, Switzerland)*, 14(2), 117. https://doi.org/10.3390/pathogens14020117.

Kong, Y., Yan, C., Liu, D., Jiang, L., Zhang, G., He, B., & Li, Y. (2022). Phylogenetic analysis of Crimean-Congo hemorrhagic fever virus in inner Mongolia, China. *Ticks and tick-borne diseases*, 13(1), 101856. https://doi.org/10.1016/j.ttbdis.2021.101856.

Kopácek, P., Hajdusek, O., Buresová, V., & Daffre, S. (2010). Tick innate immunity. *Advances in experimental medicine and biology*, 708, 137–162. doi: 10.1007/978-1-4419-8059-5_8.

Kopecky, J., Perotti, M. A., Nesvorna, M., Erban, T., & Hubert, J. (2013). Cardinium endosymbionts are widespread in synanthropic mite species (Acari: Astigmata). *Journal of invertebrate pathology*, 112(1), 20–23. https://doi.org/10.1016/j.jip.2012.11.001.

Kovryha, N., Tsyhankova, A., Zelenuchina, O., Mashchak, O., Terekhov, R., & Rogovskyy, A. S. (2021). Prevalence of Borrelia burgdorferi and Anaplasma phagocytophilum in Ixodid Ticks from Southeastern Ukraine. *Vector borne and zoonotic diseases (Larchmont, N.Y.)*, 21(4), 242–246. https://doi.org/10.1089/vbz.2020.2716.

Krause, P. J. (2019). Human babesiosis. *International journal for parasitology*, 49(2), 165–174. https://doi.org/10.1016/j.ijpara.2018.11.007.

Krawczyk, A. I., Röttjers, S., Coimbra-Dores, M. J., Heylen, D., Fonville, M., Takken, W., Faust, K., & Sprong, H. (2022). Tick microbial associations at the crossroad of horizontal and vertical transmission pathways. *Parasites & vectors*, 15(1), 380. https://doi.org/10.1186/s13071-022-05519-w.

Krige, A. S., Thompson, R. C. A., Seidlitz, A., Keatley, S., Botero, A., & Clode, P. L. (2021). 'Hook, line, and sinker': Fluorescence in situ hybridisation (FISH) uncovers Trypanosoma noyesi in Australian questing ticks. *Ticks and tick-borne diseases*, 12(1), 101596. https://doi.org/10.1016/j.ttbdis.2020.101596.

Kuan, C. Y., Lin, T. L., Ou, S. C., Chuang, S. T., Chan, J. P., Maeda, K., Mizutani, T., Wu, M. P., Lee, F., Chan, F. T., Chang, C. C., Liang, R. L., Yang, S. F., Liu, T. C., Tu, W. C., Tzeng, H. Y., Lee, C. J., Lin, C. F., Lee, H. H., Wu, J. H., … & Chou, C. C. (2023). The first nationwide surveillance of severe fever with thrombocytopenia syndrome in ruminants and wildlife in Taiwan. *Viruses*, 15(2), 441. https://doi.org/10.3390/v15020441.

Kuba, Y., Azama, Y., Kyan, H., Fukuchi, Y., Maeshiro, N., Kakita, T., Miyahira, M., Kudeken, T., & Nidaira, M. (2022). Severe fever with thrombocytopenia syndrome virus RNA in ticks from wild Mongooses in Okinawa Prefecture, Japan. *Japanese journal of infectious diseases*, 75(6), 612–615. https://doi.org/10.7883/yoken.JJID.2021.808.

Kubiak, K., Szymańska, H., Dziekońska-Rynko, J., Tylkowska, A., Dmitryjuk, M., & Dzika, E. (2024). Tick-borne pathogens in questing adults Dermacentor reticulatus from the Eastern European population (north-eastern Poland). *Scientific reports*, 14(1), 698. https://doi.org/10.1038/s41598-024-51302-x.

Kuhn, J. H., Alkhovsky, S. V., Avšič-Županc, T., Bergeron, É., Burt, F., Ergünay, K., Garrison, A. R., Marklewitz, M., Mirazimi, A., Papa, A., Pawęska, J. T., Spengler, J. R., & Palacios, G. (2024). ICTV virus taxonomy profile: Nairoviridae 2024. *The Journal of general virology*, 105(4), 001974. https://doi.org/10.1099/jgv.0.001974.

Kuivanen, S., Levanov, L., Kareinen, L., Sironen, T., Jääskeläinen, A. J., Plyusnin, I., Zakham, F., Emmerich, P., Schmidt-Chanasit, J., Hepojoki, J., Smura, T., & Vapalahti, O. (2019). Detection of novel tick-borne pathogen, Alongshan virus, in Ixodes ricinus ticks, south-eastern Finland, 2019. *Euro surveillance: bulletin Europeen sur les maladies transmissibles = European communicable disease bulletin*, 24(27), 1900394. https://doi.org/10.2807/1560-7917.ES.2019.24.27.1900394.

Kumar, D., Budachetri, K., Meyers, V. C., & Karim, S. (2016). Assessment of tick antioxidant responses to exogenous oxidative stressors and insight into the role of catalase in the reproductive fitness of the Gulf Coast tick, Amblyomma maculatum. *Insect molecular biology*, 25(3), 283–294. https://doi.org/10.1111/imb.12218.

Kuo, C. C., Huang, J. L., Chien, C. H., Shih, H. C., & Wang, H. C. (2018). First molecular detection of Anaplasma phagocytophilum in the hard tick Rhipicephalus haemaphysaloides in Taiwan. *Experimental & applied acarology*, 75(4), 437–443. https://doi.org/10.1007/s10493-018-0283-6.

Kurilshikov, A., Livanova, N. N., Fomenko, N. V., Tupikin, A. E., Rar, V. A., Kabilov, M. R., Livanov, S. G., & Tikunova, N. V. (2015). Comparative metagenomic profiling of symbiotic bacterial communities associated with Ixodes persulcatus, Ixodes pavlovskyi and Dermacentor reticulatus ticks. *PloS one*, 10(7), e0131413. https://doi.org/10.1371/journal.pone.0131413.

Kurokawa, C., Lynn, G. E., Pedra, J. H. F., Pal, U., Narasimhan, S., & Fikrig, E. (2020). Interactions between Borrelia burgdorferi and ticks. *Nature reviews*

Microbiology, 18(10), 587–600. https://doi.org/10.1038/s41579-020-0400-5.

Kurtti, T. J., Felsheim, R. F., Burkhardt, N. Y., Oliver, J. D., Heu, C. C., & Munderloh, U. G. (2015). Rickettsia buchneri sp. nov., a rickettsial endosymbiont of the blacklegged tick Ixodes scapularis. *International journal of systematic and evolutionary microbiology*, 65(Pt 3), 965–970. https://doi.org/10.1099/ijs.0.000047.

Kurtti, T. J., Palmer, A. T., & Oliver, J. H., Jr. (2002). Rickettsiella-like bacteria in Ixodes woodi (Acari: Ixodidae). *Journal of medical entomology*, 39(3), 534–540. https://doi.org/10.1603/0022-2585-39.3.534.

Kwan, J. Y., Griggs, R., Chicana, B., Miller, C., & Swei, A. (2017). Vertical vs. horizontal transmission of the microbiome in a key disease vector, Ixodes pacificus. *Molecular ecology*, 26(23), 6578–6589. https://doi.org/10.1111/mec.14391.

Laatamna, A., Strube, C., Bakkes, D. K., Schaper, S., Aziza, F. Z., Ben Chelef, H., Amrane, N. E. H., Bedraoui, R., Dobler, G., & Chitimia-Dobler, L. (2022). Molecular detection of tick-borne pathogens in Rhipicephalus sanguineus sensu stricto collected from dogs in the steppe and high plateau regions of Algeria. *Acta tropica*, 234, 106582. https://doi.org/10.1016/j.actatropica.2022.106584.

Lagunova, E. K., Liapunova, N. A., Tuul, D., Otgonsuren, G., Nomin, D., Erdenebat, N., Abmed, D., Danchinova, G. A., Sato, K., Kawabata, H., & Khasnatinov, M. A. (2022). Co-infections with multiple pathogens in natural populations of Ixodes persulcatus ticks in Mongolia. *Parasites & vectors*, 15(1), 236. https://doi.org/10.1186/s13071-022-05359-x.

Lalzar, I., Friedmann, Y., & Gottlieb, Y. (2014). Tissue tropism and vertical transmission of Coxiella in Rhipicephalus sanguineus and Rhipicephalus turanicus ticks. *Environmental microbiology*, 16(12), 3657–3668. https://doi.org/10.1111/1462-2920.12455.

Lalzar, I., Harrus, S., Mumcuoglu, K. Y., & Gottlieb, Y. (2012). Composition and seasonal variation of Rhipicephalus turanicus and Rhipicephalus sanguineus bacterial communities. *Applied and environmental microbiology*, 78(12), 4110–4116. https://doi.org/10.1128/AEM.00323-12.

Lamkanfi, M., & Dixit, V. M. (2010). Manipulation of host cell death pathways during microbial infections. *Cell host & microbe*, 8(1), 44–54. https://doi.org/10.1016/j.chom.2010.06.007.

Landesman, W. J., Mulder, K., Fredericks, L. P., & Allan, B. F. (2019). Cross-kingdom analysis of nymphal-stage Ixodes scapularis microbial communities in relation to Borrelia burgdorferi infection and load. *FEMS microbiology ecology*, 95(12), fiz167. https://doi.org/10.1093/femsec/fiz167.

Landmann, F. (2019). The Wolbachia endosymbionts. *Microbiology spectrum*, 7(2), 7(2), https://doi.org/10.1128/microbiolspec.bai-0018-2019.

Latrofa, M. S., Dantas-Torres, F., Giannelli, A., & Otranto, D. (2014). Molecular detection of tick-borne pathogens in Rhipicephalus sanguineus group ticks. *Ticks Tick Borne Dis*, 5, 943–6.

Leblebicioglu, H., Eroglu, C., Erciyas-Yavuz, K., Hokelek, M., Acici, M., & Yilmaz, H. (2014). Role of migratory birds in spreading Crimean-Congo hemorrhagic fever, Turkey. *Emerging infectious diseases*, 20(8), 1331–1334. https://doi.org/10.3201/eid2008.131547.

Leclerque, A., & Kleespies, R. G. (2012). A Rickettsiella bacterium from the hard tick, Ixodes woodi: Molecular taxonomy combining multilocus sequence typing (MLST) with significance testing. *PloS one*, 7(5), e38062. https://doi.org/10.1371/journal.pone.0038062.

Ledger, K. J., Beati, L., & Wisely, S. M. (2021). Survey of ticks and tick-borne Rickettsial and Protozoan pathogens in Eswatini. *Pathogens (Basel, Switzerland)*, 10(8), 1043. https://doi.org/10.3390/pathogens10081043.

Lee, J. S., & Chung, S. Y. (2023). The threat of climate change on tick-borne infections: Rising trend of infections and geographic distribution of climate risk factors associated with ticks. *The Journal of infectious diseases*, 227(2), 295–303. https://doi.org/10.1093/infdis/jiac300.

Lee, M. R., Kim, J. C., Park, S. E., Lee, S. J., Kim, W. J., Lee, D. H., & Kim, J. S. (2021). Interactive gene expression between Metarhizium anisopliae JEF-290 and Longhorned tick Haemaphysalis longicornis at early stage of infection. *Frontiers in physiology*, 12, 643389. https://doi.org/10.3389/fphys.2021.643389.

Lee, M. R., Li, D., Lee, S. J., Kim, J. C., Kim, S., Park, S. E., Baek, S., Shin, T. Y., Lee, D. H., & Kim, J. S. (2019). Use of Metarhizum aniopliae s.l. to control soil-dwelling longhorned tick, Haemaphysalis longicornis. *Journal of invertebrate pathology*, 166, 107230. https://doi.org/10.1016/j.jip.2019.107230.

Lee, S. H., Shin, N. R., Kim, C. M., Park, S., Yun, N. R., Kim, D. M., & Jung, D. S. (2020). First identification of Anaplasma phagocytophilum in both a biting tick Ixodes nipponensis and a patient in Korea: A case report. *BMC infectious diseases*, 20(1), 826. https://doi.org/10.1186/s12879-020-05522-5.

Lee, Y. J., Kim, B., Kwak, D., Nyamsuren, O., Guugandaa, N., & Seo, M. G. (2024). Tick-borne pathogens in Mongolian ticks: The high prevalence of Rickettsia raoultii and its public health implications. *Acta tropica*, 260, 107412. https://doi.org/10.1016/j.actatropica.2024.107412.

Lefoulon, E., Bain, O., Bourret, J., Junker, K., Guerrero, R., Cañizales, I., et al. (2015). Shaking the tree: Multi-locus sequence typing usurps current onchocercid (filarial nematode) phylogeny. *PLoS Negl Trop Dis*, 9, e0004233.

Lei, Y., Dai, Y., Guo, M., Yi, L., Cai, K., & Hu, B. (2025). Molecular survey and phylogenetic characterization of tick-borne pathogens in Hubei Province, China. *Vector borne and zoonotic diseases (Larchmont, N.Y.)*, 25(6), 377–389. https://doi.org/10.1089/vbz.2024.0126.

Lejal, E., Chiquet, J., Aubert, J., Robin, S., Estrada-Peña, A., Rue, O., Midoux, C., Mariadassou, M., Bailly, X., Cougoul, A., Gasqui, P., Cosson, J. F., Chalvet-Monfray, K., Vayssier-Taussat, M., & Pollet, T. (2021). Temporal patterns in Ixodes ricinus microbial communities: An insight into tick-borne microbe interactions. *Microbiome*, 9(1), 153. https://doi.org/10.1186/s40168-021-01051-8.

Levytska, V. A., Mushinsky, A. B., Zubrikova, D., Blanarova, L., Długosz, E., Vichova, B., Slivinska, K. A., Gajewski, Z., Gizinski, S., Liu, S., Zhou, L., & Rogovskyy, A. S. (2021). Detection of pathogens in ixodid ticks collected from animals and vegetation in five regions of Ukraine. *Ticks and tick-borne diseases*, 12(1), 101586. https://doi.org/10.1016/j.ttbdis.2020.101586.

Lewis, D. (1979). The detection of rickettsia-like microorganisms within the ovaries of female Ixodes ricinus ticks. *Zeitschrift fur Parasitenkunde (Berlin, Germany)*, 59(3), 295–298. https://doi.org/10.1007/BF00927523.

Li, J., Zhang, S., Liang, W., Zhao, S., Wang, Z., Li, H., Yang, B., Zhang, Z., Li, J., & Jia, L. (2022). Survey of tick species and molecular detection of selected tick-borne pathogens in Yanbian, China. Enquête sur les espèces de tiques et détection moléculaire de certains agents pathogènes transmis par les tiques à Yanbian, en Chine. *Parasite (Paris, France)*, 29, 38. https://doi.org/10.1051/parasite/2022039.

Li, L. H., Zhang, Y., & Zhu, D. (2018b). Effects of antibiotic treatment on the fecundity of Rhipicephalus haemaphysaloides ticks. *Parasites & vectors*, 11(1), 242. https://doi.org/10.1186/s13071-018-2807-7.

Li, L. H., Zhang, Y., Zhu, D., & Zhou, X. N. (2018a). Endosymbionts alter larva-to-nymph transstadial transmission of Babesia microti in Rhipicephalus haemaphysaloides ticks. *Frontiers in microbiology*, 9, 1415. https://doi.org/10.3389/fmicb.2018.01415.

Li, S., Zhang, L., Zhou, P., Li, Z., Song, H., Song, Y., Li, C., Xiangzheng, W., & Wu, J. (2023). Detection of Babesia Spp., Theileria Spp., and Anaplasma Ovis in Ornithodoros Lahorensis from Southern Xinjiang, China. *Journal of veterinary research*, 67(1), 79–85. https://doi.org/10.2478/jvetres-2023-0019.

Li, Y., Li, J., Xieripu, G., Rizk, M. A., Macalanda, A. M. C., Gan, L., Ren, J., Mohanta, U. K., El-Sayed, S. A. E., Chahan, B., Xuan, X., & Guo, Q. (2024). Molecular detection of Theileria ovis, Anaplasma ovis, and Rickettsia spp. in Rhipicephalus turanicus and Hyalomma anatolicum collected from sheep in Southern Xinjiang, China. *Pathogens (Basel, Switzerland)*, 13(8), 680. https://doi.org/10.3390/pathogens13080680.

Li, Y., Wen, X., Li, M., Moumouni, P. F. A., Galon, E. M., Guo, Q., Rizk, M. A., Liu, M., Li, J., Ji, S., Tumwebaze, M. A., Byamukama, B., Chahan, B., & Xuan, X. (2020). Molecular detection of tick-borne pathogens harbored by ticks collected from livestock in the Xinjiang Uygur Autonomous Region, China. *Ticks and tick-borne diseases*, 11(5), 101478. https://doi.org/10.1016/j.ttbdis.2020.101482.

Li, Z., Bao, C., Hu, J., Liu, W., Wang, X., Zhang, L., Ji, Z., Feng, Z., Li, L., Shen, A., Liu, X., Zhao, H., Tan, W., Zhou, J., Qi, X., Zhu, Y., Tang, F., Cardona, C. J., & Xing, Z. (2016). Ecology of the tick-borne phlebovirus causing severe fever with thrombocytopenia syndrome in an endemic area of China. *PLoS neglected tropical diseases*, 10(4), e0004574. https://doi.org/10.1371/journal.pntd.0004574.

Lin, T. L., Ou, S. C., Maeda, K., Shimoda, H., Chan, J. P., Tu, W. C., Hsu, W. L., & Chou, C. C. (2020). The first discovery of severe fever with thrombocytopenia syndrome virus in Taiwan. *Emerging microbes & infections*, 9(1), 148–151. https://doi.org/10.1080/22221751.2019.1710436.

Lina, T. T., Farris, T., Luo, T., Mitra, S., Zhu, B., & McBride, J. W. (2016). Hacker within! Ehrlichia chaffeensis effector driven phagocyte reprogramming strategy. *Frontiers in cellular and infection microbiology*, 6, 58. https://doi.org/10.3389/fcimb.2016.00058.

Lineberry, M. W., Sundstrom, K. D., Little, S. E., & Allen, K. E. (2021). Molecular detection of Cercopithifilaria bainae in brown dog ticks collected from dogs across the United States. *Veterinary parasitology*, 299, 109583. https://doi.org/10.1016/j.vetpar.2021.109583.

Lineberry, M. W., Sundstrom, K. D., Little, S. E., Stayton, E. M., & Allen, K. E. (2020). Detection of Cercopithifilaria bainae infection in shelter dogs and ticks in Oklahoma, USA. *Parasites & vectors*, 13(1), 216. https://doi.org/10.1186/s13071-020-04089-z.

Lipatova, I., Černevičienė, D., Griciuvienė, L., Ražanskė, I., Aleksandravičienė, A., Kibiša, A., Radzijevskaja, J., Olech, W., Anusz, K., Didkowska, A., & Paulauskas, A. (2023). Anaplasma phagocytophilum in European bison (Bison bonasus) and their ticks from Lithuania and Poland. *Ticks and tick-borne diseases*, 14(6), 102246. https://doi.org/10.1016/j.ttbdis.2023.102246.

Liu, H., Li, Z., Wang, Z., He, B., Wang, S., Wei, F., Tu, C., & Liu, Q. (2016). The first molecular evidence of severe fever with thrombocytopenia syndrome virus in ticks in Jilin, Northeastern China. *Ticks and tick-borne diseases*, 7(6), 1280–1283. https://doi.org/10.1016/j.ttbdis.2016.06.007.

Liu, L., Dai, J., Zhao, Y. O., Narasimhan, S., Yang, Y., Zhang, L., & Fikrig, E. (2012). Ixodes scapularis JAK-STAT pathway regulates tick antimicrobial peptides, thereby controlling the agent of human granulocytic anaplasmosis. *The Journal of infectious diseases*, 206(8), 1233–1241. https://doi.org/10.1093/infdis/jis484.

Liu, L., Li, L., Liu, J., Hu, Y., Liu, Z., Guo, L., & Liu, J. (2013). Coinfection of Dermacentor silvarum olenev (Acari: Ixodidae) by Coxiella-Like, Arsenophonus-like, and Rickettsia-like symbionts. *Applied and environmental microbiology*, 79(7), 2450–2454. https://doi.org/10.1128/AEM.03575-12.

Livengood, J., Hutchinson, M. L., Thirumalapura, N., & Tewari, D. (2020). Detection of Babesia, Borrelia, Anaplasma, and Rickettsia spp. in Adult Black-Legged Ticks (Ixodes scapularis) from Pennsylvania, United States, with a Luminex Multiplex Bead Assay. *Vector borne and zoonotic diseases (Larchmont, N.Y.)*, 20(6), 406–411. https://doi.org/10.1089/vbz.2019.2551.

Lo, N., Beninati, T., Sassera, D., Bouman, E. A., Santagati, S., Gern, L., Sambri, V., Masuzawa, T., Gray, J. S., Jaenson, T. G., Bouattour, A., Kenny, M. J., Guner, E. S., Kharitonenkov, I. G., Bitam, I., & Bandi, C. (2006). Widespread distribution and high prevalence of an alpha-proteobacterial symbiont in the tick Ixodes ricinus. *Environmental microbiology*, 8(7), 1280–1287. https://doi.org/10.1111/j.1462-2920.2006.01024.x.

Lopes de Carvalho, I., Toledo, A., Carvalho, C. L., Barandika, J. F., Respicio-Kingry, L. B., Garcia-Amil, C., García-Pérez, A. L., Olmeda, A. S., Zé-Zé, L., Petersen, J. M., Anda, P., Núncio, M. S., & Escudero, R. (2016). Francisella species in ticks and animals, Iberian Peninsula. *Ticks and tick-borne diseases*, 7(1), 159–165. https://doi.org/10.1016/j.ttbdis.2015.10.009.

López, Y., Miranda, J., Mattar, S. et al. (2020). First report of Lihan Tick virus (Phlebovirus, Phenuiviridae) in ticks, Colombia. *Virology journal*, 17, 63.

Lu, M., Meng, C., Li, Y., Zhou, G., Wang, L., Xu, X., Li, N., Ji, Y., Tian, J., Wang, W., & Li, K. (2023). Rickettsia sp. and Anaplasma spp. in Haemaphysalis longicornis from Shandong province of China, with evidence of a novel species "Candidatus Anaplasma Shandongensis". *Ticks and tick-borne diseases*, 14(1), 102082. https://doi.org/10.1016/j.ttbdis.2022.102083.

Lu, M., Meng, C., Zhang, B., Wang, X., Tian, J., Tang, G., Wang, W., Li, N., Li, M., Xu, X., Sun, Y., Duan, C., Qin, X., & Li, K. (2022a). Prevalence of spotted fever group Rickettsia and Candidatus Lariskella in multiple tick species from Guizhou Province, China. *Biomolecules*, 12(11), 1701. https://doi.org/10.3390/biom12111701.

Luan, Y., Gou, J., Zhong, D., Ma, L., Yin, C., Shu, M., Liu, G., & Lin, Q. (2023). The tick-borne pathogens: An overview of China's situation. *Acta parasitologica*, 68(1), 1–20. https://doi.org/10.1007/s11686-023-00658-1.

Luo, T., Hu, E., Gan, L., Yang, D., Wu, J., Gao, S., Tuo, X., Bayin, C. G., Hu, Z., & Guo, Q. (2024). Candidatus Midichloria mitochondrii can be vertically transmitted in Hyalomma anatolicum. *Experimental parasitology*, 265, 108828. https://doi.org/10.1016/j.exppara.2024.108828.

Luu, L., Bown, K. J., Palomar, A. M., Kazimírová, M., & Bell-Sakyi, L. (2020). Isolation and partial characterisation of a novel Trypanosoma from the tick Ixodes ricinus. *Ticks and tick-borne diseases*, 11(5), 101501. https://doi.org/10.1016/j.ttbdis.2020.101501.

Ma, J., Lv, X. L., Zhang, X., Han, S. Z., Wang, Z. D., Li, L., Sun, H. T., Ma, L. X., Cheng, Z. L., Shao, J. W., Chen, C., Zhao, Y. H., Sui, L., Liu, L. N., Qian, J., Wang, W., & Liu, Q. (2021). Identification of a new orthonairovirus associated with human febrile illness in China. *Nature medicine*, 27(3), 434–439. https://doi.org/10.1038/s41591-020-01228-y.

Ma, R., Shi, Y., Wu, W., Huang, C., Xue, F., Hou, R., Zhou, Y., Gu, J., Feng, F., Yu, X., Liu, J., Li, Z., Zhang, L., Lan, G., Chen, C., Bi, W., Song, X., Fu, M., Yang, H., Gu, H., … & Qi, D. (2025). The bacterial diversity and potential pathogenic risks of giant panda-infesting ticks. *Microbiology spectrum*, 13(7), e0219724. https://doi.org/10.1128/spectrum.02197-24.

Maas, L. M. I., Winter, M., Herrmann, V., Abate, S. D., Obiegala, A., Nava, S., & Sebastian, P. S. (2024). Anaplasma platys and Rickettsia massiliae in Rhipicephalus sanguineus sensu stricto ticks collected on dogs in the Patagonian region of Argentina. *Parasitology*, 151(8), 848–852. https://doi.org/10.1017/S0031182024000933.

Machado-Ferreira, E., Dietrich, G., Hojgaard, A., Levin, M., Piesman, J., Zeidner, N. S., & Soares, C. A. (2011). Coxiella symbionts in the Cayenne tick Amblyomma cajennense. *Microbial ecology*, 62(1), 134–142. https://doi.org/10.1007/s00248-011-9868-x.

Madison-Antenucci, S., Kramer, L. D., Gebhardt, L. L., & Kauffman, E. (2020). Emerging tick-borne diseases. *Clinical microbiology reviews*, 33(2), e00083-18. https://doi.org/10.1128/CMR.00083-18.

Mahdi, M., Erickson, B. R., Comer, J. A., Nichol, S. T., Rollin, P. E., AlMazroa, M. A., & Memish, Z. A.

(2011). Kyasanur forest disease virus Alkhurma subtype in ticks, Najran Province, Saudi Arabia. *Emerging infectious diseases*, 17(5), 945–947. https://doi.org/10.3201/eid1705.101824.

Maia, C., Ferreira, A., Nunes, M., Vieira, M. L., Campino, L., & Cardoso, L. (2014). Molecular detection of bacterial and parasitic pathogens in hard ticks from Portugal. *Ticks and tick-borne diseases*, 5(4), 409–414. https://doi.org/10.1016/j.ttbdis.2014.01.009.

Mani, R. J., Metcalf, J. A., & Clinkenbeard, K. D. (2015). Amblyomma americanum as a Bridging Vector for Human Infection with Francisella tularensis. *PloS one*, 10(6), e0130513. https://doi.org/10.1371/journal.pone.0130513.

Manoj, R. R. S., Iatta, R., Latrofa, M. S., Capozzi, L., Raman, M., Colella, V., & Otranto, D. (2020). Canine vector-borne pathogens from dogs and ticks from Tamil Nadu, India. *Acta tropica*, 203, 105308. https://doi.org/10.1016/j.actatropica.2019.105310.

Mansfield, K. L., Cook, C., Ellis, R. J., Bell-Sakyi, L., Johnson, N., Alberdi, P., de la Fuente, J., & Fooks, A. R. (2017). Tick-borne pathogens induce differential expression of genes promoting cell survival and host resistance in Ixodes ricinus cells. *Parasit Vectors*, 10(1), 81. https://doi.org/10.1186/s13071-017-2011-1.

Mansfield, K. L., González, E., McKay, S., Apaa, T., Kent, A. J., Cropper, P., Berry, N., Hernández-Triana, L. M., & Johnson, N. (2024). Short communication: Anaplasma phagocytophilum and Babesia spp. in ixodid ticks infesting red foxes (Vulpes vulpes) in Great Britain. *Ticks and tick-borne diseases*, 15(6), 102401. https://doi.org/10.1016/j.ttbdis.2024.102401.

Maqbool, M., Sajid, M. S., Saqib, M., Anjum, F. R., Tayyab, M. H., Rizwan, H. M., Rashid, M. I., Rashid, I., Iqbal, A., Siddique, R. M., Shamim, A., Hassan, M. A., Atif, F. A., Razzaq, A., Zeeshan, M., Hussain, K., Nisar, R. H. A., Tanveer, A., Younas, S., Kamran, K., … & Rahman, S. U. (2022). Potential mechanisms of transmission of tick-borne viruses at the virus-tick interface. *Frontiers in microbiology*, 13, 846884. https://doi.org/10.3389/fmicb.2022.846884.

Mariconti, M., Epis, S., Gaibani, P., Dalla Valle, C., Sassera, D., Tomao, P., Fabbi, M., Castelli, F., Marone, P., Sambri, V., Bazzocchi, C., & Bandi, C. (2012). Humans parasitized by the hard tick Ixodes ricinus are seropositive to Midichloria mitochondrii: Is Midichloria a novel pathogen, or just a marker of tick bite?. *Pathogens and global health*, 106(7), 391–396. https://doi.org/10.1179/2047773212Y.0000000050.

Martinez-Villegas, L., Lado, P., Klompen, H., Wang, S., Cummings, C., Pesapane, R., & Short, S. M. (2024). The microbiota of Amblyomma americanum reflects known westward expansion. *PloS one*, 19(6), e0304959. https://doi.org/10.1371/journal.pone.0304959.

Martins, L. A., Galletti, M. F. B. M., Ribeiro, J. M., Fujita, A., Costa, F. B., Labruna, M. B., Daffre, S., & Fogaça, A. C. (2017). The distinct transcriptional response of the midgut of Amblyomma sculptum and Amblyomma aureolatum ticks to Rickettsia rickettsii correlates to their differences in susceptibility to infection. *Frontiers in cellular and infection microbiology*, 7, 129. https://doi.org/10.3389/fcimb.2017.00129.

Martins, L. A., Malossi, C. D., Galletti, M. F. B. M., Ribeiro, J. M., Fujita, A., Esteves, E., Costa, F. B., Labruna, M. B., Daffre, S., & Fogaça, A. C. (2019). The transcriptome of the salivary glands of Amblyomma aureolatum reveals the antimicrobial peptide microplusin as an important factor for the tick protection against Rickettsia rickettsii infection. *Frontiers in physiology*, 10, 529. https://doi.org/10.3389/fphys.2019.00529.

Martins, L. A., Palmisano, G., Cortez, M., Kawahara, R., de Freitas Balanco, J. M., Fujita, A., Alonso, B. I., Barros-Battesti, D. M., Braz, G. R. C., Tirloni, L., Esteves, E., Daffre, S., & Fogaça, A. C. (2020). The intracellular bacterium Rickettsia rickettsii exerts an inhibitory effect on the apoptosis of tick cells. *Parasites & vectors*, 13(1), 603. https://doi.org/10.1186/s13071-020-04477-5.

Marzouk, A. S., Swelim, H. H., & Ali, A. A. B. (2020). Ultrastructural changes induced by the entomopathogenic fungus Beauveria bassiana in the ovary of the tick Argas (Persicargas) persicus (Oken). *Ticks and tick-borne diseases*, 11(6), 101507. https://doi.org/10.1016/j.ttbdis.2020.101507.

Masatani, T., Hayashi, K., Andoh, M., Tateno, M., Endo, Y., Asada, M., Kusakisako, K., Tanaka, T., Gokuden, M., Hozumi, N., Nakadohzono, F., & Matsuo, T. (2017). Detection and molecular characterization of Babesia, Theileria, and Hepatozoon species in hard ticks collected from Kagoshima, the southern region in Japan. *Ticks and tick-borne diseases*, 8(4), 581–587. https://doi.org/10.1016/j.ttbdis.2017.03.007.

Matei, I. A., Kalmár, Z., Magdaş, C., Magdaş, V., Toriay, H., Dumitrache, M. O., Ionică, A. M., D'Amico, G., Sándor, A. D., Mărcuţan, D. I., Domşa, C., Gherman, C. M., & Mihalca, A. D. (2015). Anaplasma phagocytophilum in questing Ixodes ricinus ticks from Romania. *Ticks and tick-borne diseases*, 6(3), 408–413. https://doi.org/10.1016/j.ttbdis.2015.03.010.

Matsumura, R., Kobayashi, D., Yamauchi, T., Park, E., Nishino, A., Maeda, K., Kasai, S., Itoyama, K., & Isawa, H. (2024). Isolation and characterization of Iwanai Valley virus, a new tick-borne nairovirus from Ixodes ovatus ticks in Hokkaido, Japan. *Archives of virology*, 170(1), 23. https://doi.org/10.1007/s00705-024-06195-0.

Matsuura, Y., Kikuchi, Y., Meng, X.Y., Koga, R., & Fukatsu, T. (2012). Novel clade of alphaproteobacterial endosymbionts associated with stinkbugs and other arthropods. *Applied and environmental microbiology*, 78(12), 4149–4156. https://doi.org/10.1128/AEM.00673-12.

Mays Maestas, S. E., Maestas, L. P., & Kaufman, P. E. (2025). Pathogen and host associations of soft ticks collected in South Texas. *Vector borne and zoonotic diseases (Larchmont, N.Y.)*, 25(1), 34–42. https://doi.org/10.1089/vbz.2023.0135.

McCauley, J. W., Hongo, S., Kaverin, N. V., Kochs, G., Lamb, R. A., Matrosovich, M. N., Perez, D. R., Palese, P., Presti, R. M., Rimstad, E., & Smith, G. J. D. (2009). Family: Orthomyxoviridae, international committee on taxonomy of viruses (ICTV). https://ictv.global/report_9th/RNAneg/Orthomyxoviridae. (accessed 2011).

Mediannikov, O. I. U., Ivanov, L. I., Nishikawa, M., Saito, R., Sidel'nikov, I. U. N., Zdanovskaia, N. I., Mokretsova, E. V., Tarasevich, I. V., & Suzuki, H. (2004). Mikroorganizm "Montezuma" iz poriadka Rickettsiales: Potentsial'nyĭ vozbuditel' kleshchevogo transmissivnogo zabolevaniia na Dal'nem Vostoke Rossii [Microorganism "Montezuma" of the order Rickettsiales: The potential causative agent of tick-borne disease in the Far East of Russia]. *Zhurnal mikrobiologii, epidemiologii i immunobiologii*, 1, 7–13.

Mediannikov, O., Subramanian, G., Sekeyova, Z., Bell-Sakyi, L., & Raoult, D. (2012). Isolation of Arsenophonus nasoniae from Ixodes ricinus ticks in Slovakia. *Ticks and tick-borne diseases*, 3(5-6), 367–370. https://doi.org/10.1016/j.ttbdis.2012.10.016.

Melis, S., Batisti Biffignandi, G., Olivieri, E., Galon, C., Vicari, N., Prati, P., Moutailler, S., Sassera, D., & Castelli, M. (2024). High-throughput screening of pathogens in Ixodes ricinus removed from hosts in Lombardy, northern Italy. *Ticks and tick-borne diseases*, 15(1), 102285. https://doi.org/10.1016/j.ttbdis.2023.102291.

Meng, K., Sun, W., Cheng, Z., Guo, H., Liu, J., & Chai, T. (2015). First detection of severe fever with thrombocytopenia syndrome virus in the tick species Haemaphysalis concinna in Shandong Province, China. *Parasitology research*, 114(12), 4703–4707. https://doi.org/10.1007/s00436-015-4718-2.

Mesquita, E., da Costa, D. P., Meirelles, L. N., Camargo, M. G., Corrêa, T. A., Bittencourt, V. R. E. P., da Silva Coelho, I., Santos, H. A., Humber, R. A., & Golo, P. S. (2023). Entomopathogenic fungus treatment changes the gut bacterial diversity of Rhipicephalus microplus ticks. *Parasites & vectors*, 16(1), 185. https://doi.org/10.1186/s13071-023-05790-5.

Mhamadi, M., Badji, A., Dieng, I., Gaye, A., Ndiaye, E. H., Ndiaye, M., Mhamadi, M., Touré, C. T., Mbaye, M. R., Barry, M. A., Ndiaye, O., Faye, B., Ba, F. A., Diop, B., Ndiaye, M., Fall, M., Sagne, S. N., Fall, G., Loucoubar, C., Fausther-Bovendo, H., … & Faye, O. (2022). Crimean-Congo hemorrhagic fever virus survey in humans, ticks, and livestock in Agnam (Northeastern Senegal) from february 2021 to March 2022. *Tropical medicine and infectious disease*, 7(10), 324. https://doi.org/10.3390/tropicalmed7100324.

Mohanta, U. K., Abdullah, S. M., Al-Wasef, Chikufenji, B., Ma, Z., Li, H., El-Sayed, S. A. E., Amer, M. M., Do, T. T., Islam, S., Nath, T. C., Li, Y., Umemiya-Shirafuji, R., Guo, Q., & Xuan, X. (2024). First molecular survey of tick-borne protozoan and bacterial pathogens in the questing tick population in Bangladesh. *Acta tropica*, 256, 107244. https://doi.org/10.1016/j.actatropica.2024.107248.

Molins, C. R., Delorey, M. J., Yockey, B. M., Young, J. W., Sheldon, S. W., Reese, S. M., Schriefer, M. E., & Petersen, J. M. (2010). Virulence differences among Francisella tularensis subsp. tularensis clades in mice. *PloS one*, 5(4), e10205. https://doi.org/10.1371/journal.pone.0010205.

Moming, A., Bai, Y., Wang, J., Zhang, Y., Tang, S., Fan, Z., Deng, F., & Shen, S. (2024). The known and unknown of global tick-borne viruses. *Viruses*, 16(12), 1807. https://doi.org/10.3390/v16121807.

Moming, A., Yue, X., Shen, S., Chang, C., Wang, C., Luo, T., Zhang, Y., Guo, R., Hu, Z., Zhang, Y., Deng, F., & Sun, S. (2018). Prevalence and phylogenetic analysis of Crimean-Congo hemorrhagic fever virus in ticks from different ecosystems in Xinjiang, China. *Virologica Sinica*, 33(1), 67–73. https://doi.org/10.1007/s12250-018-0016-3.

Moraga-Fernández, A., Muñoz-Hernández, C., Sánchez-Sánchez, M., Fernández de Mera, I. G., & de la Fuente, J. (2023). Exploring the diversity of tick-borne pathogens: The case of bacteria (Anaplasma, Rickettsia, Coxiella and Borrelia) protozoa (Babesia and Theileria) and viruses (Orthonairovirus, tick-borne encephalitis virus and louping ill virus) in the European continent. *Veterinary microbiology*, 286, 109892. https://doi.org/10.1016/j.vetmic.2023.109892.

Moraga-Fernández, A., Sánchez-Sánchez, M., Muñoz-Hernández, C., Pardavila, X., Sereno-Cadierno, J., Queirós, J., Vicente, J., & Fernández de Mera, I. G. (2024). Beware with the backpack! New hosts and pathogens identified for Ixodes simplex ticks collected from bats in the Iberian Peninsula. *Research in veterinary science*, 176, 105316. https://doi.org/10.1016/j.rvsc.2024.105329.

Morais, J. D., Dawson, J. E., Greene, C., Filipe, A. R., Galhardas, L. C., & Bacellar, F. (1991, September 7). First European case of ehrlichiosis. *Lancet*, 338(8767):633-4.

Morel, N., Thompson, C. S., Rossner, M. V., Mangold, A. J., & Nava, S. (2021). A Trypanosoma species detected in Rhipicephalus (Boophilus) microplus ticks from Argentina. *Ticks and tick-borne diseases*, 12(1), 101573. https://doi.org/10.1016/j.ttbdis.2020.101573.

Morozkin, E. S., Makenov, M. T., Zhurenkova, O. B., Kholodilov, I. S., Belova, O. A., Radyuk, E. V., Fyodorova, M. V., Grigoreva, Y. E., Litov, A. G., Valdokhina, A. V., Bulanenko, V. P., Samoilov, A. E., Korneenko, E. V., Voizekhovskaya, Y. A., Neverov, A. D., Karganova, G. G., & Karan, L. S. (2022). Integrated jingmenvirus polymerase gene in Ixodes ricinus genome. *Viruses*, 14(9), 1908. https://doi.org/10.3390/v14091908.

Mosha, E. T., Kuria, J. K. N., Otiende, M., & Lekolool, I. (2024). Molecular detection of Anaplasma phagocytophilum in small mammals and infesting ticks in Laikipia County, Kenya. *Veterinary medicine international*, 2024, 5575162. https://doi.org/10.1155/2024/5575162.

Moura-Martiniano, N. O., Machado-Ferreira, E., Gazêta, G. S., & Soares, C. A. G. (2017). Relative transcription of autophagy-related genes in Amblyomma sculptum and Rhipicephalus microplus ticks. *Experimental & applied acarology*, 73(3-4), 401–428. https://doi.org/10.1007/s10493-017-0193-z.

Moutailler, S., Valiente Moro, C., Vaumourin, E., Michelet, L., Tran, F. H., Devillers, E., Cosson, J. F., Gasqui, P., Van, V. T., Mavingui, P., Vourc'h, G., & Vayssier-Taussat, M. (2016). Co-infection of Ticks: The Rule Rather Than the Exception. *PLoS neglected tropical diseases*, 10(3), e0004539. https://doi.org/10.1371/journal.pntd.0004539.

Mukhacheva, T. A., & Kovalev, S. Y. (2017). Bacteria of the family 'candidatus midichloriaceae' in sympatric zones of Ixodes ticks: Genetic evidence for vertical transmission. *Microbial ecology*, 74(1), 185–193. https://doi.org/10.1007/s00248-017-0932-z.

Mumcuoglu, K. Y., Arslan-Akveran, G., Aydogdu, S., Karasartova, D., Koşar, A., Savci, U., Keskin, A., & Taylan-Ozkan, A. (2022b). Pathogens in ticks collected in Israel: II. Bacteria and protozoa found in Rhipicephalus sanguineus sensu lato and Rhipicephalus turanicus. *Ticks and tick-borne diseases*, 13(5), 101986. https://doi.org/10.1016/j.ttbdis.2022.101986.

Mumcuoglu, K. Y., Arslan-Akveran, G., Aydogdu, S., Karasartova, D., Kosar, N., Gureser, A. S., Shacham, B., & Taylan-Ozkan, A. (2022a). Pathogens in ticks collected in Israel: I. Bacteria and protozoa in Hyalomma aegyptium and Hyalomma dromedarii collected from tortoises and camels. *Ticks and tick-borne diseases*, 13(1), 101866. https://doi.org/10.1016/j.ttbdis.2021.101866.

Mumcuoglu, K. Y., Banet-Noach, C., Malkinson, M., Shalom, U., & Galun, R. (2005). Argasid ticks as possible vectors of West Nile virus in Israel. *Vector borne and zoonotic diseases (Larchmont, N.Y.)*, 5(1), 65–71. https://doi.org/10.1089/vbz.2005.5.65.

Muñoz-García, C. I., Rendón-Franco, E., Grostieta, E., Navarrete-Sotelo, M., & Sánchez-Montes, S. (2023). Novel Francisella-like endosymbiont and Anaplasma species from Amblyomma nodosum hosted by the anteater Tamandua Mexicana in Mexico. *Experimental & applied acarology*, 91(1), 111–121. https://doi.org/10.1007/s10493-023-00828-x.

Naidenova, E. V., Zakharov, K. S., Kartashov, M. Y., Agafonov, D. A., Senichkina, A. M., Magassouba, N., Nourdine, I., Nassour, A. A., Bah, M. B., Kourouma, A., Boumbali, S., Boiro, M. Y., Scherbakova, S. A., Kutyrev, V. V., & Dedkov, V. G. (2020). Prevalence of Crimean-Congo hemorrhagic fever virus in rural areas of Guinea. *Ticks and tick-borne diseases*, 11(5), 101475. https://doi.org/10.1016/j.ttbdis.2020.101475.

Nakamura, Y., Kawai, S., Yukuhiro, F., Ito, S., Gotoh, T., Kisimoto, R., Yanase, T., Matsumoto, Y., Kageyama, D., & Noda, H. (2009). Prevalence of Cardinium bacteria in planthoppers and spider mites and taxonomic revision of "Candidatus Cardinium hertigii" based on detection of a new Cardinium group from biting midges. *Applied and environmental microbiology*, 75(21), 6757–6763. https://doi.org/10.1128/AEM.01583-09.

Namina, A., Kazarina, A., Lazovska, M., Akopjana, S., Ulanova, V., Kivrane, A., Freimane, L., Sadovska, D., Kimsis, J., Bormane, A., Capligina, V., & Ranka, R. (2023). Comparative microbiome analysis of three epidemiologically important tick species in Latvia. *Microorganisms*, 11(8), 1970. https://doi.org/10.3390/microorganisms11081970.

Namrata, P., Miller, J. M., Shilpa, M., Reddy, P. R., Bandoski, C., Rossi, M. J., & Sapi, E. (2014). Filarial nematode infection in Ixodes scapularis ticks collected from Southern connecticut. *Veterinary Sciences*, 1(1), 5–15. https://doi.org/10.3390/vetsci1010005.

Narasimhan, S., & Fikrig, E. (2015). Tick microbiome: The force within. *Trends in parasitology*, 31(7), 315–323. https://doi.org/10.1016/j.pt.2015.03.010.

Narasimhan, S., Rajeevan, N., Liu, L., Zhao, Y. O., Heisig, J., Pan, J., Eppler-Epstein, R., Deponte, K., Fish, D., &

Fikrig, E. (2014). Gut microbiota of the tick vector Ixodes scapularis modulate colonization of the Lyme disease spirochete. *Cell host & microbe*, 15(1), 58–71. https://doi.org/10.1016/j.chom.2013.12.001.

Narasimhan, S., Schuijt, T. J., Abraham, N. M., Rajeevan, N., Coumou, J., Graham, M., Robson, A., Wu, M. J., Daffre, S., Hovius, J. W., & Fikrig, E. (2017). Modulation of the tick gut milieu by a secreted tick protein favors Borrelia burgdorferi colonization. *Nature communications*, 8(1), 184. https://doi.org/10.1038/s41467-017-00208-0.

Narasimhan, S., Sukumaran, B., Bozdogan, U., Thomas, V., Liang, X., DePonte, K., Marcantonio, N., Koski, R. A., Anderson, J. F., Kantor, F., & Fikrig, E. (2007). A tick antioxidant facilitates the Lyme disease agent's successful migration from the mammalian host to the arthropod vector. *Cell host & microbe*, 2(1), 7–18. https://doi.org/10.1016/j.chom.2007.06.001.

Narasimhan, S., Swei, A., Abouneameh, S., Pal, U., Pedra, J. H. F., & Fikrig, E. (2021). Grappling with the tick microbiome. *Trends in parasitology*, 37(8), 722–733. https://doi.org/10.1016/j.pt.2021.04.004.

Nardi, T., Olivieri, E., Kariuki, E., Sassera, D., & Castelli, M. (2021). Sequence of a Coxiella endosymbiont of the tick Amblyomma nuttalli suggests a pattern of convergent genome reduction in the Coxiella genus. *Genome biology and evolution*, 13(1), evaa253. https://doi.org/10.1093/gbe/evaa253.

Nava, S., Guglielmone, A. A., & Mangold, A. J. (2009). An overview of systematics and evolution of ticks. *Frontiers in bioscience (Landmark edition)*, 14(8), 2857–2877. https://doi.org/10.2741/3418.

Ndiaye, M., Badji, A., Dieng, I., Dolgova, A. S., Mhamadi, M., Kirichenko, A. D., Gladkikh, A. S., Gaye, A., Faye, O., Sall, A. A., Diallo, M., Dedkov, V. G., & Faye, O. (2024). Molecular detection and genetic characterization of two dugbe orthonairovirus isolates detected from ticks in Southern Senegal. *Viruses*, 16(6), 964. https://doi.org/10.3390/v16060964.

Nebbak, A., Dahmana, H., Almeras, L., Raoult, D., Boulanger, N., Jaulhac, B., Mediannikov, O., & Parola, P. (2019). Co-infection of bacteria and protozoan parasites in Ixodes ricinus nymphs collected in the Alsace region, France. *Ticks and tick-borne diseases*, 10(6), 101241. https://doi.org/10.1016/j.ttbdis.2019.06.001.

Neelakanta, G., Sultana, H., Fish, D., Anderson, J. F., & Fikrig, E. (2010). Anaplasma phagocytophilum induces Ixodes scapularis ticks to express an antifreeze glycoprotein gene that enhances their survival in the cold. *The Journal of clinical investigation*, 120(9), 3179–3190. https://doi.org/10.1172/JCI42868.

Negredo, A., Habela, M. Á., Ramírez de Arellano, E., Diez, F., Lasala, F., López, P., Sarriá, A., Labiod, N., Calero-Bernal, R., Arenas, M., Tenorio, A., Estrada-Peña, A., & Sánchez-Seco, M. P. (2019, June). Survey of Crimean-Congo hemorrhagic fever enzootic focus, Spain, 2011-2015. *Emerging infectious diseases*, 25(6), 1177-1184.

Newman, B. C., Sutton, W. B., Moncayo, A. C., Hughes, H. R., Taheri, A., Moore, T. C., Schweitzer, C. J., & Wang, Y. (2020). Heartland virus in lone star ticks, Alabama, USA. *Emerging infectious diseases*, 26(8), 1954–1956. https://doi.org/10.3201/eid2608.200494.

Ngnindji-Youdje, Y., Diarra, A. Z., Lontsi-Demano, M., Tchuinkam, T., & Parola, P. (2022). Detection of tick-borne pathogens in ticks from cattle in western highlands of cameroon. *Microorganisms*, 10(10), 1957. https://doi.org/10.3390/microorganisms10101957.

Nikoh, N., Hosokawa, T., Moriyama, M., Oshima, K., Hattori, M., & Fukatsu, T. (2014). Evolutionary origin of insect-Wolbachia nutritional mutualism. *Proceedings of the National Academy of Sciences of the United States of America*, 111(28), 10257–10262. https://doi.org/10.1073/pnas.1409284111.

Nooroong, P., Trinachartvanit, W., Baimai, V., & Ahantarig, A. (2018). Phylogenetic studies of bacteria (Rickettsia, Coxiella, and Anaplasma) in Amblyomma and Dermacentor ticks in Thailand and their co-infection. *Ticks and tick-borne diseases*, 9(4), 963–971. https://doi.org/10.1016/j.ttbdis.2018.03.032.

Numan, M., Alouffi, A., Almutairi, M. M., Tanaka, T., Ahmed, H., Akbar, H., Rashid, M. I., Tsai, K. H., & Ali, A. (2023). First detection of Theileria sinensis-like and Anaplasma capra in Ixodes kashmiricus: With notes on cox1-based phylogenetic position and new locality records. *Animals: an open access journal from MDPI*, 13(20), 3232. https://doi.org/10.3390/ani13203233.

Oakes, V. J., Yabsley, M. J., Schwartz, D., LeRoith, T., Bissett, C., Broaddus, C., Schlater, J. L., Todd, S. M., Boes, K. M., Brookhart, M., & Lahmers, K. K. (2019). Theileria orientalis Ikeda Genotype in Cattle, Virginia, USA. *Emerging infectious diseases*, 25(9), 1653–1659. https://doi.org/10.3201/eid2509.190088.

Oba, M., Omatsu, T., Takano, A., Fujita, H., Sato, K., Nakamoto, A., Takahashi, M., Takada, N., Kawabata, H., Ando, S., & Mizutani, T. (2016). A novel Bunyavirus from the soft tick, Argas vespertilionis, in Japan. *The Journal of veterinary medical science*, 78(3), 443–445. https://doi.org/10.1292/jvms.15-0536.

Obaid, M. K., Lan, X., Ren, Q., Zeb, J., Luo, J., Yang, J., Jia, W., Zan, X., Yin, H., Rashid, M., & Guan, G. (2025). Molecular insights into Rickettsiales in blood and ticks of two-humped camels at Gansu

Province, China: With an accidental detection of Colpodella sp. *Veterinary microbiology*, 305, 110528. https://doi.org/10.1016/j.vetmic.2025.110532.

Obanda, V., Akinyi, M., King'ori, E., Nyakundi, R., Ochola, G., Oreng, P., Mugambi, K., Waiguchu, G. M., Chege, M., Rosenbaum, W., Ylitalo, E. B., Bäck, A. T., Pettersson, L., Mukunzi, O. S., Agwanda, B., Stenberg-Lewerin, S., & Lwande, O. W. (2024). Epidemiology and ecology of the sylvatic cycle of African Swine Fever Virus in Kenya. *Virus research*, 348, 199434. https://doi.org/10.1016/j.virusres.2024.199434.

Obregón, D., Bard, E., Abrial, D., Estrada-Peña, A., & Cabezas-Cruz, A. (2019). Sex-specific linkages between taxonomic and functional profiles of tick gut microbiomes. *Frontiers in cellular and infection microbiology*, 9, 298. https://doi.org/10.3389/fcimb.2019.00298.

Ogata, S., Mohamed, W. M. A., Kusakisako, K., Thu, M. J., Qiu, Y., Moustafa, M. A. M., Matsuno, K., Katakura, K., Nonaka, N., & Nakao, R. (2021). Spiroplasma infection among Ixodid ticks exhibits species dependence and suggests a vertical pattern of transmission. *Microorganisms*, 9(2), 333. https://doi.org/10.3390/microorganisms9020333.

Ogata, S., Umemiya-Shirafuji, R., Kusakisako, K., Kakisaka, K., Chatanga, E., Hayashi, N., Taya, Y., Ohari, Y., Pandey, G. S., Abdelbaset, A. E., Qiu, Y., Matsuno, K., Nonaka, N., & Nakao, R. (2023). Investigation of vertical and horizontal transmission of Spiroplasma in ticks under laboratory conditions. *Scientific reports*, 13(1), 13265. https://doi.org/10.1038/s41598-023-39128-z.

Ogden, N. H., & Lindsay, L. R. (2016). Effects of climate and climate change on vectors and vector-borne diseases: Ticks are different. *Trends in parasitology*, 32(8), 646–656. https://doi.org/10.1016/j.pt.2016.04.015.

Oh, S. S., Chae, J. B., Kang, J. G., Kim, H. C., Chong, S.T., Shin, J. H., Hur, M. S., Suh, J. H., Oh, M. D., Jeong, S. M., Shin, N. S., Choi, K. S., & Chae, J. S. (2016, June). Detection of severe fever with thrombocytopenia syndrome virus from wild animals and ixodidae ticks in the Republic of Korea. *Vector-borne and zoonotic diseases*, 16(6), 408-414.

Olafson, P. U., Buckmeier, B. G., May, M. A., & Thomas, D. B. (2020). Molecular screening for rickettsial bacteria and piroplasms in ixodid ticks surveyed from white-tailed deer (Odocoileus virginianus) and nilgai antelope (Boselaphus tragocamelus) in southern Texas. *International journal for parasitology Parasites and wildlife*, 13, 252–260. https://doi.org/10.1016/j.ijppaw.2020.11.005.

Oliva Chávez, A. S., Shaw, D. K., Munderloh, U. G., & Pedra, J. H. (2017). Tick humoral responses: Marching to the beat of a different drummer. *Frontiers in microbiology*, 8, 223. https://doi.org/10.3389/fmicb.2017.00223.

Oliveira, L. R., Moraes-Souza, H., Maltos, A. L., Santos, K. C., Molina, R. J., & Barata, C. H. (2013). Aureobasidium pullulans infection in a patient with chronic lymphocytic leukemia. *Revista da Sociedade Brasileira de Medicina Tropical*, 46(5), 660–662. https://doi.org/10.1590/0037-8682-0117-2013.

Olivieri, E., Epis, S., Castelli, M., Varotto Boccazzi, I., Romeo, C., Desir`o, A., Bazzocchi, C., Bandi, C., & Sassera, D. (2019). Tissue tropism and metabolic pathways of Midichloria mitochondrii suggest tissue-specific functions in the symbiosis with Ixodes ricinus. *Ticks Tick Borne Diseases*, 10, 1070–1077. https://doi.org/10.1016/j.ttbdis.2019.05.019.

Olmeda-García, A. S., & Rodríguez-Rodríguez, J. A. (1994). Stage-specific development of a filarial nematode (Dipetalonema dracunculoides) in vector ticks. *Journal of helminthology*, 68(3), 231–235. https://doi.org/10.1017/s0022149x00014395.

Olsthoorn, F., Sprong, H., Fonville, M., Rocchi, M., Medlock, J., Gilbert, L., & Ghazoul, J. (2021). Occurrence of tick-borne pathogens in questing Ixodes ricinus ticks from Wester Ross, Northwest Scotland. *Parasites & vectors*, 14(1), 430. https://doi.org/10.1186/s13071-021-04946-6.

Ostfeld, R. S., & Brunner, J. L. (2015). Climate change and Ixodes tick-borne diseases of humans. *Philosophical transactions of the Royal Society of London. Series B, Biological sciences*, 370(1665), 20140051. https://doi.org/10.1098/rstb.2014.0051.

Oteo, J. A., & Portillo, A. (2012). Tick-borne rickettsioses in Europe. *Ticks and tick-borne diseases*, 3(5-6), 271–278. https://doi.org/10.1016/j.ttbdis.2012.10.035.

Otranto, D., Brianti, E., Latrofa, M. S., Annoscia, G., Weigl, S., Lia, R. P., Gaglio, G., Napoli, E., Giannetto, S., Papadopoulos, E., Mirò, G., Dantas-Torres, F., & Bain, O. (2012). On a Cercopithifilaria sp. transmitted by Rhipicephalus sanguineus: A neglected, but widespread filarioid of dogs. *Parasites & vectors*, 5(1), 1. https://doi.org/10.1186/1756-3305-5-1.

Paddock, C. D., & Childs, J. E. (2003). Ehrlichia chaffeensis: A prototypical emerging pathogen. *Clinical microbiology reviews*, 16(1), 37–64. https://doi.org/10.1128/CMR.16.1.37-64.2003.

Padgett, K. A., Kjemtrup, A., Novak, M., Velez, J. O., & Panella, N. (2022). Colorado tick fever virus in the far west: Forgotten, but not gone. *Vector borne and zoonotic diseases (Larchmont, N.Y.)*, 22(8), 443–448. https://doi.org/10.1089/vbz.2022.0018.

Pakanen, V. M., Sormunen, J. J., Sippola, E., Blomqvist, D., & Kallio, E. R. (2020). Questing abundance of adult taiga ticks Ixodes persulcatus and their Borrelia prevalence at the north-western part of their distribution. *Parasites & vectors*, 13(1), 384. https://doi.org/10.1186/s13071-020-04263-z.

Palomar, A. M., Portillo, A., Santibáñez, P., Mazuelas, D., Roncero, L., García-Álvarez, L., Santibáñez, S., Gutiérrez, Ó., & Oteo, J. A. (2015). Detection of tick-borne Anaplasma bovis, Anaplasma phagocytophilum and Anaplasma centrale in Spain. *Medical and veterinary entomology*, 29(3), 349–353. https://doi.org/10.1111/mve.12124.

Papa, A., Chaligiannis, I., Kontana, N., Sourba, T., Tsioka, K., Tsatsaris, A., & Sotiraki, S. (2014). A novel AP92-like Crimean-Congo hemorrhagic fever virus strain, Greece. *Ticks and tick-borne diseases*, 5(5), 590–593. https://doi.org/10.1016/j.ttbdis.2014.04.008.

Papa, A., Kontana, A., Tsioka, K., Chaligiannis, I., & Sotiraki, S. (2017). Molecular detection of Crimean-Congo hemorrhagic fever virus in ticks, Greece, 2012-2014. *Parasitology research*, 116(11), 3057–3063. https://doi.org/10.1007/s00436-017-5616-6.

Papa, A., Velo, E., Papadimitriou, E., Cahani, G., Kota, M., & Bino, S. (2009). Ecology of the Crimean-Congo hemorrhagic fever endemic area in Albania. *Vector borne and zoonotic diseases (Larchmont, N.Y.)*, 9(6), 713–716. https://doi.org/10.1089/vbz.2008.0141.

Park, K. B., Roh, J. Y., Kim, S. Y., Kim, N.-Y., Jo, Y. H., Jeong, R.-D., Lee, W.-G., Lee, H.-I., Cho, S.-H., Lee, Y. S., & Han, Y. S. (2021). Deep sequencing and phylogenetic analysis of severe fever with thrombocytopenia syndrome virus from the tick, *Haemaphysalis longicornis*, in Korea. *Entomological research*, 51, 3–11.

Parola, P., Paddock, C. D., Socolovschi, C., Labruna, M. B., Mediannikov, O., Kernif, T., Abdad, M. Y., Stenos, J., Bitam, I., Fournier, P. E., & Raoult, D. (2013). Update on tick-borne rickettsioses around the world: A geographic approach. *Clinical microbiology reviews*, 26(4), 657–702. https://doi.org/10.1128/CMR.00032-13.

Pavanelo, D. B., Schröder, N. C. H., Pin Viso, N. D., Martins, L. A., Malossi, C. D., Galletti, M. F. B. M., Labruna, M. B., Daffre, S., Farber, M., & Fogaça, A. C. (2020). Comparative analysis of the midgut microbiota of two natural tick vectors of Rickettsia rickettsii. *Developmental and comparative immunology*, 106, 103606. https://doi.org/10.1016/j.dci.2019.103606.

Pell, J. K., Hannam, J. J., & Steinkraus, D. C. (2010). Conservation biological control using fungal entomopathogens. *BioControl*, 55, 187–198. https://doi.org/10.1007/s10526-009-9245-6.

Peñalver, E., Arillo, A., Delclòs, X., Peris, D., Grimaldi, D. A., Anderson, S. R., Nascimbene, P. C., & Pérez-de la Fuente, R. (2017). parasitised feathered dinosaurs as revealed by Cretaceous amber assemblages. *Nature communications*, 8(1), 1924. https://doi.org/10.1038/s41467-017-01550-z.

Pereira, A., Parreira, R., Cotão, A. J., Nunes, M., Vieira, M. L., Azevedo, F., Campino, L., & Maia, C. (2018). Tick-borne bacteria and protozoa detected in ticks collected from domestic animals and wildlife in central and southern Portugal. *Ticks and tick-borne diseases*, 9(2), 225–234. https://doi.org/10.1016/j.ttbdis.2017.09.008.

Peter, E., Machuka, E., Githae, D., Okoth, E., Cleaveland, S., Shirima, G., Kusiluka, L., & Pelle, R. (2021). Detection of African swine fever virus genotype XV in a sylvatic cycle in Saadani National Park, Tanzania. *Transboundary and emerging diseases*, 68(2), 813–823. https://doi.org/10.1111/tbed.13747.

Petit, T. J. P., & Lebreton, A. (2022). Adaptations of intracellular bacteria to vacuolar or cytosolic niches. *Trends in microbiology*, 30(8), 736–748. https://doi.org/10.1016/j.tim.2022.01.015.

Pettersson, J. H., Ellström, P., Ling, J., Nilsson, I., Bergström, S., González-Acuña, D., Olsen, B., & Holmes, E. C. (2020). Circumpolar diversification of the Ixodes uriae tick virome. *PLoS pathogens*, 16(8), e1008759. https://doi.org/10.1371/journal.ppat.1008759.

Plantard, O., Bouju-Albert, A., Malard, M. A., Hermouet, A., Capron, G., & Verheyden, H. (2012). Detection of Wolbachia in the tick Ixodes ricinus is due to the presence of the hymenoptera endoparasitoid Ixodiphagus hookeri. *PloS one*, 7(1), e30692. https://doi.org/10.1371/journal.pone.0030692.

Pollet, T., Sprong, H., Lejal, E., Krawczyk, A. I., Moutailler, S., Cosson, J. F., Vayssier-Taussat, M., & Estrada-Peña, A. (2020). The scale affects our view on the identification and distribution of microbial communities in ticks. *Parasites & vectors*, 13(1), 36. https://doi.org/10.1186/s13071-020-3908-7.

Portillo, A., Santibáñez, P., Palomar, A. M., Santibáñez, S., & Oteo, J. A. (2018). 'Candidatus Neoehrlichia mikurensis' in Europe. *New microbes and new infections*, 22, 30–36. https://doi.org/10.1016/j.nmni.2017.12.011.

Pospisilova, T., Urbanova, V., Hes, O., Kopacek, P., Hajdusek, O., & Sima, R. (2019). Tracking of Borrelia afzelii transmission from infected Ixodes ricinus Nymphs to mice. *Infection and immunity*, 87(6), e00896-18. https://doi.org/10.1128/IAI.00896-18.

Price, K. J., Ayres, B. N., Maes, S. E., Witmier, B. J., Chapman, H. A., Coder, B. L., Boyer, C. N., Eisen, R. J., & Nicholson, W. L. (2022). First detection of human pathogenic variant of Anaplasma phagocytophilum in field-collected Haemaphysalis longicornis, Pennsylvania, USA. *Zoonoses and public health*, 69(2), 143–148. https://doi.org/10.1111/zph.12901.

Probst, J., Springer, A., Fingerle, V., & Strube, C. (2024). Frequency of Anaplasma phagocytophilum, Borrelia spp., and coinfections in Ixodes ricinus ticks collected from dogs and cats in Germany. *Parasites & vectors*, 17(1), 87. https://doi.org/10.1186/s13071-024-06193-w.

Qi, Y., Ai, L., Zhu, C., Lu, Y., Lv, R., Mao, Y., Lu, N., & Tan, W. (2022). Co-existence of multiple Anaplasma species and variants in ticks feeding on Hedgehogs or cattle poses potential threats of anaplasmosis to humans and livestock in Eastern China. *Frontiers in microbiology*, 13, 913650. https://doi.org/10.3389/fmicb.2022.913653.

Qin, X. R., Han, F. J., Luo, L. M., Zhao, F. M., Han, H. J., Zhang, Z. T., Liu, J. W., Xue, Z. F., Liu, M. M., Ma, D. Q., Huang, Y. T., Yue Sun, Sun, X. F., Li, W. Q., Zhao, L., Hao Yu, & Yu, X. J. (2018). Anaplasma species detected in *Haemaphysalis longicornis* tick from China. *Ticks and tick-borne diseases*, 9(4), 840–843. https://doi.org/10.1016/j.ttbdis.2018.03.014.

Qin, X. R., Han, F. J., Luo, L. M., Zhao, F. M., Han, H. J., Zhang, Z. T., Liu, J. W., Xue, Z. F., Liu, M. M., Ma, D. Q., Huang, Y. T., Yue, S., Sun, X. F., Li, W. Q., Zhao, L., Hao, Y., & Yu, X. J. (2018). Anaplasma species detected in Haemaphysalis longicornis tick from China. *Ticks and tick-borne diseases*, 9(4), 840–843. https://doi.org/10.1016/j.ttbdis.2018.03.016.

Qiu, Y., Nakao, R., Ohnuma, A., Kawamori, F., & Sugimoto, C. (2014). Microbial population analysis of the salivary glands of ticks; a possible strategy for the surveillance of bacterial pathogens. *PloS one*, 9(8), e103961. https://doi.org/10.1371/journal.pone.0103961.

Qiu, Y., Simuunza, M., Kajihara, M., Chambaro, H., Harima, H., Eto, Y., Simulundu, E., Squarre, D., Torii, S., Takada, A., Hang'ombe, B. M., Sawa, H., Sugimoto, C., & Nakao, R. (2021). Screening of tick-borne pathogens in argasid ticks in Zambia: Expansion of the geographic distribution of Rickettsia lusitaniae and Rickettsia hoogstraalii and detection of putative novel Anaplasma species. *Ticks and tick-borne diseases*, 12(4), 101720. https://doi.org/10.1016/j.ttbdis.2021.101724.

Quembo, C. J., Jori, F., Vosloo, W., & Heath, L. (2018). Genetic characterization of African swine fever virus isolates from soft ticks at the wildlife/domestic interface in Mozambique and identification of a novel genotype. *Transboundary and emerging diseases*, 65(2), 420–431. https://doi.org/10.1111/tbed.12700.

Rakov, A. V., Volchev, E. G., Petremgvdlishvili, K., & Chekanova, T. A. (2025). Comparative analysis of bacterial tick-borne pathogens in questing ticks from Sambia Peninsula, Kaliningrad Oblast, Russia: Spring and Autumn prevalence and public health risks. *Microorganisms*, 13(6), 1403. https://doi.org/10.3390/microorganisms13061408.

Ramos, R. A., Giannelli, A., Dantas-Torres, F., Mallia, E., Passantino, G., Lia, R. P., Latrofa, M. S., Mutafchiev, Y., & Otranto, D. (2013). Cercopithifilaria rugosicauda (Spirurida, Onchocercidae) in a roe deer and ticks from southern Italy. *International journal for parasitology. Parasites and wildlife*, 2, 292–296. https://doi.org/10.1016/j.ijppaw.2013.09.009.

Rar, V., & Golovljova, I. (2011). Anaplasma, Ehrlichia, and "Candidatus Neoehrlichia" bacteria: Pathogenicity, biodiversity, and molecular genetic characteristics, a review. *Infection, genetics and evolution: journal of molecular epidemiology and evolutionary genetics in infectious diseases*, 11(8), 1842–1861. https://doi.org/10.1016/j.meegid.2011.09.019.

Ravaomanana, J., Michaud, V., Jori, F., Andriatsimahavandy, A., Roger, F., Albina, E., & Vial, L. (2010). First detection of African Swine Fever Virus in Ornithodoros porcinus in Madagascar and new insights into tick distribution and taxonomy. *Parasites & vectors*, 3, 115. https://doi.org/10.1186/1756-3305-3-115.

Reeves, W. K., Mans, B. J., Durden, L. A., Miller, M. M., Gratton, E. M., & Laverty, T. M. (2020). Rickettsia hoogstraalii and a Rickettsiella from the Bat Tick Argas transgariepinus, in Namibia. *The Journal of parasitology*, 106(5), 663–669. https://doi.org/10.1645/20-46.

Remesar, S., Díaz, P., Prieto, A., García-Dios, D., Panadero, R., Fernández, G., Brianti, E., Díez-Baños, P., Morrondo, P., & López, C. M. (2021). Molecular detection and identification of piroplasms (Babesia spp. and Theileria spp.) and Anaplasma phagocytophilum in questing ticks from northwest Spain. *Medical and veterinary entomology*, 35(1), 51–58. https://doi.org/10.1111/mve.12469.

Ren, M., Sun, M., Dan, B., Tu, Y., Niu, G., Lin, W., Sun, H., & Zhu, Y. (2025). Identification and genetic characterization of Xiabuqu River virus: A novel member of the Iflaviridae family detected in soft ticks from Tibet, China. *Virus research*, 354, 199543. https://doi.org/10.1016/j.virusres.2025.199543.

Rjeibi, M. R., Amairia, S., Mhadhbi, M., Rekik, M., & Gharbi, M. (2022). Detection and molecular identification of Anaplasma phagocytophilum and Babesia spp. infections in Hyalomma aegyptium ticks in Tunisia. *Archives of microbiology*, 204(7), 385. https://doi.org/10.1007/s00203-022-02995-7.

Romano, D. M. M., Pereira, T. N., Almeida, I. B., Coelho, C. S. G., Duarte, F. C., Harakava, R., Cassiano, L. L., & Mendes, M. C. (2023). First molecular evidence of Wolbachia occurrence in Amblyomma sculptum (Acari: Ixodidae). *Veterinary parasitology*, 317, 109907. https://doi.org/10.1016/j.vetpar.2023.109907.

Rosa, R. D., Capelli-Peixoto, J., Mesquita, R. D., Kalil, S. P., Pohl, P. C., Braz, G. R., Fogaça, A. C., & Daffre, S. (2016). Exploring the immune signalling pathway-related genes of the cattle tick Rhipicephalus microplus: From molecular characterization to transcriptional profile upon microbial challenge. *Developmental and comparative immunology*, 59, 1–14. https://doi.org/10.1016/j.dci.2015.12.018.

Ross, B. D., Hayes, B., Radey, M. C., Lee, X., Josek, T., Bjork, J., Neitzel, D., Paskewitz, S., Chou, S., & Mougous, J. D. (2018). Ixodes scapularis does not harbor a stable midgut microbiome. *The ISME journal*, 12(11), 2596–2607. https://doi.org/10.1038/s41396-018-0161-6.

Sacchi, L., Bigliardi, E., Corona, S., Beninati, T., Lo, N., & Franceschi, A. (2004). A symbiont of the tick Ixodes ricinus invades and consumes mitochondria in a mode similar to that of the parasitic bacterium Bdellovibrio bacteriovorus. *Tissue & Cell*, 36, 43–53.

Sadeddine, R., Diarra, A. Z., Laroche, M., Mediannikov, O., Righi, S., Benakhla, A., Dahmana, H., Raoult, D., & Parola, P. (2020). Molecular identification of protozoal and bacterial organisms in domestic animals and their infesting ticks from north-eastern Algeria. *Ticks and tick-borne diseases*, 11(2), 101330. https://doi.org/10.1016/j.ttbdis.2019.101330.

Saghafipour, A., Mousazadeh-Mojarrad, A., Arzamani, N., Telmadarraiy, Z., Rajabzadeh, R., & Arzamani, K. (2019). Molecular and seroepidemiological survey on Crimean-Congo Hemorrhagic Fever Virus in Northeast of Iran. *Medical journal of the Islamic Republic of Iran*, 33, 41. https://doi.org/10.34171/mjiri.33.41.

Sahni, A., Fang, R., Sahni, S. K., & Walker, D. H. (2019). Pathogenesis of Rickettsial diseases: Pathogenic and immune mechanisms of an endotheliotropic infection. *Annual review of pathology*, 14, 127–152. https://doi.org/10.1146/annurev-pathmechdis-012418-012800.

Said, Y., Lahmar, S., Dhibi, M., Rjeibi, M. R., Jdidi, M., & Gharbi, M. (2021). First survey of ticks, tick-borne pathogens (Theileria, Babesia, Anaplasma and Ehrlichia) and Trypanosoma evansi in protected areas for threatened wild ruminants in Tunisia. *Parasitology international*, 81, 102275. https://doi.org/10.1016/j.parint.2020.102277.

Salje, J. (2021). Cells within cells: Rickettsiales and the obligate intracellular bacterial lifestyle. *Nature reviews. Microbiology*, 19(6), 375–390. https://doi.org/10.1038/s41579-020-00507-2.

Salvioni Recalde, O. D., Rolón, M. S., Velázquez, M. C., Kowalewski, M. M., Alfonso Ruiz Diaz, J. J., Rojas de Arias, A., Moraes, M. O., Magdinier Gomes, H., de Azevedo Baêta, B., Dias Cordeiro, M., & Vega Gómez, M. C. (2024). Diversity of anaplasmataceae transmitted by ticks (Ixodidae) and the first molecular evidence of Anaplasma phagocytophilum and Candidatus Anaplasma boleense in Paraguay. *Microorganisms*, 12(9), 1893. https://doi.org/10.3390/microorganisms12091895.

Sameroff, S., Tokarz, R., Jain, K., Oleynik, A., Carrington, C. V. F., Lipkin, W. I., & Oura, C. A. L. (2021). Novel quaranjavirus and other viral sequences identified from ticks parasitizing hunted wildlife in Trinidad and Tobago. *Ticks and tick-borne diseases*, 12(4), 101730. https://doi.org/10.1016/j.ttbdis.2021.101730.

Samish, M., & Rehacek, J. (1999). Pathogens and predators of ticks and their potential in biological control. *Annual review of entomology*, 44, 159–182. https://doi.org/10.1146/annurev.ento.44.1.159.

Samsináková, A., Kálalová, S., Daniel, M., Dusbábek, F., Honzáková, E., & Cerný, V. (1974). Entomogenous fungi associated with the tick Ixodes ricinus (L.). *Folia parasitologica*, 21(1), 39–48.

Sandhu, S. S., Sharma, A. K., Beniwal, V., Goel, G., Batra, P., Kumar, A., Jaglan, S., Sharma, A. K., & Malhotra, S. (2012). Myco-biocontrol of insect pests: Factors involved, mechanism, and regulation. *Journal of pathogens*, 2012, 126819. https://doi.org/10.1155/2012/126819.

Sang, C., Yang, M., Xu, B., Liu, G., Yang, Y., Kairullayev, K., Bauyrzhan, O., Hazihan, W., Hornok, S., & Wang, Y. (2021). Tick distribution and detection of Babesia and Theileria species in Eastern and Southern Kazakhstan. *Ticks and tick-borne diseases*, 12(6), 101817. https://doi.org/10.1016/j.ttbdis.2021.101817.

Santos, M. A. B., de Souza, I. B., de Macedo, L. O., do Nascimento Ramos, C. A., de Oliveira Rego, A. G., Alves, L. C., Ramos, R. A. N., & de Carvalho, G. A. (2017). Cercopithifilaria bainae in Rhipicephalus sanguineus sensu lato ticks from dogs in Brazil. *Ticks and tick-borne diseases*, 8(4), 623–625. https://doi.org/10.1016/j.ttbdis.2017.04.007.

Santos, M. A. B., Macedo, L. O., Otranto, D., Ramos, C. A. D. N., Rêgo, A. G. O. D., Giannelli, A., Alves, L. C., Carvalho, G. A., & Ramos, R. A. N. (2018). Screening of Cercopithifilaria bainae and Hepatozoon canis in ticks collected from dogs of Northeastern Brazil. *Acta parasitologica*, 63(3), 605–608. https://doi.org/10.1515/ap-2018-0069.

Santos-Garcia, D., Morel, O., Henri, H., El Filali, A., Buysse, M., Noel, V., McCoy, K. D., Gottlieb, Y., Klasson, L., Zener, L., Duron, O., & Vavre, F. (2023). Genomic changes during the evolution of the Coxiella genus along the parasitism-mutualism continuum. *Peer Community Journal*, 3, e41. https://doi.org/10.24072/pcjournal.269.

Sassera, D., Beninati, T., Bandi, C., Bouman, E. A. P., Sacchi, L., Fabbi, M., & Lo, N. (2006).

'Candidatus Midichloria mitochondrii', an endosymbiont of the tick Ixodes ricinus with a unique intramitochondrial lifestyle. *International journal of systematic and evolutionary microbiology*, 56(Pt 11), 2535–2540.

Sassera, D., Lo, N., Epis, S., D'Auria, G., Montagna, M., Comandatore, F., Horner, D., Peretó, J., Luciano, A. M., Franciosi, F., Ferri, E., Crotti, E., Bazzocchi, C., Daffonchio, D., Sacchi, L., Moya, A., Latorre, A., & Bandi, C. (2011). Phylogenomic evidence for the presence of a flagellum and cbb(3) oxidase in the free-living mitochondrial ancestor. *Molecular biology and evolution*, 28(12), 3285–3296. https://doi.org/10.1093/molbev/msr159.

Sato, Y., Mekata, H., Sudaryatma, P. E., Kirino, Y., Yamamoto, S., Ando, S., Sugimoto, T., & Okabayashi, T. (2021). Isolation of severe fever with thrombocytopenia syndrome virus from various tick species in area with human severe fever with thrombocytopenia syndrome cases. *Vector borne and zoonotic diseases (Larchmont, N.Y.)*, 21(5), 378–384. https://doi.org/10.1089/vbz.2020.2720.

Savage, H. M., Godsey, M. S., Jr, Tatman, J., Burkhalter, K. L., Hamm, A., Panella, N. A., Ghosh, A., & Raghavan, R. K. (2018). Surveillance for heartland and bourbon viruses in Eastern Kansas, June 2016. *Journal of medical entomology*, 55(6), 1613–1616. https://doi.org/10.1093/jme/tjy103.

Sawczyn-Domańska, A., & Wójcik-Fatla, A. (2024). Detection and prevalence of Anaplasma phagocytophilum in Ixodes ricinus ticks in eastern Poland. *Annals of agricultural and environmental medicine: AAEM*, 31(3), 439–441. https://doi.org/10.26444/aaem/192238.

Schiff, P., Cumbie, A. N., Roberts, A., Riley, J., & Eastwood, G. (2025). Eastern cottontails (Sylvilagus floridanus) as hosts for ticks infected with Borrelia burgdorferi, Anaplasma phagocytophilum, and Powassan virus in Virginia, USA. *Journal of medical entomology*, 62(3), 610–620. https://doi.org/10.1093/jme/tjaf009.

Schuh, A. J., Amman, B. R., Patel, K., Sealy, T. K., Swanepoel, R., & Towner, J. S. (2020). Human-pathogenic kasokero virus in field-collected ticks. *Emerging infectious diseases*, 26(12), 2944–2950. https://doi.org/10.3201/eid2612.202411.

Schulz, A., Barry, Y., Stoek, F., Pickin, M. J., Ba, A., Chitimia-Dobler, L., Haki, M. L., Doumbia, B. A., Eisenbarth, A., Diambar, A., Bah, M. Y., Eiden, M., & Groschup, M. H. (2021, June 29). Detection of Crimean-Congo hemorrhagic fever virus in blood-fed *Hyalomma* ticks collected from Mauritanian livestock. *Parasites vectors*, 14(1), 342.

Segura, J. A., Dibernardo, A., Manguiat, K., Waitt, B., Rueda, Z. V., Keynan, Y., Wood, H., & Gutiérrez, L. A. (2024). Molecular surveillance of microbial agents from cattle-attached and questing ticks from livestock agroecosystems of Antioquia, Colombia. *Comparative immunology, microbiology and infectious diseases*, 105, 102113. https://doi.org/10.1016/j.cimid.2023.102113.

Senbill, H., Karawia, D., Zeb, J., Alyami, N. M., Almeer, R., Rahman, S., Sparagano, O., & Baruah, A. (2024). Molecular screening and genetic diversity of tick-borne pathogens associated with dogs and livestock ticks in Egypt. *PLoS neglected tropical diseases*, 18(6), e0012185. https://doi.org/10.1371/journal.pntd.0012207.

Seo, H. J., Noh, J., Kim, H. C., Chong, S. T., Klein, T. A., Park, C. U., Choi, C. Y., Kwon, Y. S., Kim, M., Min, S., Park, Y., Yoo, M. S., & Cho, Y. S. (2021). Molecular detection and phylogenetic analysis of anaplasma and borrelia species in ticks collected from migratory birds at Heuksan, Hong, and Nan Islands, Republic of Korea. *Vector borne and zoonotic diseases (Larchmont, N.Y.)*, 21(1), 20–31. https://doi.org/10.1089/vbz.2020.2631.

Seo, J. Y., Kim, Y. J., Kim, S. Y., & Lee, H. I. (2023). Molecular detection of Anaplasma, Ehrlichia and Rickettsia pathogens in ticks collected from humans in the Republic of Korea, 2021. *Pathogens (Basel, Switzerland)*, 12(6), 802. https://doi.org/10.3390/pathogens12060802.

Seo, M. G., Kwon, O. D., & Kwak, D. (2020). Genotypic analysis of piroplasms and associated pathogens from ticks infesting Cattle in Korea. *Microorganisms*, 8(5), 728. https://doi.org/10.3390/microorganisms8050729.

Severo, M. S., Choy, A., Stephens, K. D., Sakhon, O. S., Chen, G., Chung, D. W., Le Roch, K. G., Blaha, G., & Pedra, J. H. (2013). The E3 ubiquitin ligase XIAP restricts Anaplasma phagocytophilum colonization of Ixodes scapularis ticks. *The Journal of infectious diseases*, 208(11), 1830–1840. https://doi.org/10.1093/infdis/jit380.

Sgroi, G., Iatta, R., Lia, R. P., D'Alessio, N., Manoj, R. R. S., Veneziano, V., & Otranto, D. (2021). Spotted fever group rickettsiae in Dermacentor marginatus from wild boars in Italy. *Transboundary and emerging diseases*, 68(4), 2111–2120. https://doi.org/10.1111/tbed.13859.

Shao, L., Pang, Z., Fu, H., Chang, R., Lin, Z., Lv, A., Wang, S., Kong, X., Luo, M., Liu, X., Yu, X., Liu, L., & Niu, G. (2020). Identification of recently identified tick-borne viruses (Dabieshan tick virus and SFTSV) by metagenomic analysis in ticks from Shandong Province, China. *The Journal of infection*, 81(6), 973–978. https://doi.org/10.1016/j.jinf.2020.10.022.

Shaw, E. I., & Voth, D. E. (2019). Coxiella burnetii: A Pathogenic Intracellular Acidophile.

Microbiology (Reading, England), 165(1), 1–3. https://doi.org/10.1099/mic.0.000707.

Shaw, W. R., & Catteruccia, F. (2024). Bacteria induce metabolic perturbations in ticks. *Nature microbiology*, 9(9), 2206–2207. https://doi.org/10.1038/s41564-024-01787-7.

Shen, S., Duan, X., Wang, B., Zhu, L., Zhang, Y., Zhang, J., Wang, J., Luo, T., Kou, C., Liu, D., Lv, C., Zhang, L., Chang, C., Su, Z., Tang, S., Qiao, J., Moming, A., Wang, C., Abudurexiti, A., Wang, H., … & Deng, F. (2018). A novel tick-borne phlebovirus, closely related to severe fever with thrombocytopenia syndrome virus and Heartland virus, is a potential pathogen. *Emerging microbes & infections*, 7(1), 95. https://doi.org/10.1038/s41426-018-0093-2.

Shi, H., Li, B., Li, J., Chen, S., Wang, L., Bai, Z., Zhu, L., Yan, B., & Yao, L. (2022). Molecular detection of haemophilic pathogens reveals evidence of Candidatus Mycoplasma haemobos in dogs and parasitic ticks in central China. *BMC veterinary research*, 18(1), 254. https://doi.org/10.1186/s12917-022-03364-x.

Shi, J., Hu, Z., Deng, F., & Shen, S. (2018). Tick-borne viruses. *Virologica Sinica*, 33(1), 21–43. https://doi.org/10.1007/s12250-018-0019-0.

Shibata, S., Sivakumar, T., Igarashi, I., Umemiya-Shirafuji, R., Inokuma, H., Fukumoto, S., & Yokoyama, N. (2018). Epidemiological survey of a cervine Theileria in wild deer, questing ticks, and cattle in Hokkaido, Japan. *Ticks and tick-borne diseases*, 9(5), 1235–1240. https://doi.org/10.1016/j.ttbdis.2018.05.006.

Shock, B. C., Moncayo, A., Cohen, S., Mitchell, E. A., Williamson, P. C., Lopez, G., Garrison, L. E., & Yabsley, M. J. (2014). Diversity of piroplasms detected in blood-fed and questing ticks from several states in the United States. *Ticks and tick-borne diseases*, 5(4), 373–380. https://doi.org/10.1016/j.ttbdis.2014.01.003.

Silaghi, C., Beck, R., Oteo, J. A., Pfeffer, M., & Sprong, H. (2016). Neoehrlichiosis: An emerging tick-borne zoonosis caused by Candidatus Neoehrlichia mikurensis. *Experimental & applied acarology*, 68(3), 279–297. https://doi.org/10.1007/s10493-015-9935-y.

Simo Tchetgna, H., Yousseu, F. S., Cosset, F. L., de Freitas, N. B., Kamgang, B., McCall, P. J., Ndip, R. N., Legros, V., & Wondji, C. S. (2023). Molecular and serological evidence of Crimean-Congo hemorrhagic fever orthonairovirus prevalence in livestock and ticks in Cameroon. *Frontiers in cellular and infection microbiology*, 13, 1132495. https://doi.org/10.3389/fcimb.2023.1132495.

Sivakumar, T., Tattiyapong, M., Okubo, K., Suganuma, K., Hayashida, K., Igarashi, I., Zakimi, S., Matsumoto, K., Inokuma, H., & Yokoyama, N. (2014). PCR detection of Babesia ovata from questing ticks in Japan. *Ticks and tick-borne diseases*, 5(3), 305–310. https://doi.org/10.1016/j.ttbdis.2013.12.006.

Skinner, K. M., Underwood, J., Ghosh, A., Oliva Chavez, A. S., & Brelsfoard, C. L. (2022). Wolbachia impacts Anaplasma infection in Ixodes scapularis tick cells. *International journal of environmental research and public health*, 19(3), 1051. https://doi.org/10.3390/ijerph19031051.

Smalley, R., 4th, Zafar, H., Land, J., Samour, A., Hance, D., & Brennan, R. E. (2022). Detection of Borrelia miyamotoi and Powassan Virus Lineage II (Deer Tick Virus) from Odocoileus virginianus Harvested Ixodes scapularis in Oklahoma. *Vector borne and zoonotic diseases (Larchmont, N.Y.)*, 22(4), 209–216. https://doi.org/10.1089/vbz.2021.0057.

Smith, K., Oesterle, P. T., Jardine, C. M., Dibernardo, A., Huynh, C., Lindsay, R., Pearl, D. L., Bosco-Lauth, A. M., & Nemeth, N. M. (2018). Powassan virus and other arthropod-borne viruses in wildlife and ticks in Ontario, Canada. *The American journal of tropical medicine and hygiene*, 99(2), 458–465. https://doi.org/10.4269/ajtmh.18-0098.

Smith, T. A., Driscoll, T., Gillespie, J. J., & Raghavan, R. (2015). A Coxiella-like endosymbiont is a potential vitamin source for the Lone Star tick. *Genome biology and evolution*, 7(3), 831–838. https://doi.org/10.1093/gbe/evv016.

Soares, J. F., Costa, F. B., Girotto-Soares, A., Da Silva, A. S., França, R. T., Taniwaki, S. A., Dall'Agnol, B., Reck, J., Hagiwara, M. K., & Labruna, M. B. (2018). Evaluation of the vector competence of six ixodid tick species for Rangelia vitalii (Apicomplexa, Piroplasmorida), the agent of canine rangeliosis. *Ticks and tick-borne diseases*, 9(5), 1221–1234. https://doi.org/10.1016/j.ttbdis.2018.05.004.

Soliman, A. M., Mahmoud, H. Y. A. H., Amer, M. M., Hifumi, T., & Tanaka, T. (2024a). Molecular detection and diversity of tick-borne rickettsial pathogens in ticks collected from camel (Camelus dromedarius) in Upper Egypt. *Acta tropica*, 253, 107172. https://doi.org/10.1016/j.actatropica.2024.107172.

Soliman, A. M., Mahmoud, H. Y. A. H., Amer, M. M., Mohamed, S., Hifumi, T., Tsukiyama-Kohara, K., & Tanaka, T. (2025, December). Molecular detection and phylogenetic characterization of Borrelia theileri and a putative novel Borrelia genotype in Hyalomma dromedarii ticks infesting camels in southern Egypt. *Comparative immunology, microbiology and infectious diseases*, 124, 102403.

Soliman, A. M., Mahmoud, H. Y. A. H., Hifumi, T., & Tanaka, T. (2024b). Discovery of Colpodella spp. in ticks (Hyalomma dromedarii) infesting camels in southern Egypt. *Ticks and tick-borne diseases*,

15(5), 102352. https://doi.org/10.1016/j.ttbdis.2024.102352.

Song, X., Zhong, Z., Gao, L., Weiss, B. L., & Wang, J. (2022). Metabolic interactions between disease-transmitting vectors and their microbiota. *Trends in parasitology*, 38(8), 697–708. https://doi.org/10.1016/j.pt.2022.05.002.

Sormunen, J. J., Penttinen, R., Klemola, T., Vesterinen, E. J., & Hänninen, J. (2016). Anaplasma phagocytophilum in questing Ixodes ricinus ticks in southwestern Finland. *Experimental & applied acarology*, 70(4), 491-500.

Sperling, J. L. H., Fitzgerald, D., Sperling, F. A. H., & Magor, K. E. (2020). Microbiome composition and Borrelia detection in Ixodes scapularis ticks at the Northwestern Edge of their range. *Tropical medicine and infectious disease*, 5(4), 173. https://doi.org/10.3390/tropicalmed5040173.

Špitalská, E., Boldišová, E., Štefanidesová, K., Kocianová, E., Majerčíková, Z., Tarageľová, V. R., Selyemová, D., Chvostáč, M., Derdáková, M., & Škultéty, Ľ. (2021). Pathogenic microorganisms in ticks removed from Slovakian residents over the years 2008-2018. *Ticks and tick-borne diseases*, 12(2), 101626. https://doi.org/10.1016/j.ttbdis.2020.101628.

Springer, A., Özbakış-Beceriklisoy, G., Topp, A. K., Probst, J., Fingerle, V., & Strube, C. (2025). Borrelia spp. and Anaplasma phagocytophilum in ruminant ticks – Borrelia prevalence declines only in female, but not nymphal ticks feeding on cervids. *Ticks and tick-borne diseases*, 16(3), 102476. https://doi.org/10.1016/j.ttbdis.2025.102477.

Sri-In, C., Thongmeesee, K., Wechtaisong, W., Yurayart, N., Rittisornthanoo, G., Akarapas, C., Bunphungbaramee, N., Sipraya, N., Riana, E., Bui, T. T. H., Kamkong, P., Maikaew, U., Kongmakee, P., Saedan, A., Bartholomay, L. C., & Tiawsirisup, S. (2024). Tick diversity and molecular detection of Anaplasma, Babesia, and Theileria from Khao Kheow open zoo, Chonburi Province, Thailand. *Frontiers in veterinary science*, 11, 1430892. https://doi.org/10.3389/fvets.2024.1430898.

Stafford, K. C., 3rd, & Allan, S. A. (2010). Field applications of entomopathogenic fungi Beauveria bassiana and Metarhizium anisopliae F52 (Hypocreales: Clavicipitaceae) for the control of Ixodes scapularis (Acari: Ixodidae). *Journal of medical entomology*, 47(6), 1107–1115. https://doi.org/10.1603/me10019.

Stanek, G., Wormser, G. P., Gray, J., & Strle, F. (2012). Lyme borreliosis. *Lancet (London, England)*, 379(9814), 461–473. https://doi.org/10.1016/S0140-6736(11)60103-7.

Stavru, F., Riemer, J., Jex, A., & Sassera, D., 2020. When bacteria meet mitochondria: The strange case of the tick symbiont Midichloria mitochondrii. *Cell Microbiology*, 22, e13189. https://doi.org/10.1111/cmi.13189.

Stegmüller, S., Fraefel, C., & Kubacki, J. (2023). Genome sequence of Alongshan virus from Ixodes ricinus ticks collected in Switzerland. *Microbiology resource announcements*, 12(3), e0128722. https://doi.org/10.1128/mra.01287-22.

Stewart, P. E., & Bloom, M. E. (2020). Sharing the Ride: Ixodes scapularis Symbionts and Their Interactions. *Frontiers in cellular and infection microbiology*, 10, 142. https://doi.org/10.3389/fcimb.2020.00142.

Stockman, S. (1918). Louping-ill. *J Comp Pathol Ther*, 31, 137–193. https://doi.org/10.1016/S0368-1742(18)80019-4.

Strunov, A., Kiseleva, E., & Gottlieb, Y. (2013). Spatial and temporal distribution of pathogenic Wolbachia strain wMelPop in Drosophila melanogaster central nervous system under different temperature conditions. *Journal of invertebrate pathology*, 114(1), 22–30. https://doi.org/10.1016/j.jip.2013.05.001.

Stuen, S., Granquist, E. G., & Silaghi, C. (2013). Anaplasma phagocytophilum--a widespread multi-host pathogen with highly adaptive strategies. *Frontiers in cellular and infection microbiology*, 3, 31. https://doi.org/10.3389/fcimb.2013.00031.

Su, S., Cui, M. Y., Xing, L. L., Gao, R. J., Mu, L., Hong, M., Guo, Q. Q., Ren, H., Yu, J. F., Si, X. Y., & Eerde, M. (2024). Metatranscriptomic analysis reveals the diversity of RNA viruses in ticks in Inner Mongolia, China. *PLoS neglected tropical diseases*, 18(12), e0012706. https://doi.org/10.1371/journal.pntd.0012707.

Su, S., Hong, M., Cui, M. Y., Gui, Z., Ma, S. F., Wu, L., Xing, L. L., Mu, L., Yu, J. F., Fu, S. Y., Gao, R. J., & Qi, D. D. (2023). Microbial diversity of ticks and a novel typhus group Rickettsia species (Rickettsiales bacterium Ac37b) in Inner Mongolia, China. Diversité microbienne des tiques et nouvelle espèce de Rickettsia du groupe du typhus (bactérie Rickettsiales Ac37b) en Mongolie intérieure, Chine. *Parasite (Paris, France)*, 30, 58. https://doi.org/10.1051/parasite/2023057.

Subramanian, G., Sekeyova, Z., Raoult, D., & Mediannikov, O. (2012). Multiple tick-associated bacteria in Ixodes ricinus from Slovakia. *Ticks and tick-borne diseases*, 3(5-6), 406–410. https://doi.org/10.1016/j.ttbdis.2012.10.001.

Sultankulova, K. T., Shynybekova, G. O., Kozhabergenov, N. S., Mukhami, N. N., Chervyakova, O. V., Burashev, Y. D., Zakarya, K. D., Nakhanov, A. K., Barakbayev, K. B., & Orynbayev, M. B. (2022). The prevalence and genetic variants of the CCHF virus circulating among ticks in the Southern regions of Kazakhstan.

Pathogens (Basel, Switzerland), 11(8), 841. https://doi.org/10.3390/pathogens11080841.

Sun, S., Lin, Y., Han, J., He, Z., Zhang, L., Zhou, Q., Li, R., Zhang, W., Lu, Z., & Shao, Z. (2024). Revealing the diversity of the mycobiome in different phases of ticks: ITS gene-based analysis. *Transboundary and emerging diseases*, 2024, 8814592. https://doi.org/10.1155/2024/8814592.

Swei, A., Russell, B. J., Naccache, S. N., Kabre, B., Veeraraghavan, N., Pilgard, M. A., Johnson, B. J., & Chiu, C. Y. (2013). The genome sequence of Lone Star virus, a highly divergent bunyavirus found in the Amblyomma americanum tick. *PloS one*, 8(4), e62083. https://doi.org/10.1371/journal.pone.0062083.

Tang, J., Xu, J., Liu, X. H., Lv, F. Z., Yao, Q. J., Zhou, X. F., Lu, H. Y., Yu, T. M., Jiang, Z. Z., Jin, X. Z., Guo, F., & Yu, X. J. (2024). Prevalence and genetic diversity of Anaplasma and Ehrlichia in ticks and domesticated animals in Suizhou County, Hubei Province, China. *Scientific reports*, 14(1), 12621. https://doi.org/10.1038/s41598-024-63267-6.

Taylor, M. J., Hoerauf, A., & Bockarie, M. (2010). Lymphatic filariasis and onchocerciasis. *Lancet (London, England)*, 376(9747), 1175–1185. https://doi.org/10.1016/S0140-6736(10)60586-7.

Taylor, M., Mediannikov, O., Raoult, D., & Greub, G. (2012). Endosymbiotic bacteria associated with nematodes, ticks and amoebae. *FEMS immunology and medical microbiology*, 64(1), 21–31. https://doi.org/10.1111/j.1574-695X.2011.00916.x.

Tekin, S., Bursali, A., Mutluay, N., Keskin, A., & Dundar, E. (2012). Crimean-Congo hemorrhagic fever virus in various ixodid tick species from a highly endemic area. *Veterinary parasitology*, 186(3-4), 546–552. https://doi.org/10.1016/j.vetpar.2011.11.010.

Thapa, S., Zhang, Y., & Allen, M. S. (2019). Bacterial microbiomes of Ixodes scapularis ticks collected from Massachusetts and Texas, USA. *BMC microbiology*, 19(1), 138. https://doi.org/10.1186/s12866-019-1514-7.

Thinnabut, K., Rodpai, R., Sanpool, O., Maleewong, W., & Tangkawanit, U. (2023). Genetic diversity of tick (Acari: Ixodidae) populations and molecular detection of Anaplasma and Ehrlichia infesting beef cattle from upper-northeastern Thailand. *Infection, genetics and evolution: journal of molecular epidemiology and evolutionary genetics in infectious diseases*, 107, 105394. https://doi.org/10.1016/j.meegid.2022.105394.

Thomas, R. J., Dumler, J. S., & Carlyon, J. A. (2009). Current management of human granulocytic anaplasmosis, human monocytic ehrlichiosis and *Ehrlichia ewingii* ehrlichiosis. *Expert review of anti-infective therapy*, 7(6), 709–722. https://doi.org/10.1586/eri.09.44.

Thomas, R., Santodomingo, A., Saboya-Acosta, L., Quintero-Galvis, J. F., Moreno, L., Uribe, J. E., & Muñoz-Leal, S. (2024). Hepatozoon (Eucoccidiorida: Hepatozoidae) in wild mammals of the Americas: A systematic review. *Parasites & vectors*, 17(1), 108. https://doi.org/10.1186/s13071-024-06154-3.

Thu, M. J., Qiu, Y., Matsuno, K. et al. (2019). Diversity of spotted fever group rickettsiae and their association with host ticks in Japan. *Scientific Reports*, 9, 1500.

Tian, D., Ye, R. Z., Li, Y. Y., Wang, N., Gao, W. Y., Wang, B. H., Lin, Z. T., Zhu, W. J., Wang, Q. S., Liu, Y. T., Wei, H., Wang, Y. F., Sun, Y., Shi, X. Y., Jia, N., Jiang, J. F., Cui, X. M., Tick Genome and Microbiome Consortium (TIGMIC), Cao, W. C., & Liu, Z. H. (2025). Virome specific to tick genus with distinct ecogeographical distribution. *Microbiome*, 13(1), 57. https://doi.org/10.1186/s40168-025-02061-6.

Tian, H., Yu, P., Chowell, G., Li, S., Wei, J., Tian, H., Lv, W., Han, Z., Yang, J., Huang, S., Zhou, S., Brownstein, J. S., Wang, J., & Xu, B. (2017). Severe fever with thrombocytopenia syndrome virus in humans, domesticated animals, ticks, and mosquitoes, Shaanxi Province, China. *The American journal of tropical medicine and hygiene*, 96(6), 1346–1349. https://doi.org/10.4269/ajtmh.16-0333.

Tian, J., Liu, J., Zhao, H., Chen, X., Geng, X., Lu, M., & Li, K. (2023). Molecular surveillance reveals a potential hotspot of tick-borne disease in Yakeshi City, Inner Mongolia. *BMC microbiology*, 23(1), 359. https://doi.org/10.1186/s12866-023-03110-10.

Tian, Z. C., Liu, G. Y., Yin, H., Luo, J. X., Guan, G. Q., Luo, J., Xie, J. R., Shen, H., Tian, M. Y., Zheng, J. F., Yuan, X. S., & Wang, F. F. (2013). RPS8--a new informative DNA marker for phylogeny of Babesia and Theileria parasites in China. *PloS one*, 8(11), e79860. https://doi.org/10.1371/journal.pone.0079860.

Tijsse-Klasen, E., Braks, M., Scholte, E. J., & Sprong, H. (2011). Parasites of vectors--Ixodiphagus hookeri and its Wolbachia symbionts in ticks in The Netherlands. *Parasites & vectors*, 4, 228. https://doi.org/10.1186/1756-3305-4-228.

Tokarz, R., Tagliafierro, T., Ian Lipkin, W., & Marques, A. R. (2020). Characterization of a Monanema nematode in Ixodes scapularis. *Parasites & vectors*, 13(1), 371. https://doi.org/10.1186/s13071-020-04228-6.

Tominello, T. R., Oliveira, E. R. A., Hussain, S. S., Elfert, A., Wells, J., Golden, B., & Ismail, N. (2019). Emerging roles of autophagy and inflammasome in ehrlichiosis. *Frontiers in immunology*, 10, 1011. https://doi.org/10.3389/fimmu.2019.01011.

Tonbak, S., Aktas, M., Altay, K., Azkur, A. K., Kalkan, A., Bolat, Y., Dumanli, N., & Ozdarendeli, A. (2006). Crimean-Congo hemorrhagic fever virus: Genetic analysis and tick survey in Turkey. *Journal of clinical microbiology*, 44(11), 4120–4124. https://doi.org/10.1128/JCM.00644-06.

Tong, Y., Wang, Q., Fu, Y., Li, S., Zhang, Z., Zhang, Z., & Yu, X. (2020). Molecular identification of severe fever with thrombocytopenia syndrome viruses from tick and bitten patient in Southeast China. *Virology journal*, 17(1), 122. https://doi.org/10.1186/s12985-020-01391-1.

Trevisan, G., Cinco, M., Trevisini, S., di Meo, N., Ruscio, M., Forgione, P., & Bonin, S. (2021). Borreliae part 2: Borrelia relapsing fever group and unclassified Borrelia. *Biology*, 10(11), 1117. https://doi.org/10.3390/biology10111117.

Trout Fryxell, R. T., & DeBruyn, J. M. (2016). The microbiome of ehrlichia-infected and uninfected lone star ticks (Amblyomma americanum). *PloS one*, 11(1), e0146651. https://doi.org/10.1371/journal.pone.0146651.

Tsapko, N. V., Volynkina, A. S., Evchenko, A. Y., Lisitskaya, Y. V., & Shaposhnikova, L. I. (2022). Detection of Crimean-Congo hemorrhagic fever virus in ticks collected from South Russia. *Ticks and tick-borne diseases*, 13(2), 101890. https://doi.org/10.1016/j.ttbdis.2021.101891.

Tsementzi, D., Castro Gordillo, J., Mahagna, M., Gottlieb, Y., & Konstantinidis, K. T. (2018). Comparison of closely related, uncultivated Coxiella tick endosymbiont population genomes reveals clues about the mechanisms of symbiosis. *Environmental microbiology*, 20(5), 1751–1764. https://doi.org/10.1111/1462-2920.14104.

Tuininga, A. R., Miller, J. L., Morath, S. U., Daniels, T. J., Falco, R. C., Marchese, M., Sahabi, S., Rosa, D., & Stafford, K. C., 3rd. (2009). Isolation of entomopathogenic fungi from soils and Ixodes scapularis (Acari: Ixodidae) ticks: Prevalence and methods. *Journal of medical entomology*, 46(3), 557–565. https://doi.org/10.1603/033.046.0321.

Tully, B. G., & Huntley, J. F. (2020). Mechanisms affecting the acquisition, persistence and transmission of Francisella tularensis in ticks. *Microorganisms*, 8(11), 1639. https://doi.org/10.3390/microorganisms8111639.

Tuten, H. C., Burkhalter, K. L., Noel, K. R., Hernandez, E. J., Yates, S., Wojnowski, K., Hartleb, J., Debosik, S., Holmes, A., & Stone, C. M. (2020). Heartland virus in humans and ticks, illinois, USA, 2018-2019. *Emerging infectious diseases*, 26(7), 1548–1552. https://doi.org/10.3201/eid2607.200110.

Ullah, Z., Liaqat, I., Khan, M., Alouffi, A., Almutairi, M. M., Apanaskevich, D. A., Tanaka, T., & Ali, A. (2024). Investigation of Hyalomma turanicum and Hyalomma asiaticum in Pakistan, with notes on the detection of tickborne Rickettsiales. *Frontiers in veterinary science*, 11, 1500930. https://doi.org/10.3389/fvets.2024.1500932.

Umanzor, E. F., Kelly, S. E., Ravenscraft, A., Matsuura, Y., & Hunter, M. S. (2025). The facultative intracellular symbiont Lariskella is neutral for lifetime fitness and spreads through cytoplasmic incompatibility in the leaffooted bug, Leptoglossus zonatus. *Frontiers in microbiology*, 16, 1595917. https://doi.org/10.3389/fmicb.2025.1595917.

Umemiya, R., Matsuo, T., Hatta, T., Sakakibara, S., Boldbaatar, D., & Fujisaki, K. (2008). Autophagy-related genes from a tick, Haemaphysalis longicornis. *Autophagy*, 4(1), 79–81. https://doi.org/10.4161/auto.5143.

Urbanová, V., Hajdušek, O., Hönig Mondeková, H., Šíma, R., & Kopáček, P. (2017). Tick thioester-containing proteins and phagocytosis do not affect transmission of Borrelia afzelii from the competent vector Ixodes ricinus. *Frontiers in cellular and infection microbiology*, 7, 73. https://doi.org/10.3389/fcimb.2017.00073.

Uzum, Z., Ershov, D., Pavia, M. J., Mallet, A., Gorgette, O., Plantard, O., Sassera, D., & Stavru, F. (2023). Three-dimensional images reveal the impact of the endosymbiont Midichloria mitochondrii on the host mitochondria. *Nature communications*, 14(1), 4133. https://doi.org/10.1038/s41467-023-39758-x.

Valdés, J. J., Cabezas-Cruz, A., Sima, R., Butterill, P. T., Růžek, D., & Nuttall, P. A. (2016). Substrate prediction of Ixodes ricinus salivary lipocalins differentially expressed during Borrelia afzelii infection. *Scientific reports*, 6, 32372. https://doi.org/10.1038/srep32372.

Vandegrift, K. J., & Kapoor, A. (2019). The ecology of new constituents of the tick virome and their relevance to public health. *Viruses*, 11(6), 529. https://doi.org/10.3390/v11060529.

Vaschalde, P. J., Flores, F. S., Tauro, L. B., & Monje, L. D. (2025). Wild birds as hosts of ticks (Acari: Ixodidae) and Anaplasmataceae (Rickettsiales) in the Atlantic rainforest ecoregion, Argentina. *Medical and veterinary entomology*, 39(1), 187–199. https://doi.org/10.1111/mve.12772.

Vayssier-Taussat, M., Moutailler, S., Michelet, L., Devillers, E., Bonnet, S., Cheval, J., Hébert, C., & Eloit, M. (2013). Next generation sequencing uncovers unexpected bacterial pathogens in ticks in western Europe. *PloS one*, 8(11), e81439. https://doi.org/10.1371/journal.pone.0081439.

Vega, F. E., Meyling, N. V., Luangsa-ard, J. J., & Blackwell, M. (2012). Fungal entomopathogens. In: Vega, F.E., Kaya, H.K. (Eds.), *Insect pathology*. Academic Press, London, pp. 171–220.

Venclikova, K., Mendel, J., Betasova, L., Hubalek, Z., & Rudolf, I. (2015). First evidence of Babesia venatorum and Babesia capreoli in questing Ixodes ricinus ticks in the Czech Republic. *Annals of agricultural and environmental medicine: AAEM*, 22(2), 212–214. https://doi.org/10.5604/12321966.1152067.

Vial, L., Wieland, B., Jori, F., Etter, E., Dixon, L., & Roger, F. (2007). African swine fever virus DNA in soft ticks, Senegal. *Emerging infectious diseases*, 13(12), 1928–1931. https://doi.org/10.3201/eid1312.071022.

Vikentjeva, M., Geller, J., & Bragina, O. (2024). Ticks and tick-borne pathogens in popular recreational areas in Tallinn, Estonia: The underestimated risk of tick-borne diseases. *Microorganisms*, 12(9), 1918. https://doi.org/10.3390/microorganisms12091918.

Vilcins, I. M., Old, J. M., & Deane, E. (2009). Molecular detection of Rickettsia, Coxiella and Rickettsiella DNA in three native Australian tick species. *Experimental & applied acarology*, 49(3), 229–242. https://doi.org/10.1007/s10493-009-9260-4.

von Fricken, M. E., Qurollo, B. A., Boldbaatar, B., Wang, Y. W., Jiang, R. R., Lkhagvatseren, S., Koehler, J. W., Moore, T. C., Nymadawa, P., Anderson, B. D., Matulis, G., Jiang, J. F., & Gray, G. C. (2020). Genetic diversity of Anaplasma and Ehrlichia bacteria found in Dermacentor and Ixodes ticks in Mongolia. *Ticks and tick-borne diseases*, 11(1), 101316. https://doi.org/10.1016/j.ttbdis.2019.101317.

Voorhees, M. A., Padilla, S. L., Jamsransuren, D., Koehler, J. W., Delp, K. L., Adiyadorj, D., Baasandagwa, U., Jigjav, B., Olschner, S. P., Minogue, T. D., & Schoepp, R. J. (2018). Crimean-Congo hemorrhagic fever virus, Mongolia, 2013-2014. *Emerging infectious diseases*, 24(12), 2202–2209. https://doi.org/10.3201/eid2412.180175.

Walker, D. H., Dumler, J. S. (1996). Emergence of the ehrlichioses as human health problems. *Emerging Infectious Diseases*, 2, 18–29. https://doi.org/10.3201/eid0201.960102.

Wampande, E. M., Waiswa, P., Allen, D. J., Hewson, R., Frost, S. D. W., & Stubbs, S. C. B. (2021). Phylogenetic characterization of Crimean-Congo Hemorrhagic fever virus detected in African Blue ticks feeding on cattle in a Ugandan Abattoir. *Microorganisms*, 9(2), 438. https://doi.org/10.3390/microorganisms9020438.

Wang, B., Liu, Z., Zhu, S., Zhang, J., Qi, W., Wang, J., Li, D., He, L., & Zhao, J. (2025). Molecular characterization of tick-borne bacterial and protozoan pathogens in parasitic ticks from Xinjiang, China. *Parasites & vectors*, 18(1), 207. https://doi.org/10.1186/s13071-025-06857-1.

Wang, L., Teng, Z., Wan, L., Wang, W., Yuan, S., Huang, Q., Huang, J., Zhao, N., Wang, M., Cao, K., Huang, H., Xu, J., Yuan, Y., & Qin, T. (2024). Molecular epidemiology of ticks and tick-borne pathogens in the Ta-Pa mountain area of Chongqing, China. *Pathogens (Basel, Switzerland)*, 13(11), 948. https://doi.org/10.3390/pathogens13110956.

Wang, M., Zhu, D., Dai, J., Zhong, Z., Zhang, Y., & Wang, J. (2018). Tissue localization and variation of major symbionts in Haemaphysalis longicornis, Rhipicephalus haemaphysaloides, and Dermacentor silvarum in China. *Applied and environmental microbiology*, 84(10), e00029–18. https://doi.org/10.1128/AEM.00029-18.

Wang, N., Lu, S., Ye, R. Z., Li, C., Huang, J. H., Ye, G., Li, Y. Y., Shen, S. J., Shi, X. Y., Zhu, D. Y., Shi, W., Zhao, L., Jia, N., Jiang, J. F., Cui, X. M., Sun, Y., & Cao, W. C. (2025, April 19). Virome of Hyalomma and Rhipicephalus ticks from desert of Northwestern China. *Virus Evolution*, 11(1), veaf022. doi: 10.1093/ve/veaf022.

Wang, R., Liu, S., Sun, H., Xu, C., Wen, Y., Wu, X., Zhang, W., Nie, K., Li, F., Fu, S., Yin, Q., He, Y., Xu, S., Liang, G., Deng, L., Wei, Q., & Wang, H. (2023). Metatranscriptomics reveals the RNA virome of Ixodes persulcatus in the China-North Korea border, 2017. *Viruses*, 16(1), 62. https://doi.org/10.3390/v16010063.

Wang, S., Li, J., Niu, G., Wang, X., Ding, S., Jiang, X., Li, C., Zhang, Q., Liang, M., Bi, Z., & Li, D. (2015). SFTS virus in ticks in an endemic area of China. *The American journal of tropical medicine and hygiene*, 92(4), 684–689. https://doi.org/10.4269/ajtmh.14-0008.

Wang, X. R., & Cull, B. (2022). Apoptosis and autophagy: Current understanding in tick-pathogen interactions. *Frontiers in cellular and infection microbiology*, 12, 784430. https://doi.org/10.3389/fcimb.2022.784430.

Wang, X. R., Kurtti, T. J., Oliver, J. D., & Munderloh, U. G. (2020). The identification of tick autophagy-related genes in Ixodes scapularis responding to amino acid starvation. *Ticks and tick-borne diseases*, 11(3), 101402. https://doi.org/10.1016/j.ttbdis.2020.101402.

Wang, Z. D., Wang, B., Wei, F., Han, S. Z., Zhang, L., Yang, Z. T., Yan, Y., Lv, X. L., Li, L., Wang, S. C., Song, M. X., Zhang, H. J., Huang, S. J., Chen, J., Huang, F. Q., Li, S., Liu, H. H., Hong, J., Jin, Y. L., Wang, W., … & Liu, Q. (2019). A new segmented virus associated with human febrile illness in China. *The New England journal of medicine*, 380(22), 2116–2125. https://doi.org/10.1056/NEJMoa1805068.

Wass, L., Quarsten, H., Lindgren, P. E., Forsberg, P., Skoog, E., Nilsson, K., Lingblom, C., & Wennerås, C. (2022). Cytokine responses of immunosuppressed and immunocompetent patients with Neoehrlichia mikurensis infection. *Medical microbiology and immunology*, 211(2-3), 133–141. https://doi.org/10.1007/s00430-022-00737-6.

Wassermann, M., Selzer, P., Steidle, J. L. M., & Mackenstedt, U. (2016). Biological control of Ixodes ricinus larvae and nymphs with Metarhizium anisopliae blastospores. *Ticks and tick-borne diseases*, 7(5), 768–771. https://doi.org/10.1016/j.ttbdis.2016.03.010.

Wattanamethanont, J., Kaewthamasorn, M., & Tiawsirisup, S. (2018). Natural infection of questing ixodid ticks with protozoa and bacteria in Chonburi Province, Thailand. *Ticks and tick-borne diseases*, 9(3), 749–758. https://doi.org/10.1016/j.ttbdis.2018.02.020.

Weinert, L. A., Werren, J. H., Aebi, A., Stone, G. N., & Jiggins, F. M. (2009). Evolution and diversity of Rickettsia bacteria. *BMC biology*, 7, 6. https://doi.org/10.1186/1741-7007-7-6.

Welinder-Olsson, C., Kjellin, E., Vaht, K., Jacobsson, S., & Wennerås, C. (2010). First case of human "Candidatus Neoehrlichia mikurensis" infection in a febrile patient with chronic lymphocytic leukemia. *Journal of clinical microbiology*, 48(5), 1956–1959. https://doi.org/10.1128/JCM.02423-09.

Wijnveld, M., Schötta, A. M., Stelzer, T., Duscher, G., Leschnik, M., Stockinger, H., Lindgren, P. E., & Stanek, G. (2021). Novel protozoans in Austria revealed through the use of dogs as sentinels for ticks and tick-borne pathogens. *Microorganisms*, 9(7), 1392. https://doi.org/10.3390/microorganisms9071392.

Wilhelmsson, P., Lindgren, P. E., Fransson, T., Grandi, G., Olsen, B., & Nilsson, K. (2023). Candidatus Rickettsia Vini DNA in ticks collected from nest burrows of the European Sand Martin (Riparia riparia) in Sweden. *Vector borne and zoonotic diseases (Larchmont, N.Y.)*, 23(7), 378–383. https://doi.org/10.1089/vbz.2022.0081.

Williamson, B. N., Fischer, R. J., Lopez, J. E., Ebihara, H., & Schwan, T. G. (2019). Prevalence and strains of colorado tick fever virus in rocky mountain wood ticks in the Bitterroot Valley, Montana. *Vector borne and zoonotic diseases (Larchmont, N.Y.)*, 19(9), 694–702. https://doi.org/10.1089/vbz.2018.2407.

Wójcik-Fatla, A., Zając, V., Sawczyn, A., Cisak, E., Sroka, J., & Dutkiewicz, J. (2015). Occurrence of Francisella spp. in Dermacentor reticulatus and Ixodes ricinus ticks collected in eastern Poland. *Ticks and tick-borne diseases*, 6(3), 253–257. https://doi.org/10.1016/j.ttbdis.2015.01.005.

Wu, Y., Gao, Y., Tian, C., Li, J., Wu, L., & Wang, H. (2023). Identification of Rhipicephalus sanguineus sensu lato infected with tick-borne pathogens from pet and stray dogs in Guangzhou, Southern China. *Ticks and tick-borne diseases*, 15(1), 102267. Advance online publication. https://doi.org/10.1016/j.ttbdis.2023.102267.

Wu-Chuang, A., Hodžić, A., Mateos-Hernández, L., Estrada-Peña, A., Obregon, D., & Cabezas-Cruz, A. (2021, June 6). Current debates and advances in tick microbiome research. *Current research in parasitology & vector-borne diseases*, 1, 100036. doi: 10.1016/j.crpvbd.2021.100036.

Xia, H., Li, P., Yang, J., Pan, L., Zhao, J., Wang, Z., Li, Y., Zhou, H., Dong, Y., Guo, S., Tang, S., Zhang, Z., Fan, Z., Hu, Z., Kou, Z., & Li, T. (2011). Epidemiological survey of Crimean-Congo hemorrhagic fever virus in Yunnan, China, 2008. *International journal of infectious diseases: IJID: official publication of the International Society for Infectious Diseases*, 15(7), e459–e463. https://doi.org/10.1016/j.ijid.2011.03.013.

Xiao, J., Yao, X., Guan, X., Xiong, J., Fang, Y., Zhang, J., Zhang, Y., Moming, A., Su, Z., Jin, J., Ge, Y., Wang, J., Fan, Z., Tang, S., Shen, S., & Deng, F. (2024). Viromes of Haemaphysalis longicornis reveal different viral abundance and diversity in free and engorged ticks. *Virologica Sinica*, 39(2), 194–204. https://doi.org/10.1016/j.virs.2024.02.003.

Xing, X., Guan, X., Liu, L., Zhan, J., Jiang, H., Liu, L., Li, G., Xiong, J., Tan, L., Xu, J., Jiang, Y., Yao, X., Zhan, F., & Nie, S. (2016). Natural transmission model for severe fever with thrombocytopenia syndrome Bunyavirus in villages of Hubei Province, China. *Medicine*, 95(4), e2533. https://doi.org/10.1097/MD.0000000000002533.

Xu, G., Siegel, E., Fernandez, N., Bechtold, E., Daly, T., Dupuis, A. P., 2nd, Ciota, A., & Rich, S. M. (2024). Active surveillance of powassan virus in Massachusetts Ixodes scapularis ticks, comparing detection using a new triplex real-time PCR assay with a luminex vector-borne panel. *Viruses*, 16(2), 250. https://doi.org/10.3390/v16020250.

Xu, J., Gu, X. L., Jiang, Z. Z., Cao, X. Q., Wang, R., Peng, Q. M., Li, Z. M., Zhang, L., Zhou, C. M., Qin, X. R., & Yu, X. J. (2023). Pathogenic Rickettsia, Anaplasma, and Ehrlichia in Rhipicephalus microplus ticks collected from cattle and laboratory hatched tick larvae. *PLoS neglected tropical diseases*, 17(8), e0011546. https://doi.org/10.1371/journal.pntd.0011549.

Xue, J., Ren, Q., Yang, X. L., Wang, J., Xie, G., Du, L., & Guo, W. P. (2023). Human pathogens in ticks removed from humans in Hebei, China. *Heliyon*, 9(3), e13859. https://doi.org/10.1016/j.heliyon.2023.e13859.

Yadav, P. D., Gurav, Y. K., Mistry, M., Shete, A. M., Sarkale, P., Deoshatwar, A. R., Unadkat, V. B., Kokate, P., Patil, D.Y., Raval, D. K., & Mourya, D.T. (2014). Emergence of Crimean-Congo hemorrhagic fever in Amreli district of Gujarat State, India, June to July 2013. *International journal of infectious diseases: IJID: official publication*

of the International Society for Infectious Diseases, 18, 97–100. https://doi.org/10.1016/j.ijid.2013.09.019.
Yadav, P. D., Whitmer, S. L. M., Sarkale, P., Fei Fan Ng, T., Goldsmith, C. S., Nyayanit, D. A., Esona, M. D., Shrivastava-Ranjan, P., Lakra, R., Pardeshi, P., Majumdar, T. D., Francis, A., Klena, J. D., Nichol, S. T., Ströher, U., & Mourya, D. (2019). Characterization of novel reoviruses wad medani virus (Orbivirus) and kundal virus (Coltivirus) collected from Hyalomma anatolicum ticks in India during surveillance for Crimean Congo Hemorrhagic fever. *Journal of virology*, 93(13), e00106–19. https://doi.org/10.1128/JVI.00106-19.
Yang, L., Zhao, Z., Hou, G., Zhang, C., Liu, J., Xu, L., Li, W., Tan, Z., Tu, C., & He, B. (2019). Genomes and seroprevalence of severe fever with thrombocytopenia syndrome virus and Nairobi sheep disease virus in Haemaphysalis longicornis ticks and goats in Hubei, China. *Virology*, 529, 234–245. https://doi.org/10.1016/j.virol.2019.01.026.
Yang, X., Koči, J., Smith, A. A., Zhuang, X., Sharma, K., Dutta, S., Rana, V. S., Kitsou, C., Yas, O. B., Mongodin, E. F., & Pal, U. (2021, February). A novel tick protein supports integrity of gut peritrophic matrix impacting existence of gut microbiome and Lyme disease pathogens. *Cellular Microbiology*, 23(2), e13275.
Yaser, S. A., Sadegh, C., Zakkyeh, T., Hassan, V., Maryam, M., Ali, O. M., & Mojtaba, G. S. (2011). Crimean--Congo hemorrhagic fever: A molecular survey on hard ticks (Ixodidae) in Yazd province, Iran. *Asian Pacific journal of tropical medicine*, 4(1), 61–63. https://doi.org/10.1016/S1995-7645(11)60034-5.
Yates, J. A., & Lowrie, R. J. (1984). Development of Yatesia hydrochoerus (Nematoda: Filarioidea) to the Infective stage in Ixodid ticks. *Proceedings of the Helminthological Society of Washington*, 51(2), 187–190.
Yessinou, R. E., Koumassou, A., Galadima, H. B., Nanoukon-Ahigan, H., Farougou, S., & Pfeffer, M. (2025). Tick diversity and distribution of pathogen in ticks collected from wild animals and vegetation in Africa. *Pathogens (Basel, Switzerland)*, 14(2), 116. https://doi.org/10.3390/pathogens14020116.
Yoo, J. R., Heo, S. T., Song, S. W., Bae, S. G., Lee, S., Choi, S., Lee, C., Jeong, S., Kim, M., Sa, W., Lee, Y., Choi, H., Kee, S. H., & Lee, K. H. (2020, September). Severe fever with thrombocytopenia syndrome virus in ticks and SFTS incidence in humans, South Korea. *Emerging Infectious Diseases*, 26(9), 2292-2294.
Yousefi-Behzadi, M., Moazzezy, N., Rohani, M., Naddaf, S. R., Mostafavi, E., Mohamadi, A., Shams-Ghahfarokhi, M., Pashootan, N., & Razzaghi-Abyaneh, M. (2022). Identification of intestinal fungal microflora and bacterial pathogens in the collected adult Ixodes ricinus from the Northern Provinces of Iran. *Journal of arthropod-borne diseases*, 16(2), 97–107. https://doi.org/10.18502/jad.v16i2.11801.
Yu, K. M., & Park, S. J. (2024). Tick-borne viruses: Epidemiology, pathogenesis, and animal models. *One health (Amsterdam, Netherlands)*, 19, 100903. https://doi.org/10.1016/j.onehlt.2024.100903.
Yuan, C., Wu, J., Peng, Y., Li, Y., Shen, S., Deng, F., Hu, Z., Zhou, J., Wang, M., & Zou, Z. (2020). Transcriptome analysis of the innate immune system of Hyalomma asiaticum. *Journal of invertebrate pathology*, 177, 107481. https://doi.org/10.1016/j.jip.2020.107481.
Yun, S. M., Lee, Y. J., Choi, W., Kim, H. C., Chong, S. T., Chang, K. S., Coburn, J. M., Klein, T. A., & Lee, W. J. (2016). Molecular detection of severe fever with thrombocytopenia syndrome and tick-borne encephalitis viruses in ixodid ticks collected from vegetation, Republic of Korea, 2014. *Ticks and tick-borne diseases*, 7(5), 970–978. https://doi.org/10.1016/j.ttbdis.2016.05.003.
Zaid, T., Ereqat, S., Nasereddin, A., Al-Jawabreh, A., Abdelkader, A., & Abdeen, Z. (2019). Molecular characterization of Anaplasma and Ehrlichia in ixodid ticks and reservoir hosts from Palestine: A pilot survey. *Veterinary medicine and science*, 5(2), 230–242. https://doi.org/10.1002/vms3.150.
Zając, Z., Kulisz, J., Woźniak, A., Obregón, D., Foucault-Simonin, A., Bartosik, K., Moutailler, S., & Cabezas-Cruz, A. (2024). Spatial distribution and pathogen profile of Dermacentor reticulatus ticks in Southeastern Poland: A genetic and environmental analysis. *Transboundary and emerging diseases*, 2024, 5458278. https://doi.org/10.1155/2024/5458280.
Zakham, F., Korhonen, E. M., Puonti, P. T., Castrén, R. S., Uusitalo, R., Smura, T., Kant, R., Vapalahti, O., Sironen, T., & Kinnunen, P. M. (2023). Molecular detection of pathogens from ticks collected from dogs and cats at veterinary clinics in Finland. *Parasites & vectors*, 16(1), 327. https://doi.org/10.1186/s13071-023-05864-8.
Zchori-Fein, E., & Perlman, S. J. (2004). Distribution of the bacterial symbiont Cardinium in arthropods. *Molecular ecology*, 13(7), 2009–2016. https://doi.org/10.1111/j.1365-294X.2004.02203.x.
Zchori-Fein, E., Perlman, S. J., Kelly, S. E., Katzir, N., & Hunter, M. S. (2004). Characterization of a 'Bacteroidetes' symbiont in Encarsia wasps (Hymenoptera: Aphelinidae): Proposal of 'Candidatus Cardinium hertigii'. *International journal of systematic and evolutionary microbiology*, 54(Pt 3), 961–968. https://doi.org/10.1099/ijs.0.02957-0.
Žekić, M., Kišek, T. C., Cvitković-Špik, V., Ružić-Sabljić, E., Gajdov, V., Potkonjak, A., Jurišić, A., & Savić, S. (2025). Molecular detection and characterization

of tick-borne pathogens in Ixodes ricinus ticks collected from humans. *Pathogens (Basel, Switzerland)*, 14(6), 528. https://doi.org/10.3390/pathogens14060529.

Zellner, B., & Huntley, J. F. (2019). Ticks and Tularemia: Do we know what we don't know?. *Frontiers in cellular and infection microbiology*, 9, 146. https://doi.org/10.3389/fcimb.2019.00146.

Zeng, W., Kairat, Z., Awulibieer, M., Abylay, S., Serik, K., Yang, M., Wang, Y., & Hazihan, W. (2025). Molecular detection of piroplasms, Anaplasma, and Ehrlichia species in Kazakhstan. *Frontiers in veterinary science*, 12, 1533589. https://doi.org/10.3389/fvets.2025.1533603.

Zhang, B., Wang, X., Aguli Nurland, R., Lu, M., Guan, Y., Liu, M., Gao, F., & Li, K. (2024a). Investigation of tick-borne bacterial microorganisms in Haemaphysalis ticks from Hebei, Shandong, and Qinghai provinces, China. *Ticks and tick-borne diseases*, 15(2), 102290. https://doi.org/10.1016/j.ttbdis.2023.102292.

Zhang, B., Zhang, N., Zheng, T., Lu, M., Baoli, B., Jie, R., Wang, X., & Li, K. (2024b). Tick-borne bacterial agents in Hyalomma asiaticum ticks from Xinjiang Uygur Autonomous Region, Northwest China. *Parasites & vectors*, 17(1), 167. https://doi.org/10.1186/s13071-024-06258-y.

Zhang, C. M., Li, N. X., Zhang, T. T., Qiu, Z. X., Li, Y., Li, L. W., & Liu, J. Z. (2017). Endosymbiont CLS-HI plays a role in reproduction and development of Haemaphysalis longicornis. *Experimental & applied acarology*, 73(3-4), 429–438. https://doi.org/10.1007/s10493-017-0194-y.

Zhang, M. Z., Bian, C., Ye, R. Z., Cui, X. M., Chu, Y. L., Yao, N. N., Xu, X. W., Ye, J. L., Chen, L., Yang, J. H., Su, X. L., Huang, L., Shi, X. Y., Zhao, L., Chen, Y. G., Zheng, Y. C., Zheng, X. M., Jiang, J. F., & Cao, W. C. (2025a). Human infection with a novel tickborne orthonairovirus species in China. *The New England journal of medicine*, 392(2), 200–202. https://doi.org/10.1056/NEJMc2410853.

Zhang, M. Z., Bian, C., Ye, R. Z., Cui, X. M., Yao, N. N., Yang, J. H., Chu, Y. L., Su, X. L., Wu, Y. F., Ye, J. L., Liu, S. S., Shi, X. Y., Shi, W., Jia, N., Chen, Y. G., Zhao, L., Zheng, Y. C., Zheng, X. M., Jiang, J. F., & Cao, W. C. (2025b). A series of patients infected with the emerging tick-borne Yezo virus in China: An active surveillance and genomic analysis. *The Lancet. Infectious diseases*, 25(4), 390–398. https://doi.org/10.1016/S1473-3099(24)00616-9.

Zhang, R., Zhang, Q., Yu, G., & Zhang, Z. (2021). Metagenomic deep sequencing obtains taxonomic and functional profiles of Haemaphysalis longicornis that vary in response to different developmental stages and sexes. *Experimental & applied acarology*, 83(2), 285–300. https://doi.org/10.1007/s10493-020-00582-3.

Zhang, X. A., Ma, Y. D., Zhang, Y. F., Hu, Z. Y., Zhang, J. T., Han, S., Wang, G., Li, S., Wang, X., Tang, F., Liang, W. J., Yuan, H. X., Zhao, J. Q., Jiang, L. F., Zhang, L., Si, G. Q., Peng, C., Wang, R., Ge, H. H., Li, N., … & Liu, W. (2024). A new orthonairovirus associated with human febrile illness. *The New England journal of medicine*, 391(9), 821–831. https://doi.org/10.1056/NEJMoa2313722.

Zhang, X., Norris, D. E., & Rasgon, J. L. (2011). Distribution and molecular characterization of Wolbachia endosymbionts and filarial nematodes in Maryland populations of the lone star tick (Amblyomma americanum). *FEMS microbiology ecology*, 77(1), 50–56. https://doi.org/10.1111/j.1574-6941.2011.01089.x.

Zhang, X., Zhao, C., Cheng, C., Zhang, G., Yu, T., Lawrence, K., Li, H., Sun, J., Yang, Z., Ye, L., Chu, H., Wang, Y., Han, X., Jia, Y., Fan, S., Kanuka, H., Tanaka, T., Jenkins, C., Gedye, K., Chandra, S., … & Zheng, A. (2022). Rapid spread of severe fever with thrombocytopenia syndrome virus by parthenogenetic Asian longhorned ticks. *Emerging infectious diseases*, 28(2), 363–372. https://doi.org/10.3201/eid2802.211532.

Zhang, Y. K., Yu, Z. J., Wang, D., Bronislava, V., Branislav, P., & Liu, J. Z. (2019). The bacterial microbiome of field-collected Dermacentor marginatus and Dermacentor reticulatus from Slovakia. *Parasites & vectors*, 12(1), 325. https://doi.org/10.1186/s13071-019-3582-9.

Zhang, Y., Cheng, T. Y., Liu, G. H., Liu, L., & Duan, D. Y. (2024). Metagenome reveals the midgut microbial community of Haemaphysalis qinghaiensis ticks collected from yaks and Tibetan sheep. *Parasites & vectors*, 17(1), 370. https://doi.org/10.1186/s13071-024-06442-y.

Zhao, G. P., Wang, Y. X., Fan, Z. W., Ji, Y., Liu, M. J., Zhang, W. H., Li, X. L., Zhou, S. X., Li, H., Liang, S., Liu, W., Yang, Y., & Fang, L. Q. (2021). Mapping ticks and tick-borne pathogens in China. *Nature communications*, 12(1), 1075. https://doi.org/10.1038/s41467-021-21375-1.

Zhao, L., Ma, Y. M., Yang, B., Han, W. X., Zhao, W. H., Chai, H. L., Zhang, Z. S., Zhan, Y. J., Wang, L. F., Xing, Y., Yu, L. F., Wang, J. L., Ding, Y. L., & Liu, Y. H. (2022). Comparative analysis of microbial communities in different growth stages of Dermacentor nuttalli. *Frontiers in veterinary science*, 9, 1021426. https://doi.org/10.3389/fvets.2022.1021426.

Zhao, N., Pan, K., Teng, Z., Wang, H., Zhang, X., Ren, H., Yi, L., He, J., Cai, K., & Qin, T. (2023). Molecular detection reveals diverse tick-borne bacterial and protozoan pathogens in two tick species from Yingshan County of Hubei Province, China in 2021-2022. *Frontiers in microbiology*, 14, 1298037. https://doi.org/10.3389/fmicb.2023.1298051.

Zhendong, H., Guangfu, Y., Zhong, Z., & Ruiling, Z. (2019). Phylogenetic relationships and effectiveness of four Beauveria bassiana sensu lato strains for control of Haemaphysalis longicornis (Acari: Ixodidae). *Experimental & applied acarology*, 77(1), 83–92. https://doi.org/10.1007/s10493-018-0329-9.

Zheng, H., Dietrich, C., Hongoh, Y., & Brune, A. (2016). Restriction-modification systems as mobile genetic elements in the evolution of an intracellular symbiont. *Molecular biology and evolution*, 33(3), 721–725. https://doi.org/10.1093/molbev/msv264.

Zhioua, E., Browning, M., Johnson, P. W., Ginsberg, H. S., & LeBrun, R. A. (1997). Pathogenicity of the entomopathogenic fungus Metarhizium anisopliae (Deuteromycetes) to Ixodes scapularis (Acari: Ixodidae). *The Journal of parasitology*, 83(5), 815–818.

Zhioua, E., Ginsberg, H. S., Humber, R. A., & Lebrun, R. A. (1999). Preliminary survey for entomopathogenic fungi associated with Ixodes scapularis (Acari: Ixodidae) in southern New York and New England, USA. *Journal of medical entomology*, 36(5), 635–637. https://doi.org/10.1093/jmedent/36.5.635.

Zhong, J., Jasinskas, A., & Barbour, A. G. (2007). Antibiotic treatment of the tick vector Amblyomma americanum reduced reproductive fitness. *PloS one*, 2(5), e405. https://doi.org/10.1371/journal.pone.0000405.

Zhong, Z., Zhong, T., Peng, Y., Zhou, X., Wang, Z., Tang, H., & Wang, J. (2021). Symbiont-regulated serotonin biosynthesis modulates tick feeding activity. *Cell host & microbe*, 29(10), 1545–1557.e4. https://doi.org/10.1016/j.chom.2021.08.011.

Zhou, H., Ma, Z., Hu, T., Bi, Y., Mamuti, A., Yu, R., Carr, M. J., Shi, M., Li, J., Sharshov, K., Gao, G. F., & Shi, W. (2019). Tamdy virus in Ixodid ticks infesting bactrian camels, Xinjiang, China, 2018. *Emerging infectious diseases*, 25(11), 2136–2138. https://doi.org/10.3201/eid2511.190512.

Zhou, Y., Zhang, H., Cao, J., Gong, H., & Zhou, J. (2016). Epidemiology of toxoplasmosis: Role of the tick Haemaphysalis longicornis. *Infectious diseases of poverty*, 5, 14. https://doi.org/10.1186/s40249-016-0106-0.

Zhu, C., Ai, L., Qi, Y., Liu, Y., Li, H., Ye, F., Wang, Q., Luo, Y., Tan, W., & Shi, C. (2022). Molecular detection of spotted fever group rickettsiae in hedgehogs (Erinaceus amurensis) and hedgehog-attached ticks in Xuyi County, Southeast China. *Experimental & applied acarology*, 88(1), 97–111. https://doi.org/10.1007/s10493-022-00721-y.

Zhuang, L., Du, J., Cui, X. M., Li, H., Tang, F., Zhang, P. H., Hu, J. G., Tong, Y. G., Feng, Z. C., & Liu, W. (2018). Identification of tick-borne pathogen diversity by metagenomic analysis in Haemaphysalis longicornis from Xinyang, China. *Infectious diseases of poverty*, 7(1), 45. https://doi.org/10.1186/s40249-018-0417-4.

CHAPTER SEVEN

The Locust Microbiome

Alex Muppasani and Sunil Thomas

7.1 INTRODUCTION

Desert locusts (*Schistocerca gregaria*) and migratory locusts (*Locusta migratoria*), together with other swarming grasshoppers, have shaped human history as some of the most destructive migratory pests, with recent plagues in Africa and Asia underscoring their capacity to destabilize food systems and threaten the livelihoods of hundreds of millions of people. Unlike most grasshoppers, these locusts are defined by extreme density-dependent phase polyphenism: at low population densities they persist as cryptic, solitary individuals, but when densities rise they rapidly shift into brightly colored, gregarious swarms capable of traversing continents and consuming crops at devastating scales (Foquet et al., 2021). This transformation has traditionally been explained in terms of host neuroendocrine plasticity, environmental triggers, and climate-driven habitat changes, yet a growing body of work now shows that the locust microbiome is intimately involved in almost every aspect of this process.

Across the gut, integument, oral cavity, and reproductive tract, desert locusts (*S. gregaria*), migratory locusts (*L. migratoria*), brown locusts (*Locustana pardalina*), Italian locusts (*Calliptamus italicus*), and Moroccan locusts (*Dociostaurus maroccanus*) maintain structured microbial communities dominated by Proteobacteria and Firmicutes, including core genera such as *Enterobacter*, *Klebsiella*, *Pantoea*, and *Weissella* (Lavy et al., 2020a). These symbionts ferment recalcitrant plant cell walls, synthesize vitamins, and buffer the host against plant toxins and pesticides, enabling swarms of different locust species to exploit low-quality vegetation over long distances. At the same time, specific gut and cuticular bacteria produce aggregation pheromones such as guaiacol, modulate immune defenses through colonization resistance, and can shift from mutualists to opportunistic pathogens under fungal or microsporidian infection, making the microbiome both a driver of locust success and a potential point of control (Dillon et al., 2000).

7.2 LOCUST BIOLOGY, SWARMING, AND MICROBIOME CONTEXT

Locust swarming in desert locusts (*S. gregaria*) and migratory locusts (*L. migratoria*), as well as other swarming acridids, remains one of the most persistent threats to global food security, with severe plague years affecting the livelihoods of an estimated one in ten people worldwide. These insects are short-horned grasshoppers in the family Acrididae, defined functionally by their ability to switch between a solitarious phase and a gregarious phase in response to environmental and social cues rather than by a unique taxonomic status. In *S. gregaria* and *L. migratoria*, this switch is striking: solitarious locusts are cryptic in coloration, avoid conspecifics, move relatively little, and disperse at low densities, whereas gregarious locusts are conspicuously patterned, highly active, and strongly attracted to one another, forming dense marching hopper bands and massive flying swarms that can travel hundreds to thousands of kilometers (Fürstenau et al., 2025). Central to

DOI: 10.1201/9781003639176-7

this metamorphosis is an intricate microbiome whose combined metabolic and immunological activities support key traits that make locusts such effective pests, including efficient exploitation of poor-quality vegetation, resistance to plant and anthropogenic toxins, and pheromone-based social coordination (Lavy et al., 2020b).

At low densities, solitarious desert locusts (*S. gregaria*) and migratory locusts (*L. migratoria*) typically lead a hidden existence, feeding on scattered vegetation and minimizing encounters with conspecifics and predators. When environmental conditions change—for example, when prolonged drought is followed by intense, localized rainfall that produces bursts of fresh vegetation—breeding rates increase and local population densities rise (Tandon, 2020). Physical crowding, particularly repeated contact on the hind legs, activates mechanoreceptors and initiates neurochemical cascades dominated by serotonin and dopamine, which quickly alter locomotor activity and social attraction. Within hours, behavior shifts from avoidance to attraction; over days to weeks, changes in body coloration, morphology, and physiology consolidate the gregarious phenotype (Rogers et al., 2003). In parallel, the microbiomes of the gut, integument, and reproductive tract reorganize, with increases in specific taxa such as *Weissella cibaria* on the cuticle and Enterobacteriaceae including *Pantoea agglomerans*, *Klebsiella pneumoniae*, and *Enterobacter cloacae* in the hindgut that enhance digestion, immune modulation, and the production of aggregation pheromones such as guaiacol and phenol (Shi et al., 2014).

Scientific understanding of these phase changes advanced greatly after Boris Uvarov recognized in 1921 that the solitary and gregarious forms of the desert locust (*S. gregaria*) are not separate "species" but phases of a single species. This recognition established the modern concept of density-dependent phenotypic plasticity, in which locusts adjust behavior, morphology, and physiology according to local population density, resource availability, and environmental context (Li et al., 2022). Although the majority of grasshopper species remain strictly solitary, around 20 locust species—including *S. gregaria*, *L. migratoria*, the brown locust *L. pardalina*, the Moroccan locust *D. maroccanus*, and, more rarely, the American grasshopper *Schistocerca americana*—have independently evolved this capacity for dramatic phase change. The repeated evolution of swarming across these genera and subfamilies indicates strong adaptive value in arid and semi-arid ecosystems, where resources are patchy and ephemeral and long-distance movement is often essential for survival.

Contemporary research extends this classic framework by demonstrating that host–microbe interactions are deeply integrated into the biology of phase change and swarming in *S. gregaria*, *L. migratoria*, *L. pardalina*, *C. italicus*, *D. maroccanus*, and *S. americana*. Gut communities dominated by *E. cloacae*, *K. pneumoniae*, *P. agglomerans*, and related Enterobacteriaceae enhance digestion of cellulose-rich diets, contribute to nitrogen economy and vitamin synthesis, and produce phenolic compounds such as guaiacol and phenol that function as aggregation pheromones in desert locusts (Lavy et al., 2019a). Integumentary microbes, notably *W. cibaria* on the head and abdominal cuticle of gregarious *S. gregaria*, accumulate during phase transition and may release volatiles that reinforce aggregation and group cohesion (Kang et al., 2005). At the same time, pathogens such as the fungus *Metarhizium acridum* and the microsporidian *Paranosema locustae* exploit or disrupt these microbial systems, sometimes promoting the translocation of gut symbionts into the hemocoel or suppressing pheromone-producing bacteria, which can turn mutualists into opportunists or break down swarm cohesion (Shi et al., 2014a). Together, these findings support a view of locust swarming as an emergent property of interacting genetic, neuroendocrine, environmental, and microbiome-mediated processes rather than a purely host-driven response to crowding.

7.3 GEOGRAPHIC RANGE AND MAJOR LOCUST SPECIES

Economically important locust species, led by the desert locust (*S. gregaria*) and the migratory locust (*L. migratoria*), occupy a vast geographic range that together covers nearly 20% of the global land surface (Lavy et al., 2020a). *S. gregaria* primarily inhabits arid and semi-arid regions of North Africa, the Middle East, and Southwest Asia, where rainfall is highly variable and swarms form when rare, intense rain events generate rapid, patchy vegetation growth that promotes crowding and high reproductive output. *L. migratoria* ranges widely across Africa, Europe, and Asia, while the brown locust (*L. pardalina*) and the Moroccan locust (*D. maroccanus*) are major pests in southern Africa and in parts of the western Mediterranean and Central Asia, respectively. Even the American grasshopper (*S. americana*), which only rarely forms large plagues, shows subtle behavioral phase changes, suggesting that the genetic and physiological capacity for swarming is widely distributed within the genus *Schistocerca* (Chang et al., 2025).

Climate change models and recent syntheses predict that suitable habitat for *S. gregaria* and *L. migratoria* may expand northward and eastward by several percentage points by the late twenty-first century, driven by changes in temperature, precipitation patterns, and land use (Tang et al., 2023). For desert locusts in particular, projected shifts in rainfall regimes in northern Africa and southwestern Asia are expected to increase the frequency and intensity of outbreaks in regions that are already vulnerable (Mitra et al., 2024). These trends raise direct concerns for agricultural resilience and regional food systems and underscore the need to understand how microbiome-mediated processes in *S. gregaria*, *L. migratoria*, *L. pardalina*, and *D. maroccanus* contribute to swarming, survival, and potential points of control.

7.4 ECONOMIC COSTS OF LOCUST PLAGUES

Plagues of *S. gregaria*, *L. migratoria*, and related species impose enormous economic costs that extend far beyond the immediate value of lost crops. The 2019–2021 desert locust upsurge, driven by unusual sequences of rainfall across the Arabian Peninsula and the Horn of Africa, was the worst infestation in 25 years in Ethiopia and Somalia and the worst in more than 70 years in Kenya and Uganda (Devi, 2023). For 2020 alone, World Bank estimates suggested that damages and losses in East Africa and Yemen linked to *S. gregaria* swarms could reach up to USD 8.5 billion (FAO, 2021). In contrast, Desert Locust Response operations coordinated by FAO and national governments cost about USD 230.5 million but were estimated to avert 4.5 million tons of cereal losses and 900 million liters of milk, securing food for 41.5 million people and avoiding roughly USD 1.77 billion in cereal and milk losses (FAO, 2022).

Earlier plagues show similar patterns of high damage and costly response. The 2003–2005 West Africa outbreak, dominated by *S. gregaria*, required more than USD 450 million in control spending and resulted in an estimated USD 2.5 billion in crop damage, including a 27% shortfall in cereal production in Niger that forced many villagers to abandon their land and migrate (*World Bank Group*, 2020). The 1986–1989 plague drew about USD 310 million in emergency protection efforts, while the 1954–1955 outbreak in Morocco's Souss Massa Valley, largely due to the Moroccan locust (*D. maroccanus*), caused losses exceeding USD 50 million (in 1994 dollars) in only six weeks. The historic 1874 Rocky Mountain locust (*Melanoplus spretus*) plague in the United States produced more than USD 200 million in crop damage and necessitated state relief bonds, even though this species is now extinct (*The 1874 Rocky Mountain Locust Plague That Blotted Out the Sun » Explorersweb*). Collectively, these events demonstrate that delayed or inadequate control of *S. gregaria*, *L. migratoria*, *L. pardalina*, *D. maroccanus*, and other locusts multiplies economic losses far beyond the cost of timely, well-coordinated interventions.

7.5 HUMAN AND NUTRITIONAL IMPACTS

The human costs of plagues of *S. gregaria*, *L. migratoria*, *L. pardalina*, and *D. maroccanus* extend well beyond immediate crop loss to long-term nutritional and developmental impacts. When swarms of desert locusts, migratory locusts, brown locusts, or Moroccan locusts destroy cereals, pulses, vegetables, and pasture, affected regions experience serious calorie shortfalls and micronutrient deficits, while livestock losses further reduce milk and meat availability (FAO, 2025). Data from 2008 to 2018 indicate that biological disasters, including locust invasions, accounted annually for average losses of about 6.9 trillion kilocalories, 256 billion milligrams of iron, 45 trillion milligrams of calcium, and 7 trillion micrograms of vitamin A across low- and lower middle–income countries (FAO, 2021). Over a decade, cumulative losses reached 76 trillion kilocalories and hundreds of trillions of milligrams or micrograms of key micronutrients, underscoring the depth of nutritional erosion associated with recurrent pest shocks.

These nutritional deficits translate into higher rates of child stunting, wasting, and impaired cognitive development in communities repeatedly affected by outbreaks of *S. gregaria*, *L. migratoria*, and other locust species. Historical analyses of major desert locust plagues have documented declines in school enrollment and attendance as households divert labor and resources to coping with crises and prioritize short-term survival over education (Asare et al., 2023). Locust plagues, therefore, act as multipliers of vulnerability, intersecting with poverty, conflict, and climate stress to deepen food insecurity and constrain long-term human development in regions within the ranges of *S. gregaria*, *L. migratoria*, *L. pardalina*, and *D. maroccanus*.

7.6 CONTROL STRATEGIES AND MICROBIOME-INFORMED APPROACHES

Control efforts against *S. gregaria*, *L. migratoria*, *L. pardalina*, and *D. maroccanus* have historically relied on broad-spectrum chemical insecticides, which can reduce swarm size but carry substantial environmental and health costs (Panopoulou & Tsagkarakis, 2025). In response, integrated pest management strategies now increasingly combine early warning, targeted interventions, and biological control agents that exploit species-specific vulnerabilities in desert locusts, migratory locusts, brown locusts, and Moroccan locusts (Lumumba et al., 2024). The entomopathogenic fungus *M. acridum*, deployed as the mycoinsecticide Green Muscle, infects acridids through the cuticle, colonizes the hemocoel, and kills the host via nutrient depletion and toxin production while largely sparing non-target organisms such as birds, fish, and beneficial insects (Wasuwan et al., 2021). As *M. acridum* infections progress in *S. gregaria* and *L. migratoria*, infected locusts become lethargic and easier to catch; in Niger, falcons and other raptors removed up to 26% of *M. acridum*-treated desert locust populations, demonstrating synergy between biocontrol and natural predators (Li & Xia, 2022).

Microbial control also includes the microsporidian *P. locustae*, which infects *S. gregaria* and *L. migratoria*, acidifies the hindgut, suppresses guaiacol-producing Enterobacteriaceae, and lowers serotonin and dopamine levels, pushing gregarious locusts back toward more solitary-like behavior and weakening swarm cohesion (Shi et al., 2014). Building on these tools, newer microbiome-based strategies aim to target the specific symbionts that sustain swarming and resilience in desert and migratory locusts (Kabir et al., 2024). Bacteriophages designed to infect *P. agglomerans*, *K. pneumoniae*, and *E. cloacae* may selectively disrupt guaiacol production in *S. gregaria* and *L. migratoria* without damaging beneficial environmental microbes. Small-molecule inhibitors such as 4-nitrophenol (4NP) can block host enzymes 4VPMT1 and 4VPMT2 involved in 4-vinylanisole synthesis in the migratory locust (*L. migratoria*), reducing aggregation signaling at its source (Adriaenssens et al., 2011).

These emerging tools are being developed within "Safe-by-Design" frameworks that use genetic safeguards—such as kill switches, quorum sensing–dependent replication, and metabolic auxotrophy—to limit the persistence and host range of engineered phages or symbionts in environments shared by desert locusts, migratory locusts, and other fauna (Hoffmann et al., 2023). Coupled with digital surveillance systems such as FAO's Desert Locust Information Service (DLIS), the eLocust3 field data collection system, and the Caucasus and Central Asia Locust Management (CCALM) platform, microbiome-informed interventions can support earlier, more precise actions to prevent local outbreaks of *S. gregaria*, *L. migratoria*, *L. pardalina*, and *D. maroccanus* from escalating into regional or global plagues (Cressman & Hodson, 2009). By focusing on the microbial systems that support locust nutrition, detoxification, communication, and immunity, these strategies offer a way to move away from reactive broad-spectrum spraying and toward targeted, One Health-aligned management of some of the world's most damaging insect pests.

7.7 NEUROCHEMICAL CASCADES AND THE DYNAMICS OF SOCIAL INTERACTION

The transition from the solitary to the gregarious phase in locusts is a rapid, density-dependent phenomenon driven primarily by social experience rather than gradual genetic change. This shift is initiated by local crowding, during which repeated physical contact stimulates mechanoreceptors on the hind legs, abdomen, and other body surfaces. These tactile inputs generate strong sensory signals that are transmitted via ascending neural pathways to the thoracic ganglia and brain, where they trigger a cascade of neurochemical changes that fundamentally reorganize behavior (University of Cambridge, 2009). The speed of this response highlights the primacy of neural plasticity in locust phase change, allowing individuals to rapidly adapt to fluctuating population densities.

Within approximately 2 hours of sustained crowding, serotonin concentrations increase markedly in discrete regions of the brain. This serotonergic surge functions as a molecular switch for gregarization. Experimental elevation of serotonin in solitary-reared locusts is sufficient to induce gregarious-like behaviors even in the absence of physical crowding, while pharmacological inhibition of serotonin signaling blocks the transition despite continued tactile stimulation (Ott & Rogers, 2010). Dopamine and other neuromodulators act in parallel with serotonin, shaping both the onset and persistence of the gregarious state. Current evidence suggests that serotonin primarily triggers the initial behavioral switch, whereas dopamine contributes to reinforcement, motivational drive, and the long-term stabilization of gregariousness across molts and developmental stages (Guo et al. 2013; Alessi et al. 2014~~Yong, 2009~~).

Crucially, behavioral changes precede any visible alterations in morphology or coloration, demonstrating that phase polyphenism begins as a neural and hormonal process that is only later consolidated in physiology and anatomy. Under crowded conditions, solitary-reared locusts can acquire most behavioral hallmarks of gregariousness within 4–8 hours. These include a shift from avoidance to attraction toward conspecifics, from sporadic to sustained locomotion, and from cryptic, risk-averse feeding to exposed, risk-tolerant foraging (Simpson et al., 2001). Even brief crowding events can leave lasting imprints: exposure to a small group for as little as 30 minutes can induce behavioral changes that persist for hours or days after return to isolation. Such persistent effects include increased grooming frequency, reduced resting time, and heightened attraction to social, visual, and olfactory cues, generating positive feedback loops that stabilize gregarious tendencies and lower the threshold for reaggregation (Shved et al., 2022). Over longer timescales, repeated or intermittent crowding episodes can therefore "train" the nervous system, biasing individuals toward sustained gregariousness as swarms form and propagate.

The locust nervous system encodes the social environment through the integration of multiple sensory modalities into a composite internal representation of group context. Visual cues such as neighbor density, movement, and orientation; tactile cues from direct contact or air displacement; and olfactory cues from fecal material and cuticular volatiles all contribute to estimates of local crowding (Unni et al., 2024). Gregarious locusts process these cues differently from solitary individuals, exhibiting stronger orientation toward conspecific silhouettes, heightened sensitivity to motion in the visual field, and enhanced attraction to aggregation pheromones such as guaiacol, phenol, and 4-vinylanisole. Notably, several of these compounds are produced or modulated by the locust-associated microbiome, providing a mechanistic link between microbial metabolism and host neural decision-making (Yang et al., 2023). The neuromodulatory state of the animal modulates how these cues are weighted, such that identical stimuli can elicit avoidance in a low-serotonin solitary state or attraction in a high-serotonin, dopamine-shifted gregarious state (Bertsch et al., 2025).

At the collective level, the highly coordinated movement of marching bands and flying swarms emerges from simple decision rules implemented by individual nervous systems rather than from centralized control or complex planning. Classical models described locusts as "self-propelled particles" that align their headings with near neighbors, but more recent behavioral and modeling studies suggest a richer mechanism (Montanari et al., 2021). Locusts appear to maintain minimal internal representations of neighbor position and density, applying consensus rules such as avoiding excessively high density behind the body while maintaining positions that minimize the risk of attack from conspecifics. Movement decisions are thus based on relative positioning, local density gradients, and the balance between attraction to the group and avoidance of immediate threats, rather than on heading alignment alone (Sayin et al., 2025).

This framework helps explain how coherent collective motion can arise from noisy, local interactions and why cannibalism risk and sensory constraints play central roles in shaping group structure. In crowded, resource-limited environments, conspecifics are simultaneously competitors and potential predators, and the risk of being bitten from behind strongly biases individuals toward forward movement, producing the characteristic "forced march" dynamics of hopper bands. Experimental disruption of sensory inputs—such as limiting rear visual fields or impairing mechanoreception—reduces the ability of individuals to detect these threats, decreases forward movement, and increases cannibalism risk, ultimately destabilizing group coherence. By coupling serotonergic and dopaminergic modulation with multisensory integration and simple movement rules, the locust nervous system scales local social encounters into large, directed swarms that are further shaped by microbiome-derived chemical cues and environmental conditions.

7.8 MOVEMENT AND CANNIBALISM IN LOCUST HOPPER BANDS

Locust hopper bands are dense, cohesive marching aggregations of juvenile, wingless locusts (nymphs) that can reach local densities of roughly 20–120 individuals per square meter. These formations represent a critical transitional phase in locust phase polyphenism, during which formerly solitary individuals become gregarious as a direct consequence of crowding and repeated social interactions. This shift triggers profound changes in morphology, physiology, and behavior, ultimately producing highly coordinated, mobile groups capable of extensive and rapid landscape-scale movement.

Although hopper bands are a striking example of collective motion on land, their cohesion does not primarily arise from cooperation or mutualistic alignment. Instead, their movement is largely driven by intense competition for scarce resources and by the constant risk of cannibalism. In the arid and semi-arid environments typically inhabited by locusts, essential nutrients—particularly protein and salts—are both limited and patchily distributed. Under these conditions, other locusts become one of the most reliable and concentrated sources of high-quality nutrition. This ecological context generates an autocatalytic feedback process often described as a "forced march," in which individuals are simultaneously fleeing attacks from behind while pursuing conspecifics ahead as potential prey (Cease et al., 2023).

At the level of individual decision-making, locusts move forward primarily to reduce the risk of being bitten or eaten by neighbors approaching from the rear. At the same time, they are strongly motivated to advance toward individuals in front of them, which represent accessible sources of protein and minerals. The result is a persistent pursuit-and-escape dynamic in which fear of cannibalism and hunger are tightly coupled. Without any centralized control or cooperative intent, these local interactions give rise to self-organized, group-level motion. In effect, the constant threat of rearward attack acts as a directional pressure that pushes the entire hopper band forward across the landscape.

Experimental studies on the desert locust *S. gregaria* demonstrate that the remarkable coordination observed in hopper bands and adult swarms is better explained as a survival strategy shaped by conspecific predation than as a product of social cooperation (Rogers et al., 2014). When researchers experimentally reduced a locust's ability to detect approaching individuals from behind—either by severing nerves connected to abdominal mechanoreceptors or by obstructing the rear visual field—those individuals showed a reduced tendency to move forward. As a consequence, they were significantly more likely to be attacked and cannibalized by nearby locusts. At the group level, these sensory impairments disrupted the coherence and directional persistence of hopper bands, providing strong evidence that rearward sensory input and cannibalism risk are central drivers of collective motion rather than incidental effects of high population density.

When scaled up, these individual-level survival responses produce dramatic large-scale consequences. Millions of locusts can move together over long distances at substantial speeds, consuming nearly all available green biomass in their path. This intense grazing pressure can rapidly transform vegetation structure, reduce soil cover, and alter nutrient cycling across entire regions. Thus, simple behavioral rules shaped by immediate predation risk at the individual level can cascade into profound ecological and agricultural disturbances at the landscape scale (Bazazi et al., 2008).

Local social conditions further modulate these dynamics by influencing how frequently locusts encounter one another and how those encounters unfold. As density increases, interactions between individuals tend to last longer and are more likely to escalate into biting and successful cannibalistic attacks, indicating that crowding amplifies aggression and increases the impact of each encounter (Guttal et al., 2012). At the same time, existing groups actively attract additional locusts through multiple sensory channels, including visual cues such as moving bodies and shadows, chemical cues such as odor plumes from feces, cuticular hydrocarbons, and microbiome-derived volatiles, as well as characteristic movement patterns. Together, these cues generate positive feedback in which small clusters grow into large, dense aggregations as more individuals are drawn in (Topaz et al., 2012).

This feedback loop creates a paradoxical situation in which the same crowd that provides directional cues and collective movement also increases the likelihood of being attacked and eaten. Consequently, the "gregarious signature" of a locust swarm reflects not only social attraction, but also tactical survival and opportunistic resource acquisition under extreme environmental stress. These behavioral dynamics are further intertwined with microbiome-mediated processes that influence digestion, immune function, and chemical communication, reinforcing the capacity of locusts to persist and move collectively in hostile environments (Olson et al., 2013).

7.9 ANATOMICAL DISTRIBUTION AND PHYLUM DIVERSITY OF THE LOCUST MICROBIOME

The locust microbiome is spread across several main body regions, and each region hosts bacterial communities that are adapted to its local conditions and functions. These communities support digestion, immunity, chemical signaling, and reproduction, and their composition reflects both the physical structure of the tissues and the types of nutrients and immune factors present (Rowland et al., 2018). The primary site of microbial activity is the alimentary canal, which is divided into foregut, midgut, and hindgut, and each section provides a different microhabitat shaped by its lining, secretions, and transit time of food (Ye & Rawls, 2021). Together, these regions form a longitudinal gradient in pH, oxygen levels, nutrient availability, and immune pressure, which selects for distinct bacterial groups and functional roles along the gut.

The foregut is lined with cuticle and functions as a storage and mixing chamber, where ingested plant material is combined with saliva and regurgitated enzymes to start mechanical and enzymatic breakdown. Because food passes through this region relatively quickly and the cuticular lining is periodically shed, the foregut microbiota are usually low in density and dominated by transient bacteria that closely resemble the communities found on the plants the locusts eat. These transient microbes may participate briefly in the first steps of fermentation or detoxification but generally do not establish long-term residency (Lavy et al., 2020db. The midgut, which often includes six cecal outpouchings in species such as *S. gregaria* and *L. migratoria*, is the main site of digestion and nutrient absorption, and it is bathed in digestive enzymes and immune effectors. Despite its central role in processing food, the midgut is relatively sparsely colonized compared to the hindgut, because the combination of a peritrophic membrane, high enzyme levels, and strong immune activity creates a harsh environment that limits bacterial growth and filters out many would-be colonizers (Lavy et al., 2020d).

By contrast, the hindgut, particularly the ileum and rectum, is the most densely colonized region of the digestive tract and serves as the major hub of microbial metabolism. Rectal pads increase surface area for the absorption of water, salts, and microbially produced metabolites, and they provide attachment sites for dense, biofilm-like assemblages of specialized bacteria (Sai Charan et al., 2025). Conditions in the hindgut, including slower transit, more stable pH, and lower concentrations of digestive enzymes, favor the establishment of a relatively stable, host-modulated microbiota. These hindgut communities ferment complex plant polysaccharides, produce volatile fatty acids (VFAs) and phenolic compounds, and contribute to colonization resistance against pathogens, making them central to both nutrition and chemical communication (von Engelhardt et al., 1998).

Across these gut regions and life stages, the microbiome is dominated by two major phyla, Proteobacteria and Firmicutes, with smaller but consistent contributions from Bacteroidetes and Actinobacteria. Within this broad taxonomic framework, a "core" microbiome has been described in both *S. gregaria* and *L. migratoria*, built around genera such as *E. cloacae*, *K. pneumoniae*, and *Pseudomonas aeruginosa* (and related *Pseudomonas* spp.), which belong to the family Enterobacteriaceae or closely related lineages and are known for their metabolic versatility and frequent association with plant and soil environments (Muratore et al., 2020). Other common gut residents include *P. agglomerans*, *Citrobacter freundii* (representing the genus *Citrobacter*), and *Pseudocitrobacter faecalis* (representing the genus *Pseudocitrobacter*) among the Proteobacteria, and *Lactococcus lactis* and *W. cibaria* among the Firmicutes, with lactic acid bacteria often becoming especially abundant in laboratory-reared populations where standardized diets favor their growth (Jabeen et al., 2023). Comparative studies of foregut and hindgut communities in *S. gregaria* and *L. migratoria* show that while foregut communities largely mirror those found on food plants, the hindgut strongly selects for high loads of specialized Enterobacteriaceae such as *P. agglomerans*, *K. pneumoniae*, and *E. cloacae*, reflecting the more stable, host-shaped conditions and the importance of these bacteria for fermentation and signaling (Wang et al., 2022).

Important microbial communities also exist outside the digestive tract, particularly on the external cuticle (integument) and in the reproductive system. The integumentary microbiome on the head, thorax, and abdomen includes both environmental bacteria picked up from soil and vegetation and symbionts that are repeatedly exchanged between individuals during physical contact, especially in dense aggregations (Ogunrinola et al., 2020). These surface communities can influence scent profiles, including the bouquet of volatiles that mediate recognition and attraction, and they may alter susceptibility to cuticular pathogens by occupying niches and producing antimicrobial compounds. The reproductive tract is another key site, maintaining a non-sterile environment that is essential for the vertical transmission of symbionts to offspring (Helletsgruber et al., 2017). Female reproductive tissues harbor gut-derived taxa such as *E. cloacae*, *K. pneumoniae*, and *Corynebacterium xerosis* (representing the genus *Corynebacterium*), and sequences from these genera are detectable in eggs and newly hatched nymphs, indicating direct transfer. Amplicon sequencing studies that track bacterial lineages across maternal tissues, foam plugs, eggs, and neonate guts confirm that the female reproductive tract is a critical vector for inoculating the next generation, complementing acquisition from food and the broader environment and helping ensure that offspring start life with a functional microbial toolkit (Dillon & Vennard, 2002).

7.10 TAXONOMIC DYNAMICS AND SUCCESSIONAL CHANGES ACROSS THE LIFE CYCLE

The composition and diversity of the locust gut microbiome change markedly over the life cycle, reflecting both vertically inherited microbes and those acquired from the environment. Locusts undergo incomplete metamorphosis, passing through egg, multiple nymphal (larval) instars, and an imago (adult) stage, and each of these stages involves different diets, habitats, and immune conditions that shape which bacteria can persist and thrive (Li et al., 2024). Work on *L. migratoria* shows that gut microbial diversity is often highest in the egg stage, where eggs contain rich and relatively even bacterial assemblages that likely function as reservoirs for vertical transmission. Many of the bacteria that dominate later larval and adult guts—including core taxa such as *E. cloacae*, *K. pneumoniae*, and *Enterococcus faecalis*—are already detectable in eggs, underscoring the importance of early colonization and maternal input to the assembly of the microbiome. This early seeding gives offspring a starting set of symbionts that can support digestion, immunity, and development from the moment they hatch into often harsh and unpredictable environments (Li et al., 2023).

As *L. migratoria* progresses through its nymphal instars, gut microbial diversity does not change in a simple linear fashion but instead fluctuates in characteristic ways. In early instars (first and second), species richness and the capacity to digest cellulose are significantly higher than in later instars (third to fifth), and these functional gains in cellulose breakdown correlate strongly with the relative abundance of genera such as *W. cibaria* and *P. faecalis* (Djamel Eddine et al., 2023). This pattern suggests that early larvae benefit from a broad and flexible set of microbes that help them efficiently process a wide range of plant materials while they are small, growing rapidly, and encountering diverse microhabitats. Over time, as the host immune system and digestive physiology mature, this diverse early community is "winnowed" down and replaced by a more stable, host-adapted resident microbiota that is better matched to the adult diet, gut environment, and immune filters (Basit et al., 2025). Principal coordinate analyses of Bray–Curtis distances reveal that gut community composition is more variable among early larvae than among adults, supporting a model in which the host gradually selects and stabilizes specific beneficial strains as development proceeds (Plantinga & Wu, 2021).

Environmental conditions overlay these developmental patterns and can strongly influence which microbes dominate at each stage. Temperature and humidity affect bacterial growth rates and survival, while geography determines the pool of environmental microbes available for acquisition from soil, plants, and other insects. Diet is particularly important: laboratory-reared *L. migratoria* maintained on standardized, limited diets and in relatively clean conditions often develop gut communities that differ markedly from those of wild populations (Qiu et al., 2022). Field-collected *L. migratoria* have been reported to be enriched in genera such as *P. agglomerans*, *Cutibacterium acnes* (representing *Cutibacterium*), and *Sphingomonas paucimobilis*, which likely reflect diverse plant-associated and environmental sources. In contrast, farmed individuals frequently show higher relative abundances of lactic acid bacteria such as *W. cibaria*, *E. faecalis*, and *L. lactis*, consistent with a narrower diet and more homogeneous rearing environment that favor these fermentative taxa (Cruz et al., 2007).

These comparisons highlight the dual nature of the locust microbiome as both conserved and flexible. A recognizable core—centered on taxa like *E. cloacae*, *K. pneumoniae*, and *P. agglomerans*—tends to persist across life stages, habitats, and populations, suggesting strong selection for functions such as cellulose degradation, nitrogen handling, vitamin synthesis, and pheromone precursor production (Lavy et al., 2019b). At the same time, the broader community is highly plastic and responsive to ecological context, which means that changes in climate, land use, host plant communities, or rearing practices can shift microbial composition in ways that alter nutritional efficiency, immune responsiveness, and susceptibility or resistance to pesticides and pathogens (Cavicchioli et al., 2019). This plasticity has practical implications, because it suggests that targeted changes to diet or environment could be used to steer microbiome composition toward states that either enhance locust fitness (in captive or farm settings) or reduce it (in biocontrol and management strategies).

7.11 PHYSIOLOGICAL CONTRIBUTIONS: CELLULOSE DIGESTION AND NUTRIENT CYCLING

One of the most clearly established and intensively studied functions of the locust gut microbiome is its contribution to digestion, particularly the breakdown of complex plant carbohydrates and the efficient exploitation of low-quality diets. Locusts are highly polyphagous herbivores that consume large quantities of grasses and other vascular plants dominated by cellulose and hemicellulose. These structural polysaccharides form rigid plant cell walls that are intrinsically difficult to digest, and the cellulolytic capacity encoded by the locust genome alone is insufficient for their complete degradation. As a result, endogenous insect-derived cellulases can only partially liberate nutrients from fibrous plant tissue. Gut-associated microbes substantially expand this digestive repertoire by producing a diverse array of cellulases, hemicellulases, and other carbohydrate-active enzymes that collectively dismantle plant cell walls and release fermentable sugars (Su et al., 2014).

Once liberated, these sugars enter microbial fermentation pathways that convert them into short-chain or VFAs, including acetate and propionate. These metabolites are readily absorbed across the hindgut epithelium and serve as important energy sources and carbon substrates for the host. VFAs are especially valuable under conditions where protein and simple carbohydrates are scarce, allowing locusts to maintain metabolic activity and locomotion even when feeding on nutritionally poor vegetation. In this way, microbial fermentation effectively buffers locusts against fluctuations in plant quality and enables survival on diets that would otherwise yield insufficient energetic returns (den Besten et al., 2013). The locust gut thus functions as a fermentation chamber that converts recalcitrant plant biomass into host-accessible energy.

Strain-level analyses provide concrete evidence for the enzymatic capabilities underlying this process. Targeted isolations from the gut of *L. migratoria* have identified specific cellulolytic bacteria that exhibit exceptionally high activity against model cellulose substrates. For example, *Pseudomonas* strain LM1 and *Enterobacter* strain LM5 display strong carboxymethyl cellulase and filter paper activities in vitro, confirming their capacity to efficiently degrade crystalline and amorphous cellulose (Li et al., 2024). Additional taxa commonly detected in locust guts, including *Bacillus* species (notably *Bacillus subtilis*), *Aeromonas* (e.g., *Aeromonas hydrophila*), and *Klebsiella* (particularly *K. pneumoniae*), further contribute to plant biomass conversion by secreting complementary suites of enzymes that weaken and disrupt plant cell wall integrity (Flores et al., 2025). These collective microbial activities release soluble sugars and other metabolites that are shared between microbes and hosts, enhancing overall digestive efficiency.

Independent support for this functional interpretation comes from metagenomic inference approaches. Bioinformatic predictions generated using tools such as PICRUSt applied to 16S rRNA gene datasets from *L. migratoria* reveal an enrichment of genes involved in carbohydrate metabolism and energy production, particularly within Proteobacteria-dominated gut communities (Langille et al., 2013). These predictions align closely with experimental observations of high cellulolytic activity and VFA production, reinforcing the conclusion that microbial metabolism substantially augments the host's ability to exploit fibrous plant diets.

Beyond carbon acquisition, the locust microbiome also plays a critical role in alleviating nitrogen limitation and micronutrient scarcity—two pervasive challenges in the arid and semi-arid environments where locusts typically occur. Nitrogen is often the most limiting nutrient for herbivorous insects, and plant nitrogen content declines rapidly during outbreaks as swarms strip landscapes of high-quality vegetation. The repeated detection of *K. pneumoniae* in locust guts and in the maternal foam plugs that surround egg pods suggests a potential role in nitrogen fixation or nitrogen recycling within the locust life cycle (Barnes et al., 2024). Such contributions would be particularly advantageous during long-distance migrations, when locusts are forced to subsist on nitrogen-poor plants after local resources have been depleted (Zheng et al., 2021).

In parallel, several common gut symbionts, including *B. subtilis* and related taxa, possess biosynthetic pathways for essential B vitamins such as thiamine, riboflavin, biotin, and folate. These vitamins act as indispensable cofactors in central metabolic and redox reactions, supporting growth, neuromuscular function, and stress tolerance when dietary vitamin intake is inadequate. By supplying these micronutrients endogenously, the microbiome increases metabolic flexibility and resilience, allowing locusts to maintain performance across heterogeneous and nutritionally unpredictable landscapes (Tarracchini et al., 2025).

Despite the substantial benefits conferred by gut symbionts, the locust–microbiome association is best described as facultative rather than strictly obligate. Locusts can be reared under axenic laboratory conditions and are capable of completing development and reproducing when provided with sterile, nutrient-rich diets and shielded from environmental stressors (Lavy, Gophna, et al., 2020a). However, this apparent independence disappears under ecologically realistic conditions. When diet quality declines or exposure to pathogens, plant secondary compounds, and nutritional imbalance increase, axenic locusts show markedly reduced performance relative to conventionally colonized individuals. This contrast underscores that while gut microbes may not be essential for survival in artificial settings, they provide critical ecological advantages in nature.

Taken together, these findings indicate that the locust microbiome underpins key physiological traits that enable swarming behavior: efficient extraction of energy from fibrous plants, buffering against nitrogen and vitamin limitation, and sustained locomotor activity across long distances. By enhancing digestive efficiency and nutritional resilience, microbial symbionts help support the extraordinary persistence, mobility, and population densities that define locust outbreaks, linking microscopic metabolic processes in the gut to some of the most dramatic large-scale phenomena in terrestrial ecology.

7.12 MICROBIOME ROLES IN PHEROMONE PRODUCTION AND SWARM COHESION

One of the clearest and best-studied roles of the locust microbiome is its contribution to digestion, especially the breakdown of complex plant carbohydrates and the efficient use of low-quality diets. Locusts are polyphagous herbivores that consume large amounts of grasses and other plants rich in cellulose and hemicellulose, which are tough structural polysaccharides that insect-derived cellulases alone only partly digest (Ling et al., 2022). Gut bacteria expand this limited endogenous capacity by producing a broad suite of cellulases, hemicellulases, and other carbohydrate-active enzymes that attack plant cell walls and release fermentable sugars (Lavy et al., 2020). These sugars are then fermented by microbes into VFAs such as acetate and propionate, which are absorbed across the hindgut epithelium and serve as important energy sources and carbon substrates for the host, particularly when protein and simple sugars are scarce (Ling et al., 2022; Wang et al., 2025). This microbial fermentation allows locusts to survive and maintain activity on fibrous, nutritionally poor vegetation that would otherwise be of limited value.

Targeted isolations from the gut of *L. migratoria* have identified specific cellulolytic strains that illustrate this capacity at the strain level. *Pseudomonas* strain LM1 and *Enterobacter* strain LM5, for example, show very high carboxymethyl cellulase and filter paper activities in vitro, confirming that they can efficiently break down model cellulose substrates (Wang et al., 2024). Additional important contributors to plant biomass conversion include species of *Bacillus* (e.g., *B. subtilis*), *Aeromonas* (e.g., *A. hydrophila*), and *Klebsiella* (notably *K. pneumoniae*), which together exploit plant-derived nutrients to proliferate in the gut and secrete enzymes that disrupt plant cell wall integrity (Lavy et al., 2020;

Wang et al., 2024). These activities release soluble sugars and other nutrients that can be used by both microbes and the host, effectively turning the gut into a fermentation chamber optimized for extracting value from tough plant material. Bioinformatic functional predictions using tools such as PICRUSt on 16S Ribosomal RNA (rRNA) gene datasets from *L. migratoria* show that genes involved in carbohydrate metabolism and energy production are enriched in Proteobacteria-dominated communities, which agrees well with the experimental evidence for strong microbial contributions to polysaccharide breakdown and VFA production (Wang et al., 2024).

Beyond carbon metabolism, the microbiome helps locusts cope with nitrogen limitation and micronutrient scarcity by contributing to nitrogen economy and vitamin synthesis. *K. pneumoniae*, a bacterium known to fix nitrogen in other host–microbe systems, has been detected in locust guts and in the maternal foam plugs that coat egg pods, suggesting that it may provide fixed nitrogen or participate in nitrogen recycling for developing and adult locusts (Lavy et al., 2020). This function would be especially valuable during long-distance migrations, when swarms rapidly exhaust local stands of high-protein plants and are forced to feed on suboptimal vegetation with low nitrogen content. In addition, various gut bacteria, including *B. subtilis* and related species, carry genes for synthesizing B vitamins such as thiamine, riboflavin, biotin, and folate, which support core metabolic pathways and redox balance when dietary vitamin intake is inadequate (Mazhar et al., 2022). Together, these microbial services provide metabolic flexibility that allows swarms to persist and reproduce while moving through landscapes where both macronutrients and micronutrients fluctuate and are often in short supply.

The association between locusts and their gut symbionts is therefore tight and highly beneficial, but it is not strictly obligatory in the narrow sense. Locusts can be reared axenically under laboratory conditions and can complete their life cycle and reproduce without their typical gut bacteria when they are provided with controlled, nutrient-rich diets and protected from environmental stress (Lavy et al., 2020). However, when fitness is measured in ecologically realistic contexts—including growth rates, survival under fluctuating diets, and resilience to pathogens and toxic compounds—axenic insects typically perform significantly worse than their conventionally colonized counterparts (Somerville et al., 2019; Liu et al., 2024). This performance gap underscores that in natural settings, where food quality is variable and exposure to pathogens and plant toxins is unavoidable, gut symbionts provide critical ecological advantages that help locusts maintain the extraordinary mobility, persistence, and population densities that characterize swarms.

7.13 LOCUST CUTICLE MICROBIOME AND PHASE-LINKED BEHAVIOR

The shift from solitary to gregarious phases in desert locusts involves not only visible changes in behavior and coloration but also a major reorganization of the microbiome on the cuticle and other body surfaces. Using 16S rRNA amplicon sequencing, studies comparing solitary *S. gregaria* before and after induction of gregariousness have shown that surface-associated bacterial communities change in parallel with those in the gut and reproductive tract (Dillon & Charnley, 2002). A key result is that bacteria in the genus *Weissella* (phylum Firmicutes) become much more abundant on the cuticle of the head and abdomen during the shift to the gregarious state, a pattern that is especially strong in laboratory colonies where contact rates are high and environmental conditions are controlled (Lavy, Lewin-Epstein, et al., 2020). This suggests that phase change involves a coordinated restructuring of microbial communities across multiple body compartments, with the integument playing an active role rather than simply serving as an inert barrier.

W. cibaria has already been shown in other insects, such as cockroaches, to promote aggregation by releasing volatile compounds that attract conspecifics and increase the time individuals spend together (Ma et al., 2025). By analogy, it is hypothesized that *W. cibaria* and related *Weissella* species on locust cuticles may contribute to the onset and maintenance of aggregation behavior, particularly in dense populations where bacteria are easily spread through direct contact, grooming, and shared microhabitats (Kim et al., 2023). From an evolutionary standpoint, a surface-dwelling microbe that encourages host aggregation gains a clear fitness advantage, because closer and more frequent interactions between hosts increase opportunities for horizontal transmission. Mathematical models of host–microbe dynamics support this idea, showing that microbes with high transmission potential are favored when they can induce or strengthen social clustering in their hosts, thereby facilitating the very swarming behavior that characterizes the gregarious phase (Mazmanian & Kasper, 2006).

The acquisition of *Weissella* and other Firmicutes on the integument indicates that the "gregarious signature" of the microbiome is shaped not only by host physiology and behavior but also by the reproductive and transmission interests of the microbes themselves. This perspective reframes phase transition: instead of being a purely host-driven, density-dependent process, it includes a microbial component that can stabilize or amplify particular behavioral states once they begin to emerge (Lavy et al., 2020b). The relationship between the locust and its cuticle-associated bacteria is bi-directional: host behavior, crowding, and environmental exposure determine which microbes are picked up and retained on the cuticle, while those microbes produce metabolites and chemical cues that feed back onto host sensory systems and social responses (Lavy, Gophna, et al., 2020a). This feedback loop links symbiont ecology directly to swarm dynamics, with integumentary bacteria acting as both passengers on and subtle drivers of the collective behavior that makes locust swarms so persistent and destructive.

7.14 REPRODUCTION AND MICROBIOME TRANSMISSION

The locust microbiome supports not only adult physiology but also reproductive success and early-life survival through well-developed vertical transmission mechanisms. The female reproductive tract contains a core community of Enterobacteriaceae and Enterococcaceae, including *E. cloacae*, *K. pneumoniae*, *C. xerosis* (representing *Corynebacterium*), and *E. faecalis*, which are consistently passed to offspring and form a major part of the initial microbiota in eggs and hatchlings (Canha-Gouveia et al., 2023). During oviposition, a female desert locust deposits her eggs in the soil and then secretes a protein-rich foam plug above the egg pod; this foam is heavily inoculated with maternal gut bacteria and functions as a microbial reservoir that coats and surrounds the eggs. As a result, the immediate microenvironment of the developing embryos is already enriched with symbionts that will later colonize the gut and other tissues (Mccaffery et al., 1998).

Detailed microbiological work has shown that viable *K. pneumoniae* and other gut-derived taxa can be isolated from the foam plug, which newly hatched larvae must tunnel through to reach the surface. This passage ensures that neonates ingest and are coated with maternal bacteria at the very moment they emerge into a variable and often nutrient-poor environment (Pope et al., 2019). The early acquisition of microbes that can participate in nitrogen fixation, cellulose degradation, and immune modulation gives hatchlings immediate access to metabolic and defensive tools that they would otherwise have to assemble more slowly from environmental sources alone. In arid habitats where high-quality food is patchy and unpredictable, this process acts as a form of maternal "nutritional compensation" and microbial seeding that primes offspring for growth and survival under fluctuating conditions (Laforest-Lapointe & Arrieta, 2017).

Maternal transmission is also likely to contribute to immune priming, although this has been more fully characterized in related insect systems than in locusts. In the red flour beetle, for example, movement of specific bacteria from the maternal gut to developing eggs triggers strain-specific immune responses in offspring, a phenomenon termed transgenerational immune priming (TGIP) (Freitak et al., 2014). In locusts, the detection of overlapping bacterial sequences in maternal reproductive tissues, foam plugs, eggs, and neonate guts suggests that a similar process may occur, with mothers passing on both symbionts and information about local pathogen pressures. Such transgenerational cues could help offspring mount more effective immune responses against pathogens common in their natal environment, providing a selective advantage when pathogen loads are high (Lavy et al., 2021).

Taken together, current evidence indicates that locusts acquire their bacterial symbionts through a combination of parental inoculation and environmental uptake from food and habitat. Swarming behavior adds a further layer, because large, mobile groups carry their vertically seeded microbiota into new regions, where they can acquire additional environmental strains and then pass a reshaped community on to the next generation (Lavy et al., 2020a). This dual acquisition strategy—maternal seeding plus dietary and environmental input—allows the microbiome to be both reliably transmitted and adaptively updated across generations, aligning microbial functions with the nutritional, immunological, and ecological challenges faced by successive cohorts of locusts (Delaroque & Chassaing, 2025).

7.15 ORAL MICROBIOME, SALIVA, AND EARLY COLONIZATION

The salivary and oral microbiome of locusts is an often-overlooked but important interface between the insect and its environment, shaping plant–insect interactions, pathogen entry, and the first steps of gut colonization. In many herbivorous insects, oral secretions contain effector molecules that dampen plant anti-herbivore defenses, and some of these effectors are produced or modified by oral bacteria (Engel & Moran, 2013). Genera such as *Stenotrophomonas maltophilia* (representing *Stenotrophomonas*), *P. aeruginosa* and related *Pseudomonas* spp., and *E. cloacae* have been identified in the oral secretions of multiple insect herbivores and are known to help initiate plant tissue degradation, although detailed species-level catalogs for locusts themselves are still limited. By participating in the softening and partial digestion of plant tissues at the bite site, these microbes may make it easier for locusts to ingest and process tough plant material and may also influence how plants perceive and respond to feeding damage (Alio et al., 2023).

In the migratory locust, the salivary glands play a direct role in immune defense in addition to their digestive functions. Following microbial infection, transcription of defensins and other antimicrobial peptides, such as LmDEF3, can increase 20- to 40-fold in the salivary glands compared with the fat body, indicating that the oral cavity is a key frontline in mucosal immunity (Lv et al., 2016). This pattern suggests that the salivary microbiome, together with the secretory tissues that support it, acts as an early antimicrobial barrier that can limit the establishment of harmful pathogens in the digestive tract during feeding. The presence of beneficial bacteria such as *Lactobacillus salivarius* in the oral microbiomes of some insects supports a broader "oral–gut axis" concept, in which commensal bacteria both contribute to local immune protection and assist in initial food processing and seeding of downstream gut communities (Cabral et al., 2017).

The composition of the oral microbiota is shaped by diet, habitat, and host behavior and reflects a dynamic balance between microbial colonization and shedding. With each feeding event, locusts introduce new environmental microbes from plant surfaces, soil particles, and water, while existing oral bacteria are swallowed, displaced, or outcompeted. Saliva provides a hydrated medium rich in nutrients that supports microbial survival and growth, and it helps maintain a relatively stable pH that favors certain taxa while constraining others (Lamont et al., 2018). Although the locust oral microbiome appears less diverse and less dense than hindgut communities, it has a disproportionate impact because it is the primary entry point for environmental microbes, and the first species to successfully colonize this niche can go on to seed midgut and hindgut compartments. In this way, early colonizers in the mouth and salivary system can influence future digestive efficiency, immune tone, and pathogen resistance by shaping the assembly and function of the downstream gut microbiome (Lavy et al., 2019).

7.16 MICROBIOME ROLE IN DETOXIFICATION

The locust resident gut microbiome provides a crucial defense against environmental toxicants and plant secondary metabolites, creating colonization resistance and metabolic buffering for the host. Many plants defend themselves with compounds such as tannins, alkaloids, and cardiac glycosides, which can bind proteins, disrupt membranes, or interfere with cellular signaling, and the locust microbiome helps transform and detoxify these molecules before they cause damage (Kaur et al., 2022). In other insect systems, *Bacillus pumilus* and certain *Acinetobacter* strains have been shown to degrade tannins under alkaline gut conditions, reducing their ability to precipitate proteins and injure intestinal tissues, and similar detoxification pathways are inferred in locusts based on the presence of related taxa and predicted gene functions. By metabolizing or modifying plant defensive chemistry, gut bacteria reduce the cost of feeding on chemically defended plants and expand the range of host plants that locusts can exploit (Song et al., 2025).

The microbiome also shapes how locusts respond to anthropogenic pollutants, including commonly used insecticides. Studies of pyrethroids such as deltamethrin show that the presence of a normal gut microbiota can prime and modulate the host's metabolic and transcriptional responses to these chemicals, even when bacteria do not directly break down the compounds (Krout et al., 2025). When the gut microbiota are experimentally disrupted—for example, by antibiotics or sterile rearing—locusts show altered expression of detoxification genes in the gut and other tissues and become more susceptible to chemical stressors (Ramirez et al., 2020). In the absence of a stable, functionally diverse microbiome, these coordinated detoxification responses are weakened, underscoring the systemic importance of the gut–microbe axis in maintaining physiological homeostasis under both chemical and nutritional stress.

A further contribution of the microbiome to metabolic health is the synthesis of essential vitamins and cofactors that may be limiting in natural diets. Members of the genus *Bacillus*, such as *B. subtilis*, carry genes for the production of multiple B vitamins, including thiamine (B1), riboflavin (B2), biotin (B7), and folate (B9), which are central to carbohydrate metabolism, nucleotide biosynthesis, and redox reactions. *Klebsiella* strains, including *K. pneumoniae*, have likewise been linked to the provision of vitamins and nitrogenous compounds, especially when locusts feed on low-quality vegetation with poor vitamin content (Li et al., 2023). Functional profiling of *L. migratoria* microbiomes across life stages shows that pathways related to carbohydrate metabolism, energy production, and cofactor biosynthesis are enriched in Proteobacteria-dominated communities, reinforcing the view that gut symbionts help buffer hosts against dietary deficiencies and environmental variability.

7.17 PATHOGENS, SYMBIONTS, AND OPPORTUNISTIC SHIFTS

The relationship between locusts and their microbiome is a finely balanced system that can fail under pathogen pressure, sometimes turning formerly helpful bacteria into opportunistic killers. A well-studied case involves the host-specialist fungus *M. acridum*, widely used as a biological control agent against locusts and grasshoppers. After *M. acridum* spores land on the cuticle, they germinate, penetrate the exoskeleton, and colonize the hemocoel, the internal body cavity normally kept free of gut bacteria by strong epithelial and immune barriers (Li et al., 2023). During this systemic infection, the fungus alters host physiology in ways that promote the translocation of gut symbionts into the hemocoel, moving them from their typical safe niche into a compartment where their presence is abnormal and dangerous (d'Enfert et al., 2020).

Once gut bacteria enter the hemocoel, they help drive a strong inflammatory response that can significantly accelerate host death, acting in synergy with direct fungal damage. To maintain control over this mixed infection and avoid being outcompeted by the very bacteria it has displaced, *M. acridum* produces an iron-binding protein known as MaCFEM1, which sequesters iron and restricts bacterial growth (Wang et al., 2023). When MaCFEM1 is experimentally knocked out or disrupted, translocated symbionts proliferate rapidly in the hemocoel and become lethal opportunistic pathogens, effectively turning the host's own microbiome into a secondary killer. This scenario demonstrates how beneficial symbionts can quickly become liabilities when they are displaced from their normal niche or when host immune control is compromised by a specialist pathogen (Li et al., 2024).

A different but related mechanism is seen with the microsporidian parasite *P. locustae*, which interferes with swarm behavior by reshaping gut physiology and microbial composition. Infection with *P. locustae* acidifies the hindgut, lowering pH and creating conditions that suppress key Enterobacteriaceae responsible for producing aggregation pheromones such as guaiacol (Shi et al. 2015). As the abundance of these pheromone-producing bacteria declines, levels of serotonin and dopamine in the locust nervous system also fall, reflecting the tight coupling between gut-derived chemical cues and neuromodulatory state. Infected gregarious locusts consequently show reduced attraction to conspecifics and can revert toward more solitary-like behavior, weakening swarm cohesion even before high mortality occurs. Here, the pathogen undermines locust success not only by direct pathology but by disrupting the microbial support system that underpins social behavior (Lavy et al., 2020b).

These examples highlight how pathogens and symbionts interact to shape disease outcomes and behavior, and they point to symbiont-targeted strategies as promising tools for pest management. By deliberately disturbing microbial homeostasis—whether through fungal infection, microsporidian deployment, or more precise interventions such as phages or microbiome-modifying compounds—control programs may exploit the same mechanisms that pathogens use to shift symbionts from mutualists to liabilities. Such approaches could increase the efficacy of biological control agents and reduce reliance on broad-spectrum chemical pesticides, which often have substantial non-target effects. Overall, current work supports a view of locust swarming and phase polyphenism as emergent properties of interactions among host genetics, neuroendocrine cascades, environmental conditions, and a multi-compartment microbiome that together shape nutrition, communication, immunity, and behavior. Bacterial symbionts are not passive passengers but active participants in locust ecological success and destructive impact, offering both challenges and new opportunities for predicting outbreaks and developing targeted, microbiome-informed strategies to protect global food security.

7.18 LOCUST BIOCONTROL

Biological control of locusts and grasshoppers using entomopathogenic fungi has emerged as an environmentally sustainable alternative to chemical insecticides. Among these fungi, *M. acridum* is a host-specialist pathogen widely used for locust management. However, its relatively slow killing speed and moderate virulence can limit field-level effectiveness. Enhancing the pathogenicity of *M. acridum* without compromising its host specificity is therefore a critical goal for improving locust biocontrol strategies.

Fungal virulence is strongly influenced by the composition and architecture of the cell wall, particularly polysaccharide components that mediate host adhesion, immune evasion, and stress tolerance. Cell wall-associated saccharides such as mannans and glucans play a central role in host–pathogen interactions by shaping fungal surface properties and influencing host immune recognition. Genomic analyses have revealed that *M. acridum* lacks several gene families involved in carbohydrate metabolism, including glycerate-3-kinase (GLYK), an enzyme involved in saccharide synthesis and metabolic flux.

Building on this insight, Tong et al. (2021) engineered *M. acridum* to enhance its virulence by introducing the GLYK gene from the host-generalist fungus *Metarhizium robertsii* into the *M. acridum* genome. The resulting mutant strain, designated Ma::MrGLYK, exhibited pronounced biochemical and structural changes compared with the wild-type strain. Notably, the mutant showed significantly elevated levels of fructose 6-phosphate in mycelia, indicating altered carbohydrate metabolism and enhanced saccharide synthesis.

These metabolic changes were accompanied by marked modifications to the fungal cell wall. The Ma::MrGLYK strain developed a longer and denser mannan fibril layer, which is known to contribute to fungal virulence and environmental resilience. Consistent with these structural changes, the mutant strain displayed increased resistance to heat stress, suggesting improved survival and stability under harsh environmental conditions commonly encountered in field applications.

Transcriptomic analyses further demonstrated that infection with the Ma::MrGLYK strain profoundly altered host immune responses. Compared with wild-type *M. acridum*, topical infection with the mutant induced higher expression of locust genes encoding pattern-recognition proteins, serine proteases, and cytochrome P450 enzymes, which are involved in detoxification and immune signaling. At the same time, the expression of key immune effectors, including antimicrobial peptides, was suppressed, and phenoloxidase activity—a central component of insect melanization and defense—was significantly reduced. These immune disruptions likely facilitate more efficient fungal invasion and proliferation within the host.

Functionally, these molecular and physiological changes translated into substantially enhanced pathogenicity. Both topical infection and direct injection with the Ma::MrGLYK strain resulted in significantly higher locust mortality and a pronounced reduction in lifespan compared with infections caused by the wild-type *M. acridum*. Together, these findings demonstrate that targeted metabolic engineering of cell wall polysaccharide synthesis can markedly improve fungal virulence while retaining host specificity.

Overall, this work highlights a promising strategy for advancing locust biocontrol by rationally enhancing the efficacy of *M. acridum* through genetic and metabolic modification. Such improvements could lead to faster-acting, more resilient biopesticides, strengthening sustainable and ecologically sound approaches to managing devastating locust outbreaks.

7.19 CONCLUSION

The locust microbiome represents a critical yet underexplored dimension of locust biology, with profound implications for development, physiology, behavior, and population dynamics. Composed of diverse bacterial, fungal, and viral communities inhabiting the gut, cuticle, and internal tissues, the microbiome functions as an integral extension of the locust host. Rather than being a passive assemblage of microbes, these communities actively shape nutrient acquisition, detoxification of plant secondary metabolites, immune competence, and resistance or susceptibility to pathogens.

A growing body of evidence indicates that the microbiome plays a pivotal role in locust phase polyphenism, influencing the transition between solitary and gregarious states. Microbial metabolites and microbe-mediated modulation of host neuroendocrine and immune pathways may contribute to behavioral aggregation, stress tolerance, and enhanced survival under high-density conditions. At the same time, shifts in microbial community composition can alter host immunity, either strengthening defenses against pathogens or, conversely, creating vulnerabilities that can be exploited by biological control agents.

From an applied perspective, understanding the locust microbiome opens new avenues for sustainable pest management. Interactions between resident microbes and entomopathogens, such as *M. acridum*, can strongly influence infection outcomes by modulating host immune responses and pathogen virulence. Manipulating the microbiome—through targeted disruption, probiotic interference, or synergistic combinations with microbial biocontrol agents—offers a promising strategy to enhance control efficacy while minimizing environmental impact.

Overall, the locust microbiome is a dynamic and functionally significant component of locust ecology and evolution. Integrating microbiome research with genomics, neurobiology, and biocontrol science will be essential for developing a holistic understanding of locust outbreaks and for designing next-generation, microbiome-informed strategies to mitigate one of the world's most destructive agricultural pests.

ACKNOWLEDGMENTS

The author (ST) acknowledges the Abraham Thomas Foundation for providing the resources for this work.

REFERENCES

Adriaenssens EM, Ceyssens PJ, Dunon V, Ackermann HW, Van Vaerenbergh J, Maes M, et al. Bacteriophages LIMElight and LIMEzero of Pantoea agglomerans, belonging to the "phiKMV-Like viruses" . *Appl Environ Microbiol*. 2011 May;77(10):3443–50.

Alio I, Moll R, Hoffmann T, Mamat U, Schaible UE, Pappenfort K, et al. Stenotrophomonas maltophilia affects the gene expression profiles of the major pathogens Pseudomonas aeruginosa and Staphylococcus aureus in an in vitro multispecies biofilm model. *Microbiol Spectr*. 2023;11(6):e00859–23.

Alessi AM, O'Connor V, Aonuma H, Newland PL. Dopaminergic modulation of phase reversal in desert locusts. *Front Behav Neurosci*. 2014 Nov 7;8:371.

Asare AO, Dannemann BC, Goeren E. Locust infestations and individual school dropout: Evidence from Africa. *Working Papers* [Internet]. 2023 Feb [cited 2026 Feb 4]. Available from: https://ideas.repec.org//p/old/dpaper/440.html

Barnes AJ, Bennett EF, Vezina B, Hudson AW, Hernandez GE, Nutter NA, et al. Ethanolamine metabolism through two genetically distinct loci enables Klebsiella pneumoniae to bypass nutritional competition in the gut. *PLoS Pathog*. 2024 May 7;20(5):e1012189.

Basit A, Haq IU, Hyder M, Humza M, Younas M, Akhtar MR, et al. Microbial symbiosis in lepidoptera: Analyzing the gut microbiota for sustainable pest management. *Biology (Basel)*. 2025 Jul 25;14(8):937.

Bazazi S, Buhl C, Hale JJ, Anstey ML, Sword GA, Simpson SJ, et al. Collective motion and cannibalism in locust migratory bands. *Current Biology*. 2008 May 20;18(10):735–9.

Bertsch DJ, Castillo LMP, Frye MA. Serotonin selectively modulates visual responses of object motion detectors in Drosophila. *J Neurophysiol*. 2025 Sep 1;134(3):962–84.

Cabral DJ, Wurster JI, Flokas ME, Alevizakos M, Zabat M, Korry BJ, et al. The salivary microbiome is consistent between subjects and resistant to impacts of short-term hospitalization. *Sci Rep*. 2017 Sep 8;7:11040.

Canha-Gouveia A, Pérez-Prieto I, Rodríguez CM, Escamez T, Leonés-Baños I, Salas-Espejo E, et al. The female upper reproductive tract harbors endogenous microbial profiles. *Front Endocrinol (Lausanne)*. 2023 Jun 21;14:1096050.

Cavicchioli R, Ripple WJ, Timmis KN, Azam F, Bakken LR, Baylis M, et al. Scientists' warning to humanity: Microorganisms and climate change. *Nat Rev Microbiol*. 2019;17(9):569–86.

Cease AJ, Trumper EV, Medina H, Bazán FC, Frana J, Harrison J, et al. Field bands of marching locust juveniles show carbohydrate, not protein, limitation. *Curr Res Insect Sci*. 2023 Sep 20;4:100069.

Chang X, Feng S, Ullah F, Zhang Y, Zhang Y, Qin Y, et al. Adapting distribution patterns of desert locusts, Schistocerca gregaria in response to global climate change. *Bull Entomol Res*. 2025 Feb;115(1):84–92.

Cressman K, Hodson D. Surveillance, information sharing and early warning systems for transboundary plant pests diseases: The FAO experience. *Arab Journal of Plant Protection*. 2009;27:226–232.

Cruz AT, Cazacu AC, Allen CH. Pantoea agglomerans, a plant pathogen causing human disease. *J Clin Microbiol*. 2007 Jun;45(6):1989–92.

d'Enfert C, Kaune AK, Alaban LR, Chakraborty S, Cole N, Delavy M, et al. The impact of the Fungus-Host-Microbiota interplay upon Candida albicans infections: Current knowledge and new perspectives. *FEMS Microbiol Rev*. 2020 Nov 24;45(3):fuaa060.

Delaroque C, Chassaing B. Microbiome in heritage: How maternal microbiome transmission impacts next generation health. *Microbiome*. 2025 Sep 26;13:196.

den Besten G, van Eunen K, Groen AK, Venema K, Reijngoud DJ, Bakker BM. The role of short-chain fatty acids in the interplay between diet, gut microbiota, and host energy metabolism. *J Lipid Res*. 2013 Sep;54(9):2325–40.

Devi S. Health situation deteriorating in Kenyan refugee camp. *The Lancet*. 2023 Feb 11;401(10375):425.

Dillon R, Charnley K. Mutualism between the desert locust Schistocerca gregaria and its gut microbiota. *Res Microbiol*. 2002 Oct;153(8):503–9.

Dillon R, Vennard C. Dillon RJ, Vennard CT, Charnley AK. A note: Gut bacteria produce components of a locust cohesion pheromone. *J Appl Microbiol*. 2002 Feb 1;92:759–63.

Dillon RJ, Vennard CT, Charnley AK. Pheromones: Exploitation of gut bacteria in the locust. *Nature*. 2000 Feb 1;403:851.

Djamel Eddine, L., Abdelmadjid, B., Imen, L., Badis, A., & Akkal, S. Review on Locusta migratoria cinerascens (Fabricius, 1781), *Journal of Bioresource Management* 2023 10(4).164–174.

Djamel Eddine L, Abdelmadjid B, Aouzal B, Akkal S. Review on Locusta migratoria. 2023 Jan 1;164–74.

Engel P, Moran NA. The gut microbiota of insects – diversity in structure and function. *FEMS Microbiol Rev*. 2013 Sep 1;37(5):699–735.

FAO. 2021. Desert Locust crisis | FAO Emergency and Resilience | Food and Agriculture Organization of the United Nations. Available from: www.fao.org/emergencies/where-we-work/desert-locust-crisis/

FAO. 2025. Locust management and its importance to food security. Available from: www.fao.org/transboundary-plant-pests-diseases/news/detail/locust-management-and-its-importance-to-food-security/en

FAO. Desert locust upsurge. 2022. Food and agriculture organization of the United Nations. Available from: https://doi.org/10.4060/cc0175en

Flores CAR, Siringan MAT, Relucio-San Diego MACV. Multiple plant growth-promoting activities exhibited by root-associated bacteria isolated from bamboo and corn. *Int J Microbiol*. 2025;2025: 6374935.

Foquet B, Castellanos AA, Song H. Comparative analysis of phenotypic plasticity sheds light on the evolution and molecular underpinnings of locust phase polyphenism. *Sci Rep*. 2021 Jun 7;11(1):11925.

Freitak D, Schmidtberg H, Dickel F, Lochnit G, Vogel H, Vilcinskas A. The maternal transfer of bacteria can mediate trans-generational immune priming in insects. *Virulence*. 2014 May 15;5(4): 547–54.

Fürstenau B, Nyasembe VO, Mokaya HO, Kirwa HK, Guerrero A, Torto B. Behavioral and physiological insights into cross-stage interactions in the desert locust. *Sci Rep*. 2025 Jun 27;15(1):20339.

Guo X, Ma Z, Kang L. Serotonin enhances solitariness in phase transition of the migratory locust. *Front Behav Neurosci*. 2013 Oct 7;7:129.

Guttal V, Romanczuk P, Simpson SJ, Sword GA, Couzin ID. Cannibalism can drive the evolution of behavioural phase polyphenism in locusts. *Ecol Lett*. 2012 Oct;15(10):1158–66.

Helletsgruber C, Dötterl S, Ruprecht U, Junker RR. Epiphytic bacteria alter floral scent emissions. *J Chem Ecol*. 2017;43(11):1073–7.

Hoffmann SA, Diggans J, Densmore D, Dai J, Knight T, Leproust E, et al. Safety by design: Biosafety and biosecurity in the age of synthetic genomics. *iScience*. 2023 Feb 10;26(3):106165.

Jabeen I, Islam S, Hassan AKMI, Tasnim Z, Shuvo SR. A brief insight into Citrobacter species – a growing threat to public health. *Front Antibiot*. 2023 Dec 5;2:1276982.

Kabir AH, Baki Md ZI, Ahmed B, Mostofa MG. Current, faltering, and future strategies for advancing microbiome-assisted sustainable agriculture and environmental resilience. *New Crops*. 2024;1:100013. https://doi.org/10.1016/j.ncrops.2024.100013

Kang MS, Na HS, Oh JS. Coaggregation ability of Weissella cibaria isolates with Fusobacterium nucleatum and their adhesiveness to epithelial cells. *FEMS Microbiol Lett*. 2005 Dec 15;253(2):323–9.

Kaur S, Samota MK, Choudhary M, Choudhary M, Pandey AK, Sharma A, Thakur J. How do plants defend themselves against pathogens-Biochemical mechanisms and genetic interventions. *Physiol Mol Biol Plants*. 2022 Feb;28(2):485–504.

Kim E, Yang SM, Jung DH, Kim HY. Differentiation between Weissella cibaria and Weissella confusa Using Machine-Learning-Combined MALDI-TOF MS. *Int J Mol Sci*. 2023 Jul 2;24(13):11009.

Krout IN, Matsuzaki R, White AC, Tsui S, Blackmer-Raynolds L, Kelly SD, et al. The gut microbiome promotes detoxification responses to an environmental toxicant. *bioRxiv*. 2025 Aug 19:2025.08.14.670327. (online only).

Laforest-Lapointe I, Arrieta MC. Patterns of early-life gut microbial colonization during human immune development: An ecological perspective. *Front Immunol*. 2017 Jul 10;8:788.

Lamont RJ, Koo H, Hajishengallis G. The oral microbiota: Dynamic communities and host interactions. *Nat Rev Microbiol*. 2018 Dec;16(12):745–59.

Langille MGI, Zaneveld J, Caporaso JG, McDonald D, Knights D, Reyes JA, et al. Predictive functional profiling of microbial communities using 16S rRNA marker gene sequences. *Nat Biotechnol*. 2013 Sep;31(9):814–21.

Lavy O, Gophna U, Ayali A, Gihaz S, Fishman A, Gefen E. The maternal foam plug constitutes a reservoir for the desert locust's bacterial symbionts. *Environ Microbiol*. 2021 May;23(5):2461–72.

Lavy O, Gophna U, Gefen E, Ayali A. The Effect of Density-Dependent Phase on the Locust Gut Bacterial Composition. *Front Microbiol*. 2019a Jan 18;9:3020.

Lavy O, Gophna U, Gefen E, Ayali A. Locust bacterial symbionts: An update. *Insects*. 2020 Sep 24;11(10):655.

Lavy O, Gophna U, Gefen E, Ayali A. Dynamics of bacterial composition in the locust reproductive tract are affected by the density-dependent phase. *FEMS Microbiol Ecol*. 2020b Apr 1;96(4):fiaa044.

Li G, Tong Y, Xiao Y, Huang S, Zhao T, Xia X. Probiotic Bacillus subtilis contributes to the modulation of gut microbiota and blood metabolic profile of hosts. *Comparative Biochemistry and Physiology Part C: Toxicology & Pharmacology*. 2023 Oct 1;272:109712.

Li J, Li J, Cao L, Chen Q, Ding D, Kang L. An iron-binding protein of entomopathogenic fungus suppresses the proliferation of host symbiotic bacteria. *Microbiome*. 2024 Oct 15;12:202.

Li J, Xia Y. Host–Pathogen Interactions between Metarhizium spp. and Locusts. *J Fungi (Basel)*. 2022 Jun 3;8(6):602.

Li S, Feng S qian, Ullah H, Tu X bing, Zhang Z hua. IPM–Biological and integrated management of desert locust. *Journal of Integrative Agriculture*. 2022 Dec 1;21(12):3467–87.

Ling Y, Li WJ, Li FF, Xue XB, Gao YY, Wang L, et al. Microbial gut diversity in four grasshopper species and its correlation with cellulose digestibility. *Front Microbiol*. 2022 Nov 10;13:1002532.

Liu F, Ye F, Yang Y, Kang Z, Liu Y, Chen W, et al. Gut bacteria are essential for development of an invasive bark beetle by regulating glucose transport. *Proceedings of the National Academy of Sciences*. 2024 Aug 13;121(33):e2410889121.

Lumumba BO, Otieno DJ, Nyikal RA. Farmers' perceptions of integrated desert locust management: A case study in Isiolo and Meru Counties of Kenya. *J Integr Pest Manag*. 2024 Jan 1;15(1):16.

Lv M, Mohamed AA, Zhang L, Zhang P, Zhang L. A Family of CSαβ defensins and defensin-like peptides from the migratory Locust, Locusta migratoria, and their expression dynamics during mycosis and nosemosis. *PLOS ONE*. 2016 Aug 24;11(8):e0161585.

Ma W, Liu X, Jing Y, Zhang M, Zhang X, Wang C, Khan MZ, Zhu M. Weissella: From beneficial probiotics to opportunistic pathogens—A review. *Nutrients*. 2025;17:3162. https://doi.org/10.3390/nu17193162

Mazhar S, Khokhlova E, Colom J, Simon A, Deaton J, Rea K. In vitro and in silico assessment of probiotic and functional properties of Bacillus subtilis DE111®. *Front Microbiol*. 2022;13:1101144.

Mazmanian SK, Kasper DL. The love--hate relationship between bacterial polysaccharides and the host immune system. *Nature Reviews Immunology*. 2006 Nov 1;6(11):849–59.

Mccaffery A, Simpson S. A gregarizing factor present in the egg pod foam of the desert locust Schistocerca gregaria. *J Exp Biol*. 1998;201(3):347–63.

Mitra B, Gayen A, Haque SM, Das A. Influence of climate on desert locust (Schistocerca gregaria Forskål, 1775) Plague and migration prediction in tropics. *Sci Rep*. 2024 Oct 16;14(1):24270.

Montanari D, O'Hearn WJ, Hambuckers J, Fischer J, Zinner D. Coordination during group departures and progressions in the tolerant multi-level society of wild Guinea baboons (Papio papio). *Sci Rep*. 2021 Nov 9;11:21938.

Muratore M, Prather C, Sun Y. The gut bacterial communities across six grasshopper species from a coastal tallgrass prairie. *PLoS One*. 2020 Jan 30;15(1):e0228406.

Ogunrinola GA, Oyewale JO, Oshamika OO, Olasehinde GI. The human microbiome and its impacts on health. *Int J Microbiol*. 2020 Jun 12;2020:8045646.

Olson RS, Hintze A, Dyer FC, Knoester DB, Adami C. Predator confusion is sufficient to evolve swarming behaviour. *J R Soc Interface*. 2013 Aug 6;10(85):20130305.

Ott SR, Rogers SM. Gregarious desert locusts have substantially larger brains with altered proportions compared with the solitarious phase. *Proc Biol Sci*. 2010 Oct 22;277(1697):3087–96.

Panopoulou C, Tsagkarakis A. From surveillance to sustainable control: A global review of strategies for locust management. *Agronomy*. 2025 Sep 24;15(10):2268.

Plantinga AM, Wu MC (2021). Beta Diversity and Distance-Based Analysis of Microbiome Data. In: Guha S, Datta S (Ed.) Statistical Analysis of Microbiome Data. Frontiers in Probability and the Statistical Sciences, Springer Nature, Switzerland.

Pope JL, Yang Y, Newsome RC, Sun W, Sun X, Ukhanova M, et al. Microbial colonization coordinates the pathogenesis of a Klebsiella pneumoniae infant isolate. *Sci Rep*. 2019 Mar 4;9:3380.

Qiu Y, Zhou Y, Chang Y, Liang X, Zhang H, Lin X, et al. The effects of ventilation, humidity, and temperature on bacterial growth and bacterial genera distribution. *Int J Environ Res Public Health*. 2022 Nov 20;19(22):15345.

Ramirez J, Guarner F, Bustos Fernandez L, Maruy A, Sdepanian VL, Cohen H. Antibiotics as major disruptors of gut microbiota. *Front Cell Infect Microbiol*. 2020 Nov 24;10:572912.

Rogers SM, Cullen DA, Anstey ML, Burrows M, Despland E, Dodgson T, et al. Rapid behavioural gregarization in the desert locust, Schistocerca gregaria entails synchronous changes in both activity and attraction to conspecifics. *J Insect Physiol*. 2014 Jun;65(10T0):9–26.

Rogers SM, Matheson T, Despland E, Dodgson T, Burrows M, Simpson SJ. Mechanosensory-induced behavioural gregarization in the desert locust Schistocerca gregaria. *J Exp Biol*. 2003 Nov;206(Pt 22):3991–4002.

Rowland I, Gibson G, Heinken A, Scott K, Swann J, Thiele I, et al. Gut microbiota functions: Metabolism of nutrients and other food components. *Eur J Nutr*. 2018;57(1):1–24.

Sai Charan D, Vidya Madhuri E, Rupali JS, Hariharan S, Verma S, Aravind K, et al. Harnessing insect gut microbiota: Approaches and applications for next-generation crop management and agricultural sustainability. *Discov Agric*. 2025 Dec 22;3(1):281.

Sayin S, Couzin-Fuchs E, Petelski I, Günzel Y, Salahshour M, Lee C-Y, Graving JM, Li L, Deussen O, Sword GA, Couzin ID. The behavioral mechanisms governing collective motion in swarming locusts. *Science*. 2025;387, 995–1000. https://doi.org/10.1126/science.adq7832

Shi W, Guo Y, Xu C, Tan S, Miao J, Feng Y, et al. Unveiling the mechanism by which microsporidian parasites prevent locust swarm behavior. *Proc Natl Acad Sci U S A*. 2014 Jan 28;111(4):1343–8.

Shved D, Kuznetsova P, Rozanov IA, Lebedeva SA, Vinokhodova A, Savinkina A, et al. Effects of isolation, crowding, and different psychological countermeasures on crew behavior and performance. *Front Physiol*. 2022 Nov 15;13:963301.

Simpson SJ, Despland E, Hägele BF, Dodgson T. Gregarious behavior in desert locusts is evoked by touching their back legs. *Proc Natl Acad Sci U S A*. 2001 Mar 27;98(7):3895–7.

Somerville J, Zhou L, Raymond B. Aseptic rearing and infection with gut bacteria Improve the fitness of transgenic diamondback moth, Plutella xylostella. *Insects*. 2019 Mar 27;10(4):89.

Song F, Shu J, Zhang S. Gut bacterium Acinetobacter sp. assists Camellia weevil with host plant adaptation by degrading tea saponin via the benzoate pathway. *Microbiome*. 2025 Jun 7;13:139.

Su LJ, Liu H, Li Y, Zhang HF, Chen M, Gao XH, et al. Cellulolytic activity and structure of symbiotic bacteria in locust guts. *Genet Mol Res*. 2014 Sep 29;13(3):7926–36.

Tandon A. 2020. [Commentary] Greening a desert invites locust swarms [Internet]. Mongabay-India. [cited 2026 Feb 3]. Available from: https://india.mongabay.com/2020/05/commentary-greening-a-desert-invites-locust-swarms/

Tang Q, Feng J, Zong D, Zhou J, Hu X, Wang B, et al. Potential spread of desert locust Schistocerca gregagia (Orthoptera: Acrididae) under climate change scenarios. *Diversity*. 2023 Sep 26;15(10):1038.

Tarracchini C, Bottacini F, Mancabelli L, Lugli GA, Turroni F, van Sinderen D, et al. Approaches to dissect the vitamin biosynthetic network of the gut microbiota. *Microbiome Res Rep*. 2025 Oct 20;4(4):37.

The World Bank Group (2020). The Locust Crisis: The World Bank's Response. https://www.worldbank.org/en/news/factsheet/2020/04/27/the-locust-crisis-the-world-banks-response

Wang Y, Tang J, Chen Y, Chen S, Chen S, Yu X, Wan C, Xiang G, Chen Y, Li Q. The Ecological–Evolutionary Game of the Insect Gut Microbiome: Environmental Drivers, Host Regulation, and Prospects for Cross-Cutting Applications. *Veterinary Sciences*. 2025; 12(9):866.

Tong X, Wang Y, Li J, Hu S, Yang P, Kang L. Transformation of glycerate kinase (GLYK) into Metarhizium acridum increases virulence to locust. *Pest Manag Sci*. 2021 Mar;77(3):1465–1475.

Topaz CM, D'Orsogna MR, Edelstein-Keshet L, Bernoff AJ. Locust dynamics: Behavioral phase change and swarming. *PLoS Comput Biol*. 2012 Aug 16;8(8):e1002642.

Unni AP, Knaden M, Hansson BS. Olfactory mating signals in the migratory Locust Locusta migratoria. *J Chem Ecol*. 2024;50(1–2):11–7.

von Engelhardt W, Bartels J, Kirschberger S, Meyer zu Düttingdorf HD, Busche R. Role of short-chain fatty acids in the hind gut. *Vet Q*. 1998;20 Suppl 3:S52–59.

Wang JL, Sun J, Song YJ, Zheng HH, Wang GJ, Luo WX, et al. An entomopathogenic fungus exploits its host humoral antibacterial immunity to minimize bacterial competition in the hemolymph. *Microbiome*. 2023 May 20;11:116.

Wang Q, Liu Y, Yin X. Comparison of gut bacterial communities of Locusta migratoria manilensis (Meyen) reared on different food plants. *Biology (Basel)*. 2022 Sep 13;11(9):1347.

Wasuwan R, Phosrithong N, Promdonkoy B, Sangsrakru D, Sonthirod C, Tangphatsornruang S, et al. The fungus Metarhizium sp. BCC 4849 is an effective and safe mycoinsecticide for the management of spider mites and other insect pests. *Insects*. 2021 Dec 30;13(1):42.

Yang J, Yu Q, Yu J, Kang L, Guo X. 4-Vinylanisole promotes conspecific interaction and acquisition of gregarious behavior in the migratory locust. *Proc Natl Acad Sci U S A*. 2023;120(37):e2306659120. https://doi.org/10.1073/pnas.2306659120

Ye L, Rawls JF. Microbial influences on gut development and gut-brain communication. *Development*. 2021 Nov 10;148(21):dev194936.

Zheng Y, Ding Y, Xu M, Chen H, zhang H, Liu Y, et al. Gut microbiota contributes to host defense against Klebsiella pneumoniae-induced liver abscess. *J Inflamm Res*. 2021 Oct 9;14:5215–25.

CHAPTER EIGHT

Honeybee Microbiome

Yun Sang Cho

8.1 INTRODUCTION

Honeybees (*Apis mellifera*) play a crucial role in our ecosystems and agriculture. As key pollinators, they contribute significantly to the production of fruits, vegetables, and nuts, supporting biodiversity and food security. Beyond their pollination services, honeybees are fascinating organisms in their own right, especially regarding their microbiomes, which are communities of microorganisms residing in and on them. The honeybee microbiome is essential for the health and longevity of these insects and serves several critical functions that enhance their survival. This chapter explores the honeybee microbiome's composition, functions, environmental impacts, and implications for bee health, ultimately highlighting its significance in our world.

8.2 HONEYBEE MICROBIOME COMPOSITION

The honeybee microbiome consists of diverse microbial communities, primarily dominated by bacteria, fungi, and other microorganisms that exist symbiotically with the bees (Figure 8.1). Key bacterial genera, such as *Lactobacillus* and *Bifidobacterium*, play critical roles in maintaining gut health and digestion. In addition to bacteria, fungi and yeasts, including the genus *Saccharomyces*, also contribute to the microbiome composition. The sources of microbial diversity in honeybees can be attributed to several factors. Firstly, honeybees acquire microorganisms by ingesting nectar and pollen, which are not only their food sources but also carriers of diverse microbiota (Figure 8.2). Additionally, environmental exposure, including contact with flowers and soil, introduces new microbial species into their systems. Furthermore, social interactions within the colony, such as grooming and feeding behaviors, facilitate the transfer of microorganisms among bees, enhancing the colony's microbial diversity.

Understanding the diversity of gut microbial communities in honeybees is essential for insights into their health and functionality. Recent studies highlight the complexity and ecological significance of gut symbionts in honeybees and other pollinators. Raymann *et al.* (2018a,b,c) investigated how antibiotic treatments reduce the genetic diversity of core species in the honeybee gut microbiome, potentially affecting overall microbiota function and resilience. Lim *et al.* (2015) conducted deep sequencing and ecological characterization of gut microbial communities across various bumble bee species, emphasizing the variability in microbial composition and its ecological relevance to pollinator health. Moran *et al.* (2012) assessed the distinctive gut microbiota of honeybees through deep sampling from individual workers, revealing significant intra-colony diversity that contributes to colony health. Copeland *et al.* (2022) explored how early queen development affects gut microbiota composition and associated metabolism, highlighting the interplay between social context and microbial diversity within honeybee colonies. Jones *et al.* (2025) examined the effects of microbial priority on strain-level composition within the honeybee gut community,

DOI: 10.1201/9781003639176-8

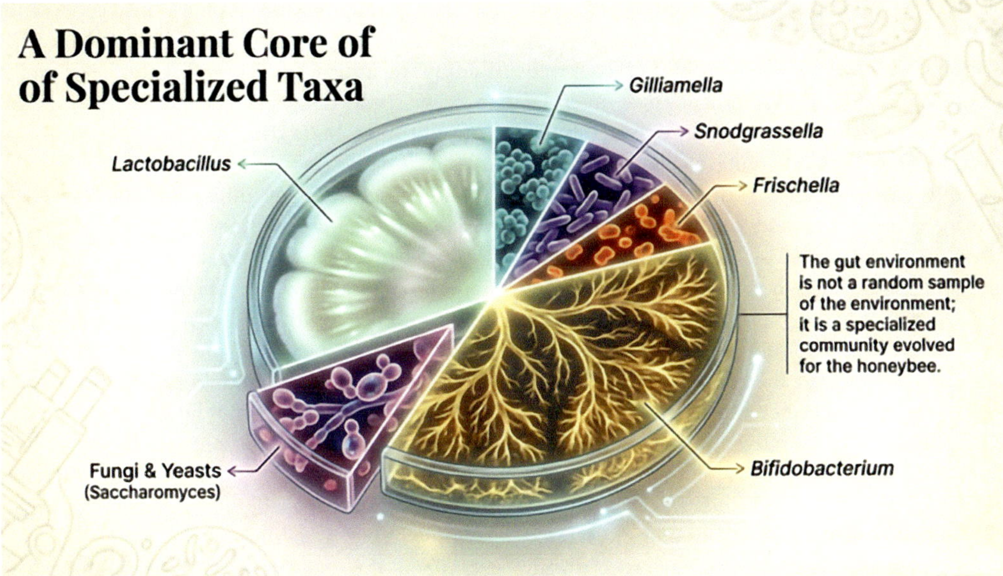

Figure 8.1 Composition of honeybee gut microbiome. Pie chart depicting the relative abundance of major microbial groups found in the honeybee gut microbiome, including the dominant bacterial genera (e.g., *Lactobacillus*, *Bifidobacterium*), fungi, and yeasts like *Saccharomyces*.

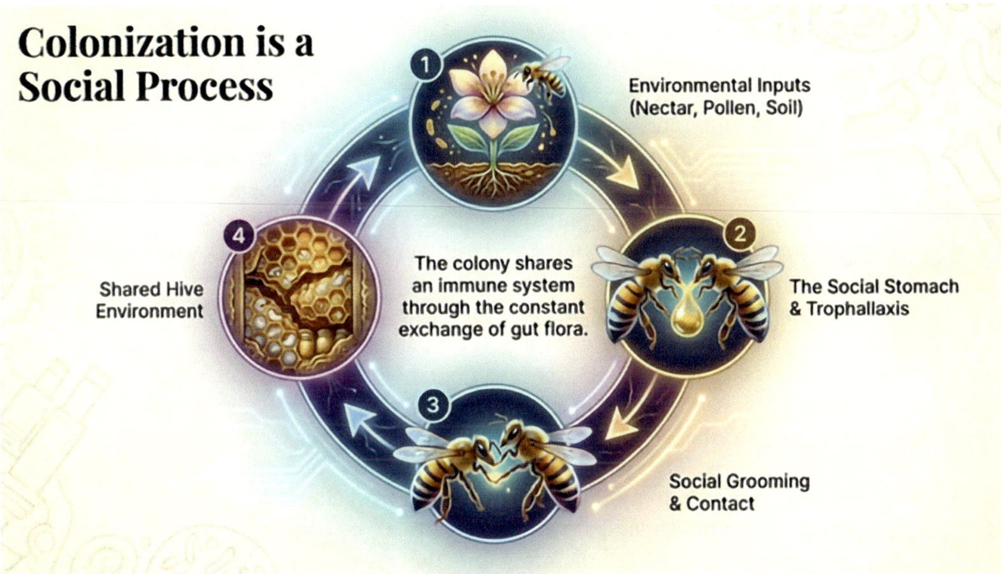

Figure 8.2 Honeybee microbial acquisition pathways. Flow diagram illustrating the various pathways through which honeybees acquire microorganisms, including the ingestion of nectar and pollen, environmental exposure (contact with flowers, soil, etc.), social interactions (grooming, trophallaxis).

providing insights into community dynamics and stability. Steele *et al.* (2021) studied the evolution of interbacterial antagonism within bee gut microbiota, reflecting host and symbiont diversification, and highlighting the ecological consequences of microbial interactions. Powell *et al.* (2016a,b) identified genetic determinants for host colonization in bacterial gut symbionts, which are crucial for understanding the establishment and maintenance of gut communities in honeybees. Engel *et al.* (2014) demonstrated that single-cell genomics reveal hidden diversity among honeybee gut symbionts, uncovering previously unrecognized genetic variation that could influence host–microbe interactions. Nazrin *et al.* (2025) investigated functional divergence in the gut microbiota of *Apis cerana* and *A. mellifera*, highlighting implications for pollination, nutrition, and sustainable apiculture practices. Powell *et al.* (2014) examined the various routes through which honeybee gut microbiota are acquired, providing insights into factors that influence microbial community assembly. Powell *et al.* (2016) conducted a detailed analysis of strain diversity and host specificity in bee gut symbionts by deep-sampling single-copy protein-coding sequences, contributing to understanding host–microbe dynamics.

The interplay between genetics and gut microbiota is a crucial area of research, especially in socially structured organisms like honeybees. Understanding how genetic factors influence microbiome composition can provide insights into the health and resilience of bee populations. Wu *et al.* (2021) demonstrated that honeybee genetics significantly shape the strain-level structure of gut microbiota through social transmission. Their study highlights the role of genetic variation in shaping the diversity and composition of microbial communities in the bee gut, underscoring the importance of host genetics in microbial interactions. Research indicates that the genetic makeup of bees not only influences gut microbiota diversity but also affects the host's immune response to pathogens. Certain genetic traits are associated with more robust microbiomes that can outcompete harmful bacteria, promoting bee health. Recent studies suggest that the microbial communities in bee guts exhibit heritable traits, meaning that specific bacterial lineages can be passed down through generations along with their host's genetics. This heritability underscores the co-evolution of bees and their gut symbionts. Investigating the functional genomics of honeybees provides insights into how genes related to digestion, metabolism, and immune response interact with the gut microbiota. Genetic variations may determine which microbial strains are favored in the gut environment, affecting nutrient utilization and overall fitness. Understanding the genetic basis of gut microbiota composition can inform breeding programs to enhance honeybee resilience. By selecting for genetic traits that promote beneficial microbiome profiles, beekeepers can improve colony health and productivity. The findings regarding honeybees can also be extended to other social insects, as genetic factors influencing gut microbiota dynamics may play a similar role in those species.

The composition of gut bacteria in honeybees varies significantly across developmental stages and seasons. Understanding these dynamics is essential for grasping how microbial communities correlate with bee health and behavior. Martinson *et al.* (2012) studied the establishment of characteristic gut bacteria during the development of honeybee workers. Their findings highlighted how specific microbial communities become established as bees transition from larvae to adults, indicating critical periods for microbial colonization that may impact lifelong health and immunity. Guo *et al.* (2015) examined the gut bacteria of Asian honeybees (*A. cerana*) at different developmental stages, revealing differences in microbial composition. This study underscored the importance of specific gut bacteria at the larval and adult stages, which may influence bees' metabolic and immunological capabilities. Ludvigsen *et al.* (2015) investigated shifts in midgut and pyloric microbiota composition within a honeybee apiary throughout the season. Their research demonstrated that environmental factors, such as food availability and climatic conditions, influence the gut microbiome, leading to seasonal fluctuations that may affect bee health and colony dynamics. Understanding how gut bacteria vary during development and across seasons can provide insights into how to enhance bee health and resilience. For instance, variations in microbiota may correlate with periods of stress or nutritional scarcity, informing management strategies to bolster colonies. The gut microbiota's ability to adapt to the diverse physiological needs of bees across life stages and seasonal conditions suggests a complex interplay between the host and the microbiome. Identifying key microbial species that confer benefits during critical developmental phases or seasonal stressors could lead to targeted interventions for bee health. Further research is needed to explore the mechanisms underlying these changes in microbiota composition. Understanding the functional roles of different bacterial populations during various stages and seasons will enhance our knowledge of host–microbe interactions and their implications for pollinator conservation.

Organic acids play a crucial role in the colonization and establishment of gut microbial communities. Organic acids can influence the adhesion properties of beneficial bacteria, facilitating their attachment to the gut epithelium and promoting colonization. Organic acids help maintain a favorable pH and nutrient availability in the gut, thereby enhancing the growth and persistence of specific symbiotic microorganisms. Quinn *et al.* (2024) highlight how host-derived organic acids enable the gut colonization of the honeybee symbiont *Snodgrassella alvi*. Their findings underscore the importance of organic acids in establishing beneficial microbial relationships in the bee gut. The ability of *S. alvi* to utilize organic acids during colonization suggests potential strategies to enhance bee health through microbial management. Understanding the mechanisms by which organic acids affect microbial colonization can have wider implications for human gut health, particularly in preventing dysbiosis and promoting the growth of beneficial microbiota. Leveraging organic acids to improve the colonization efficiency of probiotic strains in human and animal health offers a promising area for future research and applications. Further studies are needed to explore the diverse roles of organic acids across different gut environments and their interactions with various microbial communities. Exploring organic acids as tools to modulate gut microbiota composition could lead to innovative strategies to enhance health in both humans and other animals.

8.3 FUNCTIONS OF THE HONEYBEE MICROBIOME

The honeybee microbiome serves multiple vital functions essential for the health and behavior of these insects (Figure 8.3). One of its primary roles is in digestion and nutrient processing. Microbes in the honeybee gut help break down complex sugars and other nutrients in pollen and nectar, enabling effective energy uptake and supporting bee metabolism. Moreover, the microbiome plays a crucial role in supporting the immune system, enhancing disease resistance. The presence of beneficial bacteria helps maintain a balanced gut environment and inhibits pathogenic bacteria, providing a defense against infectious diseases. The microbiome also influences honeybee behavior and health, affecting foraging patterns, reproductive success, and overall fitness. In addition, the microbiome can impact honey quality and production. The microbes involved in honey fermentation can influence its flavor profile and nutritional value, underscoring the importance of the microbiome to both bee health and the quality of bee products.

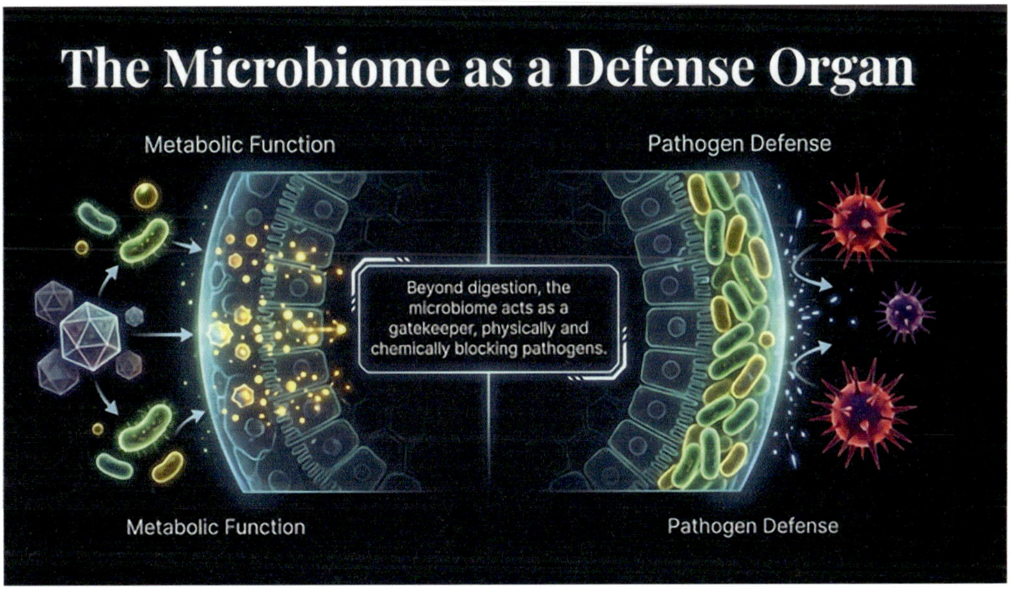

Figure 8.3 The role of microbiome in honeybee health. Schematic representation showing how the gut microbiome interacts with honeybee physiology, highlighting nutrient breakdown and digestion, immune function enhancement, protection against pathogens.

The honeybee microbiome aids in the breakdown of saccharides and fermentation processes, which are vital for nutrient absorption (Lee *et al.*, 2014; Kakumanu *et al.*, 2016). Studies have shown that gut microbiota contribute to immune responses and resilience against pathogens (Moran, 2015; Raymann *et al.*, 2018). The gut microbiome is associated with behavioral tasks in honeybees, influencing their foraging and social behaviors (Jones *et al.*, 2018; Vernier *et al.*, 2020).

Understanding the interactions between gut microbes and the immune response can provide valuable insights into bee health management. Maigoro *et al.* (2024) evaluated the gut microbiota of honeybee across selected ontogenetic stages, noting its immunogenic potential during summer. Their findings suggest that gut microbial composition can actively stimulate immune responses, thereby enhancing honeybees' overall health and defense mechanisms during peak foraging periods. Anderson *et al.* (2017) discussed the implications of gut dysbiosis in honeybees as a novel context of disease ecology. Dysbiosis can disrupt immune function and increase vulnerability to infections, underscoring the crucial link between microbial balance and the immune response. Emery *et al.* (2017) identified the gut symbiont *Frischella perrara* as a significant stimulator of the immune system in honeybees. Their research demonstrated that this symbiont enhances immune gene expression, suggesting that specific gut bacteria can play a pivotal role in modulating the host's immune defenses. Horak *et al.* (2020) revealed how symbiotic relationships shape host innate immunity in honeybees. Their study provided evidence that gut microbiota influence immune signaling pathways, demonstrating that beneficial microbes can improve resistance to pathogens and enhance the host's overall immune competence. The insights from these studies suggest potential interventions to promote beneficial gut microbiota and bolster honeybees' immune systems. Probiotics and dietary strategies can help maintain microbial diversity, which, in turn, supports robust immune responses. Investigating the interplay between gut microbiota and the immune system extends beyond honeybees to broader ecological contexts. Understanding these dynamics can inform conservation strategies for pollinators and help mitigate the impacts of diseases on wild and managed bee populations.

The relationship between hosts and their gut symbionts is a crucial aspect of understanding the health and ecology of honeybees. Host specialization refers to the adaptation of gut symbionts to their specific hosts, leading to distinct microbiota profiles that can influence host health and behavior (Figure 8.4). Kwong *et al.* (2014) provided insights into the genomics of gut symbionts, revealing how specialization varies between honeybees and

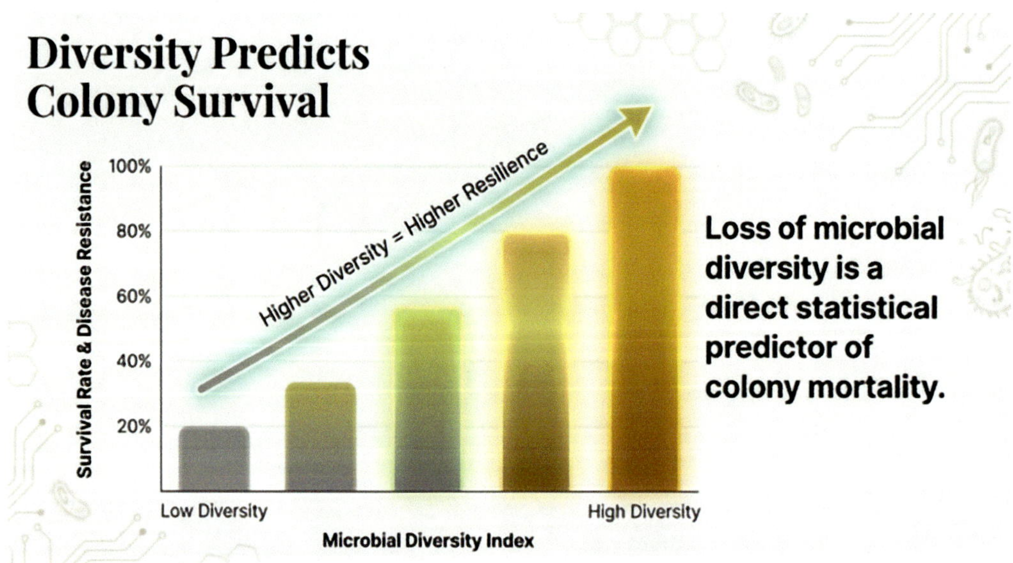

Figure 8.4 Microbial diversity and colony health. Graph demonstrating the correlation between microbial diversity in honeybee colonies and various health metrics (e.g., survival rates, resistance to diseases) expressed by a predictable data.

bumble bees. Their study highlighted distinct symbiont communities corresponding to the dietary preferences and ecological niches occupied by each bee type. Gut symbionts play a critical role in breaking down complex carbohydrates and synthesizing essential nutrients for their host, enhancing honeybee health and longevity. Specialized gut microbiota can confer resistance to pathogens by outcompeting harmful microorganisms and supporting the immune system of honeybees. The microbiota composition in honeybees may be affected by geographic location, forage availability, and exposure to different environmental pressures. Social dynamics within a honeybee colony promote the sharing of microbes, which can influence host specialization and microbiota diversity. Further research is needed to understand the genetic mechanisms behind host specialization in gut symbionts and their implications for bee health management. Investigating how agricultural practices and habitat loss affect host–symbiont relationships can provide insights for developing conservation strategies.

The gut microbiota of honeybees plays a pivotal role in their health and well-being, and the complex interactions within this microbiome, including the presence of prophages, contribute significantly to microbial dynamics. Prophages, which are dormant bacteriophages integrated into bacterial genomes, have attracted attention for their potential influence on gut microbiota composition and function. These viral elements can impact bacterial host physiology, promote horizontal gene transfer, and shape microbial community dynamics within the gut. Bueren *et al.* (2023) characterized prophages present in bacterial genomes derived from the honeybee gut microbiome. Their study provides essential insights into the diversity and prevalence of prophages and highlights their possible roles in influencing bacterial interactions, adaptation, and resilience against pathogens. The presence of prophages can alter bacterial metabolic pathways and influence nutrient and signaling availability within the gut environment. By examining the interactions between prophages and their bacterial hosts, researchers can better understand how these viral elements contribute to the overall functionality of the gut microbiota. There is growing interest in how prophages might contribute to honeybees' defenses against pathogens. By conferring resistance traits or modulating immune responses, prophages may play a critical role in enhancing the gut microbiome's resilience against infectious threats. The study of prophages in the honeybee gut microbiome is still in its early stages. Future research should focus on elucidating the specific mechanisms by which prophages interact with bacterial communities, and on the implications for honeybee health and disease management. Understanding these dynamics is crucial for developing strategies to support pollinator health. Insights gained from studying prophages in honeybees can extend to other social insects and ecological systems. The role of prophages in shaping microbiota dynamics and their potential influence on host health could have significant implications for biodiversity and ecosystem stability.

8.4 EFFECTS OF ENVIRONMENTAL FACTORS ON THE MICROBIOME

Environmental changes pose significant risks to honeybee microbiome health (Figure 8.5). The application of pesticides and exposure to pollutants can adversely affect microbial communities, leading to imbalances that may compromise the bees' immunity and overall health. Habitat loss significantly influences microbial diversity within honeybee populations, as reduced foraging options limit microbial acquisition. Air pollution can disrupt the balance of beneficial microorganisms in the honeybee gut. For instance, research by Sampson *et al.* (2025) shows that air pollution alters colonization factors in the beneficial symbiont *Snodgrassella* and disrupts the gut microbiome of bumblebees, with implications for honeybee health as well (Sampson *et al.*, 2025). Climate change also presents challenges for honeybee populations and their microbiomes. Changes in temperature and weather patterns can affect the availability of floral resources, altering bees' diets and, consequently, their microbiome composition. Understanding how these environmental factors interact with honeybee microbiomes is crucial for developing strategies to support bee health in a changing world.

The impact of insecticides and chemicals on the gut microbiota of honeybees has garnered increasing attention due to its implications for bee health and ecosystem stability (Figure 8.6). Changes in gut microbiota can influence the immune response, nutritional status, and overall resilience of bee populations, particularly in the context of chemical exposure. Blot *et al.* (2019)

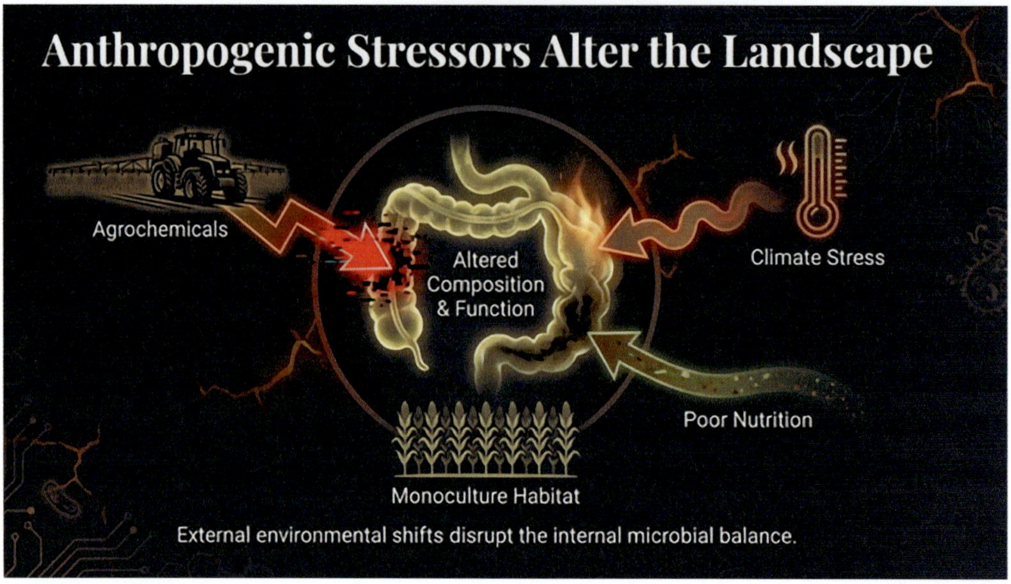

Figure 8.5 Influence of Environmental Factors on Honeybee Microbiome. Infographic summarizing the impact of different environmental factors (e.g., agricultural practices, climate change) on the composition and function of the honeybee microbiome.

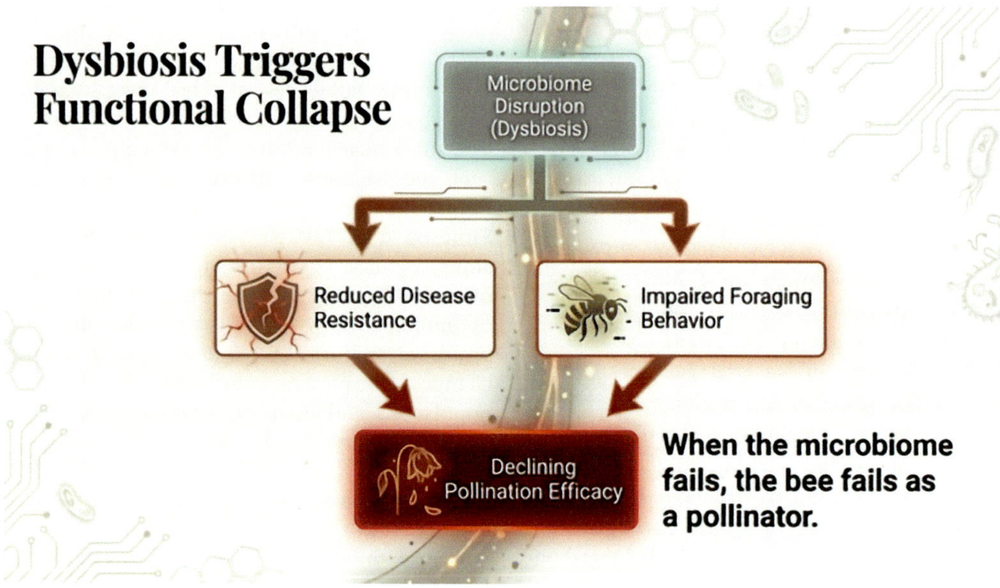

Figure 8.6 Implications of microbiome disruption. Flowchart illustrating the potential consequences of disruptions in the honeybee microbiome, including reduced disease resistance, impaired foraging behavior, declining pollination efficacy.

investigated the effects of glyphosate, a widely used herbicide, on honeybee gut microbiota. Their study found that glyphosate alters the microbial composition, which may have significant consequences for the bees' health. This research indicates that not all glyphosate metabolites, such as AMPA, exert the same effects, underscoring the need for comprehensive evaluations of different chemicals. Motta *et al.* (2024a,b) examined the effects of glyphosate on growth and biofilm formation in bee gut symbionts and their associated bacterial communities. Their findings demonstrate that glyphosate affects not only the composition of the gut microbiota but also its functional capabilities, potentially disrupting essential microbiome functions that support bee health. Motta *et al.* (2020) focused on the relationships among glyphosate exposure intensity, duration, and timing, revealing that these factors significantly modulate effects on the honeybee gut microbiota. Understanding how varying exposure levels impact microbial communities can inform management practices to mitigate risks. Castelli *et al.* (2021) examined the consequences of chronic exposure to sublethal doses of glyphosate on honeybee immunity and gut microbiota. Their study indicated that prolonged exposure not only alters microbial communities but also compromises the immune response, thereby increasing susceptibility to infections. The research by Motta *et al.* (2018) provided compelling evidence that glyphosate significantly perturbs the gut microbiota of honeybees. This alteration raises concerns about the implications for bee health and highlights the potential for glyphosate-induced dysbiosis to disrupt normal gut functions. Rouze *et al.* (2019) assessed how chronic exposure to various insecticides affects the honeybee gut microbiota, particularly in the context of infection by *Nosema ceranae*. Their findings suggest that insecticide exposure alters microbial composition and may exacerbate the effects of pathogens. Tran *et al.* (2026) identified a signature dysbiosis in honeybee gut microbiomes after exposure to neonicotinoids through metagenomic sequencing. This study underscores the profound effects of these commonly used insecticides on gut health, further emphasizing the need to consider microbial health alongside chemical exposure. The evidence highlights the critical intersection between insecticide application and gut microbiota health in honeybees. Understanding these dynamics can inform pest management strategies that reduce chemical exposure and promote healthier microbiomes, ultimately supporting the sustainability of bee populations.

Research indicates that pesticide exposure can alter the gut microbiome, raising concerns about the impact of chemical treatments on bee health (Kakumanu *et al.*, 2016). Understanding the distribution and prevalence of antibiotic resistance among gut symbionts is essential for developing effective management practices (Sun *et al.*, 2022; Raymann *et al.*, 2018). Patterns of antibiotic resistance in honeybee gut symbionts can shed light on microbial community dynamics in other ecosystems. Exploring these relationships will enhance our understanding of resistance development across various microbial populations (Duong *et al.*, 2020; Yun *et al.*, 2022).

The honeybee gut microbiome is not only essential for health but it also faces potential risks from antibiotic resistance. Research by Raymann *et al.* (2018) indicates that antibiotics can reduce the genetic diversity of core gut microbiome species, endangering bee health.

8.5 MICROBIOME AND HONEYBEE DISEASES

Investigating the relationship between gut microbiota and disease susceptibility is crucial for understanding honeybee health. The composition of gut microbial communities can influence bees' immune responses and their ability to withstand various pathogenic challenges. Schwarz *et al.* (2016) demonstrated that early gut colonizers significantly shape parasite susceptibility and microbiota composition in honeybee workers. Their study highlighted how early microbial colonization during developmental stages can affect the subsequent resilience of bees against parasites and pathogens. A diverse and balanced gut microbiome can act as a protective barrier against pathogenic infections. Specific microbial species have been shown to enhance immune responses, enabling bees to better defend against diseases such as Nosema and other microbial threats. The mechanisms by which gut microbiota confer protection against diseases include modulation of the host's immune system, competition with pathogens for resources, and production of antimicrobial compounds. These beneficial effects underscore the role of gut microbiota as an integral part of the honeybee immune system. Dysbiosis, or an imbalance in the gut microbiota, can increase disease susceptibility. Environmental stressors, nutritional deficiencies, or antibiotic treatments can disrupt the natural microbial balance, rendering bees more vulnerable to infections and diseases. Understanding the link between gut microbiota and disease susceptibility opens the door to potential interventions. Probiotic treatments or dietary modifications that enhance beneficial gut bacteria could improve overall bee health and reduce susceptibility to disease. The findings regarding gut microbiota and disease susceptibility in honeybees have implications for other social insects as well. By exploring these dynamics, researchers can gain insights into how gut microbiota influence health, survival, and productivity across various species.

The honeybee microbiome also plays a critical role in preventing pathogenic infections. A diverse and balanced microbiome can help inhibit the growth of harmful pathogens, thereby reducing the incidence of diseases such as American foulbrood and Varroa mite infestations. Case studies have shown marked shifts in microbiome composition during disease outbreaks, such as Colony Collapse Disorder (CCD), where detrimental microbial populations may outcompete beneficial ones. There is promising potential for microbiome-based therapies to enhance honeybee health. By using probiotics or promoting beneficial microbial species, researchers aim to bolster the bees' defenses against diseases, paving the way for new interventions in beekeeping practices.

Probiotics have garnered significant attention for their potential therapeutic effects on gastrointestinal disorders. Probiotics help restore dysbiotic gut microbiota, enhancing microbial diversity and functionality, which is vital for gastrointestinal health. Probiotics can modulate gut immune responses, promoting appropriate inflammation and improving barrier function. Maes *et al.* (2016) found that diet-related gut bacterial dysbiosis correlates with impaired development and increased mortality in honeybees, illustrating how disruptions in gut microbiota can lead to broader health issues such as Nosema disease. This highlights the importance of maintaining a balanced gut microbiome for overall health stability in bees. Zeng *et al.* (2025) conducted an umbrella meta-analysis that assessed the therapeutic efficacy of probiotics across a range of gastrointestinal disorders. Their findings reaffirm the potential of probiotics as an effective intervention for various gastrointestinal ailments, emphasizing their role in promoting gut health. Amalaradjou *et al.* (2012) emphasize modern approaches in probiotics research aimed at controlling foodborne pathogens, highlighting how probiotics can limit gastrointestinal illness caused by pathogenic microbes. This underscores the broader implications of probiotics, extending beyond treating existing disorders into preventive health measures. The findings on probiotics' effectiveness for gastrointestinal disorders can inform clinical applications, encouraging healthcare practitioners to consider probiotics as part of comprehensive treatment plans for managing gut health. While promising, the variability in probiotic strains and their effects necessitates careful selection and understanding of which strains are most effective for specific gastrointestinal disorders. Further research is essential to understand the mechanisms by which probiotics exert their effects on gut health and to identify the most effective strains for various clinical applications. Advancements in precision medicine could lead to the development of targeted probiotic therapies tailored to individual patient needs based on their unique gut microbiota profiles.

The use of probiotics in honeybees represents a promising approach to enhancing bee health and mitigating diseases. Probiotics are live microorganisms that, when administered in adequate amounts, confer a health benefit on the host. Probiotic bacteria have been shown to induce an immune response in honeybees, enhancing their ability to fight infections (Evans *et al.*, 2004; Abdi *et al.*, 2023). Probiotics can outcompete harmful pathogens, thereby reducing their prevalence in the gut and limiting disease outbreaks (Daisley *et al.*, 2020; Daisley *et al.*, 2023). Truong *et al.* (2023) have identified effective probiotic candidates for managing *Paenibacillus larvae*, the causative agent of American foulbrood (AFB) disease in honeybees. Commercial probiotics and prebiotics have shown potential in preventing and treating Nosema infections, which have detrimental effects on bee health (Ptaszynska *et al.*, 2016a,b; Klassen *et al.*, 2021). Recent studies have tested the effectiveness of commercial probiotics in restoring gut microbiota and improving overall health in honeybees under field conditions (Damico *et al.*, 2025; Borges *et al.*, 2021).

Research indicates that probiotic treatments can enhance therapeutic, biochemical, and immunological parameters in honeybee populations exposed to various stressors (Chmiel *et al.*, 2020; Kunat-Budzynska *et al.*, 2022). The method of probiotic delivery can significantly enhance their effectiveness against honeybee pathogens (Daisley *et al.*, 2023). Using in-hive fermentation as a source of beneficial microbes can bolster honeybees' gut health (Rodriguez *et al.*, 2023). As the importance of probiotics in honeybee health is recognized, integrating them into sustainable beekeeping practices could lead to healthier colonies and improved honey production (Smriti *et al.*, 2024). Further studies are essential to fully understand the nuances of probiotic interactions within the honeybee gut microbiome and their effectiveness against various pathogens (Garcia-Vincente *et al.*, 2023; Hwang *et al.*, 2026). Probiotics can inhibit the growth of *P. larvae* through mechanisms such as nutrient competition and niche occupation, thereby preventing infection. Probiotics may enhance honeybees' immune response, providing an additional layer of protection against infections. Guzman and Sofia (2020) highlights the efficacy of a prophylactic probiotic to fight *P. larvae* infections in honeybees, suggesting that early intervention may be crucial for success. Daisley *et al.* (2020) present a novel probiotic approach that effectively countered *P. larvae* infection in honeybees. Their findings emphasize the potential for targeted probiotic treatments to protect bee health and reduce AFB outbreaks. Yoshiyama *et al.* (2013) demonstrated that specific lactic acid bacteria isolated from fermented materials could inhibit *P. larvae*, further supporting the idea that naturally occurring probiotics can be harnessed for disease control. Understanding the efficacy of probiotics provides beekeepers with practical tools to manage AFB and promote hive health. While promising, implementing probiotics requires careful consideration of strain selection, dosage, and delivery methods to achieve optimal effectiveness. Future studies should focus on evaluating the long-term effects of probiotics on honeybee health and their potential role in integrated pest management strategies. There is a need to develop commercially viable probiotic products specifically formulated for bees to enhance their health and resilience against pathogens.

European foulbrood (EFB), caused by the bacterium *Melissococcus plutonius*, poses a significant threat to honeybee (*A. mellifera*) colonies. Recent studies have highlighted the complex interplay between honeybee microbiota and the incidence of EFB. Understanding the bacterial communities associated with healthy worker bees compared to those affected by EFB is crucial. Erban *et al.* (2017a,b) analyzed the bacterial community associated with honeybees infected with EFB, revealing distinct changes in microbiota composition associated with disease. Identifying specific microbiota members associated with resistance or susceptibility to EFB can inform management practices and therapeutic strategies. Nakamura *et al.* (2025) investigated the potential of using Nisin Z, a natural antimicrobial peptide, to protect honeybee larvae from EFB. Their findings suggest that such treatments could help mitigate the effects of the disease by influencing the microbiota dynamics in the hive. Utilizing probiotic treatments or dietary adjustments to promote beneficial microbial communities may enhance bee health and reduce EFB incidence. Further research is needed to explore interactions between the microbiota and pathogens, such as *M. plutonius*, to develop innovative control strategies for EFB. Continued exploration of antimicrobial compounds from beneficial microbes may yield new therapeutic options for combating EFB and improving overall hive health.

The microbiome plays a crucial role in shaping host health and can significantly influence the dynamics of pathogen and parasite spread within populations. The microbiome can enhance host immunity, creating a barrier against parasites. Healthy microbial communities can outcompete or inhibit pathogen growth, reducing infection risk. A disrupted microbiome (dysbiosis) may weaken the host's immune response, making it more susceptible to parasitic infections. The beetle parasite poses a significant threat to honeybee colonies, underscoring the importance of understanding its ecology and transmission dynamics. Huang *et al.* (2024a, b) investigated the relationship between honeybee microbiomes, dietary flexibility, and the spread of this beetle parasite. Their findings suggest that changes in the microbiome influence how the parasite spreads and infects bee populations. Promoting a healthy microbiome through diet, probiotics, and management practices can help mitigate the spread of parasites and enhance honeybee resilience. Regular monitoring of both bee health and microbiome diversity can inform timely interventions to control parasite spread and safeguard colonies. Future studies should focus on the complex interactions between honeybee microbiomes and various parasites, examining how these relationships evolve and influence colony dynamics. Insights from microbiome research can inform more effective management practices and conservation strategies to protect honeybee populations from parasitic threats.

The interplay between pathogens and the gut microbiota represents a critical aspect of host health and disease dynamics. Pathogen infections can lead to dysbiosis, an imbalance in gut microbial populations that may compromise gut integrity and host defenses. Pathogens can exploit host immune responses, leading to shifts in microbiota composition that favor their persistence

significantly improving honeybee survival, underscoring the potential of using beneficial microbes as an intervention against diseases such as Nosema. Shukla (2020) reviewed engineered probiotics that use synthetic biology approaches, which could enhance the efficacy of interventions designed to mitigate bee diseases. Huang *et al.* (2025) expanded on their previous work by targeting both mite and microsporidian parasites, highlighting the importance of multifaceted approaches for improving honeybee resilience. Leonard *et al.* (2020) found that engineered symbionts can activate honeybee immunity, thereby reducing pathogen loads and strengthening colony health. Lang *et al.* (2023) reported that engineered symbiotic bacteria can interfere with the Nosema redox system, inhibiting microsporidian parasitism and representing a novel target for disease management in bees. Sattayawat *et al.* explored how engineering gut symbionts could promote bee growth, suggesting that enhancing beneficial interactions between bees and their microbiota can improve health outcomes. Zhang *et al.* (2024) discussed engineering probiotics for diabetes management, which, while not directly related to bees, showcases the broad application of probiotic engineering that could extend to bee health. Kim *et al.* (2023) screened various entomopathogenic fungal culture extracts for their inhibitory activity against honeybee Nosema, demonstrating potential alternative treatments that could complement bioengineering efforts. Leonard *et al.* (2018) presented a toolkit for the genetic engineering of bacteria in the bee gut microbiome, which could facilitate the development of targeted interventions against infections such as Nosema. Lariviere *et al.* (2024) described a method for one-step genome engineering in bee gut bacterial symbionts, paving the way for precise modifications to enhance their probiotic capabilities. Gajger *et al.* (2023) studied the influence of probiotic feed supplements on Nosema infection levels and the gut microbiota of adult honeybees, suggesting that diet modification can play a role in disease management. Garcia-Vicente *et al.* (2023) examined how feeding probiotics and postbiotics affects honeybee hive strength and health during critical periods, providing insights into nutritional strategies for disease resistance. Lariviere *et al.* (2023) reported a single-step genome-engineering approach in the bee gut symbiont *S. alvi*, highlighting advancements that could enhance bee health strategies. Rebeck *et al.* (2025) discussed a yeast-based oral therapeutic for delivering immune checkpoint inhibitors, suggesting innovative avenues for exploring in bee diseases. Alberoni *et al.* (2023) studied alterations in the microbiota of caged honeybees infected with *N. ceranae*, revealing functional changes that could inform treatment strategies.

Nanotechnology has emerged as a transformative approach in the field of nutraceuticals, particularly in enhancing the efficacy and delivery of probiotics and prebiotics.This chapter explores how nanotechnology is reshaping our understanding and application of these essential components of gut health. Nanotechnology involves manipulating matter at the nanoscale to improve the bioavailability, stability, and delivery of functional ingredients. Dangi *et al.* (2023) discuss how nanotechnology is revolutionizing the formulation and delivery of probiotics, enabling enhanced colonization and survival rates within the gastrointestinal tract. Utilizing nanocarriers to encapsulate prebiotics can significantly improve their stability and controlled release, ultimately enhancing their functionality. Nanotechnology can facilitate the combination of probiotics and prebiotics, promoting symbiotic relationships that enhance gut health and improve overall efficacy. Innovations in food products incorporating nanotechnology for probiotics and prebiotics are becoming prominent, enabling the development of functional foods with enhanced health benefits. Understanding the role of nanotechnology in improving the delivery of probiotics and prebiotics can lead to novel therapeutic approaches for various gastrointestinal disorders. As nanotechnology advances, regulatory frameworks must evolve to ensure the safety and efficacy of nanotech-enhanced nutraceuticals. Further research is essential to fully explore the potential of nanotechnology in probiotics and prebiotics, focusing on long-term health outcomes and broader applications in the health sciences.

Type IV secretion systems (T4SS) are vital components in the interaction between bacteria and their hosts, facilitating the delivery of effector proteins and contributing to microbial competition. Bacteria employ T4SS to transport proteins and DNA across cell membranes, playing crucial roles in virulence, genetic exchange, and interbacterial interactions. T4SS can influence intraspecific competition by enabling bacteria to outcompete one another for resources, establish dominance, or modulate host interactions. Gut symbionts contribute to honeybee health by aiding in digestion, nutrient absorption, and immunity. Understanding their competitive interactions can provide insights into overall hive health. Motta *et al.* (2024) demonstrate that Type VI secretion systems (T6SS) facilitate intraspecific competition and host interactions in a bee gut symbiont. Their findings highlight the role of secretion systems in shaping microbial communities and competitive outcomes within the bee gut, underscoring the complexity of these interactions. Investigating T4SS and their role in intraspecific competition enhances our understanding of microbial ecology, particularly in specialized environments like the bee gut. Insights gained from studying these interactions could inform strategies to manage hive health, improve symbiotic relationships, and enhance the resilience of bee populations. T4SS play a critical role in mediating intraspecific competition among gut symbionts, influencing microbial dynamics and host interactions. Further research is essential to elucidate the complexities of these interactions and their implications for both microbial ecology and bee health.

The symbiotic relationships between honeybees and their gut microbiota play a crucial role in maintaining health and resilience against pathogens (Figure 8.8). However, the emergence of antibiotic resistance among gut symbionts has become a significant concern, affecting the efficacy of treatments and bee health. Honeybee gut symbionts are essential for digestion, nutrient absorption, and immune function. However, these beneficial microbes can also acquire antibiotic resistance traits, posing a risk to both the bees and their ability to cope with infections. Ludvigsen *et al.* (2017) investigated the antibiotic-resistant patterns of geographically widespread honeybee gut symbiont subgroups. Their study found that these subgroups exhibited locally distinct resistance profiles, highlighting the influence of environmental factors and local antibiotic use on resistance development within bee populations. The presence of antibiotic-resistant bacteria in the gut microbiota can complicate treatment strategies for honeybees. If beneficial gut symbionts carry resistance genes, it may limit the effectiveness of therapeutic antibiotics used to manage infections, necessitating a reevaluation of treatment protocols. Understanding the distribution and prevalence of antibiotic resistance among gut symbionts is essential for developing effective management practices. Continuous monitoring can provide insights into how resistance patterns change over time and in response to varying agricultural practices. Agricultural practices, including the use of antibiotics in apiculture and surrounding agricultural environments, have significant implications for the resistance patterns observed in honeybee gut symbionts. Strategies that minimize antibiotic

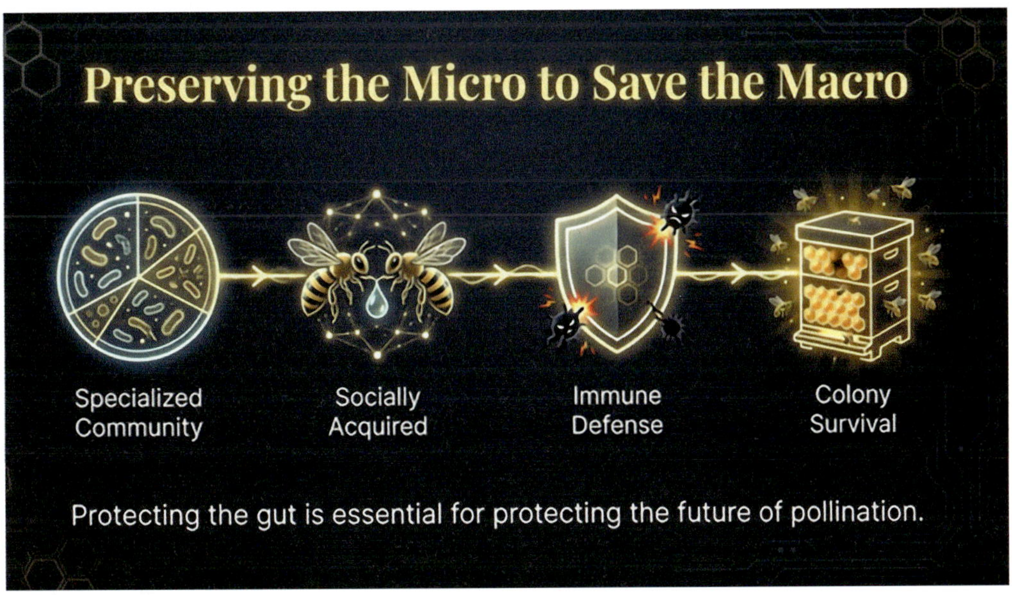

Figure 8.8 A healthy honeybee gut microbiome effects future flower pollination.

use outside the hive can reduce the selective pressure that fosters the development of resistant strains. Further research is needed to explore the mechanisms of antibiotic resistance in honeybee gut symbionts and the ecological implications of resistance genes. A deeper understanding of these dynamics can inform sustainable apiculture practices that safeguard bee health. The patterns of antibiotic resistance observed in honeybee gut symbionts have broader implications for other microbial communities in ecosystems. Investigating these relationships can contribute to our understanding of resistance development among various microorganisms.

8.7 CONCLUSION

The concept of smart probiotics has emerged as a groundbreaking advancement in probiotic research and application. These engineered probiotics are designed to enhance their effectiveness and applicability in various health-related fields. This chapter explores the definition, potential applications, and future directions for smart probiotics. Smart probiotics are engineered to exhibit specific functionalities that allow them to target particular health issues more effectively. Choudhary *et al.* (2020) discuss systematic designs for smart probiotics, emphasizing the importance of developing controlled-release systems that deliver probiotics to specific sites and at precise times within the gastrointestinal tract, thereby optimizing their therapeutic effects. Smart probiotics can be designed to deliver therapeutic compounds directly to affected tissues or cells. This targeted approach can enhance the treatment of gastrointestinal disorders, infections, and even metabolic diseases. Smart probiotics can also be utilized in food applications to inhibit pathogenic bacteria, thus contributing to food safety and public health. The development of smart probiotics faces several challenges, including regulatory hurdles, strain stability, and the complexity of interactions with the human microbiota. As research progresses, the potential for smart probiotics to revolutionize health care and nutrition continues to grow. Innovations in genetic engineering and biotechnology may facilitate the creation of more advanced probiotics tailored to individual health needs. Smart probiotics represent a significant advancement in probiotic technology, offering tailored approaches to improve health outcomes. Continued research and innovation will be vital to unlocking their full potential.

In summary, the honeybee microbiome is a vital component of honeybee biology, influencing digestion, immune function, health, behavior, and even the quality of honey. Understanding and preserving honeybee microbiomes are essential for maintaining ecological balance and ensuring the sustainability of agricultural systems reliant on these pollinators. There is a pressing need for further research and public awareness regarding honeybee health and the intricate relationships within their microbiome.

REFERENCES

Abdi K, Ben Said M, Crotti E, Masmoudi AS, Cherif A. The promise of probiotics in honeybee health and disease management. *Archives of Microbiology*, 2023 Jan 27;205(2):73. doi: 10.1007/s00203-023-03416-z

Alberoni, D, et al. (2023) Alterations in the microbiota of caged honeybees in the presence of *Nosema ceranae* infection and related changes in functionality. *Microbial Ecology*, 86:601–616. https://doi.org/10.1007/s00248-022-02050-4

Amalaradjou, MAR, et al. (2012) Modern approaches in probiotics research to control foodborne pathogens. *Advances in Food and Nutrition Research*, 67:185–239. https://dx.doi.org/10.1016/B978-0-12-394598-3.00005-8

Amalaradjou, MAR, et al. (2013) Bioengineered probiotics, a strategic approach to control enteric infections. *Bioengineered*, 4(6):379–387. https://dx.doi.org/10.4161/bioe.23574

Anderson, KE, et al. (2017) Honeybee gut dysbiosis: A novel context of disease ecology. *Current Opinion in Insect Science*, 22:125–132. http://dx.doi.org/10.1016/j.cois.2017.05.020

Blot, N, et al. (2019) Glyphosate, but not its metabolite AMPA, alters the honeybee gut microbiota. *PLOS One*, 14(4):e0215466. https://doi.org/10.1371/journal.pone.0215466

Borges, D, et al. (2021) Effects of prebiotics and probiotics on honeybees (*Apis mellifera*) infected with the microsporidian parasite *Nosema ceranae*. *Microorganisms*, 9:481. https://doi.org/10.3390/microorganisms9030481

Bueren, EK, et al. (2023) Characterization of prophages in bacterial genomes from the honeybee (*Apis mellifera*) gut microbiome. *PeerJ*, 11:e15383. https://doi.org/10.7717/peerj.15383

Castelli, L, et al. (2021) Impact of chronic exposure to Sublethal doses of Glyphosate on honeybee immunity, gut microbial and infection by pathogens. *Microorganisms*, 9:845. https://doi.org/10.3390/microorganisms9040845

Chmiel, JA, et al. (2020) Understanding the effects of sublethal pesticide exposure on honeybees: A role of probiotics as mediators of environmental stress. *Frontiers in Ecology and Evolution*, 8:22. https://doi.org/10.3389/fevo.2020.00022

Choudhary, R, et al. (2020) Toward a systematic design of smart probiotics. *Current Opinion in Biotechnology*, 64:199–209. https://doi.org/10.1016/j.copbio.2020.05.003

Coccia, M, et al. (2020) Microbiome health: Cultivation methods and implications for pollinator stress. *Insects*, 11(9):645. https://doi.org/10.3390/insects11090645

Copeland, DC, et al. (2022) Early queen development in honeybees: Social context and queen breeder source affect gut microbiota and associated metabolism. *Microbiology Spectrum*, 10(4):1–17. https://doi.org/10.1128/spectrum.00383-22

Daisley, BA, et al. (2020) Novel probiotic approach to counter *Paenibacillus larvae* infection in honeybees. *The ISME Journal*, 14:476–491. https://doi.org/10.1038/s41396-019-0541-6

Daisley, BA, et al. (2023) Delivery mechanism can enhance probiotic activity against honeybee pathogens. *ISME*, 17:1382–1395. https://doi.org/10.1038/s41396-023-01422-z

Damico, ME, et al. (2025) Testing the effectiveness of a commercially sold probiotic on restoring the gut microbiota of honeybees: A field study. *Probiotics and Antimicrobial Proteins*, 17:991–1000. https://doi.org/10.1007/s12602-023-10203-1

Dangi, P, et al. (2023) Nanotechnology impacting probiotics and prebiotics: A paradigm shift in nutraceuticals technology. *International Journal of Food Microbiology*, 388:110083. https://doi.org/10.1016/j.ijfoodmicro.2022.110083

Deng, Y, et al. (2024) Chronic bee paralysis virus exploits host antimicrobial peptides and alters gut microbiota composition to facilitate viral infection. *The ISME Journal*, 18(1):wrae051. https://doi.org/10.1093/ismejo/wrae051

Duong, BTT, et al. (2020) Investigation of the gut microbiome of *Apis cerana* honeybees from Vietnam. *Biotechnology Letters*, 42:2309–2317. https://doi.org/10.1007/s10529-020-02948-4

Emery, O, et al. (2017) Immune system stimulation by the gut symbiont *Frischella perrara* in the honeybee (*Apis mellifera*). *Molecular Ecology*, 26:2576–2590. https://doi.org/10.1111/mec.14058

Engel P, Martinson VG, Moran NA. (2012, Jul 3). Functional diversity within the simple gut microbiota of the honey bee. *Proceedings of the National Academy of Sciences of the United States of America*, 109(27): 11002-11007.

Engel, P, et al. (2014) Hidden diversity in honeybee gut symbionts detected by single-cell genomics. *PLOS Genetics*, 10(9):e1004596. https://doi.org/10.1371/journal.pgen.1004596

Erban, T, et al. (2017a) Honeybee (*Apis mellifera*)-associated bacterial community affected by American foulbrood: Detection of *Paenibacillus larvae* via microbiome analysis. *Scientific Reports*, 7:5084. https://doi.org/10.1038/s41598-017-05076-8

Erban, T, et al. (2017b) Bacterial community associated with worker honeybees (*Apis mellifera*) affected by European foulbrood. *PeerJ*, 5:e3816. https://doi.org/10.7717/peerj.3816

Evans, JD, et al. (2004) Bacterial probiotics induce an immune response in the honeybee (Hymenoptera: Apidae). *Apiculture and Social Insects*, 97(3):752–756.

Gajger, IT, et al. (2023) Influence of probiotic seed supplement on *Nosema* spp. infection level and the gut microbiota of adult honeybees (*Apis mellifera* L.). *Microorganisms*, 11:610. https://doi.org/10.3390/microorganisms11030610

Garcia-Vicente, EJ, et al. (2023) Effect of feed supplementation with probiotics and postbiotics on strength and health status of honeybee (*Apis mellifera*) hives during late spring. *Research in Veterinary Science*, 159:237–243. https://doi.org/10.1016/j.rvsc.2023.05.001

Guo, J, et al. (2015) Characterization of gut bacteria at different developmental stages of Asian honeybees, *Apis cerana*. *Journal of Invertebrate Pathology*, 127:110–114. http://dx.doi.org/10.1016/j.jip.2015.03.010

Guzman, R., Sofia, A. (2020). A prophylactic probiotic to fight paenibacillus larvae infection in honeybees. Thesis. University of British Columbia.

Horak, RD, et al. (2020) Symbionts shape host innate immunity in honeybees. *Proc.R.Soc.B*, 287:20201184. http://dx.doi.org/10.1098/rspb.2020.1184

Huang, H, et al. (2024a) The mycobiome as an integral part of the gut microbiome: Crucial roles of symbiotic fungi in health and disease. *Gut Microbes*, 16(1):2440111. https://doi.org/10.1080/19490976.2024.2440111

Huang, Q, Wei, X, Zang, L, Powell, JE, Moran, N. (2025, Dec 17). Engineered gut symbiont targets mite and microsporidian parasites and improves honey bee survival. *Proceedings of the Royal Society*, 292(2061):20251743.

Huang, Q, et al. (2023) Engineered gut symbiont inhibits microsporidian parasite and improves honeybee survival. *PNAS*, 120(25):e2220922120. https://doi.org/10.1073/pnas.2220922120

Huang, Q, et al. (2024b) Microbiomes, diet flexibility, and the spread of a beetle parasite of honeybees.

Frontiers in Microbiology, 15:1387248. https://doi.org/10.3389/fmicb.2024.1387248

Hwang, CY, et al. (2026) *In vitro* evaluation of Bacillus subtilis isolated from *Apis mellifera* for potential use as a probiotic feed additive in honeybees. *Scientific Reports*, 16:3708. https://doi.org/10.1038/s41598-025-33832-8

Jones, JC, et al. (2018) The gut microbiome is associated with behavioural task in honeybees. *Insectes Sociaux*, 65:419–429. https://doi.org/10.1007/s00040-018-0624-9

Jones, KR, et al. (2025) Effects of priority on strain-level composition of the honeybee gut community. *Applied and Environmental Microbiology*, 91(8):1–12. https://doi.org/10.1128/aem.00828-25

Kakumanu, ML, et al. (2016) honeybee gut microbiome is altered by In-Hive pesticide exposures. *Frontiers in Microbiology*, 7:1255. https://doi.org/10.3389/fmicb.2016.01255

Kim, D-J, et al. (2023) Screening of entomopathogenic fungal culture extracts with honeybee nosemosis inhibitory activity. *Insects*, 14:538. https://doi.org/10.3390/insects14060538

Klassen, SS, et al. (2021) *Nosema ceranae* infections in honeybees (*Apis mellifera*) treated with pre/probiotics and impacts on colonies in the field. *Veterinary Sciences*, 8:107. https://doi.org/10.3390/vetsci8060107

Kunat-Budzynska, M, et al. (2022) Natural substances, probiotics, and synthetic agents in the treatment and prevention of honeybee nosemosis. *Pathogens*, 11:1269. https://doi.org/10.3390/pathogens11111269

Kwong, W, et al. (2015) Evolution of host specialization in gut microbes: The bee gut as a model. *Gut Microbes*, 6(3):214–220. https://dx.doi.org/10.1080/19490976.2015.1047129

Kwong, WK, et al. (2013) Cultivation and characterization of the gut symbionts of honeybees and bumble bees: Description of *Snodgrassella alvi* gen. nov., sp. nov., a member of the family *Neisseriaceae* of the *Betaproteobacteria*, and *Gilliamella apicola* gen. nov., sp. Nov., a member of *Orbaceae* fam. nov., *Orbales* ord. nov., a sister taxon to the order '*Enterobacteriales*' of the *Gammaproteobacteria*. *International Journal of Systematic and Evolutionary Microbiology*, 63:2008–2018. https://doi.org/10.1099/ijs.0.044875-0

Kwong, WK, et al. (2014) Genomics and host specialization of honeybee and bumble bee gut symbionts. *PNAS*, 111(31):11509–11514. https://doi.org/10.1073/pnas.1405838111

Lang, H, et al. (2023) Engineered symbiotic bacteria interfering with *Nosema* redox system inhibit microsporidia parasitism in honeybees. *Nature Communications*, 14:2778. https://doi.org/10.1038/s41467-023-38498-2

Lariviere, PJ, et al. (2023) Single-step genome engineering in the bee gut symbiont *Snodgrassella alvi*. *bioRxiv preprint*. https://doi.org/10.1101/2023.09.19.558440

Lariviere, PJ, et al. (2024) One-step genome engineering in bee gut bacterial symbionts. *mBio*, 15(9):1–21. https://doi.org/10.1128/mbio.01392-24

Lee, FJ, et al. (2014) Saccharide breakdown and fermentation by the honeybee gut microbiome. *Environmental Microbiology*, 17(3):796–815. https://doi.org/10.1111/1462-2920.12526

Leonard, SP, et al. (2018) Genetic engineering of bee gut microbiome bacteria with a toolkit for modular assembly of broad-host-range plasmids. *ACS Synth Biol.*, 7(5):1279–1290. https://doi.org/10.1021/acssynbio.7b00399

Leonard, SP, et al. (2020) Engineered symbionts activate honeybee immunity and limit pathogens. *Science*, 367:573–576. https://doi.org/10.1126/science.aax9039

Lim, HC, et al. (2015) Deep sequencing and ecological characterization of gut microbial communities of diverse bumble bee species. *PLOS One*, 10(3):e0118566. https://doi.org/10.1371/journal.pone.0118566

Ludvigsen, J, Porcellato, D, Amdam, GV, Rudi, K. (2018, May) Addressing the diversity of the honeybee gut symbiont Gilliamella: description of Gilliamella apis sp. nov., isolated from the gut of honeybees (Apis mellifera). *International Journal of Systematic and Evolutionary Microbiology*, 68(5):1762-1770. doi: 10.1099/ijsem.0.002749.

Ludvigsen, J, et al. (2015) Shifts in the Midgut/Pyloric microbiota composition within a honeybee Apiary throughout a season. *Microbes and Environments*, 30(3):235–244. https://doi.org/10.1264/jsme2.ME15019

Ludvigsen, J, et al. (2017) Geographically widespread honeybee-gut symbiont subgroups show locally distinct antibiotic-resistant patterns. *Molecular Ecology*, 26:6590–6607. https://doi.org/10.1111/mec.14392

Maes, PW, et al. (2016) Diet-related gut bacterial dysbiosis correlates with impaired development, increased mortality and Nosema disease in the honeybee (*Apis mellifera*). *Molecular Ecology*, 25:5439–5450. https://doi.org/mec.13862

Maigoro, AY, et al. (2024) Gut microbiota of Apis mellifera at selected ontogenetic stages and their immunogenic potential during summer. *Pathogens*, 13:122. https://doi.org/10.3390/pathogens13020122

Martinson, VG, et al. (2012) Establishment of characteristic gut bacteria during development of the honeybee worker. *Applied and Environmental Microbiology*,

78(8):830–2840. https://doi.org/10.1128/AEM.07810-11
Martinson, VG, et al. (2019) Innovations in culturing the gut microbiota of honeybees and bumble bees. *Journal of Microbiological Methods*, 157:99–107. https://doi.org/10.1016/j.mimet.2019.04.005
McFrederick, QS, et al. (2021) Metabolomic profiles of bee gut symbionts and their roles in host health. *Environmental Microbiology*, 23(10):5890–5903. https://doi.org/10.1111/1462-2920.15441
Moran, NA, et al. (2012) Distinctive gut microbiota of honeybees assessed using deep sampling from individual worker bees. *PLOS One*, 7(4):e36393. https://doi.org/10.1371/journal.pone.0036393
Moran, NA. (2015) Genomics of the honeybee microbiome. *Current Opinion in Insect Science*, 10:22–28. http://dx.doi.org/10.1016/j.cois.2015.04.003
Motta, EVS, et al. (2018) Glyphosate perturbs the gut microbiota of honeybees. *PNAS*, 115(41):10305–10310. https://doi.org/10.1073/pnas.1803880115
Motta, EVS, et al. (2020) Impact of glyphosate on the honeybee gut microbiota: Effects of intensity, duration, and timing of exposure. *mSystems*, 5:e00268-20. https://doi.org/10.1128/mSystems.00268-20
Motta, EVS, et al. (2024a) Glyphosate effects on growth and biofilm formation in bee gut symbionts and diverse associated bacteria. *Applied and Environmental Microbiology*, 90(8):1–18. https://doi.org/10.1128/aem.00515-24
Motta, EVS, et al. (2024b) Type VI secretion systems promote intraspecific competition and host interactions in a bee gut symbiont. *PNAS*, 121(44):e241882121. https://doi.org/10.1073/pnas.2414882121
Nakamura, K, et al. (2025) Nisin Z treatment to protect honeybee larvae from European foulbrood disease. *Probiotics and Antimicrobial Proteins*, 17:3587–3594. https://doi.org/10.1007/s12602-025-10450-4
Nazrin, MRR, et al. (2025) Host-driven functional divergence in gut microbiota of honeybees *Apis cerana* and *Apis mellifera*: Implications for pollination, nutrition, and sustainable apiculture. *Current Microbiology*, 82:531. https://doi.org/10.1007/s00284-025-04525-2
Paton, AW, et al. (2012) Bioengineered microbes in disease therapy. *Trends in Molecular Medicine*, 18(7):417–425. https://dx.doi.org/10.1016/j.molmed.2012.05.006
Powell, JE, et al. (2014) Routes of acquisition of the gut microbiota of the honeybee *Apis mellifera*. *Applied and Environmental Microbiology*, 80(23):7378–7387. https://doi.org/10.1128/AEM.01861-14
Powell, JE, et al. (2016a) Genome-wide screen identifies host colonization determinants in a bacterial gut symbiont. *PNAS*, 113(48):13887–13892. https://doi.org/10.1073/pnas.1610856113
Powell, JE, et al. (2016b) Strain diversity and host specificity in bee gut symbionts revealed by deep sampling of single copy protein-coding sequences. *Molecular Ecology*, 25(18):4461–4471. https://doi.org/10.1111/mec.13787
Ptaszynska, AA, et al. (2016a) Are commercial probiotics and prebiotics effective in the treatment and prevention of honeybee nosemosis C? *Parasitol Res*, 115:397–406. https://doi.org/10.1007/s00436-015-4761-z
Ptaszynska, AA, et al. (2016b) Impact of vertebrate probiotics on honeybee yeast microbiota and on the course of nosemosis. *Med Weter*, 72(7):430–434. https://doi.org/10.21521/mw.5534
Quinn, A, et al. (2024) Host-derived organic acids enable gut colonization of the honeybee symbiont *Snodgrassella alvi*. *Nature Microbiology*, 9:477–489. https://doi.org/10.1038/s41564-023-01572-y
Raymann, K, et al. (2018a) Antibiotics reduce genetic diversity of core species in the honeybee gut microbiome. *Molecular Ecology*, 27(8):2057–2066. https://doi.org/10.1111/mec.14434
Raymann, K, et al. (2018b) The role of the gut microbiome in health and disease of adult honeybee workers. *Current Opinion in Insect Science*, 26:97–104. https://doi.org/10.1016/j.cois.2018.02.012
Raymann, K, et al. (2018c) Imidacloprid decreases honeybee survival rates but does not affect the gut microbiome. *Applied and Environmental Microbiology*, 84(13):e00545-18. https://doi.org/10.1128/AEM.00545-18
Rebeck, ON, et al. (2025) A yeast-based oral therapeutic delivers immune checkpoint inhibitors to reduce intestinal tumor burden. *Cell Chemical Biology*, 32:98–110. https://doi.org/10.1016/j.chembiol.2024.10.013
Rodriguez, MA, et al. (2023) Probiotics and in-hive fermentation as a source of beneficial microbes to support the gut microbial health of honeybees. *Journal of Insect Science*, 23(6):19. https://doi.org/10.1093/jisesa/iead093
Rouze, R, et al. (2019) The honeybee gut microbiota is altered after chronic exposure to different families of insecticides and infection by *Nosema ceranae*. *Microbes and Environments*, 34(3):226–233. https://doi.org/10.1264/jsme2.ME18169
Sampson, HR, et al. (2025) Air pollution modifies colonisation factors in beneficial symbiont *Snodgrassella* and disrupts the bumblebee gut microbiome. *Npj Biofilms and Microbiomes*, 11:2. https://doi.org/10.1038/s41522-024-00632-3
Schwarz, RS, et al. (2016) Early gut colonizers shape parasite susceptibility and microbiota composition

in honeybee workers. *PNAS*, 113(33):9345–9350. https://doi.org/10.1073/pnas.1606631113
Shukla, MYP. (2020) Efficient engineered probiotics using synthetic biology approaches: A review. *Biotechnology and Applied Biochemistry*, 1:22–29. https://doi.org/10.1002/bab.1822
Smriti, AR, et al. (2024) Prospects of probiotics in beekeeping: A review for sustainable approach to boost honeybee health. *Archives of Microbiology*, 206:205. https://doi.org/10.1007/s00203-024-03926-4
Steele, MI, et al. (2021) Evolution of interbacterial antagonism in bee gut microbiota reflects host and symbiont diversification. *mSystems*, 6(3):e00063–21. https://doi.org/10.1128/mSystems.00063-21
Sun, H, et al. (2022) Geographical resistome profiling in the honeybee microbiome reveals resistance gene transfer conferred by mobilizable plasmids. *Microbiome*, 10:69. https://doi.org/10.1186/s40168-022-01268-1
Tozkar, CO, et al. (2015) Metatranscriptomics analyses of honeybee colonies. *Frontiers in Genetics*, 6:100. https://doi.org/10.3389/fgene.2015.00100
Tran, L, et al. (2026) Neonicotinoid-induced signature dysbiosis identified via metagenomic sequencing of the honeybee gut microbiome. *Scientific Reports*, 16:1211. https://doi.org/10.1038/s41598-025-30907-4
Truong, A-T, et al. (2023) Probiotic candidates for controlling *Paenibacillus larvae*, a causative agent of American foulbrood disease in honeybee. *BMC Microbiology*, 23:150. https://doi.org/10.1186/s12866-023-02902-0
Vernier, CL, et al. (2020) The gut microbiome defines social group membership in honeybee colonies. *Science Advances*, 6:eabd3431. https://doi.org/10.1126/sciadv.abd3431
Wu, J, et al. (2021) Honeybee genetics shape the strain-level structure of gut microbiota in social transmission. *Microbiome*, 9:225. https://doi.org/10.1186/s40168-021-01174-y
Yoshiyama, M, Wu, M, Sugimura, Y, Takaya, N, Kimoto-Nira, H, Suzuki, C. (2013, Jan). Inhibition of Paenibacillus larvae by lactic acid bacteria isolated from fermented materials. *Journal of Invertebrate Pathology*, 112(1):62-67.
Yun, B-R, et al. (2022) Comparison of the gut microbiome of sacbrood virus-resistant and -susceptible *Apis cerana* from South Korea. *Scientific Reports*, 12:10010. https://doi.org/10.1038/s41598-022-13535-0
Zeng, Q, et al. (2025) Probiotics and gastrointestinal disorders: An umbrella meta-analysis of therapeutic efficacy. *European Journal of Medical Research*, 30:515. https://doi.org/10.1186/s40001-025-02908-6
Zhang, S, et al. (2024) Engineering probiotics for diabetes management: advances, challenges, and future directions in translational microbiology. *International Journal of Nanomedicine*, 19:10917–10940. https://doi.org/10.2147/IJN.S492651
Zhang, Y, et al. (2022) Microbial interactions in the honeybee gut: Implications for host health. *Frontiers in Microbiology*, 13:848–862. https://doi.org/10.3389/fmicb.2022.1087002

CHAPTER NINE

Microbiome of Silkworm (*Bombyx mori*)

Alex Muppasani and Sunil Thomas

9.1 INTRODUCTION

9.1.1 Silk Fiber

Silk is a natural protein fiber produced by the domesticated silkworm, *Bombyx mori*. It is widely celebrated for its smooth texture, characteristic luster, and exceptional strength, qualities that have made it one of the most prized materials in luxury textiles. Often referred to as the "queen of fibers," silkworm silk is extensively used as a primary textile raw material due to its distinctive sheen, high air permeability, strong moisture-absorption capacity, soft hand feel, durability, and superior mechanical performance (Wang et al., 2025).

Structurally, natural silkworm silk exhibits a sophisticated hierarchical organization. Silk proteins are synthesized within the silkworm's specialized silk glands and extruded as continuous fibers. At the microscale, each fiber is composed of multiple protofibrils; in turn, each protofibril contains on the order of a thousand nanofibrils, forming a highly organized and mechanically efficient architecture (Wang et al., 2025). As a lightweight biomaterial, silk demonstrates outstanding mechanical properties, with reported average tensile strengths ranging from approximately 400 to 600 MPa, placing it among the strongest natural protein fibers (Wang et al., 2025).

Chemically, silk fiber is a protein-based biopolymer composed primarily of fibroin, which forms two continuous filaments extruded from the larval silk glands. These twin fibroin filaments are bonded together by sericin, a second protein that acts as a natural adhesive. Sericin is more hydrophilic than fibroin and serves two principal biological functions: it reinforces the structural integrity of the cocoon and provides a protective barrier against environmental stressors and potential predators (Altman et al., 2003; Altman & Farrell, 2022).

While fibroin is well known for its biocompatibility and has been widely investigated for biomedical and tissue-engineering applications, sericin differs significantly in its biological properties. Owing to its high serine amino acid content, sericin can exhibit immunogenic and, in certain contexts, cytotoxic effects. Consequently, sericin is typically removed through a degumming process during silk preparation for medical and high-quality textile applications (Altman et al., 2003; Altman & Farrell, 2022).

The major silk-producing countries in the world are China, India, Uzbekistan, Brazil, Japan, the Republic of Korea, Thailand, Vietnam, the DPR Korea, and Iran. The global silk production in 2024 was around 100,000 mt (International Sericultural Commission, 2025).

Traditionally, silk has been closely associated with premium apparel and refined home décor, including draperies, upholstery, and bedding, where it provides both aesthetic elegance and durability (Tran et al., 2023). Beyond decorative uses, silk's unique mechanical and biochemical properties have enabled its integration into diverse industrial, technological, and biomedical domains.

DOI: 10.1201/9781003639176-9

In biomedicine, silk's biocompatibility, flexibility, and mechanical strength have made it an important material for medical applications. It has long been employed in non-absorbable surgical sutures, offering reliable wound closure with minimal tissue reaction, and in wound-care materials that provide structural support while promoting healing (Croft et al., 2022). The protein composition of silk and its relatively low immunogenicity have further stimulated extensive research into silk-based scaffolds for tissue engineering, regenerative medicine, and drug-delivery systems (Tran et al., 2023).

Silk's combination of high tensile strength and low weight has also underpinned historical and modern engineering applications. Historically, it was essential for parachutes and artillery gunpowder bags, where strength and controlled performance were critical. In contemporary contexts, silk is found in specialized products such as high-end bicycle tires, where it improves flexibility and ride quality. Additionally, silk proteins are incorporated into personal-care products, enhancing moisture retention and imparting a smooth feel to skin and hair, illustrating the material's versatility across consumer and industrial markets (Tran et al., 2023).

Recent decades have seen significant advances in silk-based materials through the reverse engineering of silkworm fibers. Native silk can now be converted into regenerated fibroin solutions, which are fabricated into diverse material formats—including films, fibers, sponges, hydrogels, and coatings—and stabilized by reformation of non-covalent β-sheet structures. This approach restores mechanical strength while preserving biocompatibility and enabling controlled degradation (Tran et al., 2023). Further enhancements through covalent crosslinking and plasticization strategies have increased elasticity, toughness, and flexibility, allowing silk to be processed with modern manufacturing techniques such as photolithography, digital light processing, and extrusion-based 3D printing, achieving precise control over micro- to nanoscale structures.

As a result, silk's application spectrum has expanded dramatically. Engineered silk is being explored for ocular prostheses, bioadhesives, and tissue-engineering matrices, as well as for environmentally friendly technologies including biodegradable electronics, such as LEDs and batteries. Its mechanical resilience and skin compatibility support emerging uses in wearable and implantable sensors, while its biodegradability positions it as a sustainable alternative for bioplastics. Collectively, these developments underscore silk fibroin's emergence as a next-generation biomaterial at the intersection of biology, materials science, and sustainable engineering (Tran et al., 2023).

9.1.2 Origin and Domestication of *Bombyx mori*

Silk production—known as sericulture—is a specialized branch of agriculture centered on the rearing of the domesticated silkworm, *B. mori*. This species of moth is native to East Asia and was domesticated in China from its wild progenitor, *Bombyx mandarina*, with which it can still hybridize under experimental and natural conditions. Molecular genetic analyses indicate that the flightless domesticated form of *B. mori* diverged from *B. mandarina* approximately 4600 years ago in China, consistent with archaeological evidence for early silk use in Neolithic societies (Arunkumar et al., 2006).

Among domesticated organisms, *B. mori* represents an extreme case of human dependence. Unlike most livestock species, which retain some capacity for survival in feral environments, *B. mori* is entirely reliant on human intervention for feeding, reproduction, and dispersal (Torres Mc Cook et al., 2021). Selective breeding over millennia has produced a pale, flightless insect with reduced sensory acuity and an impaired ability to escape predators. As a consequence, *B. mori* is widely regarded as the world's only fully domesticated insect species (Altman & Farrell, 2022). Its life cycle—from egg incubation to larval feeding, cocoon spinning, and controlled mating—is managed entirely within human-designed production systems.

Genomic research has refined this historical framework by identifying the primary domestication center in the middle and lower Yellow River basin, where early agricultural communities integrated mulberry cultivation, silkworm rearing, and textile production into a cohesive sericultural system (Zhang et al., 2024). Whole-genome resequencing demonstrates that domesticated *B. mori* forms a distinct lineage nested within Chinese *B. mandarina*, with strong selection signals in genes associated with silk gland growth, metabolism, chemosensory perception, stress tolerance, and reproductive behavior. Notably, despite clear domestication signatures, genetic diversity in *B. mori* remains comparatively higher than in many other domesticated animals, suggesting that improvement likely drew from a relatively large founding population and avoided a severe bottleneck. This pattern supports a model of sustained, directional selection acting on standing variation rather than rapid fixation from a narrow genetic base (Arunkumar et al., 2006).

Domestication profoundly reshaped both the biology and behavior of the silkworm. Relative to its wild ancestor, *B. mori* has lost the ability to fly, exhibits diminished responsiveness to environmental stimuli, and tolerates dense rearing conditions—traits advantageous under human management but disadvantageous in natural ecosystems. Long-term captive breeding has also reduced olfactory acuity and modified pheromone-mediated behaviors, making reproduction largely dependent on controlled breeding (Bisch-Knaden et al., 2014). Concurrent selection for traits such as larger cocoons, longer and more uniform filaments, rapid growth, and synchronized development further separated domesticated silkworms from wild populations. Collectively, these evolutionary changes have transformed *B. mori* into an obligate domestic insect that is effectively unable to survive without human care, exemplifying one of the most complete cases of domestication among invertebrates (Bisch-Knaden et al., 2014).

The nutritional ecology of silkworms is highly specialized. The larvae feed almost exclusively on the leaves of mulberry trees, particularly *Morus alba*, commonly known as white mulberry, a species native to China and widely cultivated across Asia. The large-scale cultivation and maintenance of mulberry plantations is therefore foundational to sericulture. Mulberry leaf yield directly determines silk output; on average, 1 mt of mulberry leaves produces approximately 25–30 kg of fresh cocoons under standard rearing conditions (Altman & Farrell, 2022).

9.1.3 The Silk Road and Global Dispersal of Sericulture

For many centuries, China maintained an effective monopoly on silk production, carefully guarding the biological and technical knowledge required for rearing the domesticated silkworm, *B. mori*. As sericulture intensified within northern and central China, silk became one of the most valuable commodities in Eurasian trade. Its desirability as a lightweight, high-value textile helped catalyze the development of the Silk Road, the vast trans-Eurasian network that facilitated the exchange not only of goods, but also of technologies, religious ideas, artistic traditions, and agricultural practices (Maro & Török, 2022).

From its Chinese heartland, sericulture spread gradually through processes of diplomatic exchange, migration, and technological diffusion. By the late centuries BCE and around the turn of the Christian Era, knowledge of mulberry cultivation and silkworm rearing had reached the Korean peninsula and the Japanese archipelago, where it was incorporated into emerging state systems through tributary relations and cultural exchange with China (Gong et al., 2016). By the early centuries CE, sericulture had also taken root in parts of the Indian subcontinent, where favorable climatic conditions and the availability of mulberry species supported further regional adaptation and diversification of silk production (Maro & Török, 2022).

Although China sought to protect the "secret of silk," the biological foundations of sericulture—silkworm eggs and mulberry germplasm—eventually moved westward. By the first millennium CE, silk textiles were widely traded across Central Asia and the Middle East. A pivotal moment occurred in the sixth century CE, when silkworm eggs and mulberry seeds were reportedly smuggled into the Byzantine Empire (Feng, 2020). This transfer enabled the establishment of independent silk production in the eastern Mediterranean, breaking China's long-standing monopoly and embedding sericulture within the economic systems of Byzantium and neighboring regions (Maro & Török, 2022).

In subsequent centuries, repeated waves of introduction and adaptation extended the geographic range of *B. mori*. Sericulture flourished in Persia and across parts of the Middle East, later spreading into southern Europe—particularly Italy, France, and Spain—where regional silkworm lineages were selected under distinct climatic regimes and disease pressures (Altman & Farrell, 2022). These European centers became major contributors to silk textile innovation during the medieval and early modern periods.

Each stage of dispersal involved not only the transport of *B. mori* strains, but also the relocation of specific mulberry cultivars and rearing technologies. Over time, local environmental conditions imposed selective pressures on silkworm populations, shaping traits such as voltinism (the number of generations produced per year), thermal tolerance, growth rate, and resistance to pathogens (Liang et al., 2025). Across roughly two millennia, these processes generated a mosaic of geographically differentiated *B. mori* populations and management systems. The result is a globally distributed yet genetically structured domesticated species, embedded in diverse agroecological contexts and associated with distinct host genetic backgrounds and microbial communities that together underpin modern sericultural diversity (Willerslev & Meltzer, 2021).

9.1.4 Contemporary Geographic Distribution of Silkworm and Rearing Systems

The domesticated silkworm survives today exclusively within human-managed environments. After millennia of artificial selection and dependence on cultivated mulberry and controlled breeding, no self-sustaining feral populations of the domesticated form are known to exist. Although the species is effectively cosmopolitan within the global sericulture sector, its distribution remains tightly constrained by two primary factors: the agroecological suitability for mulberry cultivation and the presence of technical infrastructure for egg production, disease control, cocoon processing, and silk reeling.

Modern sericulture is overwhelmingly concentrated in Asia (Warwick et al., 2024). China and India together account for the majority of global cocoon and raw silk production, supported by extensive national breeding programs, certified disease-free egg (DFE) production centers, and vertically integrated supply chains. These countries maintain a broad spectrum of rearing systems, ranging from smallholder, household-based production—often integrated into diversified farming systems—to highly industrialized operations with controlled-temperature rearing houses, mechanized feeding systems, and standardized hybrid strains. Such institutional depth allows for systematic strain improvement, including selection for yield, voltinism, thermal tolerance, and pathogen resistance (Dutta et al., 2025).

Beyond these two dominant producers, sericulture continues in several other Asian countries, including Japan, Thailand, Vietnam, and South Korea. In many of these regions, production has shifted toward high-value niche markets, heritage textiles, or specialty silk products. In parallel, some systems are increasingly embedded within circular bioeconomy frameworks, in which silkworm pupae—traditionally a byproduct of reeling—are valorized as food, animal feed, or sources of bioactive compounds.

Outside Asia, sericulture persists at smaller scales but remains economically and scientifically significant. Brazil is the leading silk producer in the Western Hemisphere, with organized contract-farming systems linked to industrial reeling facilities. Additional pockets of production survive in parts of southern and eastern Europe—particularly in Mediterranean climates historically associated with mulberry cultivation—as well as in Australia. In these contexts, *B. mori* is used not only for commercial silk production but also as a model organism in genetics, physiology, education, and biotechnology research.

Across this geographically diverse range, rearing systems differ substantially in climate regimes, mulberry cultivars, artificial diet formulations, strain selection strategies, housing design, and hygiene protocols. These environmental and managerial variables impose distinct physiological and ecological pressures on *B. mori*. Emerging research suggests that such regionally structured conditions influence patterns of gut microbiome composition, diversity, and temporal stability—factors that may affect larval growth, immunity, and silk yield (Alders et al., 2021).

Taken together, contemporary sericultural systems can be understood as parallel "natural experiments" in domestication and managed evolution. Each regional production ecology represents a distinct configuration of host genotype, diet, microbial exposure, and human intervention. These structured differences provide a powerful comparative framework for investigating host–microbe co-adaptation and for engineering the silkworm holobiont to enhance resilience, productivity, and sustainability in a changing global climate.

9.2 *BOMBYX MORI* AS A HOLOBIONT

9.2.1 Economic and Experimental Significance

B. mori serves today both as a foundation of global sericulture and as a leading experimental system for research in lepidopteran biology, genetics, and biotechnology. Its millennia-long domestication, capacity for large-scale rearing on fresh mulberry leaves or standardized artificial diets, and well-defined, synchronous developmental stages make it exceptionally amenable to laboratory study and industrial use. The availability of a fully sequenced and continually refined reference genome—including near telomere-to-telomere assemblies—together with dense genetic linkage maps and extensive mutant collections has enabled precise analyses of genes controlling silk synthesis, immune function, development, and behavior (Marzoli et al., 2022).

Within Lepidoptera, *B. mori* occupies a uniquely strategic role as a genetically tractable proxy for investigating traits shared with agriculturally important pests and pollinators that are far more difficult to rear or manipulate. Modern molecular toolkits—such as piggyBac-mediated germline transformation, CRISPR-based genome editing, and RNA interference (RNAi)—are now routine, enabling targeted functional studies and positioning silkworm larvae as efficient bioreactors for recombinant proteins and silk-derived biomaterials (Kawamoto et al., 2019). Complementing these advances, well-established infection models involving bacteria, viruses, fungi, and toxins have made the species a versatile platform for exploring host–pathogen interactions, innate immunity, and xenobiotic responses in a system that is both ethically and logistically more accessible than mammalian models (Wang et al., 2022).

9.2.2 Conceptualizing the Silkworm–Microbiome Holobiont

Moving beyond a host-centric view, the silkworm can be understood as a holobiont in which the gut microbiome functions as an auxiliary genomic system that profoundly shapes physiology and performance. This microbial community participates in nutrient assimilation, detoxification, immune modulation, and metabolic signaling, collectively influencing larval growth and the efficiency of silk biosynthesis. Comparative metagenomic and metabolomic analyses consistently show that resident bacteria expand digestive capacity by breaking down complex plant polysaccharides, contributing to amino acid and vitamin pools, and modulating metabolic fluxes that feed directly into silk protein synthesis. In this sense, the microbiome operates as a dynamic biochemical interface between diet and host, integrating environmental inputs with host developmental programs (Bordenstein & Theis, 2015).

Functional studies further illustrate how microbial metabolism extends the adaptive range of the silkworm. Gut symbionts are capable of transforming diverse plant secondary compounds into less harmful or even utilizable forms, thereby broadening dietary tolerance and stabilizing physiological homeostasis. A striking example is the microbial conversion of prenylated isoflavone (PIF) toxins from alternative host plants into glycosylated derivatives with markedly reduced toxicity, demonstrating a clear metabolic division of labor between host and microbes (Yuan et al., 2023). Such interactions highlight the holobiont as a coordinated ecological unit in which microbial processes directly influence growth trajectories, stress resilience, and cocoon quality, underscoring the importance of integrating microbiome dynamics into future studies of silkworm biology and sericultural innovation (Manzanera, 2025).

9.2.3 Microbial Taxonomic Composition and Ecological Dynamics of the Silkworm Gut

Across genetic backgrounds, feeding regimes, and husbandry conditions, the silkworm gut microbiome displays marked ecological plasticity while retaining a recognizable core taxonomic scaffold. Continuous microbial input from mulberry leaves, artificial diets, and environmental reservoirs introduces variability, particularly among low-abundance taxa; however, dominant lineages recur across independent cohorts, indicating strong host-associated filtering and partial species specificity in community assembly (Xin et al., 2024). Longitudinal analyses show pronounced developmental turnover, with early instars often enriched in Proteobacteria and later stages showing relative increases in Firmicutes, yet the same major phyla and several recurrent genera persist across life stages, strains, and experimental contexts (Hou et al., 2025). Multivariate ordination of both DNA-based (total) and RNA-based (active) communities consistently reveals clustering primarily by host species and developmental stage rather than by gut microhabitat, underscoring host identity and ontogeny as dominant structuring forces. Comparative studies with mulberry-feeding relatives such as *Acronicta major* and *Diaphania pyloalis* further demonstrate species-specific active bacterial signatures despite shared diets, supporting the view that physiological filters and immune regulation impose a conserved taxonomic template on a variable environmental species pool. Nevertheless, perturbations—including antibiotics, artificial diets, and pathogen exposure—can rapidly shift richness and evenness, highlighting considerable ecological flexibility within this core framework (Chen, Du, et al., 2018).

At finer resolution, metagenomic and 16S rRNA surveys repeatedly identify four principal bacterial phyla—Proteobacteria, Firmicutes, Actinobacteria, and Bacteroidetes—together accounting for roughly 90%–95% of classified sequences in *B. mori* (Chen, Yu, et al., 2018). Within Proteobacteria, genera such as *Pseudomonas*, *Acinetobacter*, *Aureimonas*, *Methylobacterium*, and *Rhizobium* are frequently linked to early gut colonization, nitrogen transformations, and degradation of mulberry-derived aromatic compounds, with *Pseudomonas* and *Acinetobacter* sometimes exceeding 30% and 25% relative abundance in early instars (Chen, Du, et al., 2018). Firmicutes—including *Enterococcus*, *Staphylococcus*, *Lactobacillus*, and *Weissella*—are associated with carbohydrate fermentation, short-chain fatty acid production, and immune modulation, often becoming dominant under artificial diets or physiological stress, occasionally comprising more than half of total reads (Chen et al., 2018c). Actinobacteria such as *Microbacterium*, *Rhodococcus*, *Dietzia*, and *Propionibacterium* contribute to polymer and xenobiotic degradation and frequently increase under challenging dietary or environmental conditions, whereas Bacteroidetes taxa including *Bacteroides*, *Parabacteroides*, and sphingolipid-associated lineages support polysaccharide breakdown and gut barrier stability but may decline under heavy antibiotic pressure or prolonged artificial feeding. Additional phyla—Cyanobacteria, Verrucomicrobia, and Chloroflexi—occur sporadically at low abundance and likely represent transient environmental inputs (Nalini & Prakash, 2020). Parallel surveys of the gut mycobiota show overwhelming dominance by Ascomycota and Basidiomycota, typically exceeding 99% of fungal reads (Gaspar et al., 2025), with many taxa such as *Cladosporium* and *Hannaella* traceable to the mulberry phyllosphere, emphasizing strong plant–gut coupling (Bashir et al., 2022). Transcript-based profiling indicates that subsets of fungi—including *Fusarium*, *Cryptococcus*, and *Clonostachys*—remain metabolically active, and interkingdom perturbation experiments reveal reciprocal effects on bacterial diversity, demonstrating that the silkworm gut microbiome is an integrated multi-kingdom network shaped jointly by diet, plant ecology, and host immune regulation (Huang et al., 2024).

9.4 DEVELOPMENTAL DYNAMICS AND MICROBIAL SUCCESSION IN *BOMBYX MORI*

Microbial colonization in *B. mori* is a continuous, developmentally staged process that begins at the egg surface and progresses through all five larval instars, shaped by both vertical transmission from the mother and environmental acquisition. Newly hatched larvae encounter maternal and environmental microbes on the chorion, mulberry leaves, and rearing substrates, establishing the first inoculum of gut bacteria. Several early-stage dominant genera, including *Pseudomonas* and *Escherichia*, are detectable on the egg surface and appear to persist through initial feeding, highlighting the importance of maternal and environmental seeding in establishing the larval microbiome (Kalpana et al., 1994). During the first and second instars (L1–L2), the gut remains small and physiologically immature, with rapid food transit and simple feeding patterns that favor fast-growing Proteobacteria capable of exploiting easily digestible nutrients and tolerating fluctuating gut conditions. This early colonization sets the stage for subsequent microbial succession and functional maturation of the gut ecosystem.

Across development, microbial community dynamics display a clear biphasic pattern. Early instars are dominated by Proteobacterial genera such as *Pseudomonas* and *Acinetobacter*, which together may account for over half of the total community in some cohorts. As larvae progress to later instars (≥L3), the microbiome gradually shifts toward taxa such as *Methylobacterium*, *Aureimonas*, *Thermus*, and *Geobacillus*, which are associated with the processing of more complex plant metabolites, maintenance of gut homeostasis, and stabilization of host–microbe interactions (Aljuaid et al., 2026). In a detailed longitudinal study, *Pseudomonas* and *Acinetobacter* represented roughly 31.9% and 28.2% of sequences in early stages, respectively, but their abundance declined sharply in late instars, coinciding with increases in *Methylobacterium* (19.1%) and *Aureimonas* (9.7%) as the gut matured (Celiker & Gore, 2014). Principal coordinate analyses reinforce this two-phase model, with microbial communities clustering distinctly into early and late instar groups, demonstrating that developmental stage is a primary determinant of gut microbiome composition. These shifts correlate with dramatic increases in mulberry consumption, gut expansion, and maturation of host digestive and immune functions, suggesting that changes in gut pH, oxygen gradients, peritrophic matrix

(PM) development, and microbial interactions collectively drive microbial succession and functional specialization throughout silkworm ontogeny.

9.4.1 Maturation, Peak Diversity, and Sexual Dimorphism in *Bombyx mori* Gut Microbiota

Bacterial diversity and community evenness in *B. mori* generally increase progressively from early to mid-larval instars, reaching their peak around the fifth instar, a period characterized by maximal feeding activity, rapid silk gland development, and heightened functional redundancy within the gut microbiome (Tan & Bautista, 2022). As larvae expand in size, they ingest larger quantities of mulberry leaves spanning different developmental stages, exposing themselves to a wider spectrum of phyllosphere-associated microbes and secondary plant metabolites. This environmental input provides new colonization opportunities while host-mediated filtering mechanisms select for taxa that confer metabolic or immunological benefits. Across multiple strains, mature fifth-instar larvae consistently harbor co-occurring genera such as *Pseudomonas*, *Sphingomonas*, *Delftia*, *Methylobacterium*, and *Acinetobacter*, with alpha diversity and evenness positively correlating with larval size and developmental progression. Toward the end of the fifth instar, as feeding declines and prepupation gut remodeling begins, overall diversity often experiences a transient reduction, reflecting physiological restructuring and shifts in microbial niches prior to metamorphosis.

Emerging data also reveal pronounced sexual dimorphism in gut microbial communities, particularly during late larval stages. Fifth-instar males frequently show higher relative abundances of *Enterococcus*, whereas females display greater richness and higher relative abundances of genera including *Delftia*, *Aurantimonas*, and *Staphylococcus* (De Lisle et al., 2022). These sex-specific differences may be driven by variations in feeding intensity, growth rate, hormonal regulation, or immune responsiveness, all of which shape the gut environment and selectively influence microbial colonization. Supporting this notion, experimental infections with *B. mori* cytoplasmic polyhedrosis virus (BmCPV) demonstrate that the trajectory of microbiome changes under stress differs between sexes, suggesting that males and females respond distinctively to environmental or pathogenic perturbations (Fei et al., 2025). While the precise mechanisms underlying these sex-based filters remain to be elucidated, the evidence emphasizes that host sex constitutes a meaningful axis of variation in silkworm–microbiome interactions and should be explicitly considered in future studies, particularly when designing microbiome-targeted interventions for optimized sericulture.

9.5 EVOLUTIONARY PERSPECTIVES AND DOMESTICATION EFFECTS

9.5.1 Comparisons with Wild Mulberry-Feeding Relatives

From an evolutionary and ecological perspective, domesticated *B. mori* exhibits pronounced differences in both host genetics and gut microbiome composition compared to wild mulberry-feeding moths, even when exposed to similar diets. Comparative analyses with species such as *A. major* and *D. pyloalis* reveal that despite consuming the same host plant, each species maintains a distinct, host-specific gut microbiota composed of a taxonomically constrained set of microorganisms. Clustering analyses of both DNA-based (total) and RNA-based (active) bacterial communities consistently distinguish *B. mori* from its wild relatives, with only a limited subset of genera—most notably *Pseudomonas* and *Acinetobacter*—shared across all hosts. This pattern highlights strong host species specificity in microbial community assembly, suggesting that physiological factors, immune regulation, and life-history traits exert stronger influence than diet alone in shaping gut microbiota (Lin et al., 2014).

Wild mulberry-feeding moths also harbor symbionts and endosymbionts largely absent in domesticated *B. mori*, including diverse *Wolbachia* lineages and fungal taxa such as *Botrytis*, which appear species-specific. In contrast, *B. mori* maintains a highly diverse yet distinctive gut microbiota dominated by four core bacterial phyla—Proteobacteria, Firmicutes, Actinobacteria, and Bacteroidetes—but composed primarily of host-specific operational taxonomic units (OTUs) and species-level lineages (Chen, Yu, et al., 2018). The silkworm's mycobiota similarly reflects this divergence: while *Cladosporium* and *Hannaella* are prevalent across multiple hosts, *A. major* carries unique fungal taxa such as *Botrytis*, whereas *B. mori* and *D. pyloalis* share partially overlapping fungal communities enriched with host-specific components like *Fusarium*, *Cryptococcus*, and *Clonostachys*. These observations indicate that, despite thousands of years of domestication and human-mediated management, the silkworm has not converged toward the microbial profiles of its wild relatives. Instead, *B. mori* has evolved and maintained a distinctive, host-filtered microbial assemblage shaped by both evolutionary history and the physiological constraints of domestication (Kawakita et al., 2010).

9.5.2 Domestication Bottlenecks, Host Genomics, and Microbiome Co-Evolution

The domestication of *B. mori* has been shaped by repeated population bottlenecks and strong directional selection, resulting in significant genomic divergence among strains and between domesticated silkworms and their wild ancestor, *B. mandarina*. Genome-wide resequencing of wild and domesticated populations reveals clear differentiation, yet domesticated silkworms retain surprisingly high nucleotide diversity, consistent with a relatively rapid domestication process involving a large number of founding individuals (Xia et al., 2009). Selection signals are concentrated in several hundred candidate genes enriched for expression in silk glands, midgut, and testes, encompassing loci involved in fibroin and sericin synthesis, immunity, chemosensation, and hormonal regulation. These genomic modifications not only optimize silk production and behavior but also reshape gut physiology, epithelial barrier function, and immune signaling, collectively altering the ecological and selective landscape experienced by the gut microbiota.

Recent anthropogenic selection, particularly in strains adapted to artificial diets, has introduced additional layers of genomic and microbiome interaction. Whole-genome and transcriptome analyses of lines such as Guican No.5 reveal elevated single nucleotide polymorphism (SNP) densities and adaptive changes in genes related to xenobiotic metabolism, digestive enzyme activity, and stress response compared with traditional mulberry-fed strains (Dong et al., 2017). Many of these genes retain close homology to *B. mandarina* alleles, suggesting that domesticated silkworms have repeatedly recruited ancestral variation to meet novel environmental challenges. This genomic plasticity enables *B. mori* to capitalize on both ancestral genetic diversity and contemporary microbial inputs, facilitating adaptation to high-density rearing, artificial diets, and chemical exposures. In turn, changes in host diet, immunity, and gut physiology exert selective pressures on the gut microbiota, promoting the proliferation of taxa capable of metabolizing new substrates or tolerating altered pH, redox, and nutrient conditions. These reciprocal interactions illustrate co-evolutionary dynamics at the holobiont level, where host genomic evolution and microbial community adaptation are tightly intertwined, collectively influencing growth, metabolism, and silk production (Aljuaid et al., 2026).

9.6 PHYSIOLOGICAL FILTERS OF GUT COLONIZATION

9.6.1 Physicochemical Determinants of Gut Microbial Assembly

The physicochemical environment of the *B. mori* gut exerts powerful selective pressures on microbial colonization, shaping community composition and functional potential. The midgut is among the most alkaline gut environments known in animals, with pH values typically ranging from 10.5 to 11.0, while the anterior and posterior regions remain moderately alkaline (pH 8.0–9.0) (Liang et al. 2018) ("At the right time in the right place," 2021). This extreme alkalinity enhances solubilization of leaf proteins, saponins, and other plant secondary metabolites, but it simultaneously excludes many environmental microbes that cannot tolerate such conditions. Consequently, the gut favors alkaliphilic or alkali-tolerant taxa and constrains microbial metabolic pathways. Host digestion is also adapted to this unusual chemical milieu: midgut epithelial cells primarily absorb amino acids and small peptides, and many digestive enzymes—including proteases, amylases, and lipases—function optimally under high pH, further filtering which microbes can successfully colonize.

In addition to pH, the gut environment varies along other chemical gradients, including oxygen tension, redox potential, and concentrations of plant-derived phenolics, creating a heterogeneous landscape of microhabitats. Diet composition—fresh mulberry leaves versus artificial feeds—modulates these gradients by altering carbohydrate, fiber, and micronutrient availability, which in turn selects for distinct bacterial consortia. Environmental stressors such as heavy metals, pesticides, and antibiotics can disrupt these chemical balances, often reducing overall diversity while favoring stress-tolerant taxa, such as lactic acid bacteria, illustrating how chemical constraints interact with host physiology to filter potential colonists.

9.6.2 Peritrophic Matrix Turnover and Molting-Driven Microbial Reset

The PM in *B. mori* functions as a semi-permeable barrier, protecting the gut epithelium from mechanical damage, plant-derived toxins, and pathogens while allowing nutrient absorption. Secreted by midgut cells, the PM increases in thickness as larvae grow, though its permeability decreases from the anterior to posterior regions, creating functional gradients in nutrient and particle passage (Liu et al., 2019). This structure compartmentalizes digestion and limits direct contact between microbes and epithelial cells, influencing microbial adhesion, proliferation, and interaction with the host.

During molting, the PM is periodically renewed, and gut contents are partially shed, creating a temporary reset of the luminal environment. This process often leads to transient reductions or shifts in microbial load and composition, contributing to stage-specific microbiomes observed across larval instars (Zhang et al., 2021). Ultrastructural studies show that the PM surface is dense and smooth during active feeding but becomes loose and coarse during molting and the immediate post-molting period, correlating with altered expression of peritrophins and chitin synthase. These cyclical changes, combined with reduced feeding during molts, alter nutrient flux and microbial attachment surfaces, likely underpinning the "two-phase" early–late instar separation in gut community composition (Hodžić et al., 2025).

9.6.3 Spatial versus Temporal Heterogeneity in Microbiome Assembly

Despite temporal dynamics, spatial heterogeneity along the gut appears limited in *B. mori*. Surveys show broadly similar bacterial composition across the foregut, midgut, and hindgut, with only minor differences in richness (foregut ~107 OTUs, midgut ~80 OTUs, hindgut ~74 OTUs at standardized sequencing depth) (Chen, Yu, et al., 2018). Dominant taxa at both phylum and genus levels are largely uniform, contrasting with the pronounced regional specialization observed in vertebrate guts and some other insects. This spatial uniformity highlights the primacy of temporal ecological factors—such as developmental stage, feeding status (active feeding versus molting or starvation), and environmental perturbations—over anatomical location in shaping microbial assembly. The simple, straight morphology of the silkworm gut and rapid larval development emphasize that interventions aimed at modulating the microbiome, including probiotics or dietary changes, are likely most effective when timed appropriately during development rather than targeting specific gut compartments (Chen et al., 2018d).

9.7 FUNCTIONAL METAGENOMICS OF NUTRIENT ASSIMILATION

9.7.1 Cellulose Degradation, Carbon Cycling, and Short-Chain Fatty Acids

Functional metagenomics and metabolomics have provided increasingly detailed insights into how the *B. mori* gut microbiome contributes to nutrient assimilation and energy homeostasis. Because *B. mori* lacks a complete repertoire of endogenous cellulases, it relies heavily on gut-associated bacteria such as *Enterococcus mundtii* and various *Bacillus* species to degrade complex plant cell wall polysaccharides (Puig-Castellví et al., 2023). These microbes hydrolyze cellulose, hemicellulose, and pectin into fermentable sugars, which are subsequently metabolized to short-chain fatty acids (SCFAs) including acetate, propionate, and butyrate. SCFAs serve as readily absorbable carbon sources, supplying energy for larval growth, silk gland activity, and cocoon formation. In addition, fermentation produces metabolites that can modulate gut pH, redox potential, and interspecies microbial interactions, creating feedback loops that reinforce community stability and metabolic efficiency. This cooperative carbon cycling underscores the functional interdependence between host and microbiota in enabling efficient utilization of a leaf-based diet (Weimer, 2022).

9.7.2 The Phenylalanine Axis and Silk Gland Development

Beyond general carbon metabolism, recent studies highlight the importance of amino acid-centric microbial interactions in silk production. Phenylalanine, a key aromatic amino acid, has emerged as a critical hub linking gut microbial activity to silk gland development. Probiotic supplementation with *Bacillus subtilis* not only reshapes the gut microbiota but also elevates hemolymph phenylalanine concentrations by approximately threefold, concomitant with increased digestive enzyme activity, larval growth, and improved silk traits (Ren et al., 2025). Feeding phenylalanine directly reproduces many of these benefits, indicating that microbial regulation of this metabolic axis can enhance fibroin synthesis and silk gland development. These findings suggest that strategic manipulation of microbial populations capable of promoting aromatic amino acid availability represents a promising avenue for optimizing sericultural performance.

9.7.3 Nitrogen Metabolism, Urease Activity, and Amino Acid Provisioning

Complementary approaches using nanobiotechnology demonstrate that targeted modulation of nitrogen metabolism in the gut can further influence silk production. For example, Fe_3O_4@urease nanoconjugates have been shown to enrich nitrogen-metabolizing genera such as *Pantoea* and *Klebsiella*, resulting in substantial increases in hemolymph concentrations of histidine and arginine, two amino acids essential for fibroin synthesis (Herman et al., 2025). This targeted microbial engineering illustrates how manipulating nitrogen cycling pathways in the gut can augment amino acid provisioning, directly linking microbial metabolism to silk yield and quality. Collectively, these studies underscore the multifaceted contributions of the silkworm gut microbiome to carbon and nitrogen cycling, nutrient assimilation, and silk production, highlighting a sophisticated holobiont-level network in which host physiology, diet, and microbial function are tightly integrated.

9.8 DIET, MICROBIOME, AND PERFORMANCE IN *BOMBYX MORI*

9.8.1 Mulberry Leaves versus Artificial Diets

Diet is a principal determinant of gut microbiota composition in *B. mori*, and comparisons between natural mulberry leaves and artificial diets have yielded critical insights into host–microbe interactions. Artificial diets, designed for year-round, scalable rearing and controlled nutrient delivery, offer logistical advantages but often compromise larval physiology and silk output (Park et al., 2025). Across multiple studies, larvae fed exclusively on artificial diets exhibit reduced growth rates, delayed development, lowered immune competence, and diminished silk production. These phenotypic deficits correlate strongly with altered gut microbiota, including decreased alpha diversity, reduced evenness, and a simplification of community structure dominated by a few high-abundance taxa, accompanied by a loss of numerous low-abundance lineages that may provide functional benefits (Park et al., 2025).

Artificial feed promotes enrichment of lactic acid bacteria such as *Enterococcus* and *Weissella*, reflecting the high fermentability and abundance of simple carbohydrates in these diets. In some strains, *Enterococcus* abundance rises 10–60-fold relative to mulberry-fed counterparts, coinciding with declines in genera like *Clostridium*, *Pseudomonas*, and *Escherichia*, which are characteristic of natural mulberry feeding (Li et al., 2023). This lactic acid-driven proliferation can acidify the gut, induce dysbiosis, increase susceptibility to opportunistic pathogens, and alter metabolic functional pathways such as phosphotransferase systems and starch/sucrose metabolism. In contrast, mulberry-fed larvae maintain richer, more stable microbial communities, with higher representation of Proteobacteria and Actinobacteria, including genera such as *Acinetobacter*, *Achromobacter*, *Methylobacterium*, and *Rhodococcus*. These taxa are implicated in efficient nutrient utilization, vitamin biosynthesis, and robust pathogen resistance, highlighting the functional advantages of natural diets (Ren et al., 2025).

9.8.2 Transitional Regimens and Microbiome Resilience

Transitional feeding strategies—where larvae are first reared on artificial diets and subsequently switched to mulberry leaves—demonstrate the plasticity and resilience of the silkworm gut microbiome. Microbial modeling across gut and fecal samples shows that mulberry introduces a broader spectrum of environmental microbes than artificial feed, rapidly increasing bacterial diversity and reducing the overrepresentation of lactic acid bacteria. Transitional groups exhibit intermediate diversity and selective enrichment of taxa such as Methylobacteriaceae, *Microbacterium*, and *Rhodococcus*, which likely facilitate detoxification of secondary metabolites, restore nitrogen and vitamin metabolism, and stabilize gut homeostasis (Xin et al., 2024).

Longitudinal analyses reveal that moving larvae from artificial diets to mulberry increases overall microbial richness, lowers *Enterococcus* and *Weissella* abundance, and elevates beneficial taxa such as *Achromobacter* and *Rhodococcus*. The timing of dietary transitions, for instance, during the third versus fifth instar, significantly influences growth, immune function, and silk yield. Notably, transitions during the fifth instar can enhance tolerance to thermal stress without reducing body weight or eclosion success, suggesting that even late-stage diet switches confer physiological advantages via microbiome remodeling (Wang et al., 2024). The repeated enrichment of *Lactobacillus* and *Weissella* in artificial and transitional diets positions these genera as promising candidates for host-specific synbiotic formulations, particularly when paired with *Achromobacter* and *Rhodococcus* strains that support detoxification, nutrient metabolism, and overall gut stability in artificial feeding systems (Wang et al., 2024).

This body of work underscores the importance of diet-driven microbial management in sericulture, highlighting opportunities for synbiotic development, dietary optimization, and targeted microbiome engineering to improve larval performance, silk yield, and resilience under artificial rearing conditions.

9.9 DETOXIFICATION AND RESPONSES TO ENVIRONMENTAL STRESSORS

9.9.1 Microbial Detoxification of Plant Phytochemicals

The gut microbiome of *B. mori* plays a critical role in detoxifying bioactive compounds present in mulberry and alternative host plants, effectively expanding the larval capacity to tolerate otherwise harmful dietary metabolites. When silkworms are reared on *Cudrania tricuspidata* leaves, which are rich in PIFs, gut bacteria metabolize these potentially toxic compounds into glycosylated derivatives (GPIFs) that are substantially less harmful (Wang et al., 2024). Comparative chemical profiling of leaves versus feces demonstrates that PIFs largely vanish in excreta, replaced by GPIFs that are scarcely detectable in the plant itself, indicating that glycosylation occurs in situ within the gut. In vitro assays with gut-isolated strains such as *B. subtilis*, *Staphylococcus sciuri*, and *Enterobacter hormaechei* confirm their ability to glycosylate PIFs, providing mechanistic evidence that gut microbes are active mediators of this detoxification pathway (Yuan et al., 2023). Functionally, GPIFs show greatly reduced toxicity in both cell-based and in vivo assays, and probiotic supplementation with *B. subtilis* enhances larval growth and silk traits in *C. tricuspidata*-fed silkworms. These observations illustrate a clear division of labor within the holobiont: the host provides ingestion and absorption capacity, while the microbial community extends metabolic versatility, enabling exploitation of chemically complex or otherwise inhospitable diets (Yuan et al., 2023).

9.9.2 Heavy Metals, Metabolomic Shifts, and Microbial Tolerance

Exposure to heavy metals such as chromium and copper provokes pronounced shifts in both the taxonomic composition and metabolic activity of the silkworm gut microbiome. Chromium, incorporated at graded levels into artificial diets, reduces bacterial richness and evenness in a dose-dependent manner while altering Kyoto Encyclopedia of Genes and Genomes (KEGG)-predicted pathways linked to amino acid metabolism and stress response. Low to moderate chromium concentrations favor lactic acid bacteria such as *Weissella*, whereas thermophilic Actinobacteria—including *Thermus*, *Geobacillus*, *Dietzia*, and *Propionibacterium*—decline or disappear, highlighting selective pressures on metal-tolerant taxa (Shi et al., 2024). Metabolomic analyses link these compositional changes to shifts in lipid metabolism, antioxidant capacity, and stress-responsive metabolites, suggesting that chromium-induced dysbiosis is tightly coupled with disruption of host metabolic homeostasis.

Copper exposure generates a complementary but distinct profile: high cupric chloride levels dramatically reduce Bacteroidota (from ~50.4% to 23.5%) while elevating Firmicutes (from ~0.9% to 18.9%), with lactic acid bacteria and other stress-tolerant genera emerging as dominant responders (Rong et al., 2024). Network analyses under copper stress reveal that Proteobacteria and Firmicutes occupy central roles in the reconstructed microbial interaction web, suggesting these phyla stabilize community structure against metal-induced oxidative stress (Charalambous et al., 2022). Notably, low-level copper exposure has minimal effect, indicating a threshold response: a subset of gut microbes can tolerate and adapt to metal stress up to a point, beyond which both structure and function are extensively reorganized.

9.9.3 Pesticides, Microplastics, and Nanoecotoxicology

Contemporary agrochemical and environmental pollutants, including pesticides and micro/nanoplastics, further perturb the silkworm gut ecosystem. Laboratory studies show that pyrethroids and other insecticides can damage midgut epithelium, induce oxidative stress, and disrupt the abundance of beneficial microbes. When pesticides adhere to micro- or nanoparticles, their delivery and retention in the gut increase, exacerbating microbial disruption. Such co-exposure reduces overall microbial diversity, depletes key symbionts involved in nutrient metabolism and detoxification, and promotes proliferation of opportunistic or stress-tolerant taxa, leading to increased larval mortality and compromised silk yield.

The emerging discipline of microbiome-aware nanoecotoxicology emphasizes that the effects of nanoparticles and adsorbed chemicals must be assessed not only for direct host toxicity but also for their capacity to perturb microbial communities that buffer dietary toxins, support nutrient acquisition, and maintain immune homeostasis. These findings underscore that the silkworm gut microbiome functions as both a metabolic shield against plant-derived xenobiotics and a sensitive interface for environmental stressors, with critical implications for the design of safer sericulture practices and sustainable crop protection strategies (Zhang et al., 2025; Wang et al., 2024).

9.10 THE MICROBIOME–IMMUNE SYSTEM INTERFACE

9.10.1 Innate Immune Pathways and Microbial Recognition

The silkworm immune system exemplifies a finely tuned interface between host defense and microbial ecology, where innate immune recognition integrates microbial cues with protective responses. Pattern recognition receptors (PRRs), including peptidoglycan recognition proteins (PGRPs) and β-1,3-glucan recognition proteins (βGRPs), detect conserved microbial cell wall components and initiate signaling cascades that regulate the Toll, Immune deficiency (Imd), JAK/STAT, and RNAi pathways (Muhammad et al., 2024). In *B. mori*, Gram-positive bacteria and fungi predominantly activate the Toll pathway through lysine-type peptidoglycan and β-glucan detection, leading to the production of antimicrobial peptides (AMPs) such as defensins, cecropins, and moricins. Gram-negative bacteria trigger the Imd pathway via PGRP-LC/LE recognition of diaminopimelic acid peptidoglycan, whereas DNA viruses like *B. mori* nucleopolyhedrovirus (BmNPV) activate the JAK/STAT pathway, coordinating antiviral effectors with cellular stress responses. RNAi complements these defenses by providing sequence-specific inhibition of certain viruses and bacterial transcripts, expanding the host's antiviral and antibacterial arsenal (Ghani et al., 2025).

These pathways orchestrate hemocyte-mediated cellular immunity—including phagocytosis, nodulation, and encapsulation—together with AMP production and melanization, restricting pathogen proliferation while allowing mutualistic or commensal microbes to persist. Emerging studies in Lepidoptera indicate that chronic exposure to commensals can dynamically tune immune activity; for example, Imd signaling may be downregulated during late larval stages to prevent excessive immune activation and maintain microbial homeostasis. In *B. mori*, shifts in gut microbiota composition, whether due to probiotics, diet, or environmental perturbations, are accompanied by correlated changes in immune gene expression, reflecting a bidirectional relationship in which the immune system shapes microbial assemblages while microbiota feedback modulates host defense (Eleftherianos et al., 2021).

9.10.2 Probiotic Defense, Colonization Resistance, and Pathogen Inhibition

Probiotic and commensal bacteria play a central role in reinforcing gut barrier function and conferring colonization resistance. Species such as *Lactobacillus acidophilus* and *B. subtilis* compete with pathogens for epithelial attachment, secrete antimicrobial metabolites, and modify gut pH and digestive enzyme activity, collectively stabilizing the microbial community while improving host nutrition (Guan et al., 2025). Supplementation with *B. subtilis* enhances larval growth and cocoon quality, reduces opportunistic pathogens such as *Pseudomonas*, and upregulates digestive and immune genes. By modulating α-amylase, trypsin, and lipase activity and elevating midgut pH, these probiotics indirectly favor beneficial taxa while constraining acid-tolerant opportunistic species, creating a gut environment optimized for symbiont persistence and host health.

Experimental infection models further illustrate the protective capacity of gut symbionts. Lactic acid bacteria and *Bacillus* spp. suppress BmNPV replication through competitive exclusion, acidification, and modulation of antiviral pathways, reducing viral load and enhancing survival. Similarly, the commensal *Enterococcus faecalis* LX10 protects against the microsporidian pathogen *Nosema bombycis* via secretion of bacteriocins (EntLX/EntV-like peptides) and activation of host proteases such as gelatinase (GelE) and disulfide bond-forming proteins. Colonization with *E. faecalis* LX10 upregulates host immune genes, enhances reactive oxygen species production, boosts detoxification enzyme activity, and significantly lowers microsporidian burden. These findings exemplify the potential of microbiome-informed interventions—targeted probiotics, commensal enrichment, or engineered symbionts—as effective alternatives or complements to chemical pathogen control in sericulture (Suraporn et al., 2025).

Collectively, the *B. mori* holobiont demonstrates that host immunity and gut microbiota operate as an integrated unit, where immune pathways both regulate and are shaped by microbial communities. Leveraging this interface provides a roadmap for rationally designing disease management strategies, optimizing gut microbiome composition, and enhancing resilience against environmental and pathogenic stressors in silk production systems.

9.11 ABIOTIC PERTURBATIONS AND INTERKINGDOM INTERACTIONS

9.11.1 Thermal Stress and Fermentation Disruption

Abiotic stressors, particularly elevated temperatures, exert profound effects on the *B. mori* gut microbiome, which in turn influences host physiology and development. Short-term heat exposure—for instance, maintaining larvae at 37°C for several hours—can significantly perturb gut microbial composition, with disruptions persisting for days even after return to optimal rearing conditions. High temperatures (≈35–37°C and above) reduce the abundance of key carbohydrate-fermenting taxa, including SCFA-producing genera such as *Clostridium* and unclassified members of Ruminococcaceae, thereby diminishing the pool of gut-derived SCFAs that contribute to larval energy metabolism. Dominant genera shift in response: *Enterococcus* can decline by more than 100-fold within 96 hours post-stress, while opportunistic or resilient taxa such as *Acinetobacter*, *Escherichia–Shigella*, and *Anderseniella* often rebound at later time points (Puig-Castellví et al., 2023).

These microbial perturbations have cascading effects on the host. Loss of fermentative bacteria and lactic acid producers correlates with decreased body weight, slowed growth, and heightened sensitivity to additional environmental or pathogenic challenges, suggesting that heat stress indirectly compromises gut integrity, energy balance, and overall resilience via microbiome-mediated mechanisms. Although partial recovery of the microbiota occurs over time, transient dysbiosis during critical growth windows can leave lasting impacts on larval development, cocoon quality, and silk yield. These findings underscore that thermal stress acts not only directly on the host but also indirectly, by reshaping the microbial community that mediates digestion, nutrient assimilation, and gut homeostasis (Xin et al., 2024).

9.11.2 Antibiotics, Fungal Overgrowth, and Community Stability

Chemical perturbations such as antibiotic or antifungal exposure reveal the tightly interlinked, interkingdom dynamics of the silkworm gut microbiome. Antibiotic treatment—commonly applied via mulberry leaves with agents like ampicillin and streptomycin—leads to sharp reductions in bacterial loads and diversity, often triggering compensatory overgrowth of fungi, including opportunistic taxa such as *Aspergillus*. Conversely, antifungal treatment with agents such as amphotericin B suppresses fungi but also reduces bacterial richness and evenness, highlighting the reciprocal support that fungi provide for bacterial community stability (Li et al., 2022).

Network analyses of co-occurrence patterns demonstrate that removal of one microbial kingdom rewires the remaining community, altering interactions between gut residents and leaf-associated taxa. These results suggest that microbial stability in the silkworm gut relies on a balance between bacterial and fungal populations: bacteria limit overgrowth of opportunistic fungi, while fungi maintain ecological niches, substrates, and metabolic networks that sustain bacterial diversity (Lee et al., 2022). Disruption of these relationships—whether through indiscriminate antibiotic or antifungal use—can provoke dysbiosis, reduce host performance, and increase disease susceptibility. For sericulture, these findings emphasize the importance of microbiome-aware management strategies that preserve interkingdom balance, rather than targeting a single microbial component, in order to maintain larval health, growth, and silk productivity (Zhang et al., 2025).

9.12 METHODOLOGICAL ADVANCES IN SILKWORM MICROBIOME RESEARCH

9.12.1 From Culture-Dependent Studies to Multi-Omics Integration

Research on the *B. mori* gut microbiome has undergone a profound methodological evolution, moving from traditional culture-dependent techniques to integrated multi-omics approaches that capture both taxonomic breadth and functional potential. Early studies relied on plating gut contents to isolate culturable bacteria and fungi, revealing common genera such as *Bacillus*, *Staphylococcus*, *Enterococcus*, and *Cladosporium* (Muhammad et al., 2026). While informative, these approaches captured only a small fraction of the gut ecosystem, largely missing low abundance or fastidious taxa and providing limited insight into metabolic function. The advent of high-throughput amplicon sequencing (16S rRNA for bacteria, Internal Transcribed Spacer (ITS) for fungi) expanded this view, confirming that Proteobacteria, Firmicutes, Actinobacteria, and Bacteroidetes dominate bacterial communities while Ascomycota and Basidiomycota dominate the mycobiota, exposing extensive host-, diet-, and developmental stage-specific variation in microbial composition.

More recently, shotgun metagenomics and Liquid Chromatography-Mass Spectrometry (LC/MS)-based metabolomics have been integrated to move beyond taxonomic description, linking microbial presence with functional capacities and metabolic outputs. Shotgun sequencing confirms many 16S-based taxonomic assignments but also uncovers archaeal, viral, and eukaryotic components of the gut ecosystem that amplicon surveys miss (Ji & Jung, 2025). Coupling sequencing with non-targeted metabolomics—such as studies in chromium-exposed silkworm midguts—has enabled the mapping of specific bacterial taxa (*Weissella*, *Propionibacterium*) to metabolites involved in amino acid, lipid, and antioxidant pathways, providing mechanistic insights into how environmental stressors reshape both microbiota and host metabolism. Together, these multi-omics approaches establish connections between microbial taxa, biochemical pathways, and host traits including growth, detoxification, immunity, and silk production, laying the foundation for microbiome-informed interventions in sericulture (Ma et al., 2025; Xiao et al., 2025).

9.12.2 Germ-Free and Gnotobiotic Models for Causal Inference

Germ-free and gnotobiotic *B. mori* models have emerged as critical tools for establishing causality in host–microbiome interactions. By raising silkworms on sterilized eggs and gamma-irradiated feed within isolators, researchers can generate larvae completely devoid of detectable microbes, providing a baseline for comparison with conventionally reared or probiotic-colonized cohorts. Defined microbial consortia or mono-association experiments allow direct testing of how specific taxa—such as *Enterococcus*, *Weissella*, *Bacillus*, or *Acinetobacter*—influence cellulose digestion, amino acid provisioning, immune function, and silk synthesis (Aghighi & Salami, 2024).

When combined with host transcriptomics and metabolomics, these systems illuminate the mechanisms by which the gut microbiome modulates larval physiology and environmental responses. For example, colonization with *B. subtilis* reshapes microbial networks, upregulates digestive and immune gene expression, elevates hemolymph phenylalanine levels, and enhances larval growth and cocoon yield relative to germ-free controls. Similarly, defined consortia experiments demonstrate that commensals such as *Enterococcus faecalis* LX10 can confer resistance to *N. bombycis*, while manipulation of microbial composition influences tolerance to heavy metals or pesticides. These experimental platforms transform silkworm microbiome research from descriptive ecology into mechanistic, hypothesis-driven science, enabling the rational design of microbiome-based strategies to optimize larval health, silk productivity, and resilience under industrial and environmental challenges (Xiao et al., 2025).

9.13 FUTURE DIRECTIONS: MICROBIOME ENGINEERING AND DIGITAL SERICULTURE

9.13.1 Synbiotics, Prebiotics, and Host-Specific Formulations

Looking forward, microbiome-informed sericulture offers multiple promising avenues for improving larval health, silk yield, and resilience under artificial or stress-inducing diets. Tailored synbiotic formulations—combining host-adapted probiotics such as *Weissella*, *Lactobacillus*, and *Bacillus* with mulberry-derived prebiotics—could stabilize gut microbiota in artificial diet systems and restore larval performance to levels closer to those seen with fresh mulberry feeding. Multi-omics analyses have repeatedly identified *Weissella* and *Lactobacillus* as enriched under transitional or artificial feeding regimes, highlighting their potential to support carbohydrate fermentation, SCFA production, and competitive exclusion of opportunistic pathogens, thereby maintaining gut barrier integrity and metabolic balance (Ma et al., 2025).

In practice, these probiotic strains could be paired with specific oligosaccharides or polyphenols extracted from mulberry leaves to promote beneficial fermentations and SCFA synthesis, countering the dysbiosis and acidification often observed in artificial diet–reared larvae. Optimization of synbiotic formulations must consider strain-specific compatibility, local environmental conditions, and diet composition, since "one-size-fits-all" products are unlikely to deliver consistent results. Iterative field trials integrating microbiome profiling, performance metrics, and strain-level tracking of probiotic taxa will be essential to move synbiotics from experimental validation to reliable industrial application (Wang et al., 2025).

9.13.2 CRISPR-Cas for Host and Symbiont Engineering

CRISPR-Cas technologies present a powerful toolkit for rational manipulation of both silkworm hosts and their gut symbionts. In *B. mori*, genome editing has already been employed to modify immune receptors, antiviral effectors, and silk protein genes, demonstrating the feasibility of precise host engineering and suggesting that traits such as susceptibility to BmNPV, tolerance to heat, or amino acid utilization efficiency could be directly optimized (Baci et al., 2021). Extending CRISPR approaches to commensal gut bacteria—such as *Bacillus*, *Enterococcus*, or *Weissella*—could generate strains that more efficiently degrade plant cell walls, produce essential amino acids and vitamins, or sequester heavy metals and pesticide residues, effectively transforming the microbiome into a programmable biofilter.

Engineered symbionts could be designed to respond dynamically to environmental stimuli—temperature fluctuations, chemical exposures, or nutrient availability—by modulating metabolic outputs in real time. This approach moves beyond correlative microbiome studies toward the deliberate construction of integrated host–microbe–environment modules optimized for industrial sericulture. However, implementation will require careful evaluation of ecological safety, risks of horizontal gene transfer, and regulatory compliance prior to broad deployment (Liu et al., 2024).

9.13.3 Precision Rearing and Microbiome-Informed Sericulture

Integrating microbiome data with host genomic and environmental information enables precision rearing platforms—sometimes termed digital sericulture—capable of real-time monitoring and adaptive control of holobiont health. Routine 16S rRNA profiling combined with metabolite or SCFA measurements could detect early signs of dysbiosis, triggering interventions such as diet modification, probiotic supplementation, or adjustments in temperature and stocking density. Over time, these data streams could define strain- and system-specific microbiome "reference states," providing quantitative targets for breeding and management decisions (García et al., 2025).

In addition, microbiome-aware nano ecotoxicology emphasizes the need to consider microbial responses when assessing the impact of nanoparticles, pesticides, or other agrochemicals, rather than relying solely on host survival or growth metrics. In this framework, sericulture evolves into a microbiome-conscious industry, where breeding, diet formulation, hygiene, and biosecurity are optimized for the *B. mori* holobiont as a whole. Incorporating microbial metrics alongside traditional genetic and phenotypic indices ensures that management practices sustain both host performance and microbial ecosystem stability, fostering resilient and sustainable silk production (Fometu et al., 2021).

9.14 CONCLUSIONS AND STRATEGIC RECOMMENDATIONS

The *B. mori* holobiont exemplifies a remarkable confluence of evolutionary adaptation and human-mediated selection, reflecting thousands of years of domestication, Silk Road–driven dissemination, and modern industrialization. These historical and contemporary pressures have left distinct imprints not only on the host genome but also on the gut microbiome, producing a tightly integrated system in which host physiology, diet, and microbial communities interact to optimize growth, immunity, and silk production (Yang et al., 2014). As a model organism, *B. mori* offers unparalleled opportunities to dissect the ecological and functional dynamics of gut microbiota in a tractable invertebrate host, providing insights relevant to insect agriculture, host–microbe co-evolution, and comparative immunology. Its straightforward gut morphology, well-characterized developmental stages, and extensive genomic resources make it ideal for exploring how microbial community assembly, succession, and functional outputs link to host nutrition, metabolism, and developmental physiology.

Looking forward, sustainable improvements in sericulture will require intentional incorporation of microbiome knowledge into breeding, dietary formulations, and disease management strategies. Targeted interventions—including host-specific probiotics, synbiotics, and microbiome-stabilizing feed supplements—offer the potential to enhance nutrient assimilation, silk yield, and resilience to pathogens. These approaches must be validated under large-scale production and field conditions to ensure ecological stability and functional efficacy. Integrating microbiome insights into digital and precision sericulture frameworks, where real-time data on host genetics, environmental parameters, and gut community composition inform adaptive management, will be essential. By adopting a holobiont-centered perspective, future silk production systems can achieve greater robustness, efficiency, and sustainability, leveraging the co-evolved interplay between *B. mori* and its gut microbiota to meet the challenges of a changing agricultural landscape.

ACKNOWLEDGMENTS

The author (ST) acknowledges the Abraham Thomas Foundation for providing the resources for this work.

REFERENCES

Aghighi, F., Salami, M., 2024. What we need to know about the germ-free animal models. *AIMS Microbiol* 10, 107–147. https://doi.org/10.3934/microbiol.2024007

Alders, R.G., Campbell, A., Costa, R., Guèye, E.F., Ahasanul Hoque, M., Perezgrovas-Garza, R., Rota, A., Wingett, K., 2021. Livestock across the world: Diverse animal species with complex roles in human societies and ecosystem services. *Anim Front* 11, 20–29. https://doi.org/10.1093/af/vfab047

Aljuaid, R.S., Alshareef, S.A., Jamal, B.T., Dhafeer, F.H., Alnahari, A.A., Ashy, R.A., 2026. Spatial and vertical stratification of groundwater microbial communities reveals proteobacterial dominance and redox-driven ecological transitions. *Microorganisms* 14, 232. https://doi.org/10.3390/microorganisms14010232

Altman, G.H., Diaz, F., Jakuba, C., Calabro, T., Horan, R.L., Chen, J., Lu, H., Richmond, J., Kaplan, D.L., 2003 Feb. Silk-based biomaterials. *Biomaterials*, 24(3), 401-416.

Altman, G.H., Farrell, B.D., 2022. Sericulture as a sustainable agroindustry. *Cleaner and Circular Bioeconomy* 2, 100011. https://doi.org/10.1016/j.clcb.2022.100011

Arunkumar, K.P., Metta, M., Nagaraju, J., 2006. Molecular phylogeny of silkmoths reveals the origin of domesticated silkmoth, Bombyx mori from Chinese Bombyx mandarina and paternal inheritance of Antheraea proylei mitochondrial DNA. *Mol Phylogenet Evol* 40, 419–427. https://doi.org/10.1016/j.ympev.2006.02.023

Baci, G.-M., Cucu, A.-A., Giurgiu, A.-I., Muscă, A.-S., Bagameri, L., Moise, A.R., Bobiș, O., Rațiu, A.C., Dezmirean, D.S., 2021. Advances in editing silkworms (Bombyx mori) genome by using the CRISPR-cas system. *Insects* 13, 28. https://doi.org/10.3390/insects13010028

Bashir, I., War, A.F., Rafiq, I., Reshi, Z.A., Rashid, I., Shouche, Y.S., 2022. Phyllosphere microbiome: Diversity and functions. *Microbiological Research* 254, 126888. https://doi.org/10.1016/j.micres.2021.126888

Bisch-Knaden, S., Daimon, T., Shimada, T., Hansson, B.S., Sachse, S., 2014. Anatomical and functional analysis of domestication effects on the olfactory system of the silkmoth Bombyx mori. *Proc Biol Sci* 281, 20132582. https://doi.org/10.1098/rspb.2013.2582

Bordenstein, S.R., Theis, K.R., 2015. Host biology in light of the microbiome: Ten principles of holobionts and hologenomes. *PLoS Biol* 13, e1002226. https://doi.org/10.1371/journal.pbio.1002226

Celiker, H., Gore, J., 2014. Clustering in community structure across replicate ecosystems following a long-term bacterial evolution experiment. *Nat Commun* 5, 4643. https://doi.org/10.1038/ncomms5643

Charalambous, A., Grivogiannis, E., Dieronitou, I., Michael, C., Rahme, L., Apidianakis, Y., 2022. Proteobacteria and firmicutes secreted factors exert distinct effects on Pseudomonas aeruginosa infection under normoxia or mild hypoxia. *Metabolites* 12, 449. https://doi.org/10.3390/metabo12050449

Chen, B., Du, K., Sun, C., Vimalanathan, A., Liang, X., Li, Y., Wang, B., Lu, X., Li, L., Shao, Y., 2018. Gut bacterial and fungal communities of the domesticated silkworm (Bombyx mori) and wild mulberry-feeding relatives. *ISME J* 12, 2252–2262. https://doi.org/10.1038/s41396-018-0174-1

Chen, B., Yu, T., Xie, S. et al., 2018. Comparative shotgun metagenomic data of the silkworm Bombyx mori gut microbiome. *Sci Data* 5, 180285.

Chikina, A., Matic Vignjevic, D., 2021. At the right time in the right place: How do luminal gradients position the microbiota along the gut?, *Cells & Development* 168, 203712. https://doi.org/10.1016/j.cdev.2021.203712

Croft, A.S., Spessot, E., Bhattacharjee, P., Yang, Y., Motta, A., Wöltje, M., Gantenbein, B., 2022. Biomedical applications of silk and its role for intervertebral disc repair. *JOR Spine* 5, e1225. https://doi.org/10.1002/jsp2.1225

De Lisle, S.P., Schrieber, S.J., Bolnick, D.I., 2022. Complex community-wide consequences of consumer sexual dimorphism. *J Anim Ecol* 91, 958–969. https://doi.org/10.1111/1365-2656.13685

Dong, H.-L., Zhang, S.-X., Tao, H., Chen, Z.-H., Li, X., Qiu, J.-F., Cui, W.-Z., Sima, Y.-H., Cui, W.-Z., Xu, S.-Q., 2017. Metabolomics differences between silkworms (Bombyx mori) reared on fresh mulberry (Morus) leaves or artificial diets. *Sci Rep* 7, 10972. https://doi.org/10.1038/s41598-017-11592-4

Dutta, P.L., Nath, I., Narzary, P.R., Nitish, G., Sharma, J., Rajah, A.R., Bora, N.R., 2025. Global perspectiveon silk: An economic analysisof international sericulture trade. *Agriculture Association of Textile Chemical and Critical Reviews Journal* 13, 231–238.

Eleftherianos, I., Heryanto, C., Bassal, T., Zhang, W., Tettamanti, G., Mohamed, A., 2021. Haemocyte-mediated immunity in insects: Cells, processes and associated components in the fight against pathogens and parasites. *Immunology* 164, 401–432. https://doi.org/10.1111/imm.13390

Fei, S., Awais, M.M., Zou, J., Xia, J., Wang, Y., Kong, Y., Feng, M., Sun, J., 2025. Single-nucleus RNA sequencing reveals midgut cellular heterogeneity and transcriptional profiles in Bombyx mori cytoplasmic polyhedrosis virus infection. *Insect Sci* 32, 1141–1154. https://doi.org/10.1111/1744-7917.13464

Feng, Z., 2020. The ever-changing technology and significance of silk on the silk road, in: Gaskell, I., Carter, S.A. (Eds.), *The Oxford Handbook of History and Material Culture*. Oxford University Press, pp. 222–235, https://doi.org/10.1093/oxfordhb/9780199341764.013.20

Fometu, S.S., Wu, G., Ma, L., Davids, J.S., 2021. A review on the biological effects of nanomaterials on silkworm (Bombyx mori). *Beilstein J Nanotechnol* 12, 190–202. https://doi.org/10.3762/bjnano.12.15

García, G., Carlin, M., Cano, R. de J., 2025. Holobiome harmony: Linking environmental sustainability, agriculture, and human health for a thriving planet and one health. *Microorganisms* 13, 514. https://doi.org/10.3390/microorganisms13030514

Gaspar, B.S., Roşu, O.A., Enache, R.-M., Manciulea (Profir), M., Pavelescu, L.A., Creţoiu, S.M., 2025. Gut mycobiome: Latest findings and current knowledge regarding its significance in human health and disease. *J Fungi (Basel)* 11, 333. https://doi.org/10.3390/jof11050333

Ghani, M.U., Liu, G., Zhang, Q., Liu, C., Li, S., Yang, Y., Ahmad, M., Cui, H., 2025. A Comprehensive review of humoral immune defense in the silkworm, *Bombyx mori*: Mechanistic insights and responses to microbial pathogens. *Developmental & Comparative Immunology* 170, 105418. https://doi.org/10.1016/j.dci.2025.105418

Gong, Y., Li, L., Gong, D., Yin, H., Zhang, J., 2016. Biomolecular evidence of silk from 8,500 years ago. *PLoS One* 11, e0168042. https://doi.org/10.1371/journal.pone.0168042

Guan, N., Li, C., Liu, W., Wu, Q., Zhu, J., Gao, T., Song, H., Guo, R., Yuan, F., Tian, Y., Yang, K., Zhou, D., 2025. Isolation and probiotic functions of Bacillus subtilis and its inhibitory effects on colitis. *Biology (Basel)* 14, 1786. https://doi.org/10.3390/biology14121786

Herman, R.A., Khurshid, M., Zhang, Z.-P., Yan, C.-H., Ayepa, E., Li, Y.-J.-C., You, S., Sheng, S., Wang, J., 2025. Silkworm (Bombyx mori) gut bacteria respond positively to exposure of magnetic Fe3O4@ urease-doped nanoconjugates: Potential reconstruction for nutritional efficiency. *Anim Microbiome* 8, 2. https://doi.org/10.1186/s42523-025-00500-1

Hodžić, A., Duscher, G.G., Alić, A., Beck, R., Berry, D., 2025. Peritrophic matrix: An important determinant of vector competence in hematophagous arthropods. *Trends in Parasitology* 41, 374–386. https://doi.org/10.1016/j.pt.2025.03.005

Hou, P., Liu, L., Ma, X., Men, Y., Yang, D., Wang, J., Zhang, C., 2025. Diversity characteristics and composition of gut microbiota in Antheraea pernyi (Lepidoptera: Saturniidae) larvae across different instars. *Insects* 16, 909. https://doi.org/10.3390/insects16090909

Huang, H., Wang, Q., Yang, Y., Zhong, W., He, F., Li, J., 2024. The mycobiome as integral part of the gut microbiome: Crucial role of symbiotic fungi in health and disease. *Gut Microbes* 16, 2440111. https://doi.org/10.1080/19490976.2024.2440111

Ji, J., Jung, S., 2025. PredCMB: Predicting changes in microbial metabolites based on the gene–metabolite network analysis of shotgun metagenome data. *Bioinformatics* 41, btaf020. https://doi.org/10.1093/bioinformatics/btaf020

Kalpana, S., Hatha, A.A.M., Lakshmanaperumalsamy, P., 1994. Gut microflora of the larva of silkworm, Bombyx mori. *International Journal of Tropical Insect Science* 15, 499–502. https://doi.org/10.1017/S1742758400015873

Kawakita, A., Okamoto, T., Goto, R., Kato, M., 2010. Mutualism favours higher host specificity than does antagonism in plant–herbivore interaction. *Proc Biol Sci* 277, 2765–2774. https://doi.org/10.1098/rspb.2010.0355

Kawamoto, M., Jouraku, A., Toyoda, A., Yokoi, K., Minakuchi, Y., Katsuma, S., Fujiyama, A., Kiuchi, T., Yamamoto, K., Shimada, T., 2019. High-quality genome assembly of the silkworm, *Bombyx mori*. *Insect Biochemistry and Molecular Biology* 107, 53–62. https://doi.org/10.1016/j.ibmb.2019.02.002

Lee, K.K., Kim, H., Lee, Y.-H., 2022. Cross-kingdom co-occurrence networks in the plant microbiome: Importance and ecological interpretations. *Front Microbiol* 13, 953300. https://doi.org/10.3389/fmicb.2022.953300

Li, C., Xu, S., Xiang, C., Xu, S., Zhou, Q., Zhang, J., 2022. The gut microbiota of silkworm are altered by antibiotic exposure. *FEBS Open Bio* 12, 2203–2212. https://doi.org/10.1002/2211-5463.13502

Li, J., Deng, J., Deng, X., Liu, L., Zha, X., 2023. Metabonomic analysis of silkworm midgut reveals differences between the physiological effects of an artificial and mulberry leaf diet. *Insects* 14, 347. https://doi.org/10.3390/insects14040347

Liang, J., Lu, H., Zhong, X., Qin, P., Xiang, Z., He, N., 2025. Remarkable adaptation of silkworm to mulberry. *Journal of Resource Insects*, 1, 100001. https://doi.org/10.1016/j.jori.2025.100001

Liang X, Sun C, Chen B, Du K, Yu T, Luang-In V, et al., 2018. Insect symbionts as valuable grist for the biotechnological mill: an alkaliphilic silkworm gut bacterium for efficient lactic acid production. *Applied Microbiology and Biotechnology*, 102, 4951–4962.

Lin, P., Cheng, T., Jin, S., Jiang, L., Wang, C., Xia, Q., 2014. Structural, evolutionary and functional analysis of *APN* genes in the Lepidoptera *Bombyx mori*. *Gene* 535, 303–311. https://doi.org/10.1016/j.gene.2013.11.002

Liu, X., Cooper, A.M.W., Zhang, J., Zhu, K.Y., 2019. Biosynthesis, modifications and degradation of chitin in the formation and turnover of peritrophic matrix in insects. *Journal of Insect Physiology* 114, 109–115. https://doi.org/10.1016/j.jinsphys.2019.03.006

Liu, X., Jiang, N., Wang, X., Yan, H., Guan, L., Kong, L., Chen, J., Zhang, H., Ma, H., 2024. Weissella cibaria relieves gut inflammation caused by Escherichia coli through inflammation modulation and gut microbiota regulation. *Foods* 13, 1133. https://doi.org/10.3390/foods13071133

Ma, W., Liu, X., Jing, Y., Zhang, M., Zhang, X., Wang, C., Khan, M.Z., Zhu, M., 2025. Weissella: From beneficial probiotics to opportunistic pathogens-A review. *Nutrients* 17, 3162. https://doi.org/10.3390/nu17193162

Manzanera, M., 2025. From microbial consortia to ecosystem resilience: The integrative roles of holobionts in stress biology. *Biology* 14, 1203. https://doi.org/10.3390/biology14091203

Maró, Z.M., Török, Á., 2022. China's new silk road and central and Eastern Europe—A systematic literature review. *Sustainability* 14(3), 1801.

Marzoli, F., Antonelli, P., Saviane, A., Tassoni, L., Cappellozza, S., Belluco, S., 2022. *Bombyx mori* from a food safety perspective: A systematic review. *Food Research International* 160, 111679. https://doi.org/10.1016/j.foodres.2022.111679

Muhammad, A., Sun, C., Shao, Y., 2024. The humoral immune response of the lepidopteran model insect, silkworm *Bombyx mori* L., to microbial pathogens. *Current Research in Insect Science* 6, 100097. https://doi.org/10.1016/j.cris.2024.100097

Muhammad, A., Sun, C., Shao, Y., 2026. Ecological and functional dynamics of gut microbiota in the model insect, silkworm Bombyx mori. *World J Microbiol Biotechnol* 42, 39. https://doi.org/10.1007/s11274-026-04784-6

Nalini, M.S., Prakash, H.S., 2020. Actinobacteria: Diversity, Plant Interactions and Biotechnology Applications, in: Yadav, A.N., Singh, J., Rastegari, A.A., Yadav, N. (Eds.), *Plant Microbiomes for Sustainable Agriculture*. Springer International Publishing, pp. 199–244. https://doi.org/10.1007/978-3-030-38453-1_7

Park, J.W., Park, J.S., Kook, P.R., Cho, Y.H., Park, S.K., Lee, J.H., Kang, S.K., Kim, S.-W., Kim, S.-R., 2025. Impact of artificial diet versus mulberry leaves on silkworm growth, nutrient composition, and gut microbiota. *Arch Insect Biochem Physiol* 120, e70113. https://doi.org/10.1002/arch.70113

Puig-Castellví, F., Pacheco-Tapia, R., Deslande, M., Jia, M., Andrikopoulos, P., Chechi, K., Bonnefond, A., Froguel, P., Dumas, M.-E., 2023. Advances in the integration of metabolomics and metagenomics for human gut microbiome and their clinical applications. *TrAC Trends in Analytical Chemistry* 167, 117248. https://doi.org/10.1016/j.trac.2023.117248

Ren, C., Meng, Y., Liu, Y., Wang, Y., Wang, H., Liu, Y., Liu, C., Fan, X., Zhang, S., 2025. Probiotic Bacillus subtilis enhances silkworm (Bombyx mori) growth performance and silk production via modulating gut microbiota and amino acid metabolism. *Anim Microbiome* 7, 103. https://doi.org/10.1186/s42523-025-00473-1

Rong, W., Wei, Y., Chen, Y., Huang, L., Huang, S., Lv, Y., Guan, D., Li, X., 2024. 16S rRNA Sequencing analysis uncovers dose-dependent cupric chloride effects on silkworm gut microbiome composition and diversity. *Animals* 14, 3634. https://doi.org/10.3390/ani14243634

Shi, Y.-X., Bian, D.-D., Chen, F.-X., Jin, Y., Jiang, Q., Liu, Q.-N., Tang, B.-P., Dai, L.-S., 2024. Effect of lead exposure on silkworm midgut: Insights into oxidative gene expression, digestive function, and microbial community structure. *Process Safety and Environmental Protection* 190, 481–494. https://doi.org/10.1016/j.psep.2024.07.048

Suraporn, S., Liu, J., Ren, F., Wang, L., Feng, M., Terenius, O., Swevers, L., 2025. Towards a rational basis for the selection of probiotics to improve silkworm health and performance. *Insects* 16, 162. https://doi.org/10.3390/insects16020162

Tan, I., Bautista, M., 2022. Bacterial survey in the guts of domestic silkworms, Bombyx mori L. *Insects* 13, 100. https://doi.org/10.3390/insects13010100

Torres Mc Cook, A., Caballero Fernández, B., Ruiz, A., 2021. Description of the main aspects influencing Bombyx mori L. (Lepidoptera: Bombycidae) rearing. *Cuban J Agri Sci* 55, 1–14.

Tran, H.A., Hoang, T.T., Maraldo, A., Do, T.N., Kaplan, D.L., Lim, K.S., Rnjak-Kovacina, J., 2023. Emerging silk fibroin materials and their applications: New functionality arising from innovations in silk crosslinking. *Materials Today* 65, 244-259.

Wang, D., Zhou, J., Bu, W., Ning, C., 2025. Optimization of the fermentation process for a mulberry beverage using composite microbial strains and a study on its physicochemical properties. *Foods* 14, 4312. https://doi.org/10.3390/foods14244312

Wang, G., Ding, X., Yang, J., Ma, L., Sun, X., Zhu, R., Lu, R., Xiao, Z., Xing, Z., Liu, J., Pan, Z., Xu, S., Sima, Y., 2024. Effects of habitual dietary change on the gut microbiota and health of silkworms. *International Journal of Molecular Sciences* 25, 1722. https://doi.org/10.3390/ijms25031722

Wang, Q., Sun, Z., Ma, S., Liu, X., Xia, H., Chen, K., 2022. Molecular mechanism and potential application of bacterial infection in the silkworm, *Bombyx mori*. *Developmental & Comparative Immunology* 131, 104381. https://doi.org/10.1016/j.dci.2022.104381

Warwick, C., Steedman, C., Grant, R., 2024. Is the domestication phenomenon more perceived than real? *Discov Anim* 1, 36. https://doi.org/10.1007/s44338-024-00036-9

Weimer, P.J., 2022. Degradation of cellulose and hemicellulose by ruminal microorganisms. *Microorganisms* 10, 2345. https://doi.org/10.3390/microorganisms10122345

Willerslev, E., Meltzer, D.J., 2021. Peopling of the Americas as inferred from ancient genomics. *Nature* 594, 356–364. https://doi.org/10.1038/s41586-021-03499-y

Xia, Q., Guo, Y., Zhang, Z., Li, D., Xuan, Z., Li, Z., Dai, F., Li, Y., Cheng, D., Li, R., Cheng, T., Jiang, T., Becquet, C., Xu, X., Liu, C., Zha, X., Fan, W., Lin, Y., Shen, Y., Jiang, L., Jensen, J., Hellmann, I., Tang, S., Zhao, P., Xu, H., Yu, C., Zhang, G., Li, J., Cao, J., Liu, S., He, N., Zhou, Y., Liu, H., Zhao, J., Ye, C., Du, Z., Pan, G., Zhao, A., Shao, H., Zeng, W., Wu, P., Li, C., Pan, M., Li, J., Yin, X., Li, D., Wang, J., Zheng, H., Wang, W., Zhang, X., Li, S., Yang, H., Lu, C., Nielsen, R., Zhou, Z., Wang, J., Xiang, Z., Wang, J. 2009. Complete resequencing of 40 genomes reveals domestication events and genes in silkworm (Bombyx). *Science*. 326(5951), 433-436.

Xiao, Y., Li, Q., Sun, Z., Fu, B., Yang, Q., Jiang, M., Zhang, W., Huang, X., Xing, D., 2025. Transcriptomics and metabolomics combined to analyze the response mechanism of silkworm eggs to high-temperature stress. *Insects* 16, 862. https://doi.org/10.3390/insects16080862

Xin, L., Chen, Y., Rong, W., Qin, Y., Li, X., Guan, D., 2024. Gut microbiota analysis in silkworms (Bombyx mori) provides insights into identifying key bacterials for inclusion in artificial diet formulations. *Animals (Basel)* 14, 1261. https://doi.org/10.3390/ani14091261

Yang, S.-Y., Han, M.-J., Kang, L.-F., Li, Z.-W., Shen, Y.-H., Zhang, Z., 2014. Demographic history and gene flow during silkworm domestication. *BMC Evol Biol* 14, 185. https://doi.org/10.1186/s12862-014-0185-0

Yuan, S., Sun, Y., Chang, W., Zhang, J., Sang, J., Zhao, J., Song, M., Qiao, Y., Zhang, C., Zhu, M., Tang, Y., Lou, H., 2023. The silkworm (Bombyx mori) gut microbiota is involved in metabolic detoxification by glucosylation of plant toxins. *Commun Biol* 6, 790. https://doi.org/10.1038/s42003-023-05150-0

Zhang, J., Yu, M., Abbas, Z., Liu, Q., Yin, Z., 2024. Studies on the origin and cultural value of Chinese mulberry culture. *AFJHSS* 4, 57–68. https://doi.org/10.51483/AFJHSS.4.1.2024.57-68

Zhang, M., Zhang, X., Tran, N.T., Sun, Z., Zhang, X., Ye, H., Zhang, Y., Ma, H., Aweya, J.J., Li, S., 2021. Molting alters the microbiome, immune response, and digestive enzyme activity in mud crab (Scylla paramamosain). *mSystems* 6, 10–128, https://doi.org/10.1128/msystems.00917-21

Zhang, M., Zhang, Y., Zhao, Z., Deng, F., Jiang, H., Liu, C., Li, Y., Chai, J., 2025. Bacterial–fungal interactions: Mutualism, antagonism, and competition. *Life (Basel)* 15, 1242. https://doi.org/10.3390/life15081242

CHAPTER TEN

The Mosquito Microbiome

Alex Muppasani, Olivia Lutz, and Sunil Thomas

10.1 INTRODUCTION

The mosquito is widely regarded as the deadliest animal in the world, responsible for more than 700,000 human deaths each year due to its role as a vector of numerous infectious diseases (Prudêncio, 2020). Unlike other dangerous animals that cause mortality through direct attacks, mosquitoes exert their impact indirectly by transmitting a diverse array of pathogens to humans during blood feeding. These include several medically significant viruses, such as Zika virus, yellow fever virus, chikungunya virus, dengue virus, and West Nile virus. In addition to viral pathogens, mosquitoes transmit filariasis-causing nematodes (parasitic roundworms) and protozoan parasites of the genus *Plasmodium*, the causative agents of malaria.

Among all mosquito-borne diseases, malaria remains by far the most lethal. The disease is primarily transmitted by female *Anopheles* mosquitoes and continues to pose a major global health burden, particularly in tropical and subtropical regions. According to the World Health Organization, there were an estimated 282 million malaria cases and approximately 610,000 malaria-related deaths worldwide in 2024, across 80 endemic countries (World Health Organization, 2025). These figures underscore the persistent and disproportionate impact of malaria on vulnerable populations, especially young children and pregnant women in sub-Saharan Africa.

The extraordinary mortality associated with mosquitoes reflects both the efficiency of vector-borne transmission and the global distribution of competent mosquito species. Consequently, vector control strategies—including insecticide-treated bed nets, indoor residual spraying, environmental management, and emerging genetic control approaches—remain central components of global efforts to reduce mosquito-borne disease transmission and mortality (Prudêncio, 2020).

Although mosquitoes are widely feared because of the pathogens they transmit to humans, they also play important ecological roles within natural ecosystems. Adult mosquitoes, particularly males and even many females outside of blood-feeding periods, rely on nectar as a primary energy source and in doing so act as pollinators for a variety of flowering plants. Some plant species, including certain orchids and other specialized flowers, depend in part on mosquito visitation for pollination (Foster, 2024). Beyond their role in plant reproduction, mosquitoes are integral components of food webs in aquatic and terrestrial habitats. Their larvae contribute to nutrient cycling in freshwater ecosystems by feeding on organic matter and microorganisms, while both larvae and adults serve as prey for numerous species, including bats, birds, amphibians, dragonflies, and fish. In this way, mosquitoes function not only as disease vectors but also as participants in energy transfer and trophic interactions, underscoring their embeddedness in broader ecological systems despite their substantial impact on human health (Azevedo-Santos et al., 2017; Collins et al., 2019).

DOI: 10.1201/9781003639176-10

Traditional vector control strategies have relied heavily on chemical insecticides, yet these approaches face mounting challenges from evolved resistance—documented in over 60 countries for at least one class of insecticide—environmental persistence, and non-target ecological impacts (Hemingway et al., 2016). The discovery that mosquito-associated microorganisms can profoundly modulate vector competence has opened revolutionary possibilities for biological control strategies that leverage natural host–microbe interactions rather than synthetic chemical interventions (Wei et al., 2026).

The mosquito microbiome comprises diverse and dynamic communities of bacteria, fungi, and viruses that colonize multiple tissues of the insect host, including the midgut, salivary glands, reproductive organs, and hemolymph. These microbial consortia influence mosquito development, nutrition, immunity, and vector competence (Gao et al., 2020). Among the best-studied microbial associates is the intracellular endosymbiont *Wolbachia pipientis*, which has gained significant attention for its dual capacity to manipulate host reproduction and interfere with pathogen transmission. *Wolbachia* induces cytoplasmic incompatibility (CI), a form of conditional sterility that enhances its maternal transmission through host populations, thereby facilitating its rapid spread. In addition to reproductive manipulation, *Wolbachia* can inhibit the replication of arboviruses such as dengue virus and Zika virus, as well as malaria-causing parasites of the genus *Plasmodium*. Natural *Wolbachia* infections are estimated to occur in 40–70% of all insect species, making it one of the most widespread intracellular symbionts on Earth. Although the molecular basis of CI and pathogen blocking remained elusive for decades, recent genetic and biochemical studies have demonstrated that CI is mediated by specific effector proteins encoded within prophage WO regions of the *Wolbachia* genome. Pathogen blocking appears to involve multiple, non-mutually exclusive mechanisms, including the activation of host innate immune pathways, competition for intracellular lipids and other nutrients, and the generation of reactive oxygen species (ROS) that create an inhospitable environment for pathogen replication (Ranson and Lissenden, 2016; van den Berg et al., 2012).

Beyond *Wolbachia*, the native mosquito gut microbiota includes bacterial taxa capable of directly suppressing pathogen development. Certain *Enterobacter* strains, such as Esp_Z, and species of *Chromobacterium* (e.g., Csp_P) exhibit strong anti-*Plasmodium* activity through the production of ROS and bioactive secondary metabolites (Nichols and Coon, 2025). Notably, *Chromobacterium* species produce romidepsin (FK228), a cyclic depsipeptide histone deacetylase (HDAC) inhibitor with potent antiparasitic properties. These microbiome-associated bacteria can modulate mosquito immune signaling, secrete antimicrobial compounds, and compete with invading pathogens for essential nutritional resources within the gut environment. Collectively, these findings highlight the mosquito microbiome as a complex and functionally significant component of vector biology, offering promising and mechanistically distinct avenues for malaria and arbovirus control beyond *Wolbachia*-based strategies (Dong et al., 2009; Cirimotich et al., 2011; Ramirez et al., 2014). This review therefore examines the composition, functional interactions, and vector control implications of the mosquito microbiome.

10.2 TYPES OF MOSQUITOES

Mosquitoes are dipteran insects belonging to the family Culicidae within the order Diptera and the class Insecta. Globally, more than 3500 mosquito species have been described, distributed across approximately 41 genera, reflecting their extensive evolutionary diversification and broad ecological adaptability (Zheng, 2020). Among these, three genera are of particular medical and veterinary importance: *Aedes*, *Anopheles*, and *Culex*. Species within these genera are responsible for transmitting a wide range of pathogens that cause major infectious diseases in humans and animals. For example, *Aedes* mosquitoes are primary vectors of dengue fever, chikungunya fever, and yellow fever; *Anopheles* mosquitoes transmit malaria parasites; and *Culex* species are important vectors of various forms of encephalitis and lymphatic filariasis. Collectively, mosquitoes transmit the causative agents of dengue, filariasis, malaria, yellow fever, encephalitis (affecting both humans and livestock), and chikungunya—diseases that pose significant and persistent threats to global public health (Zheng, 2020). The epidemiological importance of these mosquito genera underscores the need for continued surveillance, vector control strategies, and integrated disease management approaches worldwide. Table 10.1 summarizes

TABLE 10.1
Genus, species and disease transmitted by mosquitoes

Genus	Species	Disease
Aedes	*Aedes aegypti*	Yellow fever, dengue, Zika, chikungunya.
	Aedes albopictus	Chikungunya virus, dengue virus, yellow fever virus, Rift Valley fever virus, Japanese encephalitis virus, West Nile virus, Sindbis virus, Cache Valley virus, La Crosse virus, Eastern equine encephalitis virus, Mayaro virus, Ross River virus, Western equine encephalitis virus, Venezuelan equine encephalitis virus, Oropouche virus, Jamestown Canyon virus, San Angelo virus, Trivittatus virus.
Anopheles	*Anopheles gambiae*	Malaria, Filaria (elephantiasis), O'nyong-nyong virus.
	Anopheles arabiensis	Malaria.
	Anopheles quadrimaculatus	Malaria, Cache Valley virus, West Nile virus.
Culex	*Culex pipiens*	West Nile virus, usutu virus, Batai virus, Rift Valley fever virus, Tahyna virus, Sindbis virus, Japanese encephalitis virus, Saint Louis encephalitis virus.
	Culex quinquefasciatus	Filaria (elephantiasis), West Nile virus, Japanese encephalitis virus.

the major mosquito species implicated in disease transmission.

10.3 MOSQUITO BIOLOGY

Mosquitoes have distinct physical characteristics that aid in their identification. Their bodies are divided into three main segments: the head, thorax, and abdomen. The head contains important sensory organs, including compound eyes, antennae, and a proboscis, which is used for feeding. The thorax supports the wings and legs, enabling mosquitoes to fly swiftly and move with agility. The abdomen forms the posterior portion of the body and is primarily involved in digestion and reproduction.

Several mosquito genera can be distinguished by specific physical features. *Anopheles* mosquitoes have elongated palps located near the proboscis, which are about as long as the proboscis itself. Adult *Anopheles* can also be identified by their characteristic resting position, where the abdomen is angled upward rather than held parallel to the resting surface (Sallum et al., 2020). *Aedes* mosquitoes are recognized by their distinctive black-and-white appearance, caused by white or silver scale patches set against a black background on the legs and other parts of the body. *Aedes* mosquito feeds with its head and tail pointing down so that it makes an inverted V shape (Facchinelli et al., 2023). *Culex* mosquitoes generally have a more uniform appearance, with dark-colored bodies and narrow, elongated wings.

Mosquitoes undergo four distinct stages in their life cycle: egg, larva, pupa, and adult. Depending on the species, eggs are laid either directly on the surface of water or on moist soil and objects that may later become flooded. The larval and pupal stages are entirely aquatic and occur only in water. The adult stage, however, is terrestrial and capable of flight.

Adult mosquitoes can travel considerable distances. Both males and females feed on nectar as a source of energy. Mating typically occurs within the first three to five days after emergence, and females usually mate only once. Males generally live for about a week. Only female mosquitoes feed on blood; although blood provides little direct nourishment, it is necessary for the development of their eggs (Wellcome Sanger Institute Blog; Oklahoma State University Extension).

10.3.1 Microbiome of Mosquito Larvae

Mosquito larvae develop in small, microbe-rich freshwater habitats such as water accumulated in discarded tires, tree holes, puddles, and artificial containers. These environments are densely populated with microorganisms, including bacteria, fungi, protozoa, and algae. Unlike many organisms that inherit part of their microbiota vertically from their parents, mosquito larvae primarily acquire their microbiome horizontally from the surrounding aquatic environment. As larvae feed, they continuously ingest microorganisms suspended in the water column as well as those associated with detritus and biofilms. Although the microbial diversity present in the surrounding water is typically far greater than that found within the larval gut, the larvae selectively retain only those microbial taxa capable of surviving the extreme physiological conditions of the digestive tract. The mosquito larval midgut is highly alkaline, frequently reaching a pH of 10–11, and it contains immune defenses such as antimicrobial peptides and ROS. These conditions act as strong ecological filters that determine which microbes persist within the gut community (Schmidt and Engel, 2021).

In mosquito genera such as *Aedes* and *Culex*, the larval gut microbiome is often dominated by a relatively small number of bacterial lineages. These frequently include members of the phyla Proteobacteria, Actinobacteriota, Bacteroidota, and Firmicutes. Within these groups, genera such as *Aeromonas*, *Pseudomonas*, *Elizabethkingia*, and *Microbacterium* are commonly detected, although their relative abundance varies according to habitat conditions and mosquito species. Some bacterial taxa exhibit the ability to persist through multiple developmental stages. For example, genera such as *Microbacterium* and *Elizabethkingia* can survive the dramatic physiological changes associated with metamorphosis, persisting from the larval stage into pupae and subsequently into adult mosquitoes. This phenomenon, known as transstadial persistence, suggests that early microbial colonization during larval development can influence the microbial composition of the adult gut (Arellano and Coon, 2022).

Functionally, the larval microbiome plays a critical role in mosquito development and physiology. Experimental studies with axenic, or bacteria-free, *Aedes aegypti* larvae have demonstrated that larvae deprived of microbial partners are unable to complete normal development unless they are supplemented with live bacteria or specific bacterial metabolites. These findings indicate that microorganisms provide essential metabolic functions that the mosquito host alone cannot fully perform. Among the most important contributions of gut bacteria is the synthesis of B vitamins, including folate, riboflavin, and thiamine, which are necessary for cellular metabolism and growth. In addition, microbiomes participate in amino acid metabolism and nitrogen recycling, processes that improve the nutritional efficiency of the larval diet. Certain bacterial symbionts also influence host energy metabolism by modulating mitochondrial activity and epithelial cell turnover in the gut lining (Correa et al., 2018).

Studies using gnotobiotic systems, in which larvae are colonized with a defined microbial community, have demonstrated that even a single bacterial species can partially restore normal larval development. Mono-association experiments with isolates such as *Elizabethkingia anophelis* or *Aeromonas* species have shown that relatively simple bacterial communities can support larval growth, molting, and pupation. At the ecological level, the productivity of bacterial populations within the aquatic habitat largely determines both the nutritional resources available to larvae and the pool of microbes capable of colonizing the gut. Consequently, bacterial abundance and diversity in the environment strongly influence larval growth rate, developmental timing, and survival (Miller and Bäumler, 2021).

The composition of the mosquito larval microbiome is shaped by a combination of environmental conditions and host-specific factors. Environmental determinants include the type of organic matter present in the habitat, such as decomposing leaf litter, animal detritus, or anthropogenic waste. These substrates influence nutrient availability and microbial growth, thereby structuring the surrounding microbial community from which larvae acquire their microbiota. Additional factors such as temperature, oxygen availability, and water chemistry further influence microbial diversity and activity. At the same time, host characteristics—including mosquito species identity, gut morphology, feeding behavior, and immune responses—act as biological filters that determine which microbes persist within the digestive tract. As a result, even when larvae of different mosquito genera share the same breeding site, their gut microbiomes can differ substantially. Distinct microbial profiles have been observed among genera such as *Aedes*, *Anopheles*, and *Culex*, reflecting species-specific host–microbe interactions (Gao et al., 2020).

Experimental manipulations of larval habitats have provided additional insights into the ecological dynamics of mosquito microbiomes. Altering the composition of organic detritus, introducing specific bacterial strains, or applying low-level antimicrobial treatments can modify the microbial communities present in the water. These manipulations often produce predictable shifts in the larval gut microbiome and can significantly affect mosquito life-history traits. Changes in microbial composition have been associated with differences in the time required for larvae to reach pupation, the size and physiological condition of emerging adults, and the overall survival rate from egg to adult stage (Seabourn et al., 2023).

Importantly, the influence of the larval microbiome extends beyond the aquatic developmental stages and can affect the biology of adult mosquitoes. Microbial communities encountered during larval life can shape the gut microbiota of the emerging adult and influence immune gene expression, metabolic processes, and susceptibility to pathogens. For instance, mosquitoes reared in environments with reduced microbial diversity often emerge as smaller adults with altered gut microbial communities and modified susceptibility to infection by pathogens such as arboviruses or *Plasmodium* parasites (Dickson et al., 2017). These carryover effects highlight the ecological and epidemiological significance of the larval microbiome.

Because of its strong influence on mosquito development and disease transmission potential, the larval microbiome has become an emerging target for innovative vector control strategies. Researchers are exploring approaches that involve introducing beneficial bacterial consortia into breeding sites, manipulating organic inputs to promote microbial communities that inhibit mosquito development, or targeting essential symbiotic bacteria required for larval survival. Such microbiome-based strategies offer the possibility of environmentally sustainable mosquito management by altering the microbial ecology of breeding habitats rather than relying solely on chemical insecticides (Dickson et al., 2017).

10.3.2 Microbiome of Male and Female Mosquitoes

Male and female mosquitoes share a broadly similar core microbiome, yet important differences exist in the abundance, composition, and functional roles of their microbial communities. These differences arise from variations in physiology, nutritional requirements, reproductive biology, and feeding behavior between the two sexes. In many mosquito species, including *A. aegypti*, the gut microbiome is dominated by bacteria belonging to the phylum Proteobacteria, along with representatives from Firmicutes, Bacteroidota, and Actinobacteriota. Common genera frequently detected in both male and female mosquitoes include *Enterobacter*, *Serratia*, *Pseudomonas*, and *Asaia*. Despite this shared core community, the relative abundance of these microbes and their metabolic interactions with the host can differ significantly between the sexes.

One of the most consistent patterns observed in microbiome studies is that male mosquitoes generally harbor lower bacterial loads in the midgut compared with females. Investigations in *A. aegypti* have shown that these differences do not appear immediately after metamorphosis; newly emerged adults possess relatively similar microbial densities regardless of sex. Instead, sex-specific divergence in microbiota abundance and immune regulation develops gradually as adults mature. During this period, female mosquitoes often show higher microbial density and increased expression of immune-related genes in the gut, suggesting that host immune responses play a role in shaping the sex-specific microbial landscape (Harrison et al., 2023).

The functional consequences of these differences are particularly evident during larval development. Experimental studies using axenic, or bacteria-free, mosquito larvae have demonstrated that the microbiome supplies essential nutrients required for normal development. However, male and female larvae appear to differ in their dependence on these microbial contributions. In controlled experiments with *A. aegypti*, males were able to complete development with lower levels of microbiota-derived nutrients, especially B vitamins such as biotin (Correa et al., 2018). Female larvae, in contrast, showed a stronger requirement for bacterial metabolic products to support successful growth and metamorphosis. This increased dependence is likely linked to the greater energetic demands associated with female development, including larger adult body size and the future metabolic requirements of egg production and blood feeding (Harrison et al., 2023).

Although the bacterial communities of adult males and females are similar at broad taxonomic levels, differences often emerge in the relative abundance of particular taxa and in how microbial communities respond to environmental or biological perturbations. For example, experimental exposure of *A. aegypti* larvae to certain gut-associated fungi has been shown to alter bacterial community composition in adults in a sex-dependent manner. These findings suggest that fungal–bacterial interactions within the mosquito microbiome may operate differently in males and females, potentially influencing host physiology and immune responses (Harrison et al., 2023a).

In addition to gut-associated microbes, mosquitoes harbor microbiota in reproductive tissues, and these communities also show sex-specific patterns. Both male and female mosquitoes may contain symbiotic bacteria such as *Asaia* or *Wolbachia* within reproductive organs, but the distribution and abundance of these microbes can differ between the sexes. Some bacterial taxa exhibit tissue tropism, meaning they preferentially colonize specific organs within the reproductive tract. These symbionts can influence reproductive processes, including mating success, sperm viability, and egg development. In certain cases, symbiotic bacteria are also capable of being transmitted between mating partners or passed vertically from females to their offspring, thereby influencing microbial inheritance and population-level microbiome structure (Yin et al., 2025).

Sex-specific microbiome characteristics also intersect with mosquito immunity and disease transmission potential. Female mosquitoes, unlike males, require blood meals to support egg production. This feeding behavior exposes females to vertebrate pathogens such as arboviruses and *Plasmodium* parasites. As a result, female mosquitoes typically exhibit stronger immune activation in the gut and maintain higher microbial loads compared with nectar-feeding males. The presence of a denser microbial community in the female midgut can influence pathogen colonization, either by stimulating immune defenses or through direct microbial competition with invading pathogens (Izquierdo, 2024).

Importantly, microbial differences established during larval development may persist into adulthood and shape key life-history traits. Variations in microbial exposure during early life can influence adult body size, metabolic capacity, and reproductive output. Because body size and nutritional status affect the ability of female mosquitoes to acquire, maintain, and transmit pathogens, the larval microbiome may indirectly influence vector competence in adult females. Environmental conditions that alter microbial availability in larval habitats can therefore affect not only mosquito development but also the epidemiological dynamics of mosquito-borne diseases.

Overall, while male and female mosquitoes share a common set of core microbial taxa, the structure and functional significance of their microbiomes diverge as a consequence of sex-specific physiological demands. Female mosquitoes tend to host denser and metabolically more integrated microbial communities, reflecting their greater reliance on microbial-derived nutrients and their role as the primary vectors of mosquito-borne pathogens. Understanding these sex-based differences in mosquito microbiomes provides important insights into mosquito biology and may help guide the development of microbiome-based strategies for vector control and disease prevention (Barletta Ferreira et al., 2022).

10.3.3 Microbiome of *Aedes*, *Anopheles*, and *Culex*

Mosquitoes of the genera *Aedes*, *Anopheles*, and *Culex* host complex microbial communities that inhabit their digestive tract, reproductive organs, and other tissues. Although these genera share several bacterial taxa and broad ecological patterns, each possesses distinctive microbiome characteristics shaped by differences in habitat, feeding behavior, physiology, and evolutionary history. Across all three groups, the microbiome is predominantly bacterial and is strongly influenced by environmental exposure during larval development, dietary inputs in adulthood, and host immune regulation. These microbial communities play essential roles in nutrient metabolism, immune function, insecticide tolerance, and the capacity of mosquitoes to transmit pathogens.

In *A. aegypti*, the adult microbiome is overwhelmingly bacterial, representing approximately 99% of microbial sequence reads. Members of the phylum Proteobacteria typically dominate the community, although other groups—including Firmicutes, Bacteroidetes, and Actinobacteria—are consistently present. Within these phyla, several bacterial genera frequently occur in the mosquito midgut and other tissues, including *Bacillus*, *Chryseobacterium*, *Enterobacter*, *Klebsiella*, *Serratia*, *Acinetobacter*, and *Stenotrophomonas*.

Studies have identified a relatively stable "core microbiome" composed of bacterial species such as *Bacillus subtilis*, *Salmonella enterica*, *Staphylococcus aureus*, *Escherichia coli*, and *Enterococcus faecium* that can persist across individuals and geographic regions (Pascar et al., 2023). However, the composition of the *Aedes* microbiome changes substantially across developmental stages. Larval mosquitoes typically host a simpler microbial community with relatively low diversity, often dominated by genera such as *Microbacterium*. As mosquitoes transition to adulthood, their microbiome becomes more diverse and dynamic, reflecting exposure to environmental microbes and dietary changes. Adult feeding behavior strongly influences microbiome structure; nectar feeding favors certain bacterial taxa, while blood feeding produces temporary but pronounced shifts in microbial abundance during digestion. These changes are often reversible, with the microbiome gradually returning to a baseline composition after the blood meal has been processed (Hegde et al., 2015).

In *Anopheles gambiae*, an important vector of malaria parasites, the microbiome also undergoes significant shifts across developmental stages. Larval and pupal stages often harbor microbial communities enriched with photosynthetic Cyanobacteria, reflecting the aquatic habitats in which larvae develop and the availability of microbial biofilms associated with algae and plant material. After metamorphosis, the adult gut microbiome shifts toward dominance by Proteobacteria and Bacteroidetes, with frequent representation from families such as Enterobacteriaceae and Flavobacteriaceae. Blood feeding in adult females has a particularly strong influence on the microbiome. During blood digestion, the mosquito midgut experiences oxidative and nitrosative stress generated by the breakdown of hemoglobin and other blood components. This physiological environment tends to reduce overall microbial diversity while selecting for bacterial taxa capable of tolerating these stresses. Many of the bacteria that proliferate under these conditions appear to contribute to maintaining redox balance within the gut, thereby helping the host cope with the metabolic demands of blood digestion (Wang et al., 2011). As in *Aedes*, microbiome composition in *Anopheles* is highly variable between individuals and is strongly influenced by environmental factors such as larval habitat, diet, and geographic location. Importantly, variations in gut microbial communities have been linked to differences in susceptibility to infection by *Plasmodium* parasites, suggesting that the microbiome may influence malaria transmission dynamics (Wang et al., 2011).

The microbiome of *Culex pipiens quinquefasciatus*, a major vector of several arboviruses, also displays considerable diversity. Metagenomic sequencing studies indicate that Proteobacteria constitute nearly 70% of the microbial population in the mosquito midgut. Additional bacterial groups—including Bacteroidetes, Actinobacteria, Firmicutes, and Planctomycetes—are also commonly detected, and fungal taxa from several phyla may be present as minor community members. At the genus level, *Aeromonas* is frequently the most abundant bacterium in *Culex* mosquitoes. Other genera that commonly contribute to the gut microbiome include *Asaia*, *Morganella*, *Elizabethkingia*, *Wolbachia*, *Enterobacter*, and *Serratia*, each often representing between 5% and 10% of the microbial community. Among the most frequently identified species are *Aeromonas veronii*, *Morganella morganii*, and *Enterobacter* sp. Ag1. Functional analyses of these microbial communities indicate that many of these bacteria participate in metabolic pathways related to nutrient processing, detoxification, and resistance to environmental stressors. Notably, certain microbiome compositions have been associated with increased tolerance to insecticides, suggesting that microbial metabolism may contribute to the detoxification of chemical control agents used in mosquito management programs (Wang et al., 2021).

Despite the differences among these mosquito genera, several common patterns emerge. All three groups host microbiomes that are largely acquired from the environment during larval development and are subsequently modified by diet, physiology, and immune responses in adulthood. Many bacterial taxa—including members of the Enterobacteriaceae family as well as genera such as *Asaia*, *Elizabethkingia*, and *Serratia*—are shared across *Aedes*, *Anopheles*, and *Culex*. However, each genus displays distinctive dominant taxa and unique patterns of microbiome succession across developmental stages. These microbial communities play critical roles in mosquito biology by contributing to nutrient acquisition, modulating immune responses, influencing resistance to insecticides, and affecting the ability of mosquitoes to host and transmit pathogens. Consequently, understanding the similarities and differences among the microbiomes of these mosquito genera is increasingly important for developing microbiome-based strategies to control mosquito populations and reduce the spread of mosquito-borne diseases (Gao et al., 2020).

10.3.4 Microbiome of Mosquitoes after a Blood Meal

The ingestion of a blood meal represents one of the most dramatic physiological events in the life of a female mosquito and produces rapid and substantial changes in the gut microbiome. In species such as *A. aegypti* and *Anopheles* mosquitoes, the midgut environment is transformed within hours after feeding due to the sudden influx of nutrients, the breakdown of hemoglobin, and shifts in immune and oxidative conditions. These changes trigger a rapid expansion of specific bacterial populations while reducing overall microbial diversity. As the blood bolus enters the gut, the microbial community temporarily reorganizes, favoring bacterial groups that are well adapted to the chemically stressful environment generated during blood digestion. In many cases, members of the order Enterobacterales—such as *Enterobacter*, *Serratia*, and related enteric bacteria—proliferate rapidly during this period because they possess metabolic and stress-response systems capable of tolerating oxidative and nitrosative conditions associated with hemoglobin degradation (Salgado et al., 2024).

Metagenomic and whole-genome sequencing studies tracking individual mosquitoes during the digestive process reveal distinct temporal patterns in microbial expansion. Sampling of mosquito midguts at approximately 12, 24, and 48 hours after a blood meal demonstrates that bacterial populations undergo a pronounced bloom during the early phase of digestion. This bloom is particularly evident among Enterobacterales, which can dominate the microbial community shortly after feeding. Functional analyses of these microbial populations show enrichment in genes associated with carbohydrate metabolism, amino acid metabolism, and protein degradation, all of which contribute to the efficient breakdown of the blood meal. These microbial metabolic activities complement the host's digestive enzymes and may help process complex proteins and nutrients present in vertebrate blood (Salgado et al., 2024).

A major biochemical driver of these microbial changes is the large amount of heme released during hemoglobin digestion. Heme is a potent signaling molecule that influences cellular metabolism within the mosquito midgut. Its presence activates host signaling pathways, including protein kinase C, which in turn suppresses the production of ROS in the gut epithelium. Under normal conditions, ROS function as antimicrobial molecules that help regulate microbial populations in the gut. When ROS levels decline after a blood meal, bacterial growth becomes less constrained, allowing the microbial population to expand dramatically. In some cases, bacterial abundance can increase by as much as three orders of magnitude within the first 24 hours following feeding. Studies have demonstrated a strong inverse relationship between ROS levels and bacterial density in the mosquito midgut, highlighting the importance of oxidative regulation in shaping microbiome dynamics (Oliveira et al., 2011).

Comparative studies of mosquito feeding regimes further demonstrate that blood feeding produces a markedly different microbial profile compared with sugar feeding. Sugar-fed mosquitoes typically maintain a more diverse microbial community composed of bacteria adapted to carbohydrate-rich diets. In contrast, blood-fed mosquitoes show reduced microbial diversity but an increased dominance of bacteria capable of coping with oxidative stress and the metabolic byproducts of blood digestion. These bacteria often possess expanded genetic capacities for redox regulation, detoxification, and nutrient metabolism, suggesting that they play a beneficial role in stabilizing the gut environment during the demanding process of blood digestion (Taracena et al., 2018).

As digestion progresses and the blood bolus is gradually processed and eliminated—typically within 30–40 hours in *A. aegypti*—the gut environment begins to return to conditions similar to those seen in sugar-fed mosquitoes. During this recovery phase, the microbiome slowly shifts back toward its pre-feeding composition. However, certain bacterial taxa may persist or even become more prominent after digestion. One such example is *E. anophelis*, which often becomes highly abundant during the post-digestive stage and is consistently detected in the midguts of both *Aedes* and *Anopheles* mosquitoes. The repeated enrichment of such bacteria following blood feeding suggests that the digestive cycle periodically selects for microbial partners that are well adapted to the biochemical conditions created by blood metabolism.

The source of the blood meal can also influence microbiome dynamics. Different vertebrate hosts—such as humans, birds, or livestock—provide blood with varying biochemical compositions, including differences in proteins, lipids, and micronutrients. These variations can alter the nutrient landscape within the mosquito midgut and influence which bacterial taxa proliferate after feeding. In addition, mosquitoes that feed on multiple hosts or mixed blood sources may develop distinct microbial communities compared with those feeding on a single host species. These findings suggest that host-derived factors present in the blood meal interact with the mosquito gut environment to shape microbiome composition and function (Sarma et al., 2022).

Overall, the mosquito microbiome after a blood meal represents a highly dynamic and plastic system that temporarily reorganizes in response to the physiological demands of blood digestion. The transient bacterial bloom supports nutrient processing and helps maintain redox balance within the gut, while also interacting with the mosquito's immune system. Because many pathogens—including arboviruses and *Plasmodium* parasites—must pass through the mosquito midgut during transmission, these microbiome changes can strongly influence infection success and vector competence. Consequently, understanding how blood feeding alters the mosquito microbiome has become an important area of research, with implications for developing microbiome-based strategies to reduce the transmission of mosquito-borne diseases (Sarma et al. 2022).

10.3.5 Microbiome of Mosquitoes during Malaria Transmission

The microbiome of *Anopheles* mosquitoes plays a critical role in determining whether *Plasmodium* parasites—the causative agents of malaria—can successfully develop within the mosquito and ultimately be transmitted to humans. Within the mosquito midgut, microbial communities interact closely with both the host immune system and invading parasites, creating a complex ecological environment that can either suppress or facilitate malaria transmission. In *A. gambiae* and related malaria vectors, the midgut microbiome is typically dominated by bacteria belonging to the phylum Proteobacteria, particularly members of the family Enterobacteriaceae and other Gram-negative taxa. However, the precise composition of these microbial communities varies substantially among individual mosquitoes and populations. Environmental factors such as larval breeding sites, water chemistry, and exposure to organic material strongly influence which bacterial taxa colonize the developing mosquito and persist into adulthood (Boissière et al., 2012).

High-throughput sequencing studies have revealed that mosquito microbiomes are highly heterogeneous. Many bacterial species are detected only in a subset of mosquitoes, whereas a smaller group of bacterial genera consistently appears across most individuals, forming a core microbiome. This variation among individual mosquitoes has important implications for vectorial capacity—the ability of mosquitoes to acquire, maintain, and transmit pathogens. Differences in microbiome composition can influence immune activity, nutrient availability, and gut physiology, all of which affect whether *Plasmodium* parasites can survive and develop inside the mosquito (Bahia et al., 2014).

During malaria transmission, female mosquitoes ingest *Plasmodium* gametocytes when they take a blood meal from an infected human host. Once inside the mosquito midgut, the parasites undergo sexual reproduction and must penetrate the gut epithelium to form oocysts on the outer gut wall. This stage represents the most significant bottleneck in the parasite life cycle, with the majority of parasites failing to survive. The mosquito microbiota plays a major role in determining the outcome of this process. Experiments in which mosquitoes are treated with antibiotics to eliminate their gut bacteria consistently show higher levels of *Plasmodium* infection, indicating that a normal microbiota typically provides protection against parasite development (Salgado et al., 2024).

Several mechanisms explain how gut bacteria inhibit malaria parasites. Some bacterial species directly produce antimicrobial compounds or ROS that damage *Plasmodium* cells in the midgut lumen or at the epithelial surface. For example, certain strains of *Enterobacter* have been shown to generate reactive oxygen molecules that are toxic to developing parasites, thereby reducing the number of oocysts that successfully form within the mosquito gut. In experimental coinfection studies, mosquitoes exposed simultaneously to bacteria and *Plasmodium* typically show reduced parasite loads compared with mosquitoes infected with parasites alone (Ouologuem et al., 2023). These findings demonstrate that microbial antagonism can be a natural barrier limiting malaria transmission.

Despite these inhibitory effects, the relationship between mosquito microbiota and *Plasmodium* infection is not uniformly protective. Field studies involving wild-caught *Anopheles* mosquitoes exposed to natural *Plasmodium* falciparum isolates have revealed that certain bacterial taxa may correlate with increased parasite infection under natural conditions. For instance, higher abundance of some Enterobacteriaceae species in the midgut has been associated with particular infection patterns, suggesting that certain bacteria may either create conditions favorable for parasite survival or simply coexist with environments permissive to infection (Mitri and Vernick, 2012). Overall, the influence of the microbiome on malaria transmission is highly context dependent, reflecting complex interactions among microbial diversity, mosquito immunity, and environmental conditions. Microbiota-mediated immune priming, competition for nutrients, and direct antimicrobial activity all contribute to determining whether parasites survive within the mosquito (Boissière et al., 2012).

The mosquito immune system forms a crucial link between microbiota and parasite development. When a mosquito consumes a blood meal, the midgut microbiota often expands rapidly, stimulating immune responses throughout the insect's body. This immune activation includes the production of antimicrobial peptides, nitric oxide, and other immune effectors that can target both bacteria and *Plasmodium* parasites. Recent studies have also highlighted the importance of immune cells known as hemocytes, particularly phagocytic hemocytes, in regulating gut homeostasis. In *A. gambiae*, these cells help maintain the integrity of the midgut epithelium and control bacterial populations. When hemocytes are experimentally depleted, bacterial loads increase dramatically, gut barrier function deteriorates, and both mosquito survival and *Plasmodium* infection patterns are altered (Cardoso-Jaime & Dimopoulos, 2025). These findings illustrate how microbiota, immune regulation, and epithelial health interact to shape the mosquito's ability to host malaria parasites.

Another important microbiome-associated factor influencing malaria transmission is the presence of the endosymbiotic bacterium *Wolbachia*. For many years, *Anopheles* mosquitoes were believed to lack natural *Wolbachia* infections, but recent field studies have identified naturally occurring strains such as wAnga-Mali in *A. gambiae* populations. Mosquitoes carrying these *Wolbachia* strains often show reduced prevalence and intensity of *Plasmodium* sporozoites in their salivary glands—the stage required for transmission to humans (Gomes et al., 2017). Experimental studies suggest that *Wolbachia* has modest effects on early parasite stages in the midgut but exerts stronger inhibitory effects on later parasite development, thereby reducing overall transmission potential. Because *Wolbachia* can also spread through mosquito populations via CI, researchers are exploring the possibility of using *Wolbachia* infections in *Anopheles* mosquitoes as a novel biological strategy for malaria control (Shaw et al., 2016).

Collectively, research on mosquito microbiomes has revealed that microbial communities profoundly influence malaria transmission at multiple biological levels. Gut bacteria can directly inhibit *Plasmodium* parasites, stimulate immune responses that restrict infection, and shape the physiological environment of the mosquito midgut. At the same time, certain microbial configurations may permit or even facilitate parasite development, reflecting the ecological complexity of the mosquito gut ecosystem. These insights have stimulated growing interest in microbiome-based approaches to malaria control, including the manipulation of larval habitats to alter microbial exposure, the introduction of beneficial symbionts that block parasite development, and the use of engineered or naturally occurring bacteria capable of suppressing *Plasmodium*. Such strategies aim to complement traditional interventions such as insecticide-treated nets and indoor residual spraying, potentially offering new environmentally grounded tools to reduce malaria transmission (Cardoso-Jaime and Dimopoulos, 2025).

10.3.6 Microbiome of Mosquitoes during Dengue Transmission

The microbiome of *Aedes* mosquitoes plays a crucial role in determining whether dengue virus (DENV) can successfully infect the mosquito, replicate within its tissues, and ultimately be transmitted to humans through a subsequent bite. Dengue transmission depends on a sequence of events in which the virus must first infect the mosquito midgut after a blood meal, then disseminate to other tissues such as the hemolymph and salivary glands, and finally be released in saliva during feeding. Throughout this process, the mosquito's resident microbial community interacts with both the host immune system and the virus itself, significantly influencing the success or failure of viral infection. In *A. aegypti* and *Aedes albopictus*, the midgut microbiome is typically dominated by bacteria belonging to the phylum Proteobacteria, including members of the Enterobacteriaceae family and bacteria related to *Asaia*. In addition to bacteria, the mosquito microbiome may also include fungi, viruses, and endosymbionts such as *Wolbachia*. The composition of this community is shaped by environmental factors such as larval habitat, diet, temperature, and microbial exposure during development (Gómez et al., 2022).

The resident microbiota provides a baseline level of immune stimulation in the mosquito gut and produces microbial metabolites that help regulate the midgut environment. Together, these effects act as a first line of defense against invading pathogens, including dengue virus. When mosquitoes ingest an infectious blood meal containing DENV, the virus must overcome these microbial and immune barriers to establish infection in the gut epithelial cells. Studies in which the mosquito microbiome has been experimentally disrupted provide strong evidence for its protective role. When mosquitoes are treated with antibiotics to remove much of their gut microbiota, they become significantly more susceptible to dengue infection. Conversely, reintroducing specific bacterial species into these microbiota-depleted mosquitoes can reduce viral replication in the midgut. This protective effect is thought to occur through the stimulation of immune pathways that induce antimicrobial peptides and other antiviral molecules, which limit early viral replication and reduce the probability that the virus will disseminate beyond the gut (Zheng et al., 2023).

However, not all microbial interactions with dengue virus are inhibitory. Certain bacteria and fungi can enhance viral infection under specific conditions, demonstrating that the microbiome can influence vector competence in both protective and permissive directions. For example, some strains of Salmonella have been shown to increase the susceptibility of *A. aegypti* mosquitoes to dengue infection. In addition, a naturally occurring fungus of the genus *Talaromyces* associated with *Aedes* mosquitoes can increase the permissiveness of the mosquito midgut to dengue virus. This fungus appears to alter digestive physiology by suppressing the production of certain digestive enzymes, including trypsins, which normally contribute to the degradation of viral particles. By reducing these enzymatic defenses and modifying the physiological environment of the gut barrier, such microbes may create conditions that facilitate viral entry and replication (Gao et al., 2020). These findings highlight the complex and sometimes bidirectional effects that microbiome members can have on pathogen transmission.

The relationship between the mosquito microbiome, immune system, and dengue virus can therefore be viewed as a dynamic tripartite interaction. Many bacteria naturally present in the *Aedes* midgut stimulate low-level activation of immune signaling pathways, including those that regulate antimicrobial peptides and other defensive molecules. This basal immune activation establishes an immune "tone" that can make mosquitoes less susceptible to viral infection. At the same time, dengue virus infection itself can alter host gene expression, including genes involved in immune responses and antimicrobial peptide production. These changes may in turn influence the composition and abundance of microbial communities in the gut, creating a feedback loop between viral infection and microbiome structure. Over the course of infection, the mosquito midgut thus reaches a shifting equilibrium in which virus replication, immune activity, and microbial populations interact continuously (Zheng et al., 2023).

Comparative metagenomic studies of *A. albopictus* populations have further demonstrated that variation in microbiome composition correlates with differences in dengue susceptibility. Mosquito populations with distinct microbial communities often show different rates of viral infection and transmission potential. In some cases, functional differences in microbial genes related to metabolism, redox balance, and immune modulation appear to influence how efficiently dengue virus can replicate within the mosquito host. These findings suggest that microbial diversity and metabolic capacity are important determinants of dengue vector competence (Zheng et al., 2023).

One of the most significant microbiome components affecting dengue transmission is the intracellular bacterium *Wolbachia*. Although *A. aegypti* does not naturally carry *Wolbachia*, laboratory introduction of this symbiont has demonstrated powerful antiviral effects. When *Wolbachia* becomes established in *A. aegypti* populations, it can markedly reduce dengue virus replication within the mosquito, inhibit dissemination of the virus to secondary tissues, and decrease the likelihood that the virus reaches the salivary glands. In some cases, the presence of *Wolbachia* can completely block dengue transmission. The mechanisms underlying this antiviral effect are multifaceted and include stimulation of basal immune responses as well as alterations in host metabolism, particularly lipid and cholesterol pathways that dengue virus normally exploits for replication. Genome-scale metabolic studies have revealed that the *Aedes–Wolbachia* symbiosis modifies host cellular resources in ways that limit viral growth (Safaei et al., 2025).

Because of these strong antiviral properties, *Wolbachia*-based strategies have emerged as one of the most promising biological approaches for dengue control. Programs in several countries have released *Wolbachia*-infected *A. aegypti* mosquitoes into wild populations, where the bacterium spreads through mosquito populations via CI. As the frequency of *Wolbachia*-infected mosquitoes increases, the overall ability of the mosquito population to transmit dengue declines, effectively using the mosquito's own microbiome as a tool to interrupt disease transmission.

Overall, the mosquito microbiome influences dengue transmission through multiple biological mechanisms, including regulation of gut barrier function, modulation of immune responses, alteration of metabolic pathways, and interactions with viral replication processes. Some microbes suppress dengue virus by stimulating antiviral immunity or producing inhibitory metabolites, whereas others may enhance viral infection by weakening gut defenses or modifying digestive physiology. These insights have stimulated growing interest in microbiome-based interventions aimed at reducing dengue transmission. Strategies under investigation include introducing protective microbial symbionts, manipulating larval environments to favor beneficial microbial communities, and deploying *Wolbachia*-infected mosquitoes in endemic regions. Together, these approaches highlight the potential of microbiome-focused methods as complementary tools to traditional mosquito control strategies in the fight against dengue (Saraiva et al., 2018).

10.3.7 Microbiome of Mosquitoes during Filaria Transmission

The microbiome of mosquitoes undergoes significant and dynamic changes during the transmission of filarial parasites, and these microbial shifts can strongly influence whether infective third-stage larvae (L3) successfully develop within the vector. Filarial parasites such as *Wuchereria bancrofti*, *Brugia malayi*, and *Dirofilaria immitis* rely on mosquitoes as intermediate hosts, where they complete essential developmental stages before becoming transmissible to vertebrate hosts. Within mosquitoes, the gut microbiota forms an integral component of the internal environment encountered by these parasites. In uninfected mosquitoes, the midgut microbiome is typically dominated by bacteria belonging to the phylum Proteobacteria, including genera such as *Asaia*, *Pseudomonas*, and *Enterobacter*. Additional bacterial groups from Bacteroidetes and Actinobacteria are also present, contributing to a diverse microbial community that interacts closely with mosquito physiology and immunity. In some mosquito populations, intracellular bacteria such as *Wolbachia* may also occur naturally or be introduced experimentally, further shaping microbial ecology within the insect host (Gao et al., 2020).

The composition of the mosquito microbiome is influenced by multiple environmental and biological factors. Conditions encountered during the larval stage, including the microbial composition of breeding sites, detritus availability, and water chemistry, strongly affect the initial bacterial communities that colonize developing mosquitoes. After emergence, adult diet—particularly the balance between sugar feeding and blood feeding—continues to shape microbial abundance and diversity. Environmental exposures such as insecticides, pollutants, or microbial contaminants can further modify microbiome structure. Together, these influences determine the baseline microbial community that provides colonization resistance against invading pathogens and maintains a level of basal immune activation within the mosquito. This immune stimulation contributes to the mosquito's capacity to resist infections by a wide range of pathogens, including arboviruses, malaria parasites, and filarial nematodes (Ratcliffe et al., 2024).

When a mosquito ingests a blood meal containing microfilariae, the earliest stages of infection initiate substantial physiological and transcriptional changes in the mosquito midgut. These changes include the modulation of pattern-recognition receptors, immune signaling pathways, and antimicrobial peptide genes that collectively reshape the microbial community in the gut. As microfilariae penetrate the midgut epithelium and migrate into the hemocoel, interactions between the parasite and resident microbiota become increasingly evident. Experimental infections—for example, studies involving *A. aegypti* infected with the canine heartworm *D. immitis*—have demonstrated measurable shifts in bacterial diversity and in the relative abundance of specific microbial taxa during infection. Such findings suggest that successful establishment of filarial parasites is closely associated with remodeling of the mosquito microbiome (Erickson et al., 2009).

These observations highlight that the outcome of filarial infection is not determined solely by the genetic compatibility between parasite and mosquito host. Instead, the composition, stability, and resilience of the midgut microbiota also play crucial roles. Certain microbial communities may create conditions favorable for parasite development, while others may strengthen immune defenses that inhibit parasite survival (Erickson et al., 2009). Thus, the microbiome represents an additional ecological layer influencing vector competence for filarial transmission.

Filarial worms encounter several anatomical and immunological barriers as they develop inside the mosquito. After ingestion, microfilariae must survive the digestive environment of the midgut, penetrate the gut epithelium, migrate through the hemocoel, and eventually reach the thoracic muscles, where they develop into infective L3 larvae. Each of these stages is associated with immune responses from the mosquito host. Activation of innate immune pathways, including strong induction of antimicrobial peptides such as cecropins and defensins, can reduce bacterial populations and simultaneously affect parasite survival. In mosquito strains that are refractory to filarial infection, these immune responses may eliminate microfilariae or prevent their progression beyond early developmental stages (Erickson et al., 2009). Because many immune effectors target both microbes and parasites, immunity-driven shifts in microbiome composition can either strengthen or weaken natural barriers to infection. Consequently, the structure of the microbiome may indirectly determine the likelihood that developing larvae survive long enough to migrate to the mosquito proboscis and be transmitted during the next blood meal (Erickson et al., 2009).

An additional level of complexity arises from the bacterium *Wolbachia*, which participates in filarial transmission through multiple biological relationships. Many filarial worms harbor *Wolbachia* as an obligate intracellular symbiont essential for their normal growth, reproduction, and survival. These bacterial endosymbionts contribute to metabolic processes within the worm and regulate the expression of enzymes such as chitinases that facilitate microfilarial exsheathment and escape from the mosquito midgut. Experimental depletion of *Wolbachia* from filarial worms severely disrupts parasite development and fertility, effectively blocking transmission of the parasite through the mosquito vector (Kaur et al., 2021).

At the same time, *Wolbachia* can also exist independently as a symbiont of mosquitoes. In mosquito hosts, *Wolbachia* infections are known to alter the composition of the gut microbiota and induce a heightened basal immune state. These changes can suppress the development and transmission of various pathogens, including viruses, malaria parasites, and in some cases filarial nematodes. The interactions among *Wolbachia*, mosquito hosts, and filarial worms therefore form a complex tripartite system in which microbial symbionts influence both vector physiology and parasite survival (Kaur et al., 2021).

The interplay among mosquitoes, their microbiota, and filarial parasites has important implications for disease ecology and control. Microbiome composition can influence mosquito longevity, immune responsiveness, and tissue integrity during infection, all of which affect the mosquito's vectorial capacity. If the internal microbial environment favors parasite development, more infective L3 larvae may reach the mosquito proboscis and be transmitted to vertebrate hosts. Conversely, microbial communities that enhance immune defenses or disrupt parasite development can significantly reduce transmission potential.

These insights have stimulated growing interest in microbiome-based strategies for controlling filarial diseases. Approaches under investigation include introducing bacterial species that inhibit parasite development, manipulating larval habitats to promote protective microbial communities, and deploying *Wolbachia*-based interventions that modify mosquito immunity and microbial composition. Such strategies may eventually complement existing public health measures, including mass drug administration programs aimed at eliminating lymphatic filariasis and related diseases. By targeting the ecological interactions within the mosquito microbiome, researchers hope to create an internal environment that is hostile to filarial parasites, thereby interrupting transmission cycles and reducing disease burden in endemic regions (Bera, 2025).

10.3.8 Microbiome of Mosquito Salivary Glands

The salivary glands of mosquitoes harbor a specialized microbial community that differs markedly from the microbiota found in other organs such as the midgut. This microbiome is shaped by the anatomical isolation of the glands, the biochemical properties of saliva, and the repeated cycles of sugar and blood feeding that mosquitoes experience throughout their lifespan. Although the total bacterial load in salivary glands is generally lower than in the midgut, a stable and diverse assemblage of bacteria can still be detected. Commonly reported genera include *Serratia*, *Pseudomonas*, *Enterobacter*, *Klebsiella*, *Asaia*, and *Elizabethkingia*, which have been isolated from the salivary tissues of multiple mosquito species including *Aedes*, *Anopheles*, and *Culex* (Sharma et al., 2014). These bacteria must survive a challenging microenvironment characterized by antimicrobial components of saliva—such as lysozymes, defensins, and other antimicrobial peptides—as well as fluctuations in osmolarity and pH associated with feeding behavior. In some mosquito species, including *A. aegypti* and *A. albopictus*, intracellular symbionts such as *Wolbachia* have also been detected within salivary tissues, suggesting that parts of the microbiome occupy intracellular niches or associate closely with glandular epithelial cells rather than remaining solely within luminal spaces. Interestingly, although the midgut usually contains greater numbers of bacteria, high-throughput 16S ribosomal RNA (rRNA) surveys indicate that the salivary glands may support a more taxonomically diverse community, likely reflecting tissue-specific filtering processes and exposure to microbes circulating in the hemolymph (Onyango et al., 2024).

Functionally, the salivary gland microbiome plays a potentially important role in mosquito vector competence and in the transmission of pathogens to vertebrate hosts. During feeding, mosquitoes inject saliva into the host's skin to facilitate blood uptake by inhibiting coagulation, suppressing pain responses, and modulating local immunity. Along with salivary proteins, microbes residing in the salivary glands can also be introduced into the bite site. High-throughput sequencing of mosquito saliva has revealed that certain bacterial taxa, particularly *Serratia*, can dominate the salivary microbial community. In some *Anopheles* populations, *Serratia* species constitute more than half of the detectable bacterial sequences in saliva, and both bacteria and pathogens such as *Plasmodium* sporozoites can be delivered simultaneously during a bite. Culture-based studies have similarly demonstrated that bacteria including *Serratia marcescens*, *Serratia nematodiphila*, *Asaia* spp., *Enterobacter* spp., and *Klebsiella* spp. can be recovered directly from mosquito saliva, confirming that these microbes are actively expelled during probing and blood feeding (Sharma et al., 2014). Experimental studies further suggest that some of these bacteria influence pathogen dynamics in the salivary glands. For example, strains of *Serratia* and *Asaia* have been shown to affect arbovirus titers in saliva, potentially by interacting directly with viral particles, competing for cellular binding sites, or modifying local immune signaling pathways within gland tissues (Accoti et al., 2023).

The presence of salivary microbes may also influence the early immune response in the vertebrate host. When a mosquito bites, the host's skin is exposed to a complex mixture of salivary proteins, microbial molecules, and transmitted pathogens. Bacteria delivered during feeding carry pathogen-associated molecular patterns (PAMPs), such as lipopolysaccharides and peptidoglycan fragments, which can activate host pattern-recognition receptors and initiate inflammatory signaling. This early immune activation may either hinder or facilitate pathogen establishment depending on the specific microbial context (Guerrero et al., 2020). Mosquito salivary proteins themselves are known to modulate host physiology by altering hemostasis, vascular permeability, and immune signaling pathways. Some studies indicate that mosquito saliva can enhance the replication of certain viruses by interfering with interferon responses or by inducing autophagy pathways in host cells (Narasimhan et al., 2025). When bacterial components are introduced alongside these proteins, they add additional immunological signals that may shape the microenvironment at the bite site. Consequently, the mosquito bite site can be viewed as a transient multi-species micro-ecosystem in which salivary microbiota, vector-derived molecules, and pathogens interact to determine whether infection becomes established (Accoti et al., 2023).

Symbiotic microorganisms such as *Wolbachia* further complicate the ecological interactions within mosquito salivary glands. *Wolbachia* is widely known for its ability to influence pathogen transmission in mosquitoes. In *A. albopictus*, naturally occurring *Wolbachia* strains have been shown to reduce dengue virus replication within salivary tissues and decrease the likelihood that infectious virions appear in saliva. This phenomenon contributes to what is known as a "salivary gland escape barrier," which prevents viruses from exiting the gland and being transmitted to vertebrate hosts. Studies using transinfected *A. aegypti* mosquitoes have demonstrated that different *Wolbachia* strains—such as wMel, wAlbB, and wPip—exhibit distinct patterns of localization and density within salivary tissues. Some strains accumulate at high densities without strongly inhibiting viruses, whereas others provide substantial transmission blocking despite relatively moderate bacterial loads. These observations indicate that the antiviral effect of *Wolbachia* depends not only on its abundance but also on its precise location within the gland, the host cellular pathways it influences, and its interactions with other members of the microbial community such as *Asaia* (Alonso et al., 2019). The broader evolutionary dynamics of host–symbiont interactions further highlight how microbial inheritance and host genetics shape symbiotic relationships across generations (Rau et al., 2022).

Field studies suggest that variation in salivary microbiome composition may correlate with patterns of disease transmission. Surveys conducted in malaria- or arbovirus-endemic regions often report dominance of bacterial genera such as *Staphylococcus*, *Erwinia*, *Serratia*, and *Pantoea* in mosquitoes collected from areas with high transmission intensity. These microbial signatures could potentially serve as biomarkers of vector populations with elevated epidemiological risk. Moreover, certain bacterial taxa associated with salivary tissues have been linked to insecticide susceptibility or resistance in mosquito populations. For example, genera such as *Serratia* and *Asaia* have been implicated in metabolic interactions that influence how mosquitoes tolerate chemical insecticides in some *Anopheles* populations. Because of these connections, interventions targeting one aspect of vector control—such as insecticide application—may inadvertently alter the composition of mosquito microbiomes and thereby influence pathogen transmission dynamics. This highlights the importance of integrated monitoring programs that consider microbiome composition alongside insecticide resistance profiles and pathogen prevalence (Ruan et al., 2022).

Finally, the recognition that mosquito saliva contains viable microorganisms capable of colonizing vertebrate tissues has important implications for both disease ecology and novel control strategies. Mosquitoes may act not only as vectors for classical pathogens such as viruses and parasites but also as carriers of environmental or opportunistic bacteria introduced into host tissues during feeding. At the same time, this feature provides opportunities for innovative intervention strategies. One emerging approach involves paratransgenesis, in which symbiotic bacteria associated with mosquito tissues are genetically engineered to express molecules that block pathogen development or transmission. If such engineered microbes colonize the salivary glands, they could deliver inhibitory factors directly to the host during a mosquito bite, potentially preventing infection at its earliest stage. Viewed from this perspective, the salivary gland microbiome represents a strategic interface within the mosquito–pathogen–host network, where relatively small changes in microbial composition or function could have substantial effects on disease transmission and public health outcomes (Solfisburg et al., 2024).

10.3.9 Microbiome of Mosquito Reproductive Tissues and Vertical Transmission

Mosquito reproductive organs—including the ovaries, testes, and associated accessory glands—harbor specialized microbial communities that play an important role in symbiont maintenance and transmission across generations. Unlike the midgut microbiome, which is strongly influenced by environmental exposure and feeding behavior, the microbiota of reproductive tissues are more tightly regulated by host physiology and cellular architecture. These tissues are particularly important because they serve as the primary sites for vertical transmission, the process by which symbiotic microorganisms are passed from parent to offspring. In female mosquitoes, the ovaries commonly host bacterial symbionts such as *Wolbachia*, *Asaia*, and other taxa that can localize to germline cells, follicular epithelia, or surrounding somatic tissues. Through a process known as transovarial transmission, these bacteria are incorporated into developing oocytes and subsequently inherited by the next generation of mosquitoes (Díaz et al., 2021). This mode of inheritance allows certain microbes to persist within mosquito populations for many generations even without continual environmental reintroduction.

Male mosquitoes also possess distinct reproductive microbiomes within the testes and male accessory glands. Although some bacterial taxa overlap with those found in the gut microbiome, others appear to display strong tropism for reproductive tissues. For example, particular strains of *Asaia* and *Wolbachia* can colonize the testes and accessory glands where they become incorporated into seminal fluids or associated with the surfaces of spermatozoa. During mating, these microbes may be transferred to females through seminal fluid or through physical contact between reproductive structures. As a result, mating can serve as a secondary transmission route for certain bacteria, allowing them to colonize the female reproductive tract and occasionally other tissues such as the midgut. This process, sometimes described as venereal or sexual transmission, contributes to the spread of symbiotic bacteria within mosquito populations and may influence the microbiome composition of both sexes (Taylor et al., 2018).

Because both vertical and sexual transmission occur within reproductive tissues, these organs represent a central hub for the population-level dynamics of mosquito symbionts. The best-known example involves *Wolbachia*, an intracellular bacterium that manipulates host reproduction through mechanisms such as CI. In this system, infected males produce modified sperm that can only successfully fertilize eggs from females carrying the same *Wolbachia* strain. When infected males mate with uninfected females, embryonic development fails, whereas infected females can reproduce normally with either infected or uninfected males. This reproductive asymmetry provides a strong selective advantage for infected females and enables *Wolbachia* to spread rapidly through mosquito populations. Successful operation of this system depends on efficient colonization of the ovaries to ensure maternal inheritance, as well as colonization of male reproductive tissues where the sperm modification responsible for CI is generated (Becking et al., 2019).

Other symbiotic bacteria associated with reproductive tissues may interact with *Wolbachia* in cooperative or competitive ways. For instance, *Asaia*—a widespread bacterial symbiont in many mosquito species—can colonize reproductive organs and use similar vertical and sexual transmission pathways. Because *Asaia* and *Wolbachia* may occupy overlapping cellular niches, competition between these microbes could influence which symbionts successfully establish in particular mosquito populations. These microbial interactions are increasingly relevant to vector-control strategies that rely on symbiont manipulation. Techniques such as *Wolbachia*-based population replacement, CI programs, and paratransgenic approaches often depend on stable colonization of reproductive tissues to maintain transmission across generations. If other members of the reproductive microbiome alter symbiont density or transmission efficiency, they could either enhance or undermine these interventions (Tayyab et al., 2025).

The reproductive microbiome is also dynamic and can change across the mosquito life cycle. Symbiont densities and community composition often shift during the gonotrophic cycle, which includes blood feeding, egg development (oogenesis), and oviposition. Blood feeding triggers physiological changes that influence nutrient availability, hormone signaling, and immune responses within the ovaries, which in turn may affect microbial colonization. Similarly, mating events can introduce new bacteria into the reproductive tract, while immune responses and tissue remodeling associated with egg production may alter the abundance of resident microbes. As a result, the reproductive microbiome is not static but fluctuates in response to reproductive physiology and environmental conditions.

Spatial organization within reproductive tissues further determines how microbes are transmitted and maintained. Symbionts that colonize germline stem cells or developing oocytes are well positioned for stable vertical inheritance, while those restricted to accessory gland secretions or epithelial surfaces may rely more heavily on sexual transmission or environmental reacquisition. Localization to specific micro-niches—such as follicular epithelial cells, sperm-storage organs, or seminal fluid—can also influence interactions between different bacterial species and determine whether they compete, coexist, or cooperate within the reproductive tract. These spatial relationships highlight the importance of tissue-level ecology in shaping the evolutionary stability of mosquito–microbe symbioses (Zhang and Raikhel, 2021).

Overall, the microbiome of mosquito reproductive tissues represents a critical interface between microbial ecology, host reproduction, and vector biology. Through mechanisms of vertical and sexual transmission, reproductive symbionts influence mosquito fertility, mating compatibility, and population structure. Understanding how these microbial communities function and interact offers valuable insights into mosquito evolution and provides promising opportunities for innovative vector-control strategies aimed at reducing the transmission of mosquito-borne diseases.

10.3.10 Microbiome-Informed Mosquito Control and Future Directions

The recognition that the mosquito microbiome constitutes a tractable and influential biological system has catalyzed a paradigm shift in vector control, moving beyond traditional chemical insecticides toward strategies that exploit microbial symbioses. Among the most advanced applications are *Wolbachia*-based population replacement and suppression programs, which leverage maternally transmitted *Wolbachia* strains to reduce vector competence for arboviruses and filarial parasites. Complementing these approaches, interventions using native gut bacteria—such as *Enterobacter* Esp_Z and *Chromobacterium* Csp_P—are being explored as living biopesticides or microbial delivery platforms capable of secreting antiparasitic molecules in the mosquito midgut (Onchuru et al., 2024). These strategies aim to achieve high specificity, self-dissemination, and minimal off-target effects, while preserving essential host nutritional and developmental functions. Paratransgenesis extends this concept by engineering resident symbionts to express effector proteins that directly inhibit pathogens such as *Plasmodium* or arboviruses, offering a potentially sustainable, species-specific complement to conventional control measures. However, these advances raise complex regulatory, ecological, and ethical questions regarding the release of genetically modified microorganisms into natural ecosystems (Mahor and Gupta, 2026).

Increasingly, attention is turning to reproductive and vertically transmitted microbiota as intervention targets. Symbionts that colonize ovaries, testes, and accessory glands represent bottlenecks for microbial inheritance and are therefore well-suited to stabilize desirable traits, such as pathogen-blocking phenotypes, in wild populations. By coupling these traits with vertical or sexual transmission, fewer releases may be needed to achieve population-level effects. Integrating microbiome-informed strategies with established tools—including sterile insect technique, incompatible insect technique, larviciding, and habitat modification—offers the potential for multi-modal control, which can reduce reliance on chemical insecticides, mitigate behavioral avoidance, and buffer against insecticide resistance. Nevertheless, deploying live, evolving microbial agents introduces novel challenges, including the need to assess long-term ecological impacts, reversibility, and governance across geographic and political boundaries (Mumford and Carrasco, 2022).

Several critical challenges must be addressed to fully realize microbiome-based mosquito control. First, microbiome composition is highly context-dependent, varying by geography, season, breeding site, and exposure to insecticides, which complicates reproducibility and prediction of intervention outcomes. Second, thermal sensitivity and fitness costs associated with certain *Wolbachia* strains in *Aedes* and *Anopheles* can reduce effectiveness under field-relevant temperature regimes, motivating the search for robust strains, strain combinations, or adaptive release strategies that buffer environmental stress (Luo et al., 2025). Third, interactions among multiple symbionts—including coinfections with *Wolbachia*, *Asaia*, and native gut bacteria—can generate emergent phenotypes not predictable from single-symbiont studies. These interactions may alter pathogen-blocking efficiency, vector competence, or host life-history traits. Additional uncertainties arise from the potential for horizontal transfer of engineered traits, attenuation of effector expression over time, and indirect effects on mosquito behavior, dispersal, and host preference (Cansado-Utrilla et al., 2025).

Addressing these challenges requires longitudinal field studies, systems-level modeling of host–microbe–pathogen networks, and integration of laboratory, ecological, and social science research. Co-design with local stakeholders, robust monitoring and evaluation frameworks, and clear criteria for pausing, adapting, or rolling back deployments are essential to ensure interventions are effective, sustainable, and socially acceptable. By combining the mechanistic understanding of mosquito microbiomes with responsible implementation strategies, microbiome-informed control has the potential to transform vector management from short-term, chemical-heavy interventions toward long-term, ecologically grounded public health solutions (Cansado-Utrilla et al., 2025).

This approach reframes mosquitoes not merely as targets to be suppressed, but as complex, host–microbe ecosystems in which carefully manipulated symbionts can reduce disease transmission at multiple biological scales—from individual midgut interactions to population-level pathogen suppression—while minimizing environmental disruption.

CONCLUSION

The mosquito microbiome is a dynamic, multi-tissue ecosystem that profoundly influences mosquito biology, vector competence, and disease transmission. Across life stages and anatomical sites—from larval guts to adult midguts, salivary glands, and reproductive tissues—microbial communities interact with host physiology, immunity, and environmental factors to shape development, nutrition, pathogen susceptibility, and reproductive success. These interactions are often context-dependent, with symbionts exerting both protective and permissive effects on viruses, and parasites, while also mediating vertical and sexual transmission. Harnessing this intricate host–microbe network has opened new avenues for vector control, including *Wolbachia*-based strategies, engineered symbionts, and microbiome-informed ecological interventions. Understanding and manipulating the mosquito microbiome offers a transformative framework for sustainable disease management, highlighting the importance of integrating microbial ecology, host biology, and epidemiology in the fight against mosquito-borne pathogens.

ACKNOWLEDGMENTS

The author (ST) acknowledges the Abraham Thomas Foundation for providing resources for this work.

REFERENCES

Accoti, A., Damiani, C., Nunzi, E., Cappelli, A., Iacomelli, G., Monacchia, G., Turco, A., D'Alò, F., Peirce, M. J., Favia, G., & Spaccapelo, R. (2023). Anopheline mosquito saliva contains bacteria that are transferred to a mammalian host through blood feeding. *Frontiers in Microbiology*, 14, 1157613. https://doi.org/10.3389/fmicb.2023.1157613

Alonso, D. P., Mancini, M. V., Damiani, C., Cappelli, A., Ricci, I., Alvarez, M. V. N., Bandi, C., Ribolla, P. E. M., & Favia, G. (2019). Genome reduction in the mosquito symbiont Asaia. *Genome Biology and Evolution*, 11, 1–10. https://doi.org/10.1093/gbe/evy255

Arellano, A. A., & Coon, K. L. (2022). Bacterial communities in carnivorous pitcher plants colonize and persist in inquiline mosquitoes. *Animal Microbiome*, 4, 13. https://doi.org/10.1186/s42523-022-00164-1

Azevedo-Santos, V. M., Vitule, J. R. S., Pelicice, F. M., García-Berthou, E., & Simberloff, D. (2017). Nonnative fish to control Aedes mosquitoes: A controversial, harmful tool. *BioScience*, 67(1), 84–90.

Bahia, A. C., Dong, Y., Blumberg, B. J., Mlambo, G., Tripathi, A., BenMarzouk-Hidalgo, O. J., Chandra, R., & Dimopoulos, G. (2014). Exploring Anopheles gut bacteria for Plasmodium blocking activity. *Environmental Microbiology*, 16, 2980–2994. https://doi.org/10.1111/1462-2920.12381

Barletta Ferreira, A. B., Bahia, A. C., Pitaluga, A. N., Barros, E., Gama dos Santos, D., Bottino-Rojas, V., Kubota, M. S., Oliveira, P. L., Pimenta, P. F. P., Traub-Csekö, Y. M., & Sorgine, M. H. F. (2022). Sexual dimorphism in immune responses and infection resistance in Aedes aegypti and other hematophagous insect vectors. *Frontiers in Tropical Diseases*, 3, 847109. https://doi.org/10.3389/fitd.2022.847109

Becking, T., Chebbi, M. A., Giraud, I., Moumen, B., Laverré, T., Caubet, Y., Peccoud, J., Gilbert, C., & Cordaux, R. (2019). Sex chromosomes control vertical transmission of feminizing Wolbachia symbionts in an isopod. *PLOS Biology*, 17, e3000438. https://doi.org/10.1371/journal.pbio.3000438

Bera, B. (2025). Mosquito ecology and disease transmission: Implications for global health and vector control. *Uttar Pradesh Journal of Zoology*, 46, 64–72. https://doi.org/10.56557/upjoz/2025/v46i84894

Boissière, A., Tchioffo, M. T., Bachar, D., Abate, L., Marie, A., Nsango, S. E., Shahbazkia, H. R., Awono-Ambene, P. H., Levashina, E. A., Christen, R., & Morlais, I. (2012). Midgut microbiota of the malaria mosquito vector Anopheles gambiae and interactions with Plasmodium falciparum infection. *PLOS Pathogens*, 8, e1002742. https://doi.org/10.1371/journal.ppat.1002742

Cansado-Utrilla, C., Saldaña, M.A., Golovko, G., Khanipov, K., Watson, R.K., Wild, A.L., Brettell, L.E., Weaver, S.C., Heinz, E., Hughes, G.L. (2025). Mosquito host background impacts microbiome-Zika virus interactions in field- and laboratory-reared Aedes aegypti. *Animal Microbiome*, 7(1), 116. doi: 10.1186/s42523-025-00482-0.

Cardoso-Jaime, V., & Dimopoulos, G. (2025). Anopheles gambiae phagocytic hemocytes promote Plasmodium falciparum infection by regulating midgut epithelial integrity. *Nature Communications*, 16, 1465. https://doi.org/10.1038/s41467-025-56313-y

Cirimotich, C. M., Dong, Y., Clayton, A. M., Sandiford, S. L., Souza-Neto, J. A., Mulenga, M., & Dimopoulos, G. (2011). Natural microbe-mediated refractoriness to Plasmodium infection in Anopheles gambiae. *Science*, 332(6031), 855–858. [Mentioned via summary of gut bacteria work]

Collins, C. M., Bonds, J. A. S., Quinlan, M. M., & Mumford, J. D. (2019). Effects of the removal or reduction in density of the malaria mosquito, Anopheles gambiae s.l., on interacting predators and competitors in local ecosystems. *Medical and Veterinary Entomology*, 33(1), 1–15.

Correa, M. A., Matusovsky, B., Brackney, D. E., & Steven, B. (2018). Generation of axenic Aedes aegypti demonstrates live bacteria are not required for mosquito development. *Nature Communications*, 9, 4464. https://doi.org/10.1038/s41467-018-07014-2

Díaz, S., Camargo, C., & Avila, F. W. (2021). Characterization of the reproductive tract bacterial microbiota of virgin, mated, and blood-fed Aedes aegypti and Aedes albopictus females. *Parasites & Vectors*, 14, 592. https://doi.org/10.1186/s13071-021-05093-7

Dickson, L. B., Jiolle, D., Minard, G., Moltini-Conclois, I., Volant, S., Ghozlane, A., Bouchier, C., Ayala, D., Paupy, C., Moro, C. V., & Lambrechts, L. (2017). Carryover effects of larval exposure to different environmental bacteria drive adult trait variation in a mosquito vector. *Science Advances*, 3, e1700585. https://doi.org/10.1126/sciadv.1700585

Dong, Y., Manfredini, F., & Dimopoulos, G. (2009). Implication of the mosquito midgut microbiota in the defense against malaria parasites. *PLOS Pathogens*, 5(5), e1000423. [Summarized in gut bacteria section]

Erickson, S. M., Xi, Z., Mayhew, G. F., Ramirez, J. L., Aliota, M. T., Christensen, B. M., & Dimopoulos, G. (2009). Mosquito infection responses to developing filarial worms. *PLOS Neglected Tropical Diseases*, 3, e529. https://doi.org/10.1371/journal.pntd.0000529

Facchinelli, L., Badolo, A., & McCall, P. J. (2023). Biology and behaviour of Aedes aegypti in the human environment: Opportunities for vector control of arbovirus transmission. *Viruses*, 15, 636. https://doi.org/10.3390/v15030636

Foster, W. A. (2024). Mosquito pollination of plants: An overview of their role and an assessment of the possible contribution of disease vectors. *Transgenic Research*, 33(5), 297–322.

Gao, H., Cui, C., Wang, L., Jacobs-Lorena, M., & Wang, S. (2020). Mosquito microbiota and implications for disease control. *Trends in Parasitology*, 36, 98–111. https://doi.org/10.1016/j.pt.2019.12.001

Gomes, F. M., Hixson, B. L., Tyner, M. D. W., Ramirez, J. L., Canepa, G. E., Alves E Silva, T. L., Molina-Cruz, A., Keita, M., Kane, F., Traoré, B., Sogoba, N., & Barillas-Mury, C. (2017). Effect of naturally occurring Wolbachia in Anopheles gambiae s.l. mosquitoes from Mali on Plasmodium falciparum malaria transmission. *Proceedings of the National Academy of Sciences USA*, 114, 12566–12571. https://doi.org/10.1073/pnas.1716181114

Gómez, M., Martinez, D., Muñoz, M., & Ramírez, J. D. (2022). Aedes aegypti and Ae. albopictus microbiome/virome: New strategies for controlling arboviral transmission? *Parasites & Vectors*, 15, 287. https://doi.org/10.1186/s13071-022-05401-9

Guerrero, D., Cantaert, T., & Missé, D. (2020). Aedes mosquito salivary components and their effect on the immune response to arboviruses. *Frontiers in Cellular and Infection Microbiology*, 10, 407. https://doi.org/10.3389/fcimb.2020.00407

Harrison, R. E., Yang, X., Eum, J. H., Martinson, V. G., Dou, X., Valzania, L., Wang, Y., Boyd, B. M., Brown, M. R., & Strand, M. R. (2023). The mosquito Aedes aegypti requires a gut microbiota for normal fecundity, longevity and vector competence. *Communications Biology*, 6, 1154. https://doi.org/10.1038/s42003-023-05545-z

Hegde, S., Rasgon, J. L., & Hughes, G. L. (2015). The microbiome modulates arbovirus transmission in mosquitoes. *Current Opinion in Virology*, 15, 97–102. https://doi.org/10.1016/j.coviro.2015.08.011

Hemingway, J., Ranson, H., Magill, A., Kolaczinski, J., Fornadel, C., Gimnig, J., Coetzee, M., Simard, F., Roch, D. K., Hinzoumbe, C. K., Pickett, J., Schellenberg, D., Gething, P., Hoppé, M., & Hamon, N. (2016). Averting a malaria disaster: Will insecticide resistance derail malaria control? *The Lancet*, 387(10029), 1785–1788.

Izquierdo, J. M. (2024). Novel insights into the sexual dimorphism-associated immune response. *Cellular & Molecular Immunology*, 21, 804–806. https://doi.org/10.1038/s41423-024-01177-3

Kaur, R., Shropshire, J. D., Cross, K. L., Leigh, B., Mansueto, A. J., Stewart, V., Bordenstein, S. R., & Bordenstein, S. R. (2021). Living in the endosymbiotic world of Wolbachia: A centennial review. *Cell Host & Microbe*, 29, 879–893. https://doi.org/10.1016/j.chom.2021.03.006

Luo, P., Yang, L., Zhu, Y., Liu, M., He, Y., Liu, C., & He, W. (2025). Second genome: Rhizosphere microbiome as a key external driver of nitrogen use efficiency in maize. *Plants*, 14, 3680. https://doi.org/10.3390/plants14233680

Mahor, S., & Gupta, H. (2026). Serratia species as paratransgenic vehicles: Potential applications in

vector-borne disease control. *Clinical Microbiology Reviews*, 39(1), e00280-25. https://doi.org/10.1128/cmr.00280-25
Miller, B. M., & Bäumler, A. J. (2021). The habitat filters of microbiota-nourishing immunity. *Annual Review of Immunology*, 39, 1–18. https://doi.org/10.1146/annurev-immunol-101819-024945
Mitri, C., & Vernick, K. D. (2012). Anopheles gambiae pathogen susceptibility: The intersection of genetics, immunity and ecology. *Current Opinion in Microbiology*, 15, 285–291. https://doi.org/10.1016/j.mib.2012.04.001
Mumford, J. D., & Carrasco, L. R. (2022). Economics of transgenic insects for field release. In Benedict, M. Q., & Scott, M. J. (Eds.), *Transgenic Insects* (pp. 518–532). CABI. https://doi.org/10.1079/9781800621176.0026
Narasimhan, S., Kazimirova, M., & Wikel, S. (2025). Editorial: Skin – confluence of vertebrate host defences, arthropod vectors, and vector-borne pathogens. *Frontiers in Immunology*, 16, 1688830. https://doi.org/10.3389/fimmu.2025.1688830
Nichols, H.L., & Coon, K.L. (2025). Leveraging microbial ecology for mosquito-borne disease control. *Trends in Parasitology*, 41, 670–684. https://doi.org/10.1016/j.pt.2025.06.010
Oliveira, J. H. M., Goncalves, R. L. S., Lara, F. A., Dias, F. A., Gandara, A. C. P., Menna-Barreto, R. F. S., Edwards, M. C., Laurindo, F. R. M., Silva-Neto, M. A. C., Sorgine, M. H. F., & Oliveira, P. L. (2011). Blood meal-derived heme decreases ROS levels in the midgut of Aedes aegypti and allows proliferation of intestinal microbiota. *PLOS Pathogens*, 7, e1001320. https://doi.org/10.1371/journal.ppat.1001320
Onchuru, T. O., Makhulu, E. E., Ronnie, P. C., Mandere, S., Otieno, F. G., Gichuhi, J., & Herren, J. K. (2024). The Plasmodium transmission-blocking symbiont, Microsporidia MB, is vertically transmitted through Anopheles arabiensis germline stem cells. *PLOS Pathogens*, 20, e1012340. https://doi.org/10.1371/journal.ppat.1012340
Onyango, M. G., Payne, A. F., Stout, J., Dieme, C., Kuo, L., Kramer, L. D., & Ciota, A. T. (2024). Aedes albopictus saliva contains a richer microbial community than the midgut. *Parasites & Vectors*, 17, 267. https://doi.org/10.1186/s13071-024-06334-1
Ouologuem, D. T., Dara, A., Kone, A., Ouattara, A., & Djimde, A. A. (2023). Plasmodium falciparum development from gametocyte to oocyst: Insight from functional studies. *Microorganisms*, 11, 1966. https://doi.org/10.3390/microorganisms11081966
Pascar, J., Middleton, H., & Dorus, S. (2023). Aedes aegypti microbiome composition covaries with the density of Wolbachia infection. *Microbiome*, 11, 255. https://doi.org/10.1186/s40168-023-01678-9
Prudêncio, M. (2020). In fairness to mosquitoes. *Trends in Parasitology*, 36(11), 876–877.
Ramirez, J. L., Short, S.M., Bahia, A.C., Saraiva, R.G., Dong, Y., Kang, S., et al. (2014). Chromobacterium Csp_P reduces malaria and dengue infection in vector mosquitoes and has entomopathogenic and in vitro antipathogen activities. *PLoS Pathogens*, 10(10), e1004398. https://doi.org/10.1371/journal.ppat.1004398.
Ranson, H., & Lissenden, N. (2016). Insecticide resistance in African Anopheles mosquitoes: A worsening situation that needs urgent action to maintain malaria control. *Trends in Parasitology*, 32(3), 187–196. [Cited in intro/microbiome section]
Ratcliffe, N. A., Mello, C. B., Castro, H. C., Dyson, P., & Figueiredo, M. (2024). Immune reactions of vector insects to parasites and pathogens. *Microorganisms*, 12, 568. https://doi.org/10.3390/microorganisms12030568
Rau, J., Werner, D., Beer, M., Höper, D., & Kampen, H. (2022). The microbial RNA metagenome of Aedes albopictus (Diptera: Culicidae) from Germany. *Parasitology Research*, 121, 2587–2599. https://doi.org/10.1007/s00436-022-07576-7
Ruan, X., Luo, J., Zhang, P., & Howell, K. (2022). The salivary microbiome shows a high prevalence of core bacterial members yet variability across human populations. *NPJ Biofilms and Microbiomes*, 8, 85. https://doi.org/10.1038/s41522-022-00343-7
Safaei, S., Derakhshan-Sefidi, M., & Karimi, A. (2025). Wolbachia: A bacterial weapon against dengue fever – A narrative review of risk factors for dengue fever outbreaks. *New Microbes and New Infections*, 65, 101578. https://doi.org/10.1016/j.nmni.2025.101578
Salgado, J. F. M., Premkrishnan, B. N. V., Oliveira, E. L., Vettath, V. K., Goh, F. G., Hou, X., Drautz-Moses, D. I., Cai, Y., Schuster, S. C., & Junqueira, A. C. M. (2024). The dynamics of the midgut microbiome in Aedes aegypti during digestion reveal putative symbionts. *PNAS Nexus*, 3, pgae317. https://doi.org/10.1093/pnasnexus/pgae317
Sallum, M.A.M., Obando, R.G., Carrejo, N., & Wilkerson, R.C. (2020). Identification keys to the Anopheles mosquitoes of South America (Diptera: Culicidae). I. Introduction. *Parasit Vectors*, 13, 583. https://doi.org/10.1186/s13071-020-04298-6
Saraiva, R. G., Fang, J., Kang, S., Angleró-Rodríguez, Y. I., Dong, Y., & Dimopoulos, G. (2018). Aminopeptidase secreted by Chromobacterium sp. Panama inhibits dengue virus infection by degrading the E protein. *PLOS Neglected Tropical Diseases*, 12, e0006443. https://doi.org/10.1371/journal.pntd.0006443
Sarma, D. K., Kumar, M., Dhurve, J., Pal, N., Sharma, P., James, M. M., Das, D., Mishra, S., Shubham, S., Kumawat, M., Verma, V., Tiwari, R. R., Nagpal, R., & Marotta, F. (2022). Influence of host blood

meal source on gut microbiota of wild caught Aedes aegypti, a dominant arboviral disease vector. *Microorganisms*, 10(2), 332. https://doi.org/10.3390/microorganisms10020332

Schmidt, K., & Engel, P. (2021). Mechanisms underlying gut microbiota–host interactions in insects. *Journal of Experimental Biology*, 224, jeb207696. https://doi.org/10.1242/jeb.207696

Seabourn, P. S., Weber, D. E., Spafford, H., & Medeiros, M. C. I. (2023). Aedes albopictus microbiome derives from environmental sources and partitions across distinct host tissues. *MicrobiologyOpen*, 12, e1364. https://doi.org/10.1002/mbo3.1364

Sharma, P., Sharma, S., Maurya, R. K., De, T. D., Thomas, T., Lata, S., Singh, N., Pandey, K. C., Valecha, N., & Dixit, R. (2014). Salivary glands harbor more diverse microbial communities than gut in Anopheles culicifacies. *Parasites & Vectors*, 7, 235. https://doi.org/10.1186/1756-3305-7-235

Shaw, W. R., Marcenac, P., Childs, L. M., Buckee, C. O., Baldini, F., Sawadogo, S. P., Dabiré, R. K., Diabaté, A., & Catteruccia, F. (2016). Wolbachia infections in natural Anopheles populations affect egg laying and negatively correlate with Plasmodium development. *Nature Communications*, 7, 11772. https://doi.org/10.1038/ncomms11772

Solfisburg, Q. S., Baldini, F., Baldwin-Hunter, B., Austin, G. I., Lee, H. H., Park, H., Freedberg, D. E., Lightdale, C. J., Korem, T., & Abrams, J. A. (2024). The salivary microbiome and predicted metabolite production are associated with Barrett's esophagus and high-grade dysplasia or adenocarcinoma. *Cancer Epidemiology, Biomarkers & Prevention*, 33(3), 371-380. doi: 10.1158/1055-9965.EPI-23-0652

Taracena, M. L., Bottino-Rojas, V., Talyuli, O. A. C., Walter-Nuno, A. B., Oliveira, J. H. M., Angleró-Rodriguez, Y. I., Wells, M. B., Dimopoulos, G., Oliveira, P. L., & Paiva-Silva, G. O. (2018). Regulation of midgut cell proliferation impacts Aedes aegypti susceptibility to dengue virus. *PLOS Neglected Tropical Diseases*, 12, e0006498. https://doi.org/10.1371/journal.pntd.0006498

Taylor, M. J., Bordenstein, S. R., & Slatko, B. (2018). Microbe profile: Wolbachia: A sex selector, a viral protector and a target to treat filarial nematodes. *Microbiology*, 164, 1345–1347. https://doi.org/10.1099/mic.0.000724

Tayyab, M., Zhao, Y., & Zhang, Y. (2025). Microbiome engineering to enhance disease resistance in aquaculture: Current strategies and future directions. *Frontiers in Microbiology*, 16, 1625265. https://doi.org/10.3389/fmicb.2025.1625265

van den Berg, H., Zaim, M., Yadav, R. S., Soares, A., Ameneshewa, B., Mnzava, A., Hii J, Dash AP, Ejov M. (2012). Global trends in the use of insecticides to control vector-borne diseases. *Environmental Health Perspectives*, 120(4), 577–582.

Wang, Y., Gilbreath, T. M., Kukutla, P., Yan, G., & Xu, J. (2011). Dynamic gut microbiome across life history of the malaria mosquito Anopheles gambiae in Kenya. *PLOS ONE*, 6, e24767. https://doi.org/10.1371/journal.pone.0024767

Wang, Y.-T., Shen, R.-X., Xing, D., Zhao, C.-P., Gao, H.-T., Wu, J.-H., Zhang, N., Zhang, H.-D., Chen, Y., Zhao, T.-Y., & Li, C.-X. (2021). Metagenome sequencing reveals the midgut microbiota makeup of Culex pipiens quinquefasciatus and its possible relationship with insecticide resistance. *Frontiers in Microbiology*, 12, 625539. https://doi.org/10.3389/fmicb.2021.625539

Wei, Y. L., Wu, Z., Li, R. L., & Tang, F. (2026). Review of selected mosquito-borne diseases: arboviruses (dengue, chikungunya, Zika, West Nile, Japanese encephalitis, yellow fever) and parasitic diseases (malaria, lymphatic Filariasis). *Frontiers in Public Health*, 13, 1712094.

World Health Organization. (2025). Malaria. www.who.int/news-room/fact-sheets/detail/malaria

Yin, Y., Liu, Y., Fan, J., Yu, L., Nie, M., Zhang, Z., Han, Q., & Liao, C. (2025). Analysis of midgut bacterial communities in larvae and adult mosquitoes of Aedes aegypti invaded by three different microorganisms. *Microorganisms*, 13, 248. https://doi.org/10.3390/microorganisms13020248

Zhang, X., & Raikhel, A. S. (2021). Hormonal regulation of microRNA expression dynamics in the gut of the yellow fever mosquito Aedes aegypti. *RNA Biology*, 18, 1682–1691. https://doi.org/10.1080/15476286.2020.1864181

Zheng, R., Wang, Q., Wu, R., Paradkar, P. N., Hoffmann, A. A., & Wang, G.-H. (2023). Holobiont perspectives on tripartite interactions among microbiota, mosquitoes, and pathogens. *ISME Journal*, 17, 1143–1152. https://doi.org/10.1038/s41396-023-01436-7

Zheng, X. L. (2020). Unveiling mosquito cryptic species and their reproductive isolation. *Insect Molecular Biology*, 29(6), 499–510.

CHAPTER ELEVEN

Ecology, Evolution, and the Gut Microbiome of Honeybees (*Apis mellifera*) in a Changing Agricultural Landscape

Alex Muppasani and Sunil Thomas

11.1 INTRODUCTION

Bees (Hymenoptera: Apoidea: Anthophila) comprise one of the most diverse and ecologically indispensable radiations of insects. Current global estimates recognize at least 20,473 described species of bees, with approximately 5227 species documented from North America. This diversity spans a remarkable array of ecological strategies, morphologies, and behavioral adaptations, reflecting a long evolutionary history closely tied to the diversification of flowering plants. As primary pollinators of angiosperms, bees are fundamental to the reproductive success of both wild and cultivated plants. Their pollination services enhance fruit and seed set, promote genetic exchange among plant populations, and sustain the structure and function of terrestrial ecosystems. Consequently, maintaining wild bee species diversity is critical for ensuring crop pollination services and preserving ecosystem stability and resilience (Kilpatrick et al., 2020).

Bee species exhibit striking variation in body size and life history. At the smallest extreme is *Perdita minima*, a solitary species measuring approximately 2 mm in length and widely regarded as the smallest bee in the world. In contrast, the largest known bee, *Megachile pluto* (Wallace's giant bee), can reach lengths exceeding 35 mm. Such morphological diversity corresponds with equally diverse nesting behaviors, foraging strategies, and degrees of social organization, ranging from solitary ground-nesting species to highly eusocial taxa.

Among bees, the western honeybee, *Apis mellifera*, is arguably the most ecologically and economically significant species globally. It belongs to the genus *Apis*, which includes 11 recognized honeybee species, all within the family Apidae. As the dominant managed pollinator within the order Hymenoptera, *A. mellifera* plays a central role in modern agriculture. Its large colony sizes, perennial social structure, and capacity for long-distance foraging make it uniquely suited for large-scale crop pollination.

Animal-mediated pollination contributes substantially to global agricultural productivity, and honeybees alone are estimated to support approximately 35% of human food production through direct and indirect pollination services (Vincze et al., 2024). Beyond managed crop systems, honeybees also facilitate the reproduction of numerous wild plant species, thereby contributing to biodiversity maintenance and ecosystem resilience (Leonard et al., 2018). Although widely recognized for producing honey, beeswax, royal jelly, and other hive products, the ecological importance of honeybees extends far beyond these commodities. Through their pollination activities, they function as keystone pollinators, underpinning global food security, sustaining plant–pollinator networks, and reinforcing the ecological integrity of natural and agricultural landscapes alike.

Originally native to Africa, Europe, and parts of Western Asia, *A. mellifera* has been intentionally

 DOI: 10.1201/9781003639176-11

transported by humans to every continent except Antarctica. Today, it persists in both managed apiaries and feral populations across diverse climatic zones, from temperate to tropical regions (Han et al., 2012). This broad geographic expansion has promoted remarkable genetic and phenotypic diversification. More than 30 subspecies have been described, organized into at least five major evolutionary lineages—A (African), M (Western and Northern European), C (Central and Southeastern European), O (Near Eastern and Central Asian), and Y (Ethiopian). These lineages reflect adaptation to local environmental pressures, including temperature extremes, seasonality, pathogen load, and floral resource availability (Ruttner, 1988).

The ecological success of honeybees is rooted not only in individual-level traits but in their colony-level organization as a superorganism. A colony operates as a highly integrated biological unit in which reproductive division of labor, chemical communication (pheromonal signaling), cooperative brood care, and symbiotic microbial communities function collectively to optimize survival and productivity. This emergent social structure enhances resilience to environmental stressors and enables rapid behavioral adaptation to fluctuating ecological conditions.

Honey, the principal product of honeybees, is a high-value natural food widely consumed for its distinctive sensory qualities and nutritional profile. Its composition—primarily fructose and glucose, along with organic acids, enzymes, amino acids, vitamins, minerals, and diverse phenolic compounds—varies according to botanical and geographical origin. These compositional differences account for variation in flavor, color, antioxidant capacity, and bioactivity. The functional and health-promoting properties of honey are largely attributed to its antioxidant, antibacterial, and anti-inflammatory activities, which have supported its traditional and modern therapeutic applications (Bolat et al. 2026; Cucu et al. 2021).

Economically, honey represents a rapidly expanding global commodity. Market analyses estimate the global honey market at approximately USD 10.51 billion in 2026, with projections reaching USD 18.33 billion by 2035 (Fortune Business Insights, 2026). This growth trajectory is driven by evolving consumer preferences, particularly rising health consciousness and demand for natural, minimally processed sweeteners. As awareness of the adverse health effects of refined sugars increases, honey is increasingly positioned as a preferred natural alternative within the broader sweeteners market. Premium and region-specific varieties—such as monofloral and organic honeys—have gained strong consumer interest, supported by expanding e-commerce platforms that facilitate global access to diverse products.

Environmental and agricultural factors also significantly shape market dynamics. Climate change poses mounting challenges to apiculture by altering flowering phenology, forage availability, and pathogen pressures, all of which influence honey yield and quality. Simultaneously, there is growing emphasis on sustainable and ethical beekeeping practices, traceable sourcing, and local production systems. In markets such as the United States, consumer preference for locally sourced honey reflects both environmental awareness and support for regional producers (Zander et al., 2013).

Industry trends further indicate increasing demand for raw and unfiltered honey, value-added functional formulations (e.g., immunity-focused products), and clean-label food ingredients. The organic honey segment, in particular, is expanding due to heightened attention to sustainability, biodiversity conservation, and organic certification standards. Collectively, these drivers support a positive long-term growth forecast for the global honey industry, reinforcing honey's dual identity as both an ecologically indispensable resource and a commercially valuable natural product.

Overall, the intersection of pollination ecology, food security, human health, and global trade underscores the multidimensional importance of *A. mellifera*. Protecting honeybee populations through sustainable management and environmental stewardship remains essential not only for maintaining ecosystem services but also for sustaining the expanding global honey economy.

11.2 TAXONOMY, EVOLUTIONARY LINEAGES, AND GLOBAL DISTRIBUTION

The taxonomic framework of the western honeybee, *A. mellifera*, has undergone repeated revision as analytical approaches have shifted from classical morphology to mitochondrial and whole-genome sequencing. Early classifications were grounded in morphometrics—especially forewing venation patterns, cubital index, body size, pilosity, and coloration—most comprehensively synthesized by Ruttner. These studies recognized geographically structured "morphological lineages," but subsequent work demonstrated that many diagnostic traits exhibit homoplasy, phenotypic plasticity, and sensitivity to environmental gradients and admixture, complicating phylogenetic interpretation (Ruttner, 1988; Mitteroecker & Schaefer, 2022).

The introduction of mitochondrial DNA (mtDNA) restriction fragment length polymorphism (RFLP) analysis in the 1990s, followed by sequencing of markers such as COI–COII and ND2, microsatellites, allozymes, and ultimately complete mitochondrial and nuclear genomes, substantially increased resolution. These approaches both corroborated and challenged aspects of morphometric taxonomy, revealing cryptic structure as well as unexpected affinities among distant populations (White & Densmore 1992).

11.2.1 Major Evolutionary Lineages

Current syntheses typically recognize at least five principal evolutionary lineages within *A. mellifera*: African (A), Western and Northern European (M), Southeastern European (C), Middle Eastern/Western Asian (O), and Ethiopian (Y). Some authors further propose an S (Syrian/Lebanese) lineage and subdivide portions of A into a Z sublineage (Carr, 2023).

Lineage A comprises predominantly African subspecies, including the following:

- *A. mellifera scutellata*
- *A. mellifera capensis*
- *A. mellifera monticola*
- *A. mellifera lamarckii*
- *A. mellifera intermissa*

This lineage displays substantial genetic and phenotypic heterogeneity across latitudinal and longitudinal gradients. Mitogenomic data often reveal deep splits between northern African and sub-Saharan clades, rendering A paraphyletic in some reconstructions (Tihelka et al., 2020).

Lineage M includes the dark honeybees of Western and Northern Europe, notably

- *Apis mellifera mellifera*
- *Apis mellifera sinisxinyuan*

Despite their disjunct geographic distributions, mitochondrial affinities suggest shared ancestry. Whole mitochondrial genome analyses frequently recover M as monophyletic and, in some trees, sister to all remaining lineages (Tihelka et al., 2020).

Lineage C (Southeastern Europe) encompasses economically important subspecies widely used in global apiculture:

- *Apis mellifera ligustica*
- *Apis mellifera carnica*

These bees are central to commercial breeding due to their productivity and generally gentle temperament.

Lineages O and Y:

- The O lineage traditionally included *A. mellifera anatoliaca*, *A. mellifera meda*, *A. mellifera syriaca*, *A. mellifera cypria*, and *A. mellifera adami*. However, molecular analyses have variably reassigned some of these to A (notably the proposed Z sublineage) or C, highlighting instability at lineage boundaries (Ilyasov et al., 2020).
- The Y lineage was proposed for Ethiopian bees that deviate markedly from canonical A profiles and form distinct clusters in microsatellite and mtDNA analyses (Carr, 2023).

11.2.2 Mitogenomic and Nuclear Perspectives

Whole mitochondrial genome analyses of ≥18 subspecies strongly support the monophyly of M, C, and O, while often splitting African populations into separate clades (Tihelka et al., 2020). In several reconstructions, *A. m. intermissa* appears near the base of the tree, whereas a clade containing *A. m. mellifera* and *A. m. sinisxinyuan* is sister to other lineages, implying complex historical biogeography.

Conversely, a mitogenomic meta-analysis of 22 nominal subspecies proposed a scenario in which European *A. m. mellifera* is basal and diversification proceeded north-to-south from Europe into Africa via Anatolia and the Levant, challenging both single "Out of Africa" and "Out of Asia" hypotheses (Holzer et al., 2026). These conflicting reconstructions underscore sensitivity to taxon sampling, clock models, and genomic partitions.

Nuclear whole-genome data add further nuance. A study of 251 individuals from 18 native subspecies inferred an Asian origin with at least three expansions into Africa and Europe through Western Asia (Dogantzis et al., 2021). In that framework, lineages cluster largely as traditionally recognized, but divergence times—estimated at several hundred thousand years—predate the last glacial cycles, suggesting subsequent climatic oscillations shaped fragmentation and secondary contact. Collectively, genomic evidence portrays *A. mellifera* as an adaptive radiation with multiple dispersal routes and recurrent gene flow, rather than a simple linear expansion.

11.2.3 Trait Syndromes and Ecological Adaptation

Despite taxonomic fluidity, lineages exhibit characteristic morphological and behavioral syndromes reflecting adaptation to regional climates:

M lineage (e.g., *A. m. mellifera*): Dark coloration, broader bodies, long hairs, winter hardiness, conservative brood rearing, and defensive tendencies suited to temperate climates (Henriques et al., 2019).

C lineage (e.g., *A. m. ligustica*, *A. m. carnica*): Lighter coloration, rapid spring buildup, high productivity, gentler temperament, and strong performance in managed systems.

A lineage: High swarming and absconding rates, strong defensiveness, and resilience to heat, drought, and certain pathogens—traits advantageous in tropical and subtropical environments (Utzeri et al., 2022).

Lineage differences extend to parasite tolerance. African-derived populations (notably *A. m. scutellata*) often exhibit natural tolerance to *Varroa destructor*, associated with hygienic behavior and shorter brood development, whereas many European M and C populations historically required chemical control (Whitfield et al., 2006).

11.2.4 Human-Mediated Dispersal and Hybridization

The global distribution of *A. mellifera* has been profoundly reshaped by apiculture. From the 17th century onward, European colonists introduced M lineage bees to the Americas and Oceania, later supplementing them with C lineage stocks. In the mid-20th century, African bees imported to Brazil escaped and hybridized with European populations, giving rise to Africanized honeybees that spread across the Americas (Calfee et al., 2020; Papa et al., 2022).

Africanized populations retain adaptive African traits—rapid reproduction, disease tolerance, climatic resilience—while often exhibiting heightened defensiveness and swarming (Zárate et al., 2022). Recent work indicates that selective breeding within hybrid populations can moderate defensiveness while preserving resilience (Bianchi et al., 2023).

Extensive queen importation has generated genetic mosaics in many regions. Single Nucleotide Polymorphism (SNP), microsatellite, and mtDNA surveys reveal gradients of M, C, and A ancestry, with clinal transitions rather than sharp subspecies boundaries (Wragg et al., 2022; Techer et al., 2017). Such admixture complicates diagnosis based on morphology or single markers and necessitates integrative, multilocus approaches.

11.2.5 Conservation and Breeding Implications

Introgression of non-native lineages can erode locally adapted gene complexes—such as overwintering traits in *A. m. mellifera* or drought tolerance in certain African subspecies—threatening regional genetic resources (Frunze et al., 2025). Conversely, controlled hybridization can introduce alleles for gentleness, productivity, and parasite resistance (Henriques et al., 2025).

The central challenge for conservation genetics and apicultural breeding is to balance maintenance of local adaptation with strategic introgression of beneficial alleles. Modern genomic tools—dense SNP arrays, whole-genome sequencing, and quantitative trait locus mapping—now enable fine-scale tracking of ancestry and adaptive variation, offering unprecedented capacity to manage this balance scientifically.

11.3 BEE COLONIES AS SUPERORGANISMS

The concept of the "superorganism" is foundational to modern understanding of honeybee biology. In the western honeybee, natural selection primarily acts at the level of the colony rather than the individual. The colony functions as an integrated biological unit in which the queen, workers, and drones operate analogously to differentiated cells in a multicellular organism, each contributing to collective survival and reproduction. In this framework, colony fitness—measured in successful reproduction through swarming and queen production—supersedes individual reproductive success (Seeley, 1989; Wilson & Sober, 1989).

A typical colony contains a single reproductive queen, thousands to tens of thousands of largely sterile female workers, and a seasonal cohort of haploid males (drones) whose primary function is mating with virgin queens from other colonies. This caste system is maintained by dynamic interactions among environmental conditions, nutrition, endocrine regulation, and pheromonal signaling (Pawar et al., 2024). Workers are facultatively sterile under normal colony conditions because queen and brood pheromones suppress ovarian activation and redirect physiological investment toward cooperative tasks such as brood care, nest construction, and foraging (Hrassnigg & Crailsheim, 2005; Amdam et al., 2012). Thus, worker sterility is not an intrinsic incapacity but a socially regulated state that stabilizes colony organization.

11.3.1 Division of Labor and Age Polyethism

Division of labor among workers is primarily structured by age polyethism—a temporal progression of tasks associated with worker age. Newly emerged workers (approximately days 1–3) engage in in-hive tasks such as cleaning brood cells and thermoregulating developing larvae. Between roughly days 3–12, workers act as nurse bees, feeding larvae and the queen using protein-rich secretions from their hypopharyngeal glands. From about days 12–18, "middle-aged" workers shift to comb construction, nectar processing, food storage, and ventilation. Finally, older workers (typically from day 18 onward) transition to guarding and then to foraging outside the hive (Seeley, 1983; Hrassnigg & Crailsheim, 2005).

Importantly, this schedule is highly plastic. Colony demography, brood quantity, resource availability, disease pressure, and environmental stressors can accelerate or delay task transitions. Workers can also revert to earlier roles—for example, foragers returning to nursing behavior when young workers are scarce—demonstrating that division of labor emerges from decentralized feedback rather than rigid programming (Rueppell et al., 2007). Such flexibility allows the colony to maintain homeostasis despite fluctuating internal and external conditions.

11.3.2 Endocrine Regulation: Juvenile Hormone and Vitellogenin

The proximate regulation of age polyethism is governed by an endocrine network centered on juvenile hormone (JH) and vitellogenin, a yolk precursor protein with pleiotropic functions. In young nurse bees, vitellogenin titers are high; the protein supports brood food production and provides antioxidant protection. As workers age, vitellogenin levels decline while JH titers rise, promoting the behavioral transition to foraging (Amdam et al., 2012).

Experimental knockdown of the vitellogenin gene accelerates foraging onset and biases workers toward nectar rather than pollen collection, demonstrating that vitellogenin both delays the nurse-to-forager transition and influences foraging specialization. Conversely, elevated JH levels are associated with increased locomotor activity and risk-taking typical of foragers. Brood pheromone and queen mandibular pheromone (QMP) modulate this endocrine balance by suppressing JH titers and prolonging nursing behavior when brood demand is high (Ihle et al., 2015). Through these hormonal feedback loops, worker physiology becomes tightly aligned with colony-level needs.

11.3.3 Pheromonal Communication and Colony Integration

Pheromones serve as the principal information channels that coordinate thousands of individuals into a cohesive functional entity. QMP, a complex blend including 9-oxo-2-decenoic acid (9-ODA), 9-hydroxy-2-decenoic acid (9-HDA), methyl p-hydroxybenzoate, and 4-hydroxy-3-methoxyphenylethanol, exerts both "releaser" and "primer" effects. It elicits retinue behavior, inhibits worker ovary activation, delays foraging onset, and modulates neuroendocrine signaling pathways (Bortolotti & Costa, 2014; Ihle et al., 2015).

Brood pheromone, composed largely of fatty acid esters produced by larvae, similarly influences worker physiology by maintaining nursing behavior and suppressing premature foraging. Alarm pheromones—such as isopentyl acetate released from the sting apparatus and 2-heptanone from mandibular glands—rapidly recruit defenders and coordinate colony-level responses to predators or intruders (Bortolotti & Costa, 2014). Together, queen, brood, and worker pheromones integrate signals about reproductive status, brood demand, and environmental threats, ensuring coordinated colony responses.

11.3.4 Foraging as Colony–Environment Interface

Foraging behavior links the internal organization of the superorganism to its external ecological context. Honeybee foragers typically exploit floral resources within a radius of 1–3 km, but they can exceed this range under resource scarcity, covering areas up to tens of square kilometers (Visscher & Seeley, 1982). Individual foragers may perform 10–15 trips per day during peak activity, collectively visiting thousands of flowers and sampling nectar, pollen, water, and resins (Samuelson & Leadbeater, 2017).

Recruitment communication, particularly the waggle dance, allows for dynamic allocation of foraging effort to profitable patches. Task specialization—such as nectar versus pollen foraging—is influenced by internal physiological state, including vitellogenin titers, as well as pheromonal feedback from brood and returning foragers (Amdam et al., 2012). Through these mechanisms, the colony operates as a distributed decision-making network capable of adaptive resource exploitation.

11.3.5 Colonies as Biomonitoring Systems

During foraging, bees inevitably collect environmental contaminants alongside legitimate resources. Heavy metals, pesticides, polycyclic aromatic hydrocarbons, and airborne particulates adhere to the bees' setae or are ingested and subsequently deposited into hive matrices such as honey, wax, bee bread, propolis, and royal jelly (Perugini et al., 2011). Microorganisms—including bacteria, fungal spores, and viruses—are similarly vectored into the hive environment.

Because thousands of foragers collectively sample air, water, and vegetation across large areas over extended periods, colonies effectively integrate environmental signals across space and time. This integrative property makes honeybee colonies powerful bioindicators of landscape-level contamination in agricultural and urban systems (Donaldson-Matasci & Dornhaus, 2012). Chemical analyses of hive products can reveal spatial patterns and temporal trends in pesticide exposure and heavy metal deposition.

11.3.6 The Colony as a Self-Organizing Biological Network

Viewed through the superorganism lens, the honeybee colony is a self-organizing, chemically mediated network in which division of labor, endocrine regulation, and pheromonal communication converge to coordinate thousands of individuals into a single adaptive entity. Physiological feedback loops within workers are embedded in colony-wide information flows, allowing the superorganism simultaneously to rear brood, exploit fluctuating resources, defend against threats, and buffer environmental perturbations. In this sense, the colony is not merely a collection of insects but a higher-level biological individual—an emergent system whose coherence and fitness arise from tightly integrated social physiology.

11.4 ECONOMIC AND ECOLOGICAL VALUE OF HONEYBEE POLLINATION

The economic contributions of *A. mellifera* can be divided into two broad categories: marketable hive products (honey, beeswax, propolis, royal jelly) and the ecosystem service of pollination. Although hive products generate substantial revenue, the monetary value of pollination services far exceeds that of honey production at both regional and global scales.

11.4.1 Global Economic Significance

According to assessments by the Intergovernmental Science-Policy Platform on Biodiversity and Ecosystem Services, approximately 75% of the world's leading food crops benefit to some extent from animal pollination, and nearly 90% of wild flowering plant species depend at least partially on animal pollinators for sexual reproduction (Imperatriz-Fonseca et al., 2016). When translated into economic terms, an estimated 5–8% of global crop production volume is directly attributable to animal pollination, corresponding to an annual market value of approximately 235–577 billion USD (2015 dollars) (Khalifa et al., 2021). These figures likely underestimate total value because they do not fully account for improvements in crop quality, yield stability, and resilience.

Within the broader guild of pollinators—including wild bees, butterflies, flies, beetles, birds, and bats—managed honeybees provide the majority of commercial pollination services worldwide. Their colonies can be transported over long distances, split and propagated, and deployed at densities tailored to crop requirements, making them uniquely suited to intensive agricultural systems.

11.4.2 National-Scale Economics: The United States

At the national level, the scale of pollination services becomes particularly clear in countries with extensive pollinator-dependent specialty crops. In the United States, pollinators collectively contribute tens of billions of dollars annually to agricultural production, with managed honeybee colonies supplying most commercial pollination for almonds, apples, berries, melons, and vegetable seed crops (U.S. Fish & Wildlife Service, 2025).

Recent economic assessments indicate a structural shift in beekeeper revenue streams. In 2024, U.S. beekeepers earned approximately 400 million USD from pollination services, surpassing the roughly 361–362 million USD generated from domestic honey production in the same year (USDA, 2025). This inversion underscores that pollination services—rather than honey—have become the primary source of income for many commercial operations, tightly linking apiculture to the economic performance of pollinator-dependent crops.

11.4.3 Almonds: Industrial-Scale Pollination

Almond production in California exemplifies the industrial scale and economic centrality of honeybee pollination. Almonds represent the largest single pollination market in the United States, accounting for approximately 81% of total pollination service revenue for U.S. beekeepers in 2024. Almond pollination alone generated an estimated 325.8 million USD that year (USDA, 2025).

Between 2022 and 2024, average almond pollination fees were approximately 188 USD per colony—more than double the roughly 70 USD per colony typical for other pollinator-dependent crops. For 2024 specifically, fees averaged near 181 USD per colony (USDA, 2024). For almond growers, pollination costs can represent around 11% of variable production expenses, compared with less than 3% for most other specialty crops (Hitaj et al., 2018). These figures highlight the mutual economic dependence of growers and beekeepers on reliable, large-scale honeybee pollination.

Comparable, though smaller-scale, dynamics occur in blueberries, cherries, apples, cucurbits, and other specialty crops, where colony stocking densities and rental fees are calibrated to optimize fruit set and yield.

11.4.4 Crop Quality, Nutrition, and Human Health

Beyond increasing yield quantity, animal pollination enhances multiple quality parameters, including fruit size, symmetry, shelf life, and nutrient density (Gazzea et al., 2023). Many pollinator-dependent crops—particularly fruits, nuts, and oilseeds—contribute disproportionately to micronutrient availability in human diets, even if they supply a smaller share of total caloric intake relative to staple cereals.

A global analysis integrating FAO production data, nutrient composition databases, and pollinator dependency categories found that fully or partially animal-pollinated crops provide more than 90% of global dietary vitamin C, nearly all lycopene, the majority of β-cryptoxanthin and β-tocopherol (vitamin E), most dietary lipids derived from plants, substantial portions of vitamin A precursors, calcium, fluoride, and large fractions of folate (Blili et al., 2025). Consequently, pollinator declines are projected to disproportionately reduce access to essential micronutrients and healthy fats, even if total caloric production remains relatively stable.

Modeling studies further suggest that reductions in pollinator populations could exacerbate micronutrient deficiencies—particularly in low- and middle-income countries where diets are already constrained and access to supplementation is limited (Smith et al., 2022). Thus, pollination services are not merely an agricultural input but a determinant of global nutritional security.

11.4.5 Ecological Importance and Biodiversity

From an ecological perspective, pollination underpins the structure and persistence of terrestrial ecosystems. A global synthesis by Ollerton et al. (2011) estimated that approximately 87.5% of wild flowering plant species—roughly 308,000 species—depend to some extent on animal pollination, with dependence reaching about 94% in tropical communities and approximately 78% in temperate regions. Successful pollination influences seed set, seed quality, genetic diversity, and gene flow, thereby shaping plant population viability and evolutionary trajectories.

Declines in pollinator abundance or diversity can reduce seed production and disrupt pollen transfer among plant populations, leading to altered recruitment patterns and shifts in community composition. These plant-level effects cascade through food webs. Reduced fruit and seed availability diminishes resources for frugivorous birds and mammals, while declines in nectar and pollen affect specialist pollinators and nectarivores, amplifying biodiversity loss.

Animal-pollinated shrubs and trees also contribute to ecosystem functions such as carbon sequestration, soil stabilization, and hydrological regulation. Their decline can exacerbate erosion, reduce habitat complexity, and alter nutrient cycling—particularly in fragmented or degraded landscapes (Vahidi et al., 2025).

11.4.6 Complementarity with Wild Pollinators

Although managed honeybees dominate commercial pollination markets, diverse wild pollinator communities often complement honeybee activity. Wild bees may forage under cooler or cloudier conditions, visit flowers at different times of day, or more effectively pollinate certain floral morphologies. In agricultural mosaics, this functional complementarity enhances yield stability and resilience to environmental fluctuations (Werling, 2015). Consequently, maintaining pollinator diversity is essential not only for biodiversity conservation but also for stabilizing agricultural production.

11.4.7 Pollination as a Food Security Imperative

Taken together, the economic, nutritional, and ecological evidence demonstrates that pollination is foundational to modern food systems. Pollinator-dependent crops are major sources of vitamins A and C, folate, iron, and plant-derived lipids essential for cardiovascular and neurological health (Khan & Amanullah, 2024). When nutritional vulnerability is combined with the economic exposure of specialty crop sectors and the ecological consequences of disrupted plant–pollinator networks, safeguarding honeybees and other pollinators emerges as a public health, conservation, and food security priority.

In this broader perspective, the value of pollination transcends its contribution to crop yield. It sustains biodiversity, supports ecosystem stability, enhances dietary quality, and underwrites significant sectors of the global agricultural economy. Protecting pollinator populations—including managed *A. mellifera* and diverse wild species—is therefore integral to ensuring resilient ecosystems and nutritionally adequate food systems in the face of environmental change (Katumo et al., 2022).

11.5 COMMERCIAL BEEKEEPING, MIGRATORY POLLINATION, AND COLONY STRESS

Modern commercial beekeeping in many industrialized agricultural systems is organized around large-scale migratory pollination. In this model, colonies of the western honeybee, *A. mellifera*, are transported repeatedly over hundreds to thousands of kilometers to service sequential crop blooms. This practice is particularly prominent in the United States, where pollination contracts—especially for almonds—constitute a major portion of commercial beekeepers' annual income (Patidar et al., 2017). While migratory pollination is indispensable for sustaining high yields in pollinator-dependent crops, it exposes colonies to a suite of chronic and acute stressors that can compromise health and survival.

11.5.1 Transport Stress and Environmental Transitions

Frequent long-distance transport imposes mechanical, thermal, and physiological stress. Colonies are typically palletized, entrances screened, and trucked overnight. During transit, bees experience vibration, disrupted cluster structure, elevated CO_2 concentrations, limited ventilation, and extended confinement that restricts foraging and normal thermoregulation. Such conditions can alter brood care dynamics and impair queen performance (Kulikowski, 2016).

Upon arrival at new pollination sites, colonies must rapidly acclimate to unfamiliar floral resources, climatic regimes, and landscape contexts. These abrupt environmental transitions can create short-term mismatches between colony nutritional demands and available forage. Repeated cycles of transport followed by rapid adjustment to new foraging environments have been associated with shifts in pathogen prevalence and immune status, although outcomes vary depending on management practices and landscape quality.

11.5.2 High Stocking Densities and Pathogen Transmission

Major pollination events often aggregate thousands of colonies from geographically distant operations into concentrated areas. In almond orchards in California—the largest pollination market in North America—recommended stocking densities are commonly around two colonies per acre. This aggregation dramatically increases local densities of bees, parasites, and pathogens (Cavigli et al., 2016).

Among the most consequential parasites is *V. destructor*, an ectoparasitic mite that vectors multiple RNA viruses and feeds on developing and adult bees. Surveys of migratory colonies participating in almond pollination have documented frequent co-infections involving deformed wing virus, black queen cell virus, Lake Sinai viruses, Kashmir bee virus, Israeli acute paralysis virus, the microsporidian *Nosema ceranae*, and bacterial pathogens such as *Paenibacillus larvae* and *Melissococcus plutonius* (Glenny et al., 2017).

In one longitudinal study of commercially managed colonies involved in almond pollination, colonies harbored an average of five to six pathogens simultaneously. Higher infestation levels of *V. destructor* were strongly associated with elevated deformed wing virus titers and poorer colony health assessments (Glenny et al., 2017). Dense aggregations of colonies, combined with drifting, robbing behavior, and shared floral resources, create conditions conducive to rapid horizontal transmission and regional amplification of pathogens.

11.5.3 Nutritional Stress in Monoculture Landscapes

The nutritional landscape in intensive agricultural systems is frequently suboptimal for honeybee health. Large monocultures such as almonds, canola, or sunflower provide abundant nectar and pollen during a brief bloom window, followed by sharp declines in resource availability. Even during bloom, pollen derived from a single plant species can lack sufficient diversity in essential amino acids, lipids, sterols, vitamins, and micronutrients required for optimal immune function and brood rearing (Kaur et al., 2024).

Experimental comparisons of bees fed polyfloral pollen versus nutritionally limited pollen (e.g., *Eucalyptus grandis*) demonstrate that nutritionally stressed bees exhibit altered gut microbiota composition, including changes in key symbionts such as *Bartonella apis*. These bees also show reduced expression of immune-related genes (e.g., hymenoptaecin, glucose oxidase, lysozyme), lower vitellogenin levels, and increased proliferation of *N. ceranae* (Castelli et al., 2020). Such findings indicate that nutritionally restricted diets—typical of monoculture-dominated landscapes—can simultaneously disrupt microbial homeostasis, suppress immunity, and facilitate pathogen development.

11.5.4 Synergistic Stressors: Parasites, Pesticides, and Physiology

Nutritional stress rarely acts in isolation. Migratory colonies are also exposed to agrochemicals, including insecticides, fungicides, and herbicides applied within or adjacent to pollinated crops. While some compounds may have low acute toxicity at field-relevant doses, sublethal exposures can interact synergistically with parasites and pathogens, exacerbating physiological stress (Siviter et al., 2021).

Combined stressors have been linked to reduced worker longevity, impaired hypopharyngeal gland development (affecting brood food production), decreased brood viability, and diminished queen fecundity. Chronic stress may increase the probability of queen failure—a leading cause of colony loss in commercial operations. Colonies returning from almond pollination frequently exhibit elevated viral or parasite loads and may require intensive interventions, including supplemental feeding, miticide treatments, and queen replacement, before being strong enough for subsequent pollination contracts (Mayack et al., 2025).

11.5.5 Economic Benefits and Biological Costs

Large-scale migratory pollination systems thus embody a fundamental trade-off. Economically, they are indispensable for maintaining high yields in pollinator-dependent crops and constitute a primary revenue source for commercial beekeepers. Biologically, however, they impose cumulative stress through repeated transport, high-density pathogen exposure, nutritional imbalance, and chemical contact.

From a systems perspective, migratory pollination amplifies connectivity among colonies across regions, effectively creating a continent-scale epidemiological network. While this interconnectedness enhances agricultural productivity, it also facilitates the spread of parasites and pathogens and magnifies the consequences of management failures. Understanding and mitigating these compounded stressors—through improved nutrition, integrated pest management, reduced pesticide exposure, and strategic apiary placement—remains central to sustaining both commercial apiculture and the crop production systems that depend upon it.

11.6 HIGH-VALUE BIOACTIVE HIVE PRODUCTS: PROPOLIS AND ROYAL JELLY

Beyond honey and beeswax, the western honeybee produces several biologically active substances of ecological and biomedical importance, most notably propolis and royal jelly (Sarapa et al., 2025). These materials serve critical colony-level functions while also attracting increasing attention for their pharmacological potential in human health.

11.6.1 Propolis: The Colony's Chemical Shield

Propolis—often called "bee glue"—is a resinous composite material that worker bees collect from plant exudates (buds, bark fissures, leaf petioles, and wounds) and modify with beeswax and salivary enzymes. Inside the hive, propolis is used to seal cracks, reinforce comb structures, smooth interior surfaces, and encapsulate intruders or debris too large to remove physically. In natural tree cavities, colonies construct a continuous "propolis envelope" along interior walls, forming a chemically active barrier between the nest environment and the external world.

11.6.2 Chemical Composition and Variability

Chemically, propolis is rich in polyphenolic compounds, especially flavonoids and phenolic acids, along with terpenoids, aromatic aldehydes, esters, and waxes. Its composition varies geographically according to local flora. In temperate regions, "poplar-type" propolis—derived primarily from *Populus* spp.—is common, whereas in tropical areas, propolis may originate from genera such as *Baccharis* or *Clusia* (Wagh, 2013).

Among the best-characterized constituents are flavonoids such as pinocembrin, galangin, chrysin, and pinostrobin, as well as phenolic derivatives including caffeic acid and caffeic acid phenethyl ester (CAPE). These compounds exhibit antimicrobial properties through mechanisms such as disruption of microbial membranes, inhibition of nucleic acid and protein synthesis, and interference with enzymatic systems (Al-Khayri et al., 2022).

11.6.3 Antimicrobial and Immunomodulatory Properties

In vitro and in vivo studies demonstrate that propolis extracts inhibit the growth of numerous Gram-positive bacteria, including *Staphylococcus aureus* and *Streptococcus* spp., fungi such as *Candida albicans*, and certain protozoa (El Adaouia Taleb et al., 2020). Antimicrobial effects are frequently dose-dependent and linked to specific polyphenols, including pinocembrin, galangin, p-coumaric acid, and CAPE (Zulhendri et al., 2021).

Beyond antimicrobial activity, propolis exhibits antioxidant, anti-inflammatory, and immunomodulatory effects in vertebrate systems. Reported actions include modulation of oxidative stress pathways, suppression of pro-inflammatory mediators, and regulation of cell cycle and apoptotic signaling. Experimental studies suggest cardioprotective, neuroprotective, antidiabetic, and anticancer properties, though high-quality clinical trials remain limited (Zullkiflee et al., 2022). These findings have stimulated interest in propolis as a functional food ingredient and complementary therapeutic agent.

11.6.4 Propolis and Social Immunity

Within the hive, propolis contributes to what is often termed "social immunity." Colonies housed in environments where a propolis envelope is present show reduced baseline expression of immune-related genes such as hymenoptaecin and abaecin in young workers compared with colonies lacking such an envelope (Shanahan et al., 2024). Simultaneously, bees from propolis-rich environments often harbor lower surface loads of opportunistic bacteria and a more stable microbiota.

This pattern suggests that propolis acts as an external antimicrobial barrier, lowering environmental pathogen pressure and reducing the need for costly constitutive immune activation. By decreasing baseline immune expenditure, colonies may conserve energy for brood production, foraging, and longevity. In this way, propolis operates at the interface between nest architecture, microbial ecology, and colony-level health—an emergent defense strategy shaped by coevolution among bees, plants, and microbes.

11.6.5 Royal Jelly: A Nutritional and Developmental Signal

Royal jelly is a protein- and lipid-rich secretion produced by the hypopharyngeal and mandibular glands of young nurse workers. All larvae receive royal jelly during their first days of development, but larvae destined to become queens are fed exclusively and abundantly with royal jelly throughout larval life and into adulthood.

11.6.6 Composition and Bioactive Molecules

Royal jelly typically contains 9–18% proteins, 3–8% lipids, 7–18% carbohydrates, free amino acids, vitamins, minerals, and diverse bioactive peptides. A distinctive feature is the presence of major royal jelly proteins (MRJP1–MRJP9), which dominate its protein fraction.

Its lipid fraction includes unusual medium-chain fatty acids rarely found elsewhere in nature. Among the most studied is trans-10-hydroxy-2-decenoic acid (10-HDA), along with related compounds such as 10-hydroxydecanoic acid (10-H2DA) and 1,10-decanedioic acid (Xue et al., 2025). These fatty acids are considered biochemical markers of royal jelly authenticity and contribute to its biological activity.

11.6.7 Biomedical Potential

Experimental research indicates that royal jelly and 10-HDA possess antibacterial, anti-inflammatory, antioxidant, and immunomodulatory properties. In cultured human dermal fibroblasts, royal jelly and 10-HDA accelerate wound closure in scratch assays, reduce UVA-induced cytotoxicity and reactive oxygen species production, inhibit matrix metalloproteinases (MMP-1 and MMP-3), and stimulate collagen synthesis. These effects are mediated in part through modulation of JNK and p38 MAPK signaling pathways (Yang et al., 2018).

More recently, royal jelly–derived extracellular vesicles (RJEVs) have been shown to enhance re-epithelialization and modulate inflammatory responses in murine wound models, suggesting that vesicle-associated proteins, lipids, and small RNAs may contribute to its regenerative properties (Álvarez et al., 2023). Although promising, translational applications in dermatology and regenerative medicine require further controlled clinical validation.

11.6.8 Royal Jelly and Caste Determination

Within honeybee biology, royal jelly plays a central role in caste differentiation. Larvae genetically capable of developing into either queens or workers follow divergent developmental trajectories based on diet. Queen-destined larvae receive continuous, high-volume royal jelly feeding, whereas worker-destined larvae are switched after several days to a diet richer in pollen and lower in royal jelly (Alhosin, 2023).

This nutritional divergence triggers profound changes in gene expression, endocrine signaling, and epigenetic regulation. Altered DNA methylation patterns, modulation of JH levels, and shifts in insulin/insulin-like and TOR signaling pathways collectively drive the development of a queen phenotype characterized by larger body size, fully developed ovaries, high fecundity, and dramatically extended lifespan relative to workers of identical genetic background (Sedley, 2020; Guo et al., 2021).

Components of royal jelly—including MRJPs and 10-HDA—have been implicated in regulating these developmental pathways, although the precise molecular mechanisms remain under active investigation. In this context, royal jelly functions not merely as nutrition but as a developmental signal that orchestrates phenotypic plasticity within the honeybee superorganism.

Together, propolis and royal jelly illustrate how products of *A. mellifera* serve dual ecological and biomedical roles. Propolis enhances colony-level disease resistance through chemically mediated environmental modification, while royal jelly governs caste differentiation and exhibits bioactive properties with translational potential. These hive products exemplify the biochemical sophistication of social insects and highlight the intersection of evolutionary ecology, chemical biology, and applied biomedical research.

11.7 STRUCTURE OF THE HONEYBEE GUT MICROBIOME

The gut microbiome of adult worker honeybees is one of the most streamlined and host-specialized microbial communities described in animals, providing a powerful framework for studying host–microbe coevolution. In contrast to the hundreds of taxa commonly observed in mammalian intestines, healthy workers harbor a relatively small but stable "core" consortium typically consisting of five to nine dominant bacterial lineages. These include *Snodgrassella*, *Gilliamella*, *Frischella*, *Bombilactobacillus* (Firm-4), Firm-5 *Lactobacillus* (notably the *Lactobacillus melliventris* clade), *Bifidobacterium*, and less abundant yet characteristic members such as *Apibacter* and *B. apis*. Comparative surveys across continents consistently recover these taxa, underscoring their global conservation and functional importance (Raymann & Moran, 2018). Phylogenomic evidence further indicates that most of these bacteria are largely restricted to social corbiculate bees, rarely persisting outside the host, which supports a history of long-term co-diversification, ecological specialization, and transmission primarily through social contact rather than environmental acquisition (Cerqueira et al., 2024).

Anatomical compartmentalization of the digestive tract plays a central role in structuring this microbiome. The alimentary canal is divided into the crop (honey stomach), midgut, ileum, and rectum, each characterized by distinct physicochemical conditions such as nutrient availability, pH gradients, oxygen diffusion, and transit dynamics (De Paula et al., 2022). The crop, functioning mainly as a temporary storage organ for nectar and water, typically contains low microbial biomass and a transient community resembling that of hive substrates like honey and bee bread, reflecting frequent emptying and high osmotic pressure. Similarly, the midgut—dominated by digestive enzymes, the peritrophic matrix, and rapid flow—supports only sparse and unstable populations. In contrast, the hindgut provides a comparatively stable environment where microbial densities can reach 10^8–10^9 cells per gut, forming spatially organized communities that perform the majority of metabolic interactions (Agriculture Institute, 2023).

Within the ileum, microbial organization is highly structured. *Snodgrassella alvi* and *Gilliamella apicola* form dense, stratified biofilms along the epithelial surface, supported by the ileum's narrow lumen and longitudinal folds. *S. alvi* typically occupies the innermost layer adjacent to host tissue, where it consumes residual oxygen and helps maintain a low-oxygen environment conducive to obligately fermentative partners. *G. apicola* is more luminal and specializes in degrading complex plant polysaccharides derived from pollen, including pectins and hemicelluloses. Metabolomic and genomic studies reveal extensive metabolic cross-feeding within this niche; for example, carbohydrate fermentation by *G. apicola* generates organic acids and intermediates such as pyruvate that can be utilized by *S. alvi* and other community members, illustrating a tightly integrated metabolic network (Luo et al., 2024).

The rectum represents the primary fermentation chamber of the bee gut. Here, Firm-4 (*Bombilactobacillus*), Firm-5 *Lactobacillus*, and *Bifidobacterium* species dominate, collectively transforming residual dietary carbohydrates, pollen wall components, and host-derived substrates into lactic acid and other short-chain fatty acids. These metabolites contribute to host nutrition, reduce luminal pH, and create an environment that suppresses opportunistic pathogens. Because fecal material accumulates in the rectum prior to defecation flights, this compartment also serves as a major reservoir for microbial transmission within colonies, enabling newly emerged workers to acquire their microbiota through social interactions and fecal–oral exposure (Zheng et al., 2019; Luo et al., 2024).

Although present at lower abundance, several non-core taxa appear to play key roles in immune modulation and community stability. *Frischella perrara*, localized near the pylorus, induces a melanized epithelial "scab," a host immune response thought to prime systemic defenses and shape gut immune homeostasis (Engel et al., 2015). Members of the Acetobacteraceae, including *Commensalibacter* and *Bombella*, are frequently detected in the crop and midgut and may contribute to organic acid production and early colonization processes. *Apibacter* species are implicated in nitrogen recycling, while *B. apis*, typically associated with the rectum, has been linked to nitrogen metabolism and potential modulation of host physiology, though its functional role remains under investigation (Li et al., 2022).

Zheng et al. (2018) determined the honeybee microbiome. The bacteria in the honeybee gut include *Lactobacillus apis*, *Lactobacillus helsingborgensis*, *Lactobacillus kimbladii*, *Lactobacillus kullabergensis*, *L. melliventris*, *Lactobacillus mellifer*, *Lactobacillus mellis*, *Bifidobacterium* sp., *Bifidobacterium asteroides*, *Bifidobacterium coryneforme*, *Bifidobacterium indicum*, *S. alvi*, *G. apicola*, *F. perrara*, *B. apis*, *Commensalibacter* sp., *Apibacter adventoris*, *Parasaccharibacter apium*, *Lactobacillus kunkeei*, *Fructobacillus fructosus*, *Saccharibacter* spp.

Winter poses a substantial challenge for honeybee colonies in temperate regions, where prolonged periods of low temperature and limited floral resources threaten colony survival. Over the past two decades, beekeepers have reported consistently high overwinter colony losses. These losses have been linked to multiple interacting stressors, including cold temperature stress, geographic relocation associated with climate change, genetic background (genotype), pathogens and parasites, pesticide exposure, and inadequate nutrition (Hristov et al., 2020).

To survive winter, a colony of the western honeybee undergoes coordinated behavioral and physiological adjustments. Worker bees reduce foraging and general activity, while endocrine-regulated behavioral changes shift colony organization toward winter survival. Individuals accumulate greater nutrient reserves, particularly lipids and proteins, and exhibit extended longevity compared with summer workers. At the colony level, bees form a thermoregulatory cluster, generating and conserving heat through collective metabolic activity. These adaptations allow the colony to maintain internal temperatures necessary for survival despite harsh external conditions.

Building on this understanding of overwintering biology, Brar et al. (2025) examined the gut microbiota dynamics of overwintering adult worker bees from two commercial strains: Bolton bees and Mann Lake bees. These strains of *A. mellifera* were maintained under two storage conditions—indoor storage at 6°C and outdoor storage under natural environmental conditions—across the early winter months of October, November, and December. Across all samples, Lactobacilli dominated the gut communities of overwintering workers. An unclassified *Lactobacillus* species showed significantly higher abundance in November, whereas *L. apis* was significantly more abundant in October. Strain-specific differences were also detected. Bolton bees exhibited significantly higher abundances of uncultured *Bartonella* and *Bifidobacterium*, along with an unexpected detection of *Wolbachia*. In contrast, Mann Lake bees showed increased abundance of *Commensalibacter*. Diversity analyses indicated that microbial community structure was influenced primarily by sampling month rather than bee strain or storage condition. Shannon diversity varied significantly across months, and Bray–Curtis dissimilarity also revealed significant temporal differences. Overall, hive storage conditions (indoor constant temperature versus outdoor exposure) did not significantly affect taxonomic abundance. However, several bacterial taxa exhibited month-dependent fluctuations, with only minor variations observed between the two commercial bee strains (Brar et al., 2025).

Together, these findings illustrate that the honeybee gut microbiome is not merely a simplified community but a highly organized symbiotic system shaped by host anatomy, social behavior, and evolutionary history. Its low complexity, predictable composition, and strong functional partitioning across gut regions make it an exceptional model for dissecting principles of microbial community assembly, metabolic cooperation, and host–microbe mutualism.

11.8 KEY BACTERIAL SYMBIONTS AND FUNCTIONAL ROLES

The core microbiota of the western honeybee represents a tightly integrated functional consortium in which each dominant lineage occupies a distinct ecological niche while contributing to collective metabolism, detoxification, and host defense. Together, these symbionts enable efficient exploitation of pollen- and nectar-based diets while stabilizing gut conditions and limiting pathogen invasion. Rather than acting independently, they form a metabolically interdependent network shaped by spatial organization within the gut and by host physiology.

S. alvi is a Betaproteobacteria (family Neisseriaceae) specialized for life on the ileal epithelium, where it forms dense biofilms directly on host tissue. By consuming oxygen diffusing across the gut wall, it maintains the microaerophilic conditions required by neighboring anaerobic symbionts, effectively acting as a physiological gatekeeper of the ileal environment (Quinn et al., 2024). Genome-wide mutant screens demonstrate that pili and adhesins are essential for stable colonization; loss of these structures reduces persistence and disrupts community composition, underscoring biofilm formation as a mutualistic trait that stabilizes the microbiome (Cheung et al., 2023). Metabolically, *S. alvi* relies primarily on host-derived organic acids such as citrate and malate rather than directly degrading dietary polymers, highlighting its adaptation to a niche defined by host metabolism. By occupying epithelial attachment sites and modulating local oxygen and metabolite gradients, it also contributes to competitive exclusion of opportunistic microbes.

G. apicola, a Gammaproteobacterium in the Orbaceae, complements this role as the principal degrader of complex plant polysaccharides. Its genomes encode extensive repertoires of glycoside hydrolases, pectate lyases, and phosphotransferase systems that enable the breakdown of pollen cell-wall components such as pectin and hemicellulose (Tang et al., 2025). Experimental work shows that *G. apicola* can metabolize sugars toxic to the host, including mannose and rhamnose, illustrating its function as a dietary detoxifier (Zheng et al., 2016). Fermentation products—pyruvate, acetate, and other organic acids—serve as substrates for neighboring microbes such as *S. alvi*, demonstrating extensive cross-feeding and metabolic coupling within the ileal biofilm.

The lactic acid–producing clades formerly known as Firm-4 and Firm-5 have been taxonomically reassigned to *Bombilactobacillus* and the *L. melliventris* group, respectively (Heczko et al., 2023). These bacteria dominate the rectum, where they ferment pollen-derived carbohydrates into lactic acid and short-chain fatty acids that acidify the gut lumen and provide additional energy substrates for the host. Mono-colonization studies show that these taxa generate a substantial portion of the hindgut metabolic output, even in isolation, producing organic acids and aromatic metabolites (Han et al., 2024). Intriguingly, behavioral experiments indicate that higher abundances of *Bombilactobacillus* and *L. melliventris* are associated with nurse bees, and experimental inoculation can reduce foraging activity, suggesting links between microbial metabolism, energy balance, and division of labor within colonies (Vernier et al., 2024).

B. asteroides and related species add another functional dimension by influencing both digestion and host signaling pathways. These Actinobacteria degrade hemicelluloses and other complex carbohydrates while producing metabolites that affect host endocrine systems. Metabolomic analyses of mono-colonized bees show that *B. asteroides* uniquely stimulates production of JH derivatives, which regulate development and age-related behavioral transitions (Pino et al., 2022). In behavioral assays, colonization increases foraging intensity, contrasting with the inhibitory effects observed for lactic acid bacteria, supporting a model in which relative abundances of these taxa help tune colony task allocation (Vernier et al., 2024).

F. perrara, another Orbaceae member, occupies the pylorus and anterior ileum and induces the characteristic melanized "scab" phenotype on the gut epithelium. This lesion reflects localized immune activation and is associated with strong upregulation of antimicrobial and pattern-recognition pathways, suggesting that *F. perrara* plays a role in immune priming and barrier maintenance despite its moderate abundance (Schmidt et al., 2023). Its impact illustrates how low-abundance taxa can exert disproportionate effects on host physiology.

Several additional symbionts provide auxiliary metabolic capabilities. *Commensalibacter* and *Bombella* (Acetobacteraceae) are acetic-acid bacteria common in the crop, midgut, and hive substrates, where they participate in organic-acid production and early processing of ingested nectar and pollen (Botero et al., 2023). *Apibacter* species, enriched in the hindgut, are implicated in nitrogen recycling and fermentation pathways, while *B. apis* is associated with rectal contents and may contribute to amino-acid metabolism and host physiological modulation (Kwong et al., 2018). Although individually less abundant, these taxa likely enhance the resilience of the microbial network under fluctuating dietary or environmental conditions.

Recent studies reveal that honeybee symbionts also produce specialized metabolites that directly inhibit competitors and pathogens. Comparative genomics has identified type VI secretion systems and colibactin-like biosynthetic clusters in certain *Gilliamella* and *Snodgrassella* strains, suggesting mechanisms for interbacterial antagonism that help maintain the exclusivity of the core community (Zhang et al., 2022). Such defensive chemistry underscores that the bee gut microbiome is not merely commensal but actively protective.

Using laboratory-generated gnotobiotic bees, Zhang et al. (2022) demonstrated that an intact gut microbiota is essential for proper olfactory learning and memory. Comparative brain transcriptomic analyses revealed clear differences in gene expression between microbiota-free and conventionally colonized bees. Follow-up metabolomic profiling of hemolymph and gut samples indicated that the microbiota primarily influences tryptophan metabolism. The study further showed that host-specific *Lactobacillus* strains enhance memory performance by converting tryptophan into indole derivatives, which activate the host aryl hydrocarbon receptor. These findings underscore the role of specific gut bacteria in regulating honeybee neural function (Boucias et al., 2018).

Collectively, these findings portray a highly integrated system in which functional roles are partitioned yet interdependent: *S. alvi* engineers the physicochemical niche, *G. apicola* expands digestive capacity and detoxifies diet, *Lactobacillus* bacteria and *Bifidobacterium* generate energy-rich and signaling metabolites, and accessory taxa contribute immune modulation and metabolic flexibility. This cooperative network exemplifies how a low-diversity microbiome can achieve high functional complexity, reinforcing the honeybee gut as a premier model for studying metabolic cooperation and host–microbe mutualism (Motta & Moran, 2024).

11.9 INTERACTIONS WITH *NOSEMA CERANAE* AND OTHER PARASITES

The microsporidian parasite *N. ceranae* (often reassigned to *Vairimorpha ceranae*) is among the most widespread and consequential pathogens affecting honeybees. The parasite completes its intracellular life cycle within midgut epithelial cells—the primary site of nutrient absorption—where it disrupts tissue integrity and host physiology (Chen et al., 2009). Infection damages the peritrophic membrane, interferes with epithelial renewal pathways such as Wnt signaling, and leads to lesions that reduce digestive efficiency and shorten worker lifespan. At the organismal level, infected bees often exhibit increased sucrose consumption, reduced trophallaxis, oxidative stress, and altered immune responses, all of which can impair colony productivity even in the absence of overt clinical signs (Dussaubat et al., 2012). Because *N. ceranae* is confined to the gut, its pathophysiological effects are inseparable from changes in the resident microbiome.

High-resolution metagenomic studies demonstrate that *N. ceranae* infection consistently reshapes gut community composition. Experimental infections have revealed increases in taxa such as *Proteus mirabilis*, *F. perrara*, and *G. apicola*, alongside declines in *B. asteroides*, *F. fructosus*, and various *Lactobacillus* species (Sbaghdi et al., 2024). At broader taxonomic scales, these shifts reflect enrichment of Proteobacteria and depletion of Firmicutes and Actinobacteria, a pattern interpreted as parasite-associated dysbiosis. Independent studies report similar trends—elevated *G. apicola* and *S. alvi* coupled with reduced Alphaproteobacteria and lactic acid bacteria—suggesting that microsporidian infection reproducibly destabilizes the normal microbial equilibrium and may increase susceptibility to opportunistic pathogens (Jabal-Uriel et al., 2022).

These compositional changes are closely linked to alterations in host immune and oxidative stress pathways. *N. ceranae* modulates antimicrobial peptide genes such as apidaecin and defensin and perturbs antioxidant systems, including catalase and glutathione peroxidase expression, reflecting elevated reactive oxygen species in infected midguts (Prasad et al., 2025). Damage to epithelial cells also alters nutrient flux, pH, and immune effector levels, thereby reshaping the ecological niche available to gut bacteria. The result is a bidirectional feedback loop: parasite-induced tissue and immune changes restructure the microbiome, and the resulting dysbiosis can further influence parasite proliferation and host resilience (Bombaça et al., 2021).

Probiotic interventions have been explored as a strategy to mitigate these effects. Supplementation with *Pediococcus acidilactici* MA18/5M, for example, partially restored bacterial community structure in infected bees, increasing the relative abundance of taxa such as *Bifidobacterium* and *Lactobacillus* that are typically reduced by infection (Woo et al., 2025). Functional analyses showed normalization of several immune and antioxidant gene expression patterns (Maftei et al., 2024). However, improvements in survival were modest and not statistically significant in that study, highlighting that microbiome restoration does not always translate into clear fitness gains under controlled conditions and that host–microbe–parasite interactions remain complex.

Environmental stressors further modulate these interactions. Chronic exposure to field-realistic concentrations of the neonicotinoid thiamethoxam has been reported to alter gut microbial composition and, intriguingly, to counteract some dysbiosis signatures associated with *N. ceranae* infection. Both thiamethoxam and probiotic treatment independently reshaped microbial communities, and their combined effects varied depending on infection status, illustrating that chemical stressors, symbionts, and parasites can interact in non-additive and sometimes antagonistic ways (Stanley & Raine, 2017; Sbaghdi et al., 2024). In some contexts, co-exposure reduced survival in uninfected bees while slightly improving outcomes in infected individuals, underscoring strong context dependence.

Taken together, these findings reinforce the concept of the honeybee as a holobiont in which parasites, microbiota, and environmental factors form an interconnected network governing health. *N. ceranae* infection not only damages midgut tissues but also drives reproducible microbiome shifts and immune modulation; probiotics and pesticides can either buffer or exacerbate these effects depending on conditions (Lau et al., 2024). A mechanistic understanding of these multilayered interactions is essential for designing effective disease-management strategies—such as targeted probiotic deployment or mitigation of pesticide exposure—and for interpreting microbiome dysbiosis as both a consequence and potential driver of parasite-associated colony declines.

11.10 STRAIN-DEPENDENT ANTIMICROBIAL ACTIVITY OF CORE HONEYBEE GUT SYMBIONTS

Host-adapted members of the genera *Lactobacillus* and *Bifidobacterium* form the functional core of the honeybee gut microbiota and are widely regarded as key probiotic partners of their host. These symbionts contribute to digestion, immune modulation, and—critically—colonization resistance against pathogens. Among their most important protective roles is the suppression of major brood diseases caused by *P. larvae* and *M. plutonius*, the etiological agents of American and European foulbrood, respectively. In addition to these well-established pathogens, the opportunistic bacterium *Serratia marcescens* has gained attention as a contributor to colony morbidity, particularly under stress conditions. Although prior investigations have demonstrated inhibitory effects of individual gut symbionts on specific bee pathogens, comprehensive assessments comparing multiple clinical isolates across several pathogen species within a unified geographic context have been limited.

Addressing this gap, Dengiz et al. (2025) systematically evaluated the antimicrobial capacity of a diverse collection of *Lactobacillus* and *Bifidobacterium* strains isolated from different regions within a single country. Following taxonomic identification through 16S rRNA gene sequencing, isolates were screened in vitro in modified brain–heart infusion medium against four *P. larvae* isolates, one *M. plutonius* isolate, and the *S. marcescens* sicaria strain. The study revealed pronounced strain-level variability: 28 strains strongly inhibited at least two *P. larvae* isolates, while 12 displayed moderate but consistent inhibition across all four. Activity against *M. plutonius* was comparatively rare, detected in three strains, whereas inhibition of *S. marcescens* was more frequent, observed in 20 strains.

Notably, the most potent and broad-spectrum antagonistic effects were associated with strains belonging to *B. asteroides*, *Bifidobacterium choladohabitans*, *B. polysaccharolyticum*, *L. apis*, *L. helsingborgensis*, *L. kullabergensis*, and *L. melliventris*. Collectively, these findings emphasize that antimicrobial efficacy is highly strain dependent rather than species uniform, underscoring the necessity of targeted probiotic selection. By identifying symbionts capable of inhibiting multiple pathogens simultaneously, the work advances the development of microbiome-based strategies to enhance colony resilience and offers a framework for precision probiotic interventions in apiculture.

11.11 AGROCHEMICALS, DYSBIOSIS, AND MICROBIOME DISRUPTION

Modern agricultural landscapes expose honeybees to a complex mixture of herbicides, insecticides, antibiotics, and nutritional stressors that can disrupt the gut microbiome even at sublethal levels. Because the honeybee gut community is relatively low in diversity yet highly specialized, perturbations to a few keystone taxa can cascade into functional changes affecting digestion, immunity, and pathogen resistance. Increasing evidence therefore frames chemical exposure as a driver of microbiome-mediated health effects rather than solely a source of direct toxicity.

11.11.1 Glyphosate and Shikimate Pathway–Dependent Dysbiosis

Glyphosate, the most widely used herbicide globally, inhibits 5-enolpyruvylshikimate-3-phosphate synthase (EPSPS), a key enzyme in the bacterial shikimate pathway that is absent in animals but present in many bee gut symbionts. Comparative genomics shows that strains of *S. alvi* harbor glyphosate-sensitive class I EPSPS variants, rendering this keystone species particularly susceptible (Leino et al., 2021). Experimental exposures demonstrate dose-dependent reductions in *S. alvi*, shifts in community composition, and increased

and produce antimicrobial compounds that limit pathogen establishment. Experimental perturbations of this community through antibiotics, pesticide exposure, or microbiota-free (gnotobiotic) rearing consistently alter host transcriptional programs linked to immunity, nutrient metabolism, neuroendocrine signaling, and behavior, and often increase susceptibility to bacterial, fungal, and microsporidian infections (Zhang et al., 2025). These findings support a systems view in which microbiome integrity acts as a buffering layer that stabilizes host physiology; when environmental pressures exceed this buffering capacity, dysbiosis can trigger cascading effects including immune dysregulation, heightened pathogen loads, and reduced worker longevity (Minaud et al., 2024).

Modern anthropogenic stressors increasingly challenge this symbiotic equilibrium. Global assessments identify land-use change, intensive monoculture agriculture, and pesticide exposure as dominant drivers of pollinator decline, with climate change compounding risk by shifting floral phenology and resource availability (Farooq et al., 2025). In managed honeybees, herbicides, insecticides, and veterinary antibiotics can directly reshape gut microbial communities, while habitat simplification reduces pollen diversity and nutritional quality, constraining the metabolic niches that sustain symbiont diversity (Motta et al., 2022). The convergence of chemical, nutritional, and climatic stressors elevates the likelihood of microbiome instability, which can manifest at the colony level as impaired brood development, queen failure, reduced disease resistance, and ultimately colony collapse.

Because honeybees underpin both agricultural productivity and ecosystem functioning, safeguarding their microbiome has implications that extend well beyond apiculture. The bee gut community is increasingly viewed as a tractable model for understanding environment–microbiota–host interactions, offering insights relevant to pollinator conservation, disease ecology, and even antimicrobial stewardship (Hai et al., 2026). Practical conservation strategies therefore emphasize reducing broad-spectrum chemical inputs through integrated pest management, restoring heterogeneous floral habitats, and ensuring season-long forage availability. Such landscape-level interventions not only support honeybee nutrition but also promote microbial functional diversity, benefiting wild pollinators and strengthening ecosystem resilience (Navarro-Escalante et al., 2025).

Microbiome-informed interventions represent a complementary and rapidly developing approach. Probiotic formulations derived from native bee symbionts—such as *L. melliventris*, *S. alvi*, *G. apicola*, and *Bifidobacterium* spp.—have demonstrated potential to re-establish disrupted communities, enhance resistance to pathogens like *P. larvae* and *S. marcescens*, and improve survival or foraging performance in bees and related pollinators (Gulliver et al., 2022). However, outcomes remain context-dependent: studies using strains such as *P. acidilactici* or commercial formulations highlight variability in efficacy depending on strain compatibility, dosage, and concurrent stressors including pesticides (Aziz et al., 2025). Future applications will likely rely on well-characterized, host-adapted microbial consortia, alongside standardized protocols for storage, delivery, and evaluation of microbiome restoration.

Viewing the honeybee colony as a superorganism—or holobiont—underscores that conservation must address the integrated environment–microbiota–host system. Interventions focusing solely on chemical regulation or habitat restoration may be insufficient if microbial partners remain compromised. Instead, a holistic strategy that combines sustainable land management, reduced chemical burden, improved nutrition, and targeted microbiome support can strengthen the symbiotic capacity of bees to withstand environmental change. Such an approach not only preserves pollination services and biodiversity but also contributes to global food security by maintaining the ecological functions on which agricultural systems depend.

11.14 CONCLUSION

The honeybee microbiome represents a highly specialized and relatively simple microbial community that plays a foundational role in colony health. Core bacterial taxa, consistently found across geographic regions and bee species, support digestion of pollen and nectar, synthesize essential nutrients, and protect against opportunistic pathogens through competitive exclusion and antimicrobial metabolite production. This tight coevolution between bees and their symbionts underscores how microbial stability contributes directly to individual fitness and, by extension, colony productivity.

Disruptions to this microbial balance—driven by pesticides, antibiotics, poor nutrition, climate stress, and pathogen pressure—are increasingly linked to impaired immunity, reduced lifespan, and heightened susceptibility to diseases such as foulbrood and Nosema infections. Because the honeybee gut community is relatively low in diversity yet functionally interdependent, even subtle compositional shifts can cascade into measurable physiological and behavioral consequences. These findings position the microbiome as both a sensitive biomarker of environmental stress and a potential target for interventions such as probiotic supplementation and habitat management.

Looking forward, integrating microbiome science into apiculture and pollinator conservation offers a promising pathway to enhance resilience in managed and wild bee populations. Future research combining metagenomics, metabolomics, and ecological field studies will be critical to clarify functional mechanisms and to design evidence-based strategies that stabilize beneficial microbes under real-world conditions. Ultimately, protecting the honeybee microbiome is not only central to sustaining colony health but also to safeguarding global pollination services and agricultural ecosystems.

ACKNOWLEDGMENTS

The author (ST) acknowledges the Abraham Thomas Foundation for providing the resources for this work.

REFERENCES

Agriculture Institute, 2023. Understanding the Digestive System of Honey Bees: From Nectar to Honey. https://agriculture.institute/beekeeping-introduction/digestive-system-of-honey-bees/ (accessed 2.13.26).

Alarcón, M., Castelli, L., Branchiccela, B., Invernizzi, C., Antúnez, K., 2026. Mitigating the impacts of nutritional stress on honey bee colonies located in monoculture Eucalyptus plantations. *Apidologie* 57, 7. https://doi.org/10.1007/s13592-025-01242-4

Alberoni, D., DiGioia, D., 2025. Editorial: Impact of landscape and feeding on the bees' gut microbiome shaping and pathogens development. *Front Cell Infect Microbiol* 15, 1688273. https://doi.org/10.3389/fcimb.2025.1688273

Alhosin, M., 2023. Epigenetics mechanisms of honeybees: Secrets of royal jelly. *Epigenet Insights* 16, 25168657231213717. https://doi.org/10.1177/25168657231213717

Al-Khayri, J.M., Sahana, G.R., Nagella, P., Joseph, B.V., Alessa, F.M., Al-Mssallem, M.Q., 2022. Flavonoids as potential anti-inflammatory molecules: A review. *Molecules* 27, 2901. https://doi.org/10.3390/molecules27092901

Álvarez, S., Contreras-Kallens, P., Aguayo, S., Ramírez, O., Vallejos, C., Ruiz, J., Carrasco-Gallardo, E., Troncoso-Vera, S., Morales, B., Schuh, C.M.A.P., 2023. Royal jelly extracellular vesicles promote wound healing by modulating underlying cellular responses. *Molecular Therapy – Nucleic Acids* 31, 541–552. https://doi.org/10.1016/j.omtn.2023.02.008

Amdam, G.V., Fennern, E., Havukainen, H., 2012. Vitellogenin in Honey Bee Behavior and Lifespan, in: Galizia, C.G., Eisenhardt, D., Giurfa, M. (Eds.), *Honeybee Neurobiology and Behavior: A Tribute to Randolf Menzel*. Springer Netherlands, Dordrecht, pp. 17–29. https://doi.org/10.1007/978-94-007-2099-2_2

Aziz, T., Naveed, M., Shabbir, M.A., Naseeb, J., Sarwar, A., Zhao, L., Yang, Z., Haiying, C., Lin, L., Shami, A., Al-Asmari, F., 2025. Genomic profiling of pediococcus acidilactici BCB1H and identification of its key features for Biotechnological innovation, food technology and medicine. *Sci Rep* 15, 6050. https://doi.org/10.1038/s41598-025-90633-9

Bianchi, E.M., Ferrari, C., Aguirre, N.C., Filippi, C.V., Vera, P.A., Puebla, A.F., Gennari, G.P., Rodríguez, G.A., Scannapieco, A.C., Acuña, C.V., Lanzavecchia, S.B., 2023. Phenotypic and genetic characterization of Africanized Apis mellifera colonies with natural tolerance to Varroa destructor and contrasting defensive behavior. *Front Insect Sci* 3, 1175760. https://doi.org/10.3389/finsc.2023.1175760

Blili, Y., Abou Nader, E., Pavez, I., Prosperi, P., Harbouze, R., Kyrgiakos, L.S., Kleisiari, C., Vasileiou, M., Angelopoulos, V., Vlontzos, G., Kleftodimos, G., 2025. Assessing the economic impact of insect pollination on the agricultural sector: A department-level case study in France. *Environmental and Sustainability Indicators* 28, 100944. https://doi.org/10.1016/j.indic.2025.100944

Blot, N., Veillat, L., Rouzé, R., Delatte, H., 2019. Glyphosate, but not its metabolite AMPA, alters the honeybee gut microbiota. *PLoS One* 14, e0215466. https://doi.org/10.1371/journal.pone.0215466

Bolat, E., A. A.Canbolat, S.Sarıtaş, et al. 2026. Honey as a natural flavorful product: A comprehensive review of its potential biological activities and recent studies. *Food Science & Nutrition* 14 (4), e71645. https://doi.org/10.1002/fsn3.71645

Bombaça, A.C.S., Gandara, A.C.P., Ennes-Vidal, V., Bottino-Rojas, V., Dias, F.A., Farnesi, L.C., Sorgine,

M.H., Bahia, A.C., Bruno, R.V., Menna-Barreto, R.F.S., 2021. Aedes aegypti infection with trypanosomatid strigomonas culicis alters midgut redox metabolism and reduces mosquito reproductive fitness. *Front Cell Infect Microbiol* 11, 732925. https://doi.org/10.3389/fcimb.2021.732925

Bortolotti, L., Costa, C., 2014. Chemical Communication in the Honey Bee Society, in: Mucignat-Caretta, C. (Ed.), *Neurobiology of Chemical Communication, Frontiers in Neuroscience*. CRC Press/Taylor & Francis, Boca Raton (FL).

Botero, J., Sombolestani, A.S., Cnockaert, M., Peeters, C., Borremans, W., De Vuyst, L., Vereecken, N.J., Michez, D., Smagghe, G., Bonilla-Rosso, G., Engel, P., Vandamme, P., 2023. A phylogenomic and comparative genomic analysis of Commensalibacter, a versatile insect symbiont. *Anim Microbiome* 5, 25. https://doi.org/10.1186/s42523-023-00248-6

Boucias, D.G., Zhou, Y., Huang, S., Keyhani, N.O., 2018. Microbiota in insect fungal pathology. *Appl Microbiol Biotechnol* 102, 5873–5888. https://doi.org/10.1007/s00253-018-9089-z

Brar, G., Ngor, L., McFrederick, Q.S., Torson, A.S., Rajamohan, A., Rinehart, J., Singh, P., Bowsher, J.H. 2025. High abundance of lactobacilli in the gut microbiome of honey bees during winter. *Scientific Reports* 15(1), 7409. doi: 10.1038/s41598-025-90763-0.

Calfee, E., Agra, M.N., Palacio, M.A., Ramírez, S.R., Coop, G., 2020. Selection and hybridization shaped the rapid spread of African honey bee ancestry in the Americas. *PLoS Genet* 16, e1009038. https://doi.org/10.1371/journal.pgen.1009038

Carr, S.M., 2023. Multiple mitogenomes indicate things fall apart with out of Africa or Asia hypotheses for the phylogeographic evolution of Honey Bees (Apis mellifera). *Sci Rep* 13, 9386. https://doi.org/10.1038/s41598-023-35937-4

Castelli, L., Branchiccela, B., Garrido, M., Invernizzi, C., Porrini, M., Romero, H., Santos, E., Zunino, P., Antúnez, K., 2020. Impact of nutritional stress on honeybee gut microbiota, immunity, and Nosema ceranae infection. *Microb Ecol* 80, 908–919. https://doi.org/10.1007/s00248-020-01538-1

Cavigli, I., Daughenbaugh, K.F., Martin, M., Lerch, M., Banner, K., Garcia, E., Brutscher, L.M., Flenniken, M.L., 2016. Pathogen prevalence and abundance in honey bee colonies involved in almond pollination. *Apidologie* 47, 251–266. https://doi.org/10.1007/s13592-015-0395-5

Cerqueira, A.E.S., Lima, H.S., Silva, L.C.F., Veloso, T.G.R., de Paula, S.O., Santana, W.C., da Silva, C.C., 2024. Melipona stingless bees and honey microbiota reveal the diversity, composition, and modes of symbionts transmission. *FEMS Microbiol Ecol* 100, fiae063. https://doi.org/10.1093/femsec/fiae063

Chakraborty, A., Basu, P., 2025. Intensive agriculture influences functional diversity, redundancy and trait profile of bee community and interacting plant community in a tropical agricultural landscape. *Agriculture, Ecosystems & Environment* 383, 109544. https://doi.org/10.1016/j.agee.2025.109544

Chen, Y.P., Evans, J.D., Murphy, C., Gutell, R., Zuker, M., Gundensen-Rindal, D., Pettis, J.S., 2009. Morphological, molecular, and phylogenetic characterization of Nosema ceranae, a microsporidian parasite isolated from the European honey bee, Apis mellifera. *J Eukaryot Microbiol* 56, 142–147. https://doi.org/10.1111/j.1550-7408.2008.00374.x

Cheung, B.H., Alisoltani, A., Kochan, T.J., Lebrun-Corbin, M., Nozick, S.H., Axline, C.M.R., Bachta, K.E.R., Ozer, E.A., Hauser, A.R. 2023. Genome-wide screens reveal shared and strain-specific genes that facilitate enteric colonization by Klebsiella pneumoniae. *mBio* 14(6), e0212823. doi: 10.1128/mbio.02128-23.

Cucu, A-A., Baci, G-M., Moise, A.R., Dezsi, Ş., Marc, B.D., Stângaciu, Ş., Dezmirean, D.S., 2021. Towards a better understanding of nutritional and therapeutic effects of honey and their applications in apitherapy. *Applied Sciences* 11 (9), 4190. https://doi.org/10.3390/app11094190

Cunningham, M.M., Tran, L., McKee, C.G., Ortega Polo, R., Newman, T., Lansing, L., Griffiths, J.S., Bilodeau, G.J., Rott, M., Marta Guarna, M., 2022. Honey bees as biomonitors of environmental contaminants, pathogens, and climate change. *Ecological Indicators* 134, 108457. https://doi.org/10.1016/j.ecolind.2021.108457

de Jongh, E.J., Harper, S.L., Yamamoto, S.S., Wright, C.J., Wilkinson, C.W., Ghosh, S., Otto, S.J.G., 2022. One health, one hive: A scoping review of honey bees, climate change, pollutants, and antimicrobial resistance. *PLoS One* 17, e0242393. https://doi.org/10.1371/journal.pone.0242393

De Paula, J.C., Doello, K., Mesas, C., Kapravelou, G., Cornet-Gómez, A., Orantes, F.J., Martínez, R., Linares, F., Prados, J.C., Porres, J.M., Osuna, A., de Pablos, L.M., 2022. Exploring honeybee abdominal anatomy through micro-CT and novel multi-staining approaches. *Insects* 13, 556. https://doi.org/10.3390/insects13060556

Dengiz, B., Killer, J., Havlík, J., Dobeš, P., Hyršl, P. 2025. Selection of probiotics for honey bees: The in vitro inhibition of paenibacillus larvae, melissococcus plutonius, and serratia marcescens strain sicaria by host-specific lactobacilli and bifidobacteria. *Microorganisms* 13(5), 1159. https://doi.org/10.3390/microorganisms13051159

Dogantzis, K.A., Tiwari, T., Conflitti, I.M., Dey, A., Patch, H.M., Muli, E.M., Garnery, L., Whitfield, C.W.,

Stolle, E., Alqarni, A.S., Allsopp, M.H., Zayed, A., 2021. Thrice out of Asia and the adaptive radiation of the western honey bee. *Science Advances* 7, eabj2151. https://doi.org/10.1126/sciadv.abj2151
Donaldson-Matasci, M.C., Dornhaus, A., 2012. How habitat affects the benefits of communication in collectively foraging honey bees. *Behav Ecol Sociobiol* 66, 583–592. https://doi.org/10.1007/s00265-011-1306-z
Dussaubat, C., Brunet, J.-L., Higes, M., Colbourne, J.K., Lopez, J., Choi, J.-H., Martín-Hernández, R., Botías, C., Cousin, M., McDonnell, C., Bonnet, M., Belzunces, L.P., Moritz, R.F.A., Le Conte, Y., Alaux, C., 2012. Gut pathology and responses to the microsporidium Nosema ceranae in the honey bee Apis mellifera. *PLoS One* 7, e37017. https://doi.org/10.1371/journal.pone.0037017
El Adaouia Taleb, R., Djebli, N., Chenini, H., Sahin, H., Kolayli, S., 2020. In vivo and in vitro anti-diabetic activity of ethanolic propolis extract. *Journal of Food Biochemistry* 44, e13267. https://doi.org/10.1111/jfbc.13267
El Khoury, S., Gauthier, J., Bouslama, S., Cheaib, B., Giovenazzo, P., Derome, N., 2021. Dietary contamination with a Neonicotinoid (Clothianidin) gradient triggers specific dysbiosis signatures of microbiota activity along the honeybee (Apis mellifera) digestive tract. *Microorganisms* 9, 2283. https://doi.org/10.3390/microorganisms9112283
Encerrado-Manriquez, A.M., Pouv, A.K., Fine, J.D., Nicklisch, S.C.T., 2024. Enhancing knowledge of chemical exposures and fate in honey bee hives: Insights from colony structure and interactions. *Science of The Total Environment* 916, 170193. https://doi.org/10.1016/j.scitotenv.2024.170193
Engel, P., Bartlett, K.D., Moran, N.A., 2015. The bacterium Frischella perrara causes scab formation in the gut of its honeybee host. *mBio* 6, e00193–15. https://doi.org/10.1128/mBio.00193-15
Farooq, M., Khan, U., Adnan, M., Younas, W., Qureshi, N.B., Yasir, Q.M., Cai, Q., Chiu, M.-C., 2025. Global decline of aquatic and terrestrial insects driven by climate change and anthropogenic impacts. *Biological Conservation* 308, 111181. https://doi.org/10.1016/j.biocon.2025.111181
Fernandes, K.E., Frost, E.A., Kratz, M., Carter, D.A., 2024. Pollen products collected from honey bee hives experiencing minor stress have altered fungal communities and reduced antimicrobial properties. *FEMS Microbiol Ecol* 100, fiae091. https://doi.org/10.1093/femsec/fiae091
Fortune Business Insights, 2026. Honey Market Size, Share & Industry Analysis, By Type (Alfalfa, Buckwheat, Wildflower, Clover, Acacia, and Others), By Application (Food & Beverages, Personal Care & Cosmetics, Pharmaceuticals, and Others), By Packaging (Glass Jar, Bottle, Tub, Tube, and Others), and Regional Forecast, 2026-2034. Food Processing & Processed Food/Honey Market. www.fortunebusinessinsights.com/industry-reports/honey-market-100551.
Frunze, O., Petukhov, A.V., Brandorf, A.Z., Simankov, M.K., Kim, H., Kwon, H.-W., 2025. Conservation of Apis mellifera mellifera L. in the Middle Ural. *Insects* 16, 512. https://doi.org/10.3390/insects16050512
Gazzea, E., Batáry, P., Marini, L., 2023. Global meta-analysis shows reduced quality of food crops under inadequate animal pollination. *Nat Commun* 14, 4463. https://doi.org/10.1038/s41467-023-40231-y
Glenny, W., Cavigli, I., Daughenbaugh, K.F., Radford, R., Kegley, S.E., Flenniken, M.L., 2017. Honey bee (Apis mellifera) colony health and pathogen composition in migratory beekeeping operations. *PLOS ONE* 12, e0182814. https://doi.org/10.1371/journal.pone.0182814
Godebo, T.R., Stoner, H., Taylor, P., Jeuland, M., 2025. Metals in honey from bees as a proxy for environmental contamination in the United States. *Environmental Pollution* 364, 125221. https://doi.org/10.1016/j.envpol.2024.125221
Gopal, R.K., Ganesh, P.S., Pathoor, N.N., Viswanathan, A., 2025. Enhancing microbiome-based biomarkers: Challenges and opportunities. *The Lancet Microbe* 6(3), 101032. https://doi.org/10.1016/j.lanmic.2024.101032
Gulliver, E.L., et al., 2022. Review article: The future of microbiome-based therapeutics. *Aliment Pharmacol Ther* 56, 192–208. https://doi.org/10.1111/apt.17049
Guo, J., Wang, Z., Chen, Y., Cao, J., Tian, W., Ma, B., Dong, Y., 2021. Active components and biological functions of royal jelly. *Journal of Functional Foods* 82, 104514. https://doi.org/10.1016/j.jff.2021.104514
Hai, Q., Li, D., Huang, T., Dang, X., Xu, J., Ma, Z., Zhou, Z., 2026. The honeybee gut microbiome: A novel multidimensional model of antimicrobial resistance transmission and immune homeostasis. *FEMS Microbiol Rev* 50, fuag001. https://doi.org/10.1093/femsre/fuag001
Han, B., et al., 2024. Lactobacillus firm-5-derived succinate prevents honeybees from having diabetes-like symptoms. *Proc Natl Acad Sci U.S.A* 121, e2405410121. https://doi.org/10.1073/pnas.2405410121
Han, F., Wallberg, A., Webster, M.T., 2012. From where did the Western honeybee (Apis mellifera) originate? *Ecol Evol* 2, 1949–1957. https://doi.org/10.1002/ece3.312
Heczko, P., Kozień, Ł., Strus, M., 2023. Special issue "an update on lactobacillus": editorial. *Microorganisms*

11, 1400. https://doi.org/10.3390/microorganisms11061400

Henriques, D., Lopes, A.R., Low, M., Pinto, M.A., 2025. Human-mediated introgression and Varroa destructor shaped the genetic structure of honey bee populations in the Azores. *Sci Rep* 15, 25217. https://doi.org/10.1038/s41598-025-08950-y

Henriques, D., Parejo, M., Lopes, A.R., Pinto, M.A., 2019. Mitochondrial SNP markers to monitor evolutionary lineage ancestry in Apis melliferamellifera conservation programs. *Apidologie* 50, 538–541. https://doi.org/10.1007/s13592-019-00666-z

Hitaj, C., Smith, D., Hunt, K., 2018. Honeybees on the move: Pollination services and honey production. 2018 Agricultural & Applied Economics Association Annual Meeting, Washington, D.C., August 5–7.

Holzer, E., Malabusini, S., Savoldelli, S., Lupi, D., 2026. The genus apis in a changing world: Distribution, conservation, climate, and anthropogenic stressors. *Insects* 17, 185. https://doi.org/10.3390/insects17020185

Hrassnigg, N., Crailsheim, K., 2005. Differences in drone and worker physiology in honeybees (Apis mellifera). *Apidologie* 36, 255–277 https://doi.org/10.1051/apido:2005015

Hristov, P., Shumkova, R., Palova, N., Neov, B. (2015). Factors associated with honey bee colony losses: A mini-review. *Vet Science* 7(4),166. doi: 10.3390/vetsci7040166.

Ihle, K.E., Fondrk, M.K., Page, R.E., Amdam, G.V., 2015. Genotype effect on lifespan following vitellogenin knockdown. *Exp Gerontol* 61, 113–122. https://doi.org/10.1016/j.exger.2014.12.007

Ilyasov, R.A., Lee, M., Takahashi, J., Kwon, H.W., Nikolenko, A.G., 2020. A revision of subspecies structure of western honey bee Apis mellifera. *Saudi Journal of Biological Sciences* 27, 3615–3621. https://doi.org/10.1016/j.sjbs.2020.08.001

Imperatriz-Fonseca, et al., 2016. The assessment report of the intergovernmental science-policy platform on biodiversity and ecosystem services on pollinators. Secretariat of the Intergovernmental Science-Policy Platform on Biodiversity and Ecosystem Services, Bonn, Germany.

Jabal-Uriel, C., Alba, C., Higes, M., Rodríguez, J.M., Martín-Hernández, R., 2022. Effect of Nosema ceranae infection and season on the gut bacteriome composition. *Sci Rep* 12, 9326. https://doi.org/10.1038/s41598-022-13337-4

Katumo, D.M., Liang, H., Ochola, A.C., Lv, M., Wang, Q.-F., Yang, C.-F., 2022. Pollinator diversity benefits natural and agricultural ecosystems, environmental health, and human welfare. *Plant Divers* 44, 429–435. https://doi.org/10.1016/j.pld.2022.01.005

Kaur, M., et al., 2024. Monoculture of crops: A challenge in attaining food security, in: Sharma, A., Kumar, M., Sharma, P, (Eds.), *Advances in Food Security and Sustainability*. Elsevier, pp. 197–213. https://doi.org/10.1016/bs.af2s.2024.07.008

Khalifa, S.A.M., et al., 2021. Overview of bee pollination and its economic value for crop production. *Insects* 12, 688. https://doi.org/10.3390/insects12080688

Khan, M.N., Bilal, H., Khan, S., Shafiq, M., Jiao, X., 2025. Unintended effects on gut microbiota of non-target organisms. *Comparative Biochemistry and Physiology Part C: Toxicology & Pharmacology* 298, 110330. https://doi.org/10.1016/j.cbpc.2025.110330

Khan, U., Amanullah, 2024. Enhancing public health through modern agronomy: Sustainable and nutrient-rich practices. *Journal of Agriculture, Food Systems, and Community Development* 13(2), 11–18.

Kilpatrick, S.K., Gibbs, J., Mikulas, M.M., Spichiger, S.-E., Ostiguy, N., Biddinger, D.J., López-Uribe, M.M. 2020. An updated checklist of the bees (Hymenoptera, Apoidea, Anthophila) of Pennsylvania, United States of America. *Journal of Hymenoptera Research* 77, 1–86.

Klein, S., Pasquaretta, C., He, X.J., Perry, C., Søvik, E., Devaud, J.-M., Barron, A.B., Lihoreau, M., 2019. Honey bees increase their foraging performance and frequency of pollen trips through experience. *Sci Rep* 9, 6778. https://doi.org/10.1038/s41598-019-42677-x

Kulikowski, M., 2016. Stress from migratory beekeeping may be eased by access to food. *NC State News*. https://news.ncsu.edu/2016/08/commercial-bee-study/ (accessed 2.13.26).

Kwong, W.K., Steele, M.I., Moran, N.A., 2018. Genome sequences of Apibacter spp., gut symbionts of Asian honey bees. *Genome Biol Evol* 10, 1174–1179. https://doi.org/10.1093/gbe/evy076

Lau, E., Maccaro, J., McFrederick, Q.S., Nieh, J.C., 2024. Exploring the interactions between Nosema ceranae infection and the honey bee gut microbiome. *Sci Rep* 14, 20037. https://doi.org/10.1038/s41598-024-67796-y

Leino, L., et al., 2021. Classification of the glyphosate target enzyme for assessing sensitivity of organisms to the herbicide. *Journal of Hazardous Materials* 408, 124556. https://doi.org/10.1016/j.jhazmat.2020.124556

Leonard, R.J., Wat, K.K.Y., McArthur, C., Hochuli, D.F., 2018. Urbanisation and wing asymmetry in the western honey bee. *PeerJ* 6, e5940. https://doi.org/10.7717/peerj.5940

Li, C., Tang, M., Li, X., Zhou, X., 2022. Community dynamics in structure and function of honey bee gut bacteria in response to winter dietary shift. *mBio* Oct 26, 13 (5), e0113122. doi: 10.1128/mbio.01131-22

Luo, S., Zhang, X., Zhou, X., 2024. Temporospatial dynamics and host specificity of honeybee gut bacteria. *Cell Reports* 43, 114408. https://doi.org/10.1016/j.celrep.2024.114408

Maftei, N.-M., et al., 2024. The potential impact of probiotics on human health: An update on their health-promoting properties. *Microorganisms* 12, 234. https://doi.org/10.3390/microorganisms12020234

Mayack, C., Carlson, M., Niño, B.D., Niño, E.L., Seshadri, A., 2025. Impacts of almond pollination service and inter-row cover cropping on honey bee colony strength. *Science of the Total Environment* 983, 179703. https://doi.org/10.1016/j.scitotenv.2025.179703

Minaud, É., et al., 2024. How stressors disrupt honey bee biological traits and overwintering mechanisms. *Heliyon* 10, e34390. https://doi.org/10.1016/j.heliyon.2024.e34390

Mitteroecker, P., Schaefer, K., 2022. Geometric morphometrics: Achievements and challenges. *Am J Biol Anthropol* 178, 181–210. https://doi.org/10.1002/ajpa.24531

Motta, E.V.S., Moran, N.A., 2023. The effects of glyphosate on bumble bees and their gut microbial communities. *Sci Total Environ* 872, 162102. https://doi.org/10.1016/j.scitotenv.2023.162102

Motta, E.V.S., Moran, N.A., 2024.The honeybee microbiota and its impact on health and disease. *Nat Rev Microbiol* 22, 122–137. https://doi.org/10.1038/s41579-023-00990-3

Motta, E.V.S., Powell, J.E., Moran, N.A., 2022. Glyphosate induces immune dysregulation in honey bees. *Anim Microbiome* 4, 16. https://doi.org/10.1186/s42523-022-00165-0

Motta, E.V.S., Raymann, K., Moran, N.A., 2018. Glyphosate perturbs the gut microbiota of honey bees. *Proc Natl Acad Sci U S A* 115, 10305–10310. https://doi.org/10.1073/pnas.1803880115

Navarro-Escalante, L., Ashraf, A.H.M.Z., Leonard, S.P., Barrick, J.E., 2025. Protecting honey bees through microbiome engineering. *Current Opinion in Insect Science* 72, 101416. https://doi.org/10.1016/j.cois.2025.101416

Ollerton, J., Winfree, R., Tarrant, S., 2011. How many flowering plants are pollinated by animals? *Oikos* 120, 321–326. https://doi.org/10.1111/j.1600-0706.2010.18644.x

Papa, G., et al., 2022.The honey bee Apis mellifera: An insect at the interface between human and ecosystem health. *Biology (Basel)* 11, 233. https://doi.org/10.3390/biology11020233

Patidar, B.K., Ojha, K.N., Khan, I.U., 2017. Role of honeybee (Apis mellifera) in enhancing yield of mustard. *Int J Curr Microbiol App Sci* 6, 1879–1882. https://doi.org/10.20546/ijcmas.2017.607.224

Perugini, M., Manera, M., Grotta, L., Abete, M.C., Tarasco, R., Amorena, M., 2011 May. Heavy metal (Hg, Cr, Cd, and Pb) contamination in urban areas and wildlife reserves: honeybees as bioindicators. *Biol Trace Elem Res.* 140 (2), 170-176. doi: 10.1007/s12011-010-8688-z

Pino, A., et al., 2022. Characterization of Bifidobacterium asteroides Isolates. *Microorganisms* 10, 655. https://doi.org/10.3390/microorganisms10030655

Powell, J.E., Carver, Z., Leonard, S.P., Moran, N.A., 2021. Field-realistic tylosin exposure impacts honey bee microbiota and pathogen susceptibility. *Microbiol Spectr* 9, e0010321. https://doi.org/10.1128/Spectrum.00103-21

Prasad, A., et al., 2025. Evolution of gut microbiota across honeybee species revealed by comparative metagenomics. *Nat Commun* 16, 9069. https://doi.org/10.1038/s41467-025-64115-5

Quinn, A., et al., 2024. Host-derived organic acids enable gut colonization of the honey bee symbiont Snodgrassella alvi. *Nat Microbiol* 9, 477–489. https://doi.org/10.1038/s41564-023-01572-y

Raymann, K., Moran, N.A., 2018. The role of the gut microbiome in health and disease of adult honey bee workers. *Curr Opin Insect Sci* 26, 97–104. https://doi.org/10.1016/j.cois.2018.02.012

Rueppell, O., Bachelier, C., Fondrk, M.K., Page, R.E., 2007. Regulation of life history determines lifespan of worker honey bees. *Exp Gerontol* 42, 1020–1032. https://doi.org/10.1016/j.exger.2007.06.002

Ruttner, F. (1988) *Biogeography and Taxonomy of Honeybees*. Springer-Verlag, Berlin, Heidelberg and New York, p. 284.

Samuelson, A., Leadbeater, E., 2017. Foraging by honeybees, in: Vonk, J., Shackelford, T. (Eds.), *Encyclopedia of Animal Cognition and Behavior*. Springer, Cham, pp. 2782–2790. https://doi.org/10.1007/978-3-319-47829-6_918-1

Pawar, P., Jarpla, M., Narwade, D.K., et al. 2024. A review on the sublethal effects of pure and formulated glyphosate on bees. *Archives of Current Research International* 24(12), 28–42. https://doi.org/10.9734/ACRI/2024/V24I12995

Sarapa, A., Peter, A., Buettner, A., Loos, H.M., 2025. Organoleptic and chemical properties of propolis: A review. *Eur Food Res Technol* 251, 1331–1352. https://doi.org/10.1007/s00217-025-04708-y

Sbaghdi, T., et al., 2024. The response of the honey bee gut microbiota to Nosema ceranae. *Microorganisms* 12, 192. https://doi.org/10.3390/microorganisms12010192

Schmidt, K., et al., 2023. Integration host factor regulates colonization factors in the bee gut symbiont

Frischella perrara. *eLife* 12, e76182. https://doi.org/10.7554/eLife.76182

Sedley, L., 2020. Advances in nutritional epigenetics—A fresh perspective for an old idea. *Epigenet Insights* 13, 2516865720981924. https://doi.org/10.1177/2516865720981924

Seeley, T.D. 1983. Division of labor between scouts and recruits in honeybee foraging. *Behavioral Ecology and Sociobiology* 12, 253–259. https://doi.org/10.1007/BF00290778

Seeley, T.D. 1989. The honey bee colony as a superorganism. *American Scientist* 77, 546–553.

Shanahan, M., et al., 2024. Thinking inside the box: Restoring the propolis envelope facilitates honey bee social immunity. *PLoS One* 19, e0291744. https://doi.org/10.1371/journal.pone.0291744

Shen, X., et al., 2024. Short-term probiotic supplementation affects the diversity and interactions of the native gut microbiome. *Imeta* 3, e253. https://doi.org/10.1002/imt2.253

Siviter, H., et al., 2021. Agrochemicals interact synergistically to increase bee mortality. *Nature* 596, 389–392. https://doi.org/10.1038/s41586-021-03787-7

Smith, M.R., et al., 2022. Pollinator deficits, food consumption, and consequences for human health: A modeling study. *Environ Health Perspect* 130, 127003. https://doi.org/10.1289/EHP10947

Stanley, D.A., Raine, N.E., 2017. Bumblebee colony development following chronic exposure to thiamethoxam. *Scientific Reports* 7, 8005. https://doi.org/10.1038/s41598-017-08752-x

Tang, M., et al., 2025. Honeybee-Gilliamella synergy in carbohydrate metabolism enhances host thermogenesis in cold acclimation. *npj Biofilms Microbiomes* 11, 172. https://doi.org/10.1038/s41522-025-00798-4

Techer, M.A., et al., 2017. Genetic diversity and differentiation among insular honey bee populations in the southwest Indian Ocean. *PLOS ONE* 12, e0189234. https://doi.org/10.1371/journal.pone.0189234

Tihelka, E., Cai, C., Pisani, D., Donoghue, P.C.J., 2020. Mitochondrial genomes illuminate the evolutionary history of the Western honey bee. *Sci Rep* 10, 14515. https://doi.org/10.1038/s41598-020-71393-0

Tran, L., et al., 2025. Neonicotinoid-induced signature dysbiosis identified via metagenomic sequencing of the honey bee gut microbiome. *Sci Rep* 16, 1211. https://doi.org/10.1038/s41598-025-30907-4

U.S. Fish & Wildlife Service, 2025. Pollinators benefit economies. www.fws.gov/initiative/pollinators/pollinators-benefit-economies (accessed 2.12.26).

USDA, 2024. Cost of Pollination (December 2024). National Agricultural Statistics Service. https://esmis.nal.usda.gov/sites/default/release-files/d504rk335/3t947j47n/xs55p7492/cospol24.pdf

USDA, 2025. Pollination fees and honey production data. www.ers.usda.gov/data-products/charts-of-note/chart-detail?chartId=112782 (accessed 2.12.26b).

Utzeri, V.J., Ribani, A., Taurisano, V., Fontanesi, L., 2022. DNA method that distinguishes Apis mellifera mitochondrial C mitotypes. *Food Control* 134, 108713. https://doi.org/10.1016/j.foodcont.2021.108713

Vahidi, M., Shafian, S., Frame, W.H., 2025. Depth-specific soil moisture estimation in vegetated corn fields using a canopy-informed model. *Agricultural Water Management* 307, 109213. https://doi.org/10.1016/j.agwat.2024.109213

Vernier, C.L., et al., 2024. Gut microbiota contribute to variations in honey bee foraging intensity. *ISME J* 18, wrae030. https://doi.org/10.1093/ismejo/wrae030

Vincze, C., Leelőssy, Á., Zajácz, E., Mészáros, R. 2025. A review of short-term weather impacts on honey production. *International Journal of Biometeorology* 69(2), 303–317. doi: 10.1007/s00484-024-02824-0.

Visscher, P.K., Seeley, T.D. 1982. Foraging strategy of honeybee colonies in a temperate deciduous forest. *Ecology* 63, 1790–1801. https://doi.org/10.2307/1940121

Wagh, V.D., 2013. Propolis: A wonder bees product and its pharmacological potentials. *Adv Pharmacol Sci* 2013, 308249. https://doi.org/10.1155/2013/308249

Werling, B., 2015. Biodiversity services and bioenergy landscapes. *MSU Extension*. www.canr.msu.edu/resources/biodiversity_services_and_bioenergy_landscapes_e3164 (accessed 2.13.26).

White P.S., Densmore L.D. 1992. *Mitochondrial DNA Isolation: Molecular Genetic Analysis of Populations. A Practical Approach*. Oxford University Press. Oxford, UK.

Whitfield, C.W., et al., 2006. Thrice out of Africa: Ancient and recent expansions of the honey bee. *Science* 314, 642–645. https://doi.org/10.1126/science.1132772

Wilson, D.S., Sober, E., 1989. Colony fitness and superorganism selection. *Journal of Theoretical Biology* 36(3), 337–356. doi: 10.1016/s0022-5193(89)80169-9.

Woo, I.-K., et al., 2025. Probiotic Pediococcus acidilactici strains exert anti-inflammatory effects. *Probiotics Antimicrob Proteins* 17, 3258–3269. https://doi.org/10.1007/s12602-024-10263-x

Wragg, D., et al., 2022. Complex population structure and haplotype patterns in the Western European honey bee. *Mol Ecol Resour* 22, 3068–3086. https://doi.org/10.1111/1755-0998.13665

Xue, X., Wu, L., Wang, K., 2025. Chemical composition of royal jelly, in: Alvarez-Suarez, J.M. (Ed.), *Bee Products – Chemical and Biological Properties*. Springer Nature Switzerland, pp. 273–285. https://doi.org/10.1007/978-3-031-89049-9_8

Yang, Y.-C., Chou, W.-M., Widowati, D.A., Lin, I.-P., Peng, C.-C., 2018. 10-hydroxy-2-decenoic acid of royal jelly exhibits bactericide and anti-inflammatory activity. *BMC Complement Altern Med* 18, 202. https://doi.org/10.1186/s12906-018-2267-9

Zander, K., Stolz, H., Hamm, U., 2013. Promising ethical arguments for product differentiation in the organic food sector. *Appetite* 62, 133–142. https://doi.org/10.1016/j.appet.2012.11.015

Zárate, D., et al., 2022. Admixture in Africanized honey bees (Apis mellifera) from Panamá to San Diego. *Ecol Evol* 12, e8580. https://doi.org/10.1002/ece3.8580

Zhang, W., et al., 2022. Genomic features underlying the evolutionary transitions of Apibacter to honey bee gut symbionts. *Insect Sci* 29, 259–275. https://doi.org/10.1111/1744-7917.12912

Zhang, Z., Mu, X., Cao, Q., Zhai, Y., Zheng, L., Liu, Y., Zheng, H., Zhang, X., 2025. Antibiotic exposure alters the honeybee gut microbiota and behavioral caste transition. *Insect Sci* 32, 260–276. https://doi.org/10.1111/1744-7917.13374

Zheng, H., et al., 2016. Metabolism of toxic sugars by strains of the bee gut symbiont Gilliamella apicola. *mBio* 7, 10–128. https://doi.org/10.1128/mbio.01326-16

Zheng, H., Steele, M.I., Leonard, S.P. et al. 2018. Honey bees as models for gut microbiota research. *Lab Animal* 47, 317–325. https://doi.org/10.1038/s41684-018-0173-x

Zheng, H., Perreau, J., Powell, J.E., Han, B., Zhang, Z., Kwong, W.K., Tringe, S.G., Moran, N.A. 2019. Division of labor in honey bee gut microbiota for plant polysaccharide digestion. *Proceedings of the National Academy of Sciences of the USA* 116, 25909–25916. https://doi.org/10.1073/pnas.1916224116

Zhong, H., et al., 2025. Multilayer networks decoding Apis mellifera health and resilience. *J Proteome Res* 24, 5305–5318. https://doi.org/10.1021/acs.jproteome.5c00294

Zulhendri, F., et al., 2021. Antiviral, antibacterial, antifungal, and antiparasitic properties of propolis: A review. *Foods* 10, 1360. https://doi.org/10.3390/foods10061360

Zullkiflee, N., Taha, H., Usman, A., 2022. Propolis: Its role and efficacy in human health and diseases. *Molecules* 27, 6120. https://doi.org/10.3390/molecules27186120

INDEX